NEGATIVE-REFRACTION METAMATERIALS

NEGATIVE-REFRACTION METAMATERIALS

Fundamental Principles and Applications

Edited by

G. V. Eleftheriades

K. G. Balmain

IEEE PRESS

A JOHN WILEY & SONS, INC., PUBLICATION

Copyright © 2005 by Institute of Electrical and Electronics Engineers. All rights reserved.

Published by John Wiley & Sons, Inc., Hoboken, New Jersey.
Published simultaneously in Canada.

No part of this publication may be reproduced, stored in a retrieval system, or transmitted in any form or by any means, electronic, mechanical, photocopying, recording, scanning, or otherwise, except as permitted under Section 107 or 108 of the 1976 United States Copyright Act, without either the prior written permission of the Publisher, or authorization through payment of the appropriate per-copy fee to the Copyright Clearance Center, Inc., 222 Rosewood Drive, Danvers, MA 01923, (978) 750-8400, fax (978) 750-4470, or on the web at www.copyright.com. Requests to the Publisher for permission should be addressed to the Permissions Department, John Wiley & Sons, Inc., 111 River Street, Hoboken, NJ 07030, (201) 748-6011, fax (201) 748-6008, or online at http://www.wiley.com/go/permission.

Limit of Liability/Disclaimer of Warranty: While the publisher and author have used their best efforts in preparing this book, they make no representations or warranties with respect to the accuracy or completeness of the contents of this book and specifically disclaim any implied warranties of merchantability or fitness for a particular purpose. No warranty may be created or extended by sales representatives or written sales materials. The advice and strategies contained herein may not be suitable for your situation. You should consult with a professional where appropriate. Neither the publisher nor author shall be liable for any loss of profit or any other commercial damages, including but not limited to special, incidental, consequential, or other damages.

For general information on our other products and services or for technical support, please contact our Customer Care Department within the United States at (800) 762-2974, outside the United States at (317) 572-3993 or fax (317) 572-4002.

Wiley also publishes its books in a variety of electronic formats. Some content that appears in print may not be available in electronic format. For information about Wiley products, visit our web site at www.wiley.com.

Library of Congress Cataloging-in-Publication Data:

Negative refraction metamaterials : fundamental properties and applications / edited by
G. V. Eleftheriades and K. G. Balmain.
p. cm.
Includes bibliographical references and index.
ISBN 13: 978-0-471-60146-3 (cloth)

1. Magnetic materials. 2. Electromagnetism. I. Eleftheriades, G. V. (George V.) II. Balmain, K. G. (Keith G.)

TK454.4.M3N493 2005
620.1'1297—dc22 2005045967

10 9 8 7 6 5 4 3 2

Contents

Contributors

ANDREA ALÙ, Department of Electrical and Systems Engineering, University of Pennsylvania, Philadelphia, PA 19104, USA; Department of Applied Electronics, Universita di Roma Tre, 84-00146, Rome, Italy

KEITH G. BALMAIN, The Edward S. Rogers Sr. Department of Electrical and Computer Engineering, University of Toronto, Toronto, Ontario M5S 3G4, Canada

GEORGE V. ELEFTHERIADES, The Edward S. Rogers Sr. Department of Electrical and Computer Engineering, University of Toronto, Toronto, Ontario M5S 3G4, Canada

NADER ENGHETA, Department of Electrical and Systems Engineering, University of Pennsylvania, Philadelphia, PA 19104, United States

ANTHONY GRBIC, The Edward S. Rogers Sr. Department of Electrical and Computer Engineering, University of Toronto, Toronto, Ontario M5S 3G4, Canada

ASHWIN K. IYER, The Edward S. Rogers Sr. Department of Electrical and Computer Engineering, University of Toronto, Toronto, Ontario M5S 3G4, Canada

JOHN D. JOANNOPOULOS, Department of Physics and Center for Materials Science and Engineering, Massachusetts Institute of Technology, Cambridge, MA 02139, United States

CHIYAN LUO, Department of Physics and Center for Materials Science and Engineering, Massachusetts Institute of Technology, Cambridge, MA 02139, United States

ANDREA A. E. LÜTTGEN, The Edward S. Rogers Sr. Department of Electrical and Computer Engineering, University of Toronto, Toronto, Ontario M5S 3G4, Canada

MOHAMMAD MOJAHEDI, The Edward S. Rogers Sr. Department of Electrical and Computer Engineering, University of Toronto, Toronto, Ontario M5S 3G4, Canada

ANDREY K. SARYCHEV, School of Electrical and Computer Engineering, Purdue University, West Lafayette, IN 47907, United States

DAVID SCHURIG, Department of Physics, University of California at San Diego, La Jolla, CA 92093, United States

VLADIMIR M. SHALAEV, School of Electrical and Computer Engineering, Purdue University, West Lafayette, IN 47907, United States

DAVID R. SMITH, Department of Physics, University of California at San Diego, La Jolla, CA 92093, United States. *Present address*: Department of Electrical and Computer Engineering, Duke University, Durham, NC 27708, United States

RICHARD W. ZIOLKOWSKI, Department of Electrical and Computer Engineering, University of Arizona, Tucson, AZ 85721, USA

Preface

"... [The conception of an idea] does not at all admit of exposition like other branches of knowledge; but as a result of continued application to the subject itself and communion therewith, it is brought to birth in the soul on a sudden, as light that is kindled by a leaping spark, and thereafter it nourishes itself."

—*Plato (427–347 B.C.), 7th Epistle, 341.*

Metamaterials represent an exciting emerging research area that promises to bring about important technological and scientific advancements in diverse areas such as telecommunications, radars and defense, nanolithography with light, microelectronics, medical imaging, and so on. This book includes contributions from some of the top experts in the field in an effort to document in an authoritative, but understandable, way the most important and most recent developments. Presently, a universally accepted definition of metamaterials does not exist. Broadly speaking, metamaterials are artificial media with unusual electromagnetic properties. Some researchers restrict metamaterials to be artificially structured periodic media (in fact, effective material parameters can be defined even for nonperiodic media in analogy to amorphous materials) in which the periodicity is much smaller than the wavelength of the impinging electromagnetic wave. The underlying nature of the subwavelength periodic inclusions enables them to act as artificial "molecules" that scatter back the impinging electromagnetic fields in a prescribed manner. This process can be macroscopically characterized by means of effective material parameters such as permittivity, a permeability, and a refractive index. This definition of metamaterials is directly related to the classic work in artificial dielectrics that has been carried out at microwave frequencies in the 1950s and 1960s. Yet others do not impose strict limits to the size of the constituent unit cells, thus extending the definition of metamaterials to include structures such as photonic crystals.

In this book, both artificial dielectric and photonic crystal types of metamaterials are covered. Its scope, however, is restricted to metamaterials that support the "unusual" electromagnetic property of negative refraction. Negative refraction can be supported in isotropic media for which a negative permittivity and a negative permeability, hence a negative refractive index, can be defined. These latter materials are referred to in the book as "left-handed," "negative-refractive-index (or negative index)," and "double-negative" materials. However, negative refraction need not be limited to isotropic media, nor does it have to be associated with a negative index of refraction. The underlying physical properties of several such classes of negative-refraction metamaterials are presented and are related to corresponding emerging applications such as lenses and antennas, imaging with super-resolution, microwave passive devices, interconnects for wireless telecommunications, and radar/defense applications. The implementation of negative-refraction metamaterials at optical frequencies is also covered in this book.

Chapter 1, by Iyer and Eleftheriades, describes the fundamentals of isotropic metamaterials in which a simultaneous negative permittivity and permeability, hence a negative refractive index, can be defined. The emphasis is placed on the theory, design, and experiments involving planar transmission-line based metamaterials, although bulk split-ring-resonator/wire metamaterials are also described. Furthermore, this chapter attempts to historically link metamaterials with the classic body of work in artificial dielectrics carried out in the fifties and sixties. Moreover, in this opening chapter the reader can find a comprehensive summary of the various terms that are used throughout the text by various authors to describe these kind of metamaterials. Chapter 2, by Eleftheriades, builds on these fundamentals in order to describe a range of useful microwave devices and antennas. Chapter 3, by Grbic and Eleftheriades, describes in a comprehensive manner the theory and the experiments behind a super-resolving negative-refractive-index planar transmission-line lens. Furthermore, it explains how to extend the transmission-line-based metamaterial to three dimensions. Chapter 4, by Ziolkowski, describes numerical simulation studies of negative refraction of Gaussian beams and associated focusing phenomena. Chapter 5, by Schurig and Smith, describes in an exhaustive way the theory and the unique advantages of shaped lenses made out of negative-refractive-index metamaterials. Chapter 6, by Balmain and Lüttgen, introduces a new kind of transmission-line metamaterial that is anisotropic and supports the formation of sharp beams called resonance cones. This chapter describes the theory and some of the microwave applications of these unique negative-refraction metamaterials. The next two chapters are devoted to the potential implementations of negative-refraction metamaterials at optical frequencies. Specifically, Chapter 7, by Luo and Joannopoulos, explains how to obtain negative refraction and associated super-resolving imaging effects using dielectric photonic crystals. This chapter is unique in that these photonic crystals support negative refraction (of power flow), but without an underlying effective negative refractive index. On the other hand, in Chapter 8, Sarychev and Shalaev prescribe a method for realizing negative-refractive-index metamaterials at optical frequencies using plasmonic (metallic) nanowires. Chapter 9, by Alù and Engheta, deals with another interesting topic, namely the unusual propagation phenomena in metallic waveguides partially filled with negative-refractive-index metamaterials. Finally, Chapter 10, by Mojahedi and Eleftheriades, introduces metamaterials in which the refractive index and the underlying group velocity are both negative. Such metamaterials and associated electrical interconnects could find applications in dispersion compensation.

Lastly, a note on the notation: Since this book includes contributions from both engineers and physicists, some of the notation is inconsistent. In particular, Chapters 5, 7, and 8 use a time-harmonic variation of $\exp(-j\omega t)$ whereas the rest use $\exp(+j\omega t)$.

We would like to thank Rohin Iyer, Ashwin Iyer, and Suzanne Erickson for their tireless help in editing this material. We would also like to acknowledge our associate publisher at Wiley, Mr. George Telecki, and editorial assistant, Ms. Rachel Witmer, as well as Amy Hendrickson at TeXnology Inc. for their cheerful assistance. One of the editors, G. V. Eleftheriades, would like to dedicate this book to his wife, Maria, for her inspiring courage during times of extreme personal hardship while preparing this book.

G. V. ELEFTHERIADES AND K. G. BALMAIN

Toronto, January 2005

1 Negative-Refractive-Index Transmission-Line Metamaterials

ASHWIN K. IYER and GEORGE V. ELEFTHERIADES

The Edward S. Rogers, Sr. Department of Electrical and Computer Engineering
University of Toronto
Toronto, Ontario, M5S 3G4
Canada

1.1 INTRODUCTION

1.1.1 Veselago and the Left-Handed Medium (LHM)

In the 1960s, Victor Veselago of Moscow's P. N. Lebedev Institute of Physics examined the feasibility of media characterized by a simultaneously negative permittivity ϵ and permeability μ [1]. He concluded that such media are allowed by Maxwell's equations and that plane waves propagating inside them could be described by an electric field intensity vector $\mathbf{E}$, magnetic field intensity vector $\mathbf{H}$, and wavevector $\mathbf{k}$, forming a left-handed triplet, in seeming opposition to wave propagation in conventional media, in which these three quantities form a right-handed triplet, and accordingly labeled these materials left-handed media (LHM) and right-handed media (RHM), respectively. The two arrangements are illustrated in Fig. 1.1. Moreover, although $\mathbf{E}$, $\mathbf{H}$, and $\mathbf{k}$ form a left-handed triplet, $\mathbf{E}$, $\mathbf{H}$, and the Poynting vector $\mathbf{S}$ maintain a right-handed relationship; thus, in LHM the wavevector $\mathbf{k}$ is *antiparallel* to the Poynting vector $\mathbf{S}$. Viewed in retrospect, it seems what Veselago was describing was the backward wave. For this reason, some researchers use the term "backward wave media" to describe left-handed materials [2]. Certainly, one-dimensional backward wave lines are not new to the microwave community; in fact, through backward waves, Veselago's left-handed medium is intimately tied to many familiar concepts and known one-dimensional structures, including the backward wave amplifier/oscillator (for example, see Refs. 3, 4) and backfire antennas that operate in a higher-order negative spatial harmonic (see Ref. 5). However, what is remarkable in Veselago's work is his realization that two- or three-dimensional isotropic and homogeneous media supporting backward waves ought to be characterized by a *negative*

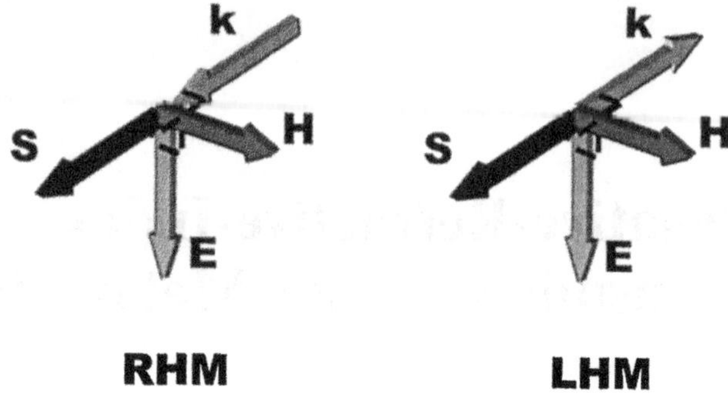

Fig. 1.1 Orientation of field quantities **E**, **H**, Poynting vector **S**, and wavevector **k** in right-handed media (RHM) and left-handed media (LHM).

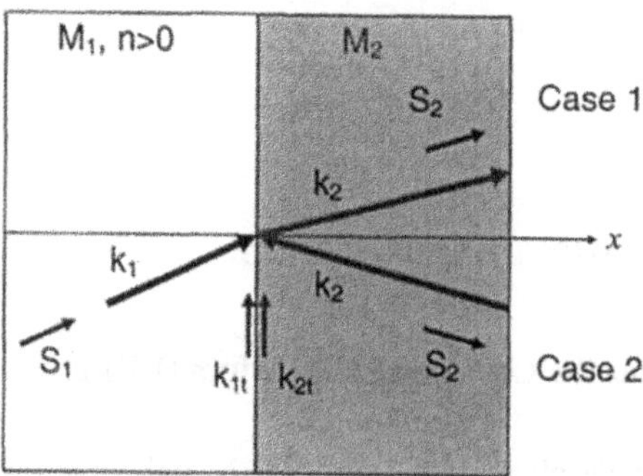

Fig. 1.2 Refraction at a two-medium interface as determined by phase matching. The support of backward waves by an LHM insists on negative refraction (case 2). After Refs. [26, 27]. Copyright © 2002 IEEE.

refractive index, which is defined by taking the negative branch of the square root in the definition $n = \pm\sqrt{\epsilon\mu}$. Consequently, when such media are interfaced with conventional dielectrics, Snell's Law is reversed, leading to the negative refraction of an incident electromagnetic plane wave. Such a material, in realized form, could appropriately be called a *metamaterial*, where the prefix *meta*, Greek for "beyond" or "after," suggests that it possesses properties that transcend those available in nature.

1.1.2 Negative Refraction at a Planar Interface

One way to understand negative refraction is through the idea of phase matching. To illustrate, consider the two-medium interface of Fig. 1.2, where medium 1 (M_1) is an RHM and medium 2 (M_2) is unspecified for the moment. A plane wave originating in M_1 is incident on the interface with wavevector $\mathbf{k}_1$, and it establishes a refracted wave in M_2 with wavevector $\mathbf{k}_2$ such that their tangential components $\mathbf{k}_{1t}$ and $\mathbf{k}_{2t}$

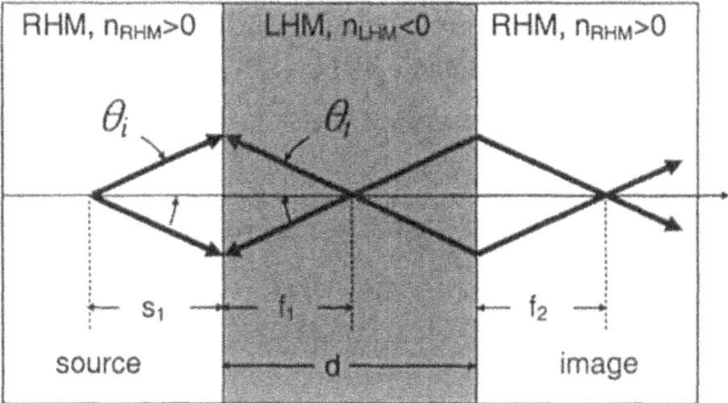

Fig. 1.3 Focusing of the rays of a cylindrical excitation by an LHM slab embedded in an RHM.

are equal, according to the conservation of the wave momentum. Having specified the tangential components, we immediately recognize that there are two possibilities for the normal component of $\mathbf{k}_2$: the first case, in which $\mathbf{k}_2$ is directed away from the interface, and the second case, usually describing reflected waves, in which $\mathbf{k}_2$ is directed towards the interface. These two cases are represented as Case 1 and Case 2 in Fig. 1.2. By the conservation of energy, the normal components of the Poynting vectors $\mathbf{S}_1$ and $\mathbf{S}_2$ must remain in the positive x-direction through both media. Thus, Case 1 depicts the usual situation in which M_2 is a conventional positive-index medium; however, if M_2 is a medium supporting propagating backward waves (LHM), then the wavevector $\mathbf{k}_2$ must be directed oppositely to the Poynting Vector $\mathbf{S}_2$ (i.e., with a normal component in the negative x-direction). Therefore, refraction in media that support backward waves must be described by the second case, in which power is propagated along the direction of phase *advance*, and so is directed through a negative angle of refraction. Thus, M_2 can be seen to possess an effectively negative refractive index.

1.1.3 Flat Lenses and Focusing

Harnessing the phenomenon of negative refraction, entirely new refractive devices can be envisioned, such as a flat "lens" without an optical axis, also proposed by Veselago. This is shown in Fig. 1.3. In general, nonparaxial rays experience spherical aberration; that is, for an arbitrary selection of positive and negative refractive indices, each ray intercepts the principal axis at a different point. However, it is not fortuitous that Veselago chose, as his example, the case of an LHM slab with refractive index $n_{LHM} = -1$, embedded in vacuum ($n_{RHM} = +1$), such that the *relative* index of refraction is $n_{REL} = n_{LHM}/n_{RHM} = -1$. In this special case, $|\theta_i|$ and $|\theta_t|$ become equal, and all the component rays are focused to the same point. Furthermore, for $n_{REL} = -1$, the slab thickness d, source distance s_1, and external focal length f_2 in Fig. 1.3 are related through $h = s_1 + f_2$, so that the phase lag (positive optical path length) incurred in the two RHM regions is fully compensated by the phase advance (negative optical path length) incurred in the LHM slab. Thus, the phase of the source is exactly restored at the image plane.

1.2 BACKGROUND

1.2.1 Artificial Dielectrics

In natural dielectrics, local electromagnetic interactions at the atomic or molecular level produced by an applied field result in a macroscopic response that may be described by an electric permittivity and magnetic permeability. These constitutive parameters acquire meaning only when the lattice exhibits some degree of spatial order and the wavelength of the impressed field is much longer than the lattice spacing. It would seem, therefore, that any effort to *synthesize* particular material parameters requires access to the scatterer itself, the atom or molecule in question, a degree of precision that makes the task prohibitively difficult. However, it is apparent that the long-wavelength condition can be met at scales far more accessible than those of atoms and molecules; indeed, for sufficiently long wavelengths—for example, those corresponding to RF/microwave frequencies—electromagnetic scatterers may be fabricated with entirely practicable dimensions, and would react to applied fields much like the atoms and molecules of a crystal lattice. Furthermore, at wavelengths substantially longer than the obstacle spacing, ordered arrays of these inclusions would behave like effective media and exhibit dielectric properties. These are hardly novel ideas; indeed, they were intensely scrutinized a half-century ago during the investigation of synthetic media known as artificial dielectrics. The term "artificial dielectric" was introduced in 1948 by Winston E. Kock of Bell Laboratories to describe electromagnetic structures of practicable dimension that could be designed to mimic the response of natural solids to electromagnetic radiation [6]. His ideas were motivated by the pressing need at the time for lightweight, low-loss substitutes for natural dielectrics when large devices were required. His earlier research examined the design of antennas using large, metallic lenses comprising parallel metal plates, as substitutes for heavy dielectric lens aerials [7, 8]. Kock's metallic lenses, resembling stacked electromagnetic waveguides, exhibited superluminal phase velocities and accordingly advanced the phase of the incoming radiation relative to propagation in vacuum. Furthermore, at wavelengths substantially longer than the plate spacing, his lenses behaved like media with an effective, positive index of refraction less than unity. It wasn't long before Kock realized that his phase-advance metallic lenses were merely a subset of a much wider category of electromagnetic structures whose behaviour, at certain frequencies, boasted an irrefutable likeness to their dielectric counterparts. Aiming toward the design of phase-delay lenses, Kock established numerous analogies between the electromagnetic response of his metallic structures and the electrodynamics of natural dielectrics. First, he noted that any analogy attempting to capture the refraction characteristics of artificial dielectrics must, as in natural dielectrics, require that the wavelength in the (effective) medium be much longer than the lattice spacing. Operated in a long-wavelength regime, the periodic perturbations of artificial dielectrics appear diminutive in contrast to the wavelength, allowing effective macroscopic material parameters to be defined. When the wavelength in the medium becomes comparable to the lattice spacing, diffraction effects are expected, analogous to the diffraction of X rays by solids. In either case, it is implicitly established that a lattice of conducting plates acts much like the crystalline lattices of a solid. Kock noted that the individual conducting elements of his lattice, indeed, behaved like electric dipoles under the influence of an impressed field, in becoming polarized and establishing a net dipole moment. Drawing on his background in microwaves, Kock acknowledged that his conducting elements could be said to capacitively load free space, much like the discrete capacitive shunt loading of waveguides to reduce the phase velocity of the guided waves. The associated displacement of charge is reflected in the charge on the capacitor plates, and so the capacitance is effectively modified by the conducting obstacle; for a large array of conducting scatterers, this change may be modeled by an effective positive permittivity greater than ϵ_0. Kock also recognized that, at higher frequencies, the existence of eddy currents on the surface of the conductors prevents the penetration of magnetic field lines into the conductors, and the consequent condensation of magnetic field lines (viewed, similarly, as a change in inductance) results in a diamagnetic response. Kock's experimentation on artificial dielectrics consisting of conducting strips, spheres, and disks showed that numerous other phenomena related to the crystalline nature of solids, including anisotropy, scattering due to diffraction, and frequency dispersion (including anomalous dispersion), are evident, suggesting that the analogy between artificial and natural dielectrics is not only sound, but seems to transcend even their most fundamental dissociations. Essentially, Kock's ideas acknowledged that natural dielectrics, or, equivalently, substances with corrugations on the atomic scale, are superfluous for many applications in the microwave or even terahertz or far infrared range, where the wavelengths are long enough to make the fabrication of an artificial dielectric practicable. In this sense, artificial dielectrics are also intimately connected to today's photonic crystals, in which the refractive index is periodically modulated in space so as to achieve stopbands in desirable frequency regions and for preferred angles of propagation, mimicking the energy band structure of solids. These devices have found numerous uses as filters and waveguides at optical frequencies (see Chapter 7). For a thorough treatment of artificial dielectrics, the interested reader is referred to Chapter 12 of Ref. 9 and to the list of early works provided therein.

1.2.2 Negative Permittivity

It is well known that plasmas are described by a permittivity function that becomes negative below a plasma frequency ω_P, causing the propagation constant in the plasma to become imaginary. In this frequency region, electromagnetic waves incident on the plasma suffer reactive attenuation and are reflected. Thus, the plasma frequency bears a resemblance to the modal cutoff frequencies of particular electromagnetic waveguides, below which the waveguide environment can be perceived as an inductively loaded free space, as observed in 1954 by R. N. Bracewell [10]. The idea of modeling plasmas using artificial dielectrics was examined as early as 1962 by Walter Rotman [11]. Rotman considered Kock's artificial dielectrics, and he employed the well-known theory for the analysis of periodic microwave networks (for example, see Ref. 4) to determine their dispersion characteristics. His analysis, however, could not explicitly consider the permittivity of the media and, instead, was limited to the consideration of their index of refraction. Rotman noted that an isotropic electrical plasma could be modeled by a medium with an index of refraction below unity, provided that its permeability was near that of free space. Consequently, the sphere- and disk-type media were excluded, since the finite dimensions of these conducting inclusions transverse to the applied electric field give the effective medium a diamagnetic response. What remained was the "rodded" dielectric medium, or conducting strip medium, consisting of thin wire rods oriented along the incident electric field. The dispersion characteristics of this medium showed that it does, indeed, behave like a plasma. The idea of a *negative* permittivity was implicit in many such works, but it was not until nearly a quarter-century later, when Rotman's rodded dielectric was rediscovered, that it was made clear exactly how a wire medium resembled a plasma.

It is evident that in the construction of electromagnetic structures of any sort in the microwave range, we rely on the properties of metals. Essentially, metals are plasmas, since they consist of an ionized "gas" of free electrons. Below their plasma frequency, the real component of the permittivity of bulk metals can be said to be negative. However, the natural plasma frequencies of metals normally occur in the ultraviolet region of the electromagnetic spectrum, in which wavelengths are extremely short. This condition certainly precludes the use of realizable artificial dielectrics in the microwave range, which, moreover, must operate in the long-wavelength regime. Although the permittivity is negative at frequencies below the plasma frequency, the approach toward absorptive resonances at lower frequencies increases the dissipation, hence the complex nature of ϵ. Thus, to observe a negative permittivity with low absorption at microwave frequencies, it would be necessary to somehow depress the plasma frequency of the metal.

This problem was addressed by Pendry et al. [12] (and simultaneously by Sievenpiper et al. [13]), who proposed the familiar structure of Rotman consisting of a mesh of very thin conducting wires arranged in a periodic lattice, but approached the problem from a novel standpoint. Due to the spatial confinement of the electrons to thin wires, the effective electron concentration in the volume of the structure is decreased, which also decreases the plasma frequency. More significant, however, is that the self-inductance of the wire array manifests itself as a greatly enhanced effective mass of the electrons confined to the wires. This enhancement reduces the effective plasma frequency of the structure by many orders of magnitude, placing it well into the gigahertz range. Thus, an array of thin metallic wires, by virtue of its macroscopic plasma-like behaviour, produces an effectively negative permittivity at microwave frequencies.

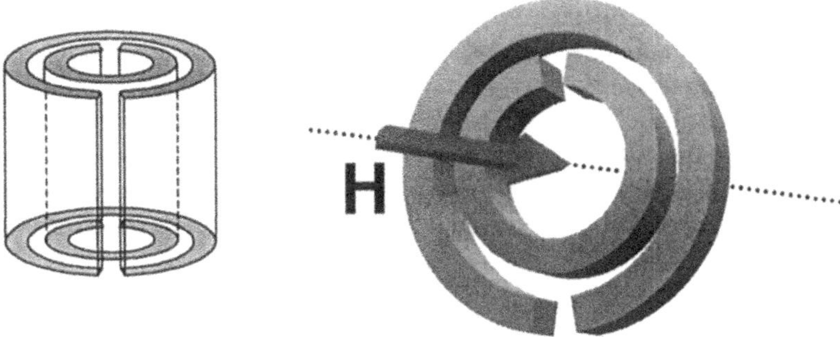

Fig. 1.4 The split-ring resonator (SRR) of Pendry et al. [14] in cylindrical and planar form, activated by a magnetic field normal to the plane of the rings.

1.2.3 Negative Permeability

Before dismissing the possibility of achieving $\mu < 0$ using natural isotropic substances, Veselago momentarily contemplated the nature of such a substance. He imagined a gas of magnetic "charges" exhibiting a magnetic plasma frequency, below which the permeability would assume negative values. The obstacle, of course, was the constitutive particle itself, the hypothetical magnetic charge. It is important to note that in the effort to synthesize a negative effective permittivity, Rotman and Pendry relied on the analogies their structures shared with the simplified electrodynamics of natural substances. Indeed, as acknowledged by Veselago himself [1], it is a much more difficult task to synthesize an isotropic negative permeability, for which there exists no known electrodynamic precedent.

In 1999, Pendry et al. [14] claimed to have developed microstructured artificial materials exhibiting strange magnetic properties. The work first developed expressions for the magnetic properties of materials resembling the wire mesh, in which the fields and currents are oriented along the wire axis. Ultimately, the work concluded that such materials are strictly diamagnetic and that the permeability approaches the free-space value as the radius of the wires is decreased, a response which may be expected of simple artificial dielectrics [9]. However, by giving the cylinders an internal electromagnetic structure resembling a parallel-plate capacitor wrapped around a central axis, Pendry et al. noticed a very different behaviour. The resulting split-ring resonator (SRR), depicted in Fig. 1.4, exhibits strong electric fields, supported by a very large capacitance, between the rings. Furthermore, although currents cannot traverse the gaps, the application of magnetic fields oriented normal to the plane of the rings induces simultaneous currents in both rings. This synthesized capacitance, along with the natural inductance of the cylindrical structure, results in a resonant response characterized by an effective relative permeability of the form

$$\mu_{eff} = 1 - \frac{\pi r^2/a^2}{1 - \frac{2lj\rho}{\omega r \mu_0} - \frac{3l}{\pi^2 \mu_0 \omega^2 C r^3}} \tag{1.1}$$

which has been appropriately modified from the original to refer to a $e^{+j\omega t}$ time variation. Here, r represents the radius of the SRR, a is the lattice spacing of SRRs lying in the same plane, l is the spacing between planes, ρ represents the resistive losses of the metal sheets, and C is the sheet capacitance between the two sheets. It is clear from the resonant form of μ_{eff} that an artificial medium composed of SRR arrays would exhibit an effective permeability that attains large values near the resonance, limited only by the amount of resistive loss. This resonance frequency is given by

$$\omega_0 = \sqrt{\frac{3l}{\pi^2 \mu_0 C r^3}} \tag{1.2}$$

However, μ_{eff} seems to possess another, more familiar trait: When $\sigma \to 0$, the permeability expression can become negative if the second term of (1.1) is greater than one. This occurs at an effective magnetic plasma frequency ω_{mp}, given by

$$\omega_{mp} = \sqrt{\frac{3l}{\pi^2 \mu_0 C r^3 (1 - \frac{\pi r^2}{a^2})}} \tag{1.3}$$

The quantity $\pi r^2/a^2$, which we shall denote F, is the fractional area occupied by the rings, or filling factor. When embedded in air, arrays of SRRs appear to exhibit a stopband in the frequency region enclosed by ω_0 and ω_{mp}, suggesting that the permeability is negative in this region. Although the phenomenon is evidently narrowband, the magnetic plasma frequency can be happily placed in the gigahertz range. Thus, comprising purely nonmagnetic materials, the SRR array of Pendry et al. had successfully simulated an artificial magnetic plasma, the substance hypothetically envisioned by Veselago, for which the effective permeability assumes negative values at microwave frequencies.

1.2.4 The First LHM

The work of Pendry et al. had yielded two distinct electromagnetic structures: Rotman's rodded artificial dielectric, unearthed and recast as a microwave plasma with $\epsilon < 0$, and the SRR, which exhibits $\mu < 0$ at microwave frequencies. The inevitable connection to Veselago was quickly made by Smith and Schultz et al. at the University of California at San Diego, in association with Pendry, who immediately engaged themselves in realizing the first LHM as a composite of conducting wires and SRRs [15–17]. For easy fabrication using standard microwave materials, the copper SRRs were implemented in planar form on fiberglass substrates, and numerous such boards were assembled into a periodic array. The UCSD team approached the task of characterizing the SRR array quite systematically, first through numerical simulations and then through simple transmission experiments in both one and two dimensions, confirming that the SRR provides a negative permeability for magnetic fields oriented normal to the plane of the rings. However, that an array of SRRs embedded in air creates a stopband in the vicinity of the SRR resonance frequency is not conclusively indicative of a negative effective permeability, since the SRR medium could possess

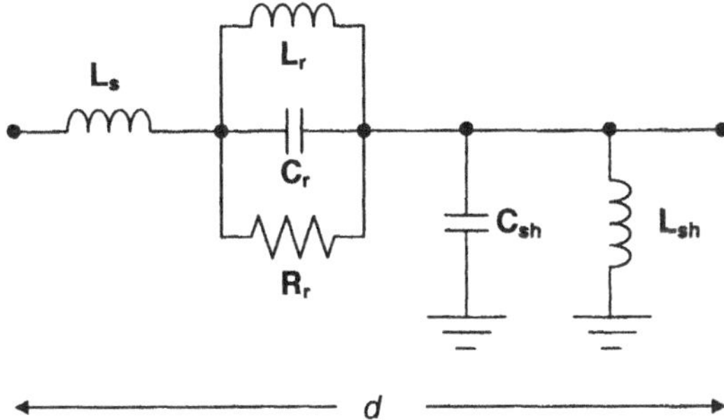

Fig. 1.5 Transmission-line model for Wire/SRR medium with $d \ll \lambda$.

an electric response as well (see Refs. 18 and 19). The approach taken by Smith et al. was to insert the SRR array into a printed wire medium, known to exhibit a negative permittivity below its effective plasma frequency, anticipating frequency bands in this range in which propagation had been restored by a negative permeability. Indeed, their simulations and experiments revealed a region of propagation enclosed by the resonance and magnetic plasma frequencies of Pendry's SRR, suggesting that the SRR array had, in fact, provided a negative effective permeability [20]. More intriguing, however, would be the nature of this region of propagation, in which the wire/SRR composite medium possesses simultaneously negative effective permittivity and permeability, the indelible signature of left-handedness.

It should be noted that Smith et al. employed a slightly different expression for the resonant form of μ_{eff} than (1.1), which ensured that the effective relative permeability approached unity in the infinite-frequency limit. This expression, along with the expression typically employed for the effective relative permittivity, is shown below:

$$\mu_{eff} = 1 - \frac{F\omega_0^2}{\omega^2 - \omega_0^2 - j\omega\Gamma} \tag{1.4}$$

$$\epsilon_{eff} = 1 - \frac{\omega_{ep}^2}{\omega^2} \tag{1.5}$$

Based on these expressions, it was shown in Ref. 22 that the wire/SRR medium embedded in vacuum possesses a direct L–C transmission-line analogue. The unit cell for this equivalent transmission line is shown in Fig. 1.5. The vacuum permeability and permittivity are represented by the series inductor $L_s = \mu_0 d$ and shunt capacitor $C_{sh} = \epsilon_0 d$, and the cell dimension d is assumed to be much smaller than the applied wavelength. The parallel RLC circuit in the series branch models the resonant response of the SRR and the inductor in the shunt branch models the effect of the wire array. The complex propagation constant for an equivalent transmission line consisting of an infinite cascade of such unit cells is given by

$$\gamma_t = \frac{j\omega}{c}\sqrt{\left[\left(\frac{L_s}{\mu_0 d}\right) - \frac{\frac{L_r}{\mu_0 d}\frac{1}{L_r C_r}}{\omega^2 - \frac{1}{L_r C_r} - \frac{j\omega}{R_r C_r}}\right]\left[\left(\frac{C_{sh}}{\epsilon_0 d}\right) - \frac{1}{\epsilon_0 \omega^2 L_{sh} d}\right]} \tag{1.6}$$

A comparison between (1.6) and the propagation constant produced by the product of (1.4) and (1.5), reveals the transmission-line model interpretation of the SRR and wire resonance frequencies:

$$\begin{aligned} \omega_0 &= \frac{1}{\sqrt{L_r C_r}} \\ \omega_{mp} &= \frac{1}{\sqrt{L_r C_r}}\sqrt{1+F} \\ \omega_{ep} &= \frac{1}{\sqrt{L_{sh} C_{sh}}} \end{aligned} \tag{1.7}$$

where $F = L_r/L_s$ is the filling factor. The value of this transmission-line model in describing the characteristics of negative-refractive-index media will be considered in the following sections.

Once it had been shown that propagation is restored in the medium when the effective material parameters are simultaneously negative, it remained only to apply the composite medium in experiment to verify whether it possessed a negative effective index of refraction, as Veselago hypothesized nearly thirty years earlier. The long-anticipated results were finally reported in *Science* [21]. R. A. Shelby, D. R. Smith, and S. Schultz at UCSD had experimentally verified negative refraction using a composite wire/SRR negative-refractive-index medium. Square SRRs printed on one side of a fiberglass substrate were coupled with wires printed on the reverse side, and the individual boards were assembled into a two-dimensional periodic, prism-shaped square lattice and embedded in air, as depicted in Fig. 1.6. The rings were designed for resonance around 10.5 GHz, where an LHM passband had previously been observed. The sample was irradiated by a microwave beam at 10.5 GHz incident at 18.43°, and a microwave detector was scanned azimuthally around the exit point in the plane of incidence. A control sample made of Teflon reported a positive angle of refraction of 27°, corresponding to the well-known refractive index of Teflon of +1.4, and calibrating the apparatus. Using the wire/SRR metamaterial, the same beam was observed to exit at an angle of −61°, which, applied to Snell's Law, yields an effectively negative index of refraction of −2.7. Resonant at 10.5 GHz, the wire and SRR media exhibited a bandwidth of approximately 500 MHz, or 5%, over which the refractive index was negative and in approximate agreement with the dispersion characteristics predicted by the product of the material parameter expressions of the wire and SRR media. The UCSD experiment had realized the LHM, and Veselago's seminal work was to be gloriously resurrected.

Fig. 1.6 Depiction of the wire/SRR metamaterial of Shelby et al. [21] used to verify negative refraction.

1.2.5 Terminology

Before concluding this section, it is perhaps useful to summarize the nomenclature found in the literature that has evolved for the description of metamaterials possessing these unique properties. The most frequently used terms are "left-handed," "negative-refractive-index" (or simply "negative-index"), "backward wave," and "double-negative" materials. We have already mentioned all terms and justified their origin except the last. The term "double-negative" media [23] originates from the fact that these materials are characterized by simultaneously negative permittivity and permeability. In our opinion, all four names are justified and have their advantages and disadvantages. However, we will not engage here in a further discussion on nomenclature because we believe it to be of secondary importance. In our research group at the University of Toronto, we have used the self-contained term "negative-refractive-index" to describe these metamaterials because this term conveys one of their most fundamental and surprising aspects and, moreover, is able to capture the imagination of the nonspecialist. Nevertheless, we also liberally use the term "left-handed" for historical reasons.

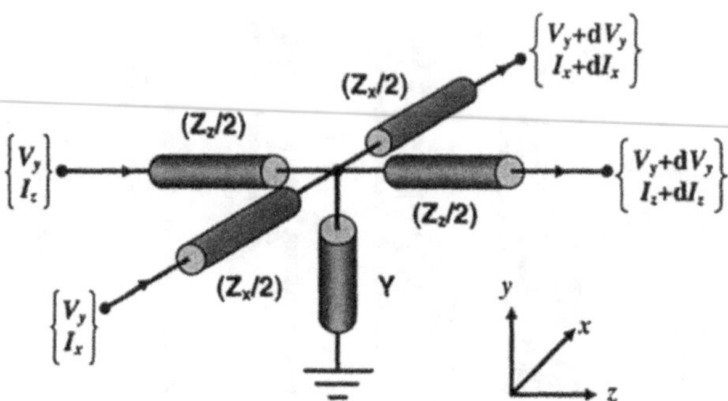

Fig. 1.7 Unit cell for a distributed transmission-line network model of a planar homogeneous medium. After Ref. [27]. Copyright © 2002 IEEE.

1.3 TRANSMISSION-LINE THEORY OF NEGATIVE-REFRACTIVE-INDEX MEDIA

The essential paradigm governing the synthesis of artificial media was the establishment of direct analogies with natural media. Such artificial dielectrics consisted of discrete electromagnetic scatterers periodically arranged into ordered arrays, analogous to the atoms and molecules in a crystal lattice. At wavelengths on the order of the lattice constant, these structures, like solids, exhibited diffraction effects, and at longer wavelengths, an effective refractive index could be defined. Central to each of these developments is the notion that all materials, natural or artificial, are granular at some level of scale; hence, artificial dielectrics were naturally and exclusively studied according to methods used to characterize natural dielectrics. For example, the investigation of such materials has traditionally begun with the determination of the electric polarizability and magnetization of the scatterer, followed by the application of the Lorentz theory for dielectrics, or they have been treated macroscopically as scattering problems.

The above methods of analysis, however rigorous, do not directly provide any insight into the *synthesis* of artificial media to possess specific effective material parameters, let alone the exotic parameters associated with negative-refractive-index metamaterials. For this we may look to the familiar transmission-line model, which represents natural media using a distributed network of reactances, a unit cell of which is shown in Fig. 1.7 for the planar case. Modeling natural media in this fashion interprets the notion of granularity through the process of discretization; that is, Maxwell's equations are solved in discrete spatial increments, or unit cells, in which the relevant field quantities are regarded to be quasi-static. The nature of the impedances and admittances in the 2-D unit cell of Fig. 1.7 can be determined using this approach, which we present herein. We begin with the statement of Maxwell's differential curl equations in the frequency domain, and we assume time-harmonic

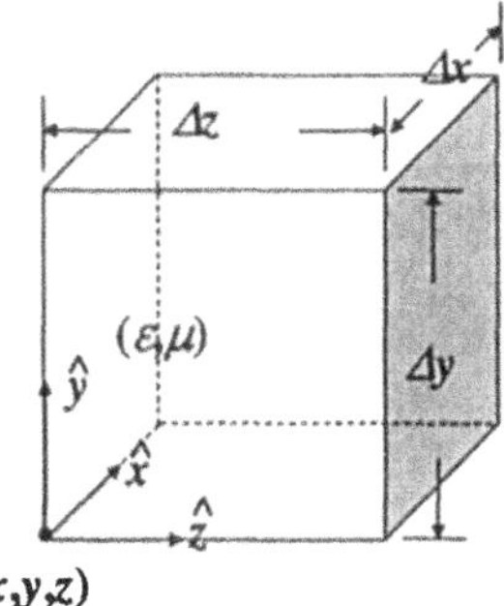

Fig. 1.8 Volume of space representing region in which fields may be considered quasi-static.

fields with time variation $e^{+j\omega t}$:

$$\nabla \times \mathbf{E} = -j\omega\mathbf{B} \qquad \text{(Faraday's Law)} \tag{1.8}$$

$$\nabla \times \mathbf{H} = \mathbf{J} + j\omega\mathbf{D} \qquad \text{(Ampère–Maxwell's Law)} \tag{1.9}$$

In homogeneous, isotropic media, these are supplemented by material constitutive relations of the form

$$\mathbf{B} = \mu(\omega)\mathbf{H} \tag{1.10}$$

$$\mathbf{D} = \epsilon(\omega)\mathbf{E} \tag{1.11}$$

where the dispersive nature of the material parameters $\mu(\omega)$ and $\epsilon(\omega)$ has been emphasized for generality.

We now suppose there exists a volume of space represented as a three-dimensional unit cell of dimensions Δx, Δy, and Δz, each of which is diminutive in comparison to the wavelength of the impressed field, thus enforcing the quasi-static field condition. This is depicted in Fig. 1.8. In the interest of ultimately studying a thin planar geometry, we may assume that there is no field variation in the y-direction so that $\partial/\partial y \to 0$, and the electromagnetic interactions supported by the resulting planar geometry may then be described by a combination of TE_y and TM_y modes. In anticipation of modeling dielectrics using distributed circuit analogies, we now develop expressions for the quasi-TM_y case, in which the predominant electric and magnetic field components are E_y, H_x, and H_z, related by the following expressions in the lossless case:

$$\frac{\partial E_y}{\partial x} = -j\omega\mu(\omega)H_z \tag{1.12}$$

$$\frac{\partial E_y}{\partial z} = +j\omega\mu(\omega)H_x \tag{1.13}$$

$$\frac{\partial H_x}{\partial z} - \frac{\partial H_z}{\partial x} = +j\omega\epsilon(\omega)E_y \tag{1.14}$$

Spatial discretization of Maxwell's equations over this cube results in the following expressions:

$$E_y(x_0 + \Delta x, z_0) - E_y(x_0, z_0) = -j\omega\mu(\omega)H_z\Delta x \quad (1.15)$$
$$E_y(x_0, z_0 + \Delta z) - E_y(x_0, z_0) = +j\omega\mu(\omega)H_x\Delta z \quad (1.16)$$

and

$$\begin{aligned}&[H_x(x_0, z_0 + \Delta z) - H_x(x_0, z_0)]\,\Delta x \\ &\quad - [H_z(x_0 + \Delta x, z_0) - H_z(x_0, z_0)]\,\Delta z = +j\omega\epsilon(\omega)E_y(x_0, z_0)\Delta x\Delta z\end{aligned} \quad (1.17)$$

The definitions of potential difference and current using field quantities are as follows:

$$V_{a'} - V_a = -\int_a^{a'} \mathbf{E} \cdot d\mathbf{l} \quad (1.18)$$
$$I = \oint_C \mathbf{H} \cdot d\mathbf{l} \quad (1.19)$$

where $a - a'$ is any path connecting the bottom and top faces of the cube, and C is a suitably chosen closed contour slicing its bottom or top face. Since the fields are quasi-static within the volume of the unit cell, the integrals degenerate into the simple products $V_y = E_y\Delta y$ (assuming the bottom face of the unit cell is taken as the zero reference potential), $I_z = -H_x\Delta x$ and $I_x = H_z\Delta z$. Furthermore, defining the impedance and admittance quantities $Z_x = j\omega\mu(\omega)\Delta x\Delta y/\Delta z$, $Z_z = j\omega\mu(\omega)\Delta y\Delta z/\Delta x$, $Y = j\omega\epsilon(\omega)\Delta x\Delta z/\Delta y$ and rearranging, (1.15)–(1.17) reduce to

$$V_y(x_0 + \Delta x, z_0) - V_y(x_0, z_0) = -Z_x I_x \quad (1.20)$$
$$V_y(x_0, z_0 + \Delta z) - V_y(x_0, z_0) = -Z_z I_z \quad (1.21)$$

and

$$\begin{aligned}&[I_z(x_0, z_0 + \Delta z) - I_z(x_0, z_0)] + [I_x(x_0 + \Delta x, z_0) - I_x(x_0, z_0)] \\ &\qquad = -YV_y(x_0, z_0)\end{aligned} \quad (1.22)$$

The first (second) equation suggests that the potential difference between the front and back (left and right) faces of the cube of Fig. 1.8 results from a current drawn by an effective impedance Z_x (Z_z). The third equation suggests that the potential difference between the top and bottom faces is given by the current drawn by an admittance Y. These results are familiar, because they are merely a two-dimensional representation of the transmission-line telegrapher's equations. Equivalently, these are Kirchhoff's voltage and current laws for the per-unit-length lumped-element model of the symmetric transmission-line unit cell of Fig. 1.7. The nature of the impedances and admittances at a particular frequency $\omega = \omega_0$ is evident in the expressions $Z_x = j\omega\mu(\omega_0)\Delta x\Delta y/\Delta z$, $Z_z = j\omega\mu(\omega_0)\Delta y\Delta z/\Delta x$, $Y = j\omega\epsilon(\omega_0)\Delta x\Delta z/\Delta y$,

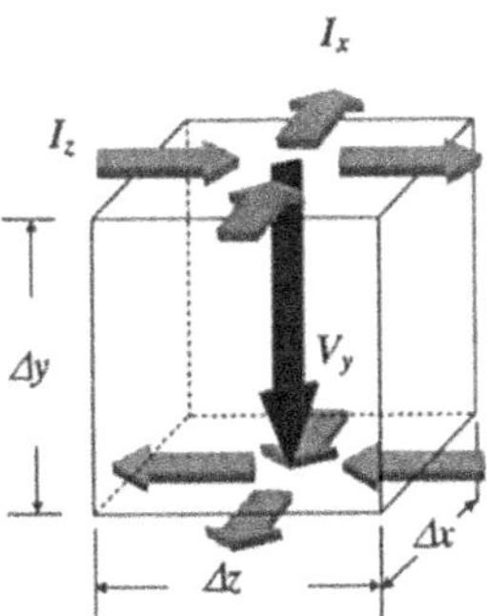

Fig. 1.9 Determination of the impedance and admittance quantities in Fig. 1.7 by mapping quasi-static field quantities in the unit cell of Fig. 1.8 to currents and voltages.

and this is illustrated in Fig. 1.9. The presence of uniform electric fields throughout the area $\Delta x \Delta z$ of the unit cell over its thickness Δy describes a parallel-plate capacitor filled with the medium $\epsilon(\omega_0)$, $\mu(\omega_0)$ whose capacitance is given by the familiar relation $C = \epsilon(\omega_0)\Delta x \Delta z/\Delta y$. The presence of quasi-static magnetic fields is akin to oppositely directed currents in the parallel plates whose flux contributions are linked through an area $\Delta y \Delta z$ for currents flowing across a distance of Δx and through $\Delta x \Delta y$ for currents flowing across Δz. This yields an inductance within the unit cell of $L_x = \mu(\omega_0)\Delta x \Delta y/\Delta z$ in the x-direction and $L_z = \mu(\omega_0)\Delta y \Delta z/\Delta x$ in the z-direction. More interesting however, are the distributed capacitance and inductance, which are, evidently, related to $\epsilon(\omega_0)$ and $\mu(\omega_0)$ through a constant term given by the geometry of the unit cell; for the isotropic medium, when the dimensions of the unit cell are infinitesimal compared to the wavelength (the continuous limit), the distributed capacitance and inductance are identically equal to the isotropic material parameters. The result, therefore, is that any homogeneous and lossless dielectric can be modeled at a particular frequency ω_0 by discrete unit cells containing only inductors and capacitors, apportioned such that the per-unit-length capacitance and inductance (i.e., these are distributed capacitances and inductances), through the geometry of the unit cell, represent the effective material parameters $\epsilon(\omega_0)$ and $\mu(\omega_0)$.

Under the assumption of a two-dimensionally isotropic, cubic unit cell ($Z = Z_x = Z_z$, $d = \Delta x = \Delta y = \Delta z$) the above impedance and admittance expressions become $Z = j\omega\mu(\omega)d$, $Y = j\omega\epsilon(\omega)d$, so that the effective material parameters modeled by the transmission-line network are expressed by

$$\mu(\omega) = \frac{Z(\omega)/d}{j\omega} \tag{1.23}$$

$$\epsilon(\omega) = \frac{Y(\omega)/d}{j\omega} \tag{1.24}$$

For any conventional medium (or RHM) that is isotropic, nonmagnetic, and possesses a relative permittivity ϵ_r, we require $\mu(\omega) = \mu_0$ and $\epsilon(\omega) = \epsilon_r\epsilon_0$, and so we must

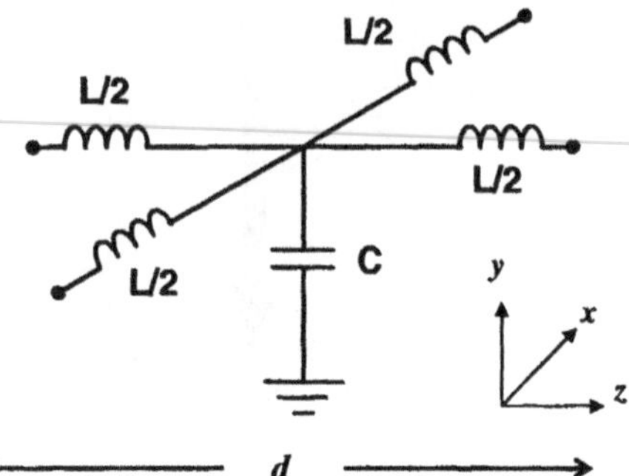

Fig. 1.10 2-D transmission-line unit cell describing a medium with $\mu(\omega) = \mu_0$ and $\epsilon(\omega) = \epsilon_r \epsilon_0$ using a per-unit-length series inductance and shunt capacitance.

choose $Z = j\omega\mu_0 d$ and $Y = j\omega\epsilon_r\epsilon_0 d$. This implies a two-dimensional medium in a low-pass topology whose unit cell possesses a series inductance $L = \mu_0 d$ (H) and shunt capacitance $C = \epsilon_r\epsilon_0 d$ (F), as shown in Fig. 1.10. Here, $L' = \mu_0 = L/d$ (H/m) and $C' = \epsilon_r\epsilon_0 = C/d$ (F/m) are the corresponding distributed quantities, both of which, it should be noted, are positive and real. The reader will also recall that the development that yielded this unit cell assumed no losses, which could, otherwise, have been modeled by a resistance in series and a conductance in shunt.

In the continuous limit, with $d/\lambda \to 0$, the corresponding propagation constant β, which is obtained from the circuit wave equation,

$$\frac{\partial^2 V_y}{\partial x^2} + \frac{\partial^2 V_y}{\partial z^2} + \beta^2 V_y = 0, \qquad \beta = \pm\sqrt{-ZY} \tag{1.25}$$

reduces to that of a standard transmission line, filled with a nonmagnetic dielectric with relative permittivity ϵ_r,

$$\beta = \pm\sqrt{-ZY} = \omega\sqrt{L'C'} = \omega\sqrt{\mu_0\epsilon_r\epsilon_0} = \omega/v_\phi \tag{1.26}$$

This dispersion relation reveals the variation of the propagation constant with frequency, and is presented in two formats for the continuous medium case in Fig. 1.11. The ω–β curve of Fig. 1.11a shows the variation of the propagation constant along a particular axis of propagation in the x–z plane as a function of frequency, and Fig. 1.11b shows the variation of the propagation constant as a function of propagation direction at a particular frequency. Accordingly, the latter diagram is known as an equifrequency surface (EFS). The ω–β curve provides the magnitude of the phase and group velocities in the medium, and the EFS provides their specific directions. When the unit cell is cubic and electrically infinitesimal, the distributed structures model isotropic media, and the EFS of the planar network is circular. The phase velocity is defined as the ratio $v_\phi = \omega/\beta$, whose magnitude is given by the slope of the line from the origin of the ω–β curve to a point (ω_0, β_0) in Fig. 1.11a and whose direction is given by the line connecting the origin to a point $(\beta_{0,x}, \beta_{0,z})$ on

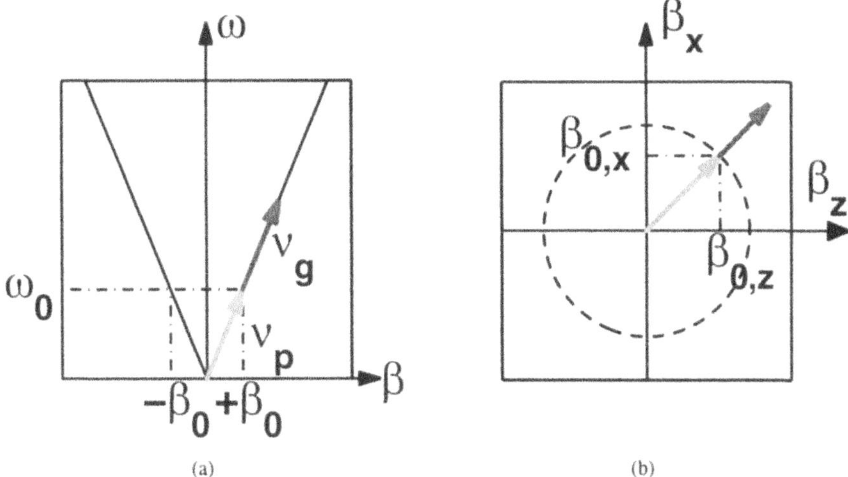

Fig. 1.11 Dispersion relation for medium with $\mu = \mu_0$ and $\epsilon = \epsilon_r\epsilon_0$ described using (a) an ω–β curve and (b) an equifrequency surface (EFS).

the EFS of Fig. 1.11b. The group velocity is defined as $v_g = (\partial\beta/\partial\omega)^{-1}$, which is the slope of the tangent to the ω–β curve at (ω_0, β_0) and which is directed along the gradient to the EFS, taken in the direction of increasing frequency. It is evident from Fig. 1.11a that the propagation constant of conventional isotropic RHM modeled by a distributed series inductance and shunt capacitance varies proportionally with frequency, as would be expected in conventional dielectrics at low frequencies. It is also clear from Fig. 1.11b that the resulting phase and group velocities are parallel and equal (dispersionless medium) and are given by

$$v_\phi = \frac{\omega}{\beta} = \frac{1}{\sqrt{L'C'}} = \frac{1}{\sqrt{\mu_0\epsilon_r\epsilon_0}} = \left(\frac{\partial\beta}{\partial\omega}\right)^{-1} = v_g \tag{1.27}$$

That the phase and group velocities are both positive results from the choice of the positive root in (1.27). The choice is arbitrary since it serves only to select one of two solutions related through a space reversal or, equivalently, one of the two branches of the ω–β curve or diametrically opposite directions of the EFS. However, to avoid the needless confusion associated with the term "negative group velocity," it is quite logical to define the phase velocity according to the branch of the ω–β curve that supports a "positive group velocity," which represents power flow in the reference direction. Accordingly, in RHM, a positive phase velocity means that the phase lags in the direction of the group velocity (in this case, parallel to the Poynting vector), a fact that is invariant to the sign selected for the root. Thus, the refractive index, which can be defined as the ratio between the speed of light in vacuum and the phase velocity in the medium, is positive:

$$n = \frac{c}{v_\phi} = \frac{\sqrt{L'C'}}{\sqrt{\mu_0\epsilon_0}} = \frac{\sqrt{\mu_0\epsilon_r\epsilon_0}}{\sqrt{\mu_0\epsilon_0}} = \sqrt{\epsilon_r} \tag{1.28}$$

Furthermore, the wave impedance of the effective medium is exactly equal to the characteristic impedance of the distributed network in the continuous limit, as expected:

$$\eta_r = \sqrt{\frac{\mu_0}{\epsilon_r\epsilon_0}} = \sqrt{\frac{L'}{C'}} = Z_0 \tag{1.29}$$

1.3.1 Application of the Transmission-Line Theory of Dielectrics to the Synthesis of LHM

That Maxwell's equations can be represented in entirety by appropriate circuit equations, and consequently that natural media could be represented by distributed circuit networks, was recognized as early as 1944 by G. Kron [24], who employed the method of spatial discretization of Maxwell's equations to arrive at Kirchhoff's voltage and current laws for three-dimensional media, and by J. R. Whinnery and S. Ramo [25], who treated two-dimensional media, although it was only four years later that W. Kock introduced the term 'artificial dielectric'. This trend suggests that, having availed of our familiarity with circuits to model the behaviour of natural media, we may now apply the same principles to synthesize the behaviour of artificial media. That is, the elegance of the distributed circuit concept lies in the fact that it is applicable to any homogeneous dielectric, since we are free to specify L' and C', and it therefore allows us to examine the scope of distributed L–C networks for reproducing the exotic material parameters associated with LHM. Specifically, Veselago's postulation of a negative permittivity and permeability prompts us to ask whether the L' and C' parameters in a network representation can also be made negative. Naturally, from an impedance perspective, imposing a negative L' and C', or equivalently a negative series impedance $-j\omega L'd$ and shunt admittance $-j\omega C'd$, essentially exchanges their reactive and susceptive roles, so that the series inductor becomes a series capacitor, and the shunt capacitor becomes a shunt inductor. The unit cell of the emerging *dual* structure is shown in Fig. 1.12, and it is easily recognized as having the topology of a two-dimensional high-pass filter network. The effective permittivity and permeability represented by this topology can, once again, be obtained using (1.23) and (1.24), which result in

$$\mu(\omega) = \frac{1/j\omega Cd}{j\omega} = -\frac{1}{\omega^2 Cd} \tag{1.30}$$

$$\epsilon(\omega) = \frac{1/j\omega Ld}{j\omega} = -\frac{1}{\omega^2 Ld} \tag{1.31}$$

Contrary to the results for the RHM unit cell, the effective material parameters of the dual network are prominently negative. However, they are no longer con-

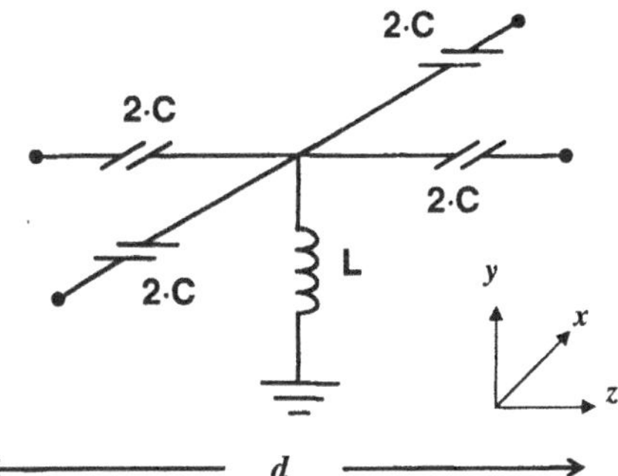

Fig. 1.12 2-D dual transmission-line unit cell describing a medium with simultaneously negative, dispersive parameters $\mu = -|\mu(\omega)|$ and $\epsilon = -|\epsilon(\omega)|$ using a per-unit-length series capacitance and shunt inductance.

stants, and are, instead, explicit functions of frequency; in fact, their particular dispersive forms ensure that the time-averaged stored electric and magnetic energies associated with this medium are positive, so that the conservation of energy is not violated [1, 9]. Thus, the simple dual high-pass network, with distributed series capacitance $C' = Cd$ (F·m) and shunt inductance $L' = Ld$ (H·m), satisfies the principal requirement for left-handed behaviour: The effective material parameters are simultaneously negative. Therefore, the network may legitimately be described as an LHM. This novel conception of the negative LHM permeability and permittivity in terms of an equivalent series capacitance and shunt inductance was investigated in Refs. 26 and 27, which presented the above dispersion analysis as well as simulations of negative refraction and focusing in two-dimensional dual L–C arrays for both continuous implementations ($d/\lambda \to 0$) and periodic implementations using finite-length transmission-line segments, in which the size of the unit cell was shown to affect the dispersion properties of the network (described in Section 1.4). The latter work also presented an experimental verification of these ideas using a small, transmission-line based, planar negative-refractive-index lens with which focusing was demonstrated. A similar theory was followed in Ref. 28, which discussed the continuous implementation of the one-dimensional dual L–C line, and in Ref. 29, which described some interesting devices and circuits. A stripline implementation of the one- and two-dimensional dual L–C medium was also suggested in Ref. 30.

The propagation constant associated with the dual structure, found through the application of (1.30) and (1.31), boasts a peculiar inverse relationship with frequency,

$$\beta = -\sqrt{-ZY} = -\frac{1}{\omega\sqrt{L'C'}} \tag{1.32}$$

and the corresponding ω–β curve and EFS are as shown in Fig. 1.13. In this case, the phase and group velocities are antiparallel and are given by

$$v_\phi = \frac{\omega}{\beta} = -\omega^2\sqrt{L'C'} = -\left(\frac{\partial\beta}{\partial\omega}\right)^{-1} = -v_g \tag{1.33}$$

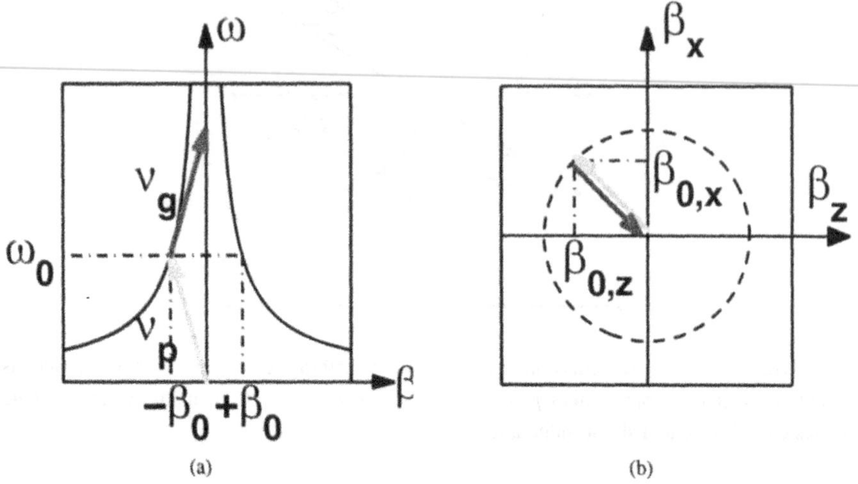

Fig. 1.13 Dispersion relation for medium with simultaneously negative, dispersive parameters $\mu = -|\mu(\omega)|$ and $\epsilon = -|\epsilon(\omega)|$ described using (a) an ω–β curve and (b) an equifrequency surface (EFS).

where the choice of the negative root in (1.33) has ensured a positive group velocity (in this case also parallel to the Poynting vector), in accordance with our adopted convention. These results are familiar because they return us to the phenomenon of the backward wave; that is, since the phase and group velocities (Poynting vector) are antiparallel, negative refraction can be expected at an interface between an RHM and LHM constructed from a dual L–C network by virtue of phase matching at the interface. Indeed, the one-dimensional simplification of the above two-dimensional analysis leads to the familiar high-pass filter, which is known to support a fundamental spatial harmonic in which the Poynting vector and wavevector are oppositely directed [3]. The backward wave manifests itself in the fact that, in LHM, the phase leads in the direction of positive group velocity, or power flow, and so the index of refraction should, accordingly, be negative, and it is clear that the relationship between the effective wave impedance and network characteristic impedance is also preserved:

$$n = \frac{c}{v_\phi} = \frac{-\sqrt{\mu(\omega)\epsilon(\omega)}}{\sqrt{\mu_0\epsilon_0}} = -\frac{1}{\omega^2\sqrt{L'C'\mu_0\epsilon_0}} \tag{1.34}$$

$$\eta = \sqrt{\frac{\mu(\omega)}{\epsilon(\omega)}} = \sqrt{\frac{L'}{C'}} = Z_0 \tag{1.35}$$

It is noteworthy that the above development has, once again, assumed no losses; it is, therefore, evident that the dual L–C representation of the LHM does not require losses to synthesize simultaneously negative effective material parameters. Furthermore, unlike the wire/SRR negative-refractive-index media described in Ref. 21, the dual

L–C medium does not rely explicitly on resonators to synthesize the required negative material parameters [22, 31]. On this note, let us revisit the transmission-line model of the wire/SRR medium that was described earlier in this chapter (Fig. 1.5). It was previously noted that the effective permeability function of (1.1) assumed negative values in the frequency region enclosed by ω_0 and ω_{mp}. In the transmission-line model of the wire/SRR medium, this range corresponds exactly to the frequency region in which the series branch is capacitive. Similary, the effective permittivity is negated below ω_{ep}, where the shunt branch is inductive. Thus, in the frequency region in which a negative refractive index is achieved, the wire/SRR medium can be said to reduce to the dual *L–C* circuit topology described herein. In this sense, the SRR introduces an "excess" resonance that may be avoided by implementing the series capacitance directly. In doing so, we are able to realize extremely large bandwidths over which the negative refractive index property is maintained. Indeed, it is evident from Fig. 1.13 that the continuous, dual *L–C* medium exhibits a backward-wave characteristic over an infinite range of frequencies.

The above results describe an isotropic dual *L–C* medium; Balmain et al. [32] have developed transmission-line based *anisotropic* metamaterials, which have demonstrated negative refraction and focusing into a spot on the order of $\lambda/25$. These metamaterials, consisting of a two-dimensional periodic *L–C* grid over ground with series capacitors loading one grid axis and series inductors loading the other (and shunt inductors to ground in some implementations [33–35]), are excited by a point source between ground and the anisotropic grid, which forms a resonant path that can be made to scan with frequency; this phenomenon is akin to the "resonance cones" observed in anisotropic plasmas, and such anisotropy is described by hyperbolic equifrequency surfaces. More on these fascinating anisotropic media can be found in Chapter 6.

1.4 PERIODICALLY LOADED NEGATIVE-REFRACTIVE-INDEX TRANSMISSION-LINE (NRI-TL) METAMATERIALS

The practical realization of the dual *L–C* unit cell of the previous section requires us to endow the continuous medium with small physical dimensions provided by a host transmission-line medium and a unit cell inductance and capacitance that must be realized using either lumped printed or discrete (chip) inductors and capacitors. Many such cells must then be arranged periodically into a two-dimensional lattice, and the array must be excited at guide wavelengths longer than the cell dimension to enforce the homogeneity condition, so that the resulting periodic structure can be regarded as distributed. Under these conditions, we may expect that such a structure will exhibit the unusual left-handed properties proven for its continuous counterpart (albeit limited in some quantifiable sense by the imposed periodicity), including a negative refractive index. We therefore consider a design that periodically loads a two-dimensional host transmission-line network (for which the relevant distributed parameters are known) with discrete or printed reactive elements. We shall hereinafter refer to this structure

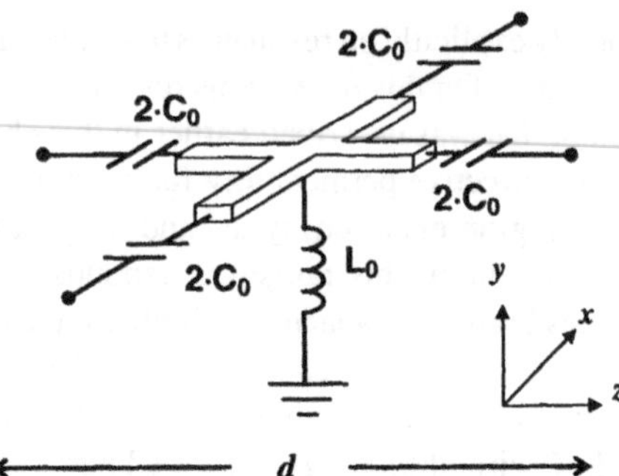

Fig. 1.14 Practical, periodic implementation of the continuous 2-D unit cell of Fig. 1.12 consisting of a transmission-line host medium periodically loaded with inductors and capacitors in a dual configuration. After Ref. [36]. Copyright © 2003 Optical Society of America, Inc.

as a negative-refractive-index transmission-line (NRI-TL) metamaterial. The unit cell for the periodic, two-dimensional NRI-TL metamaterial is depicted in Fig. 1.14, and it shows the host transmission lines loaded with series capacitors and a shunt inductor [27,36]. The host transmission line, which possesses a propagation constant k and provides the dimension d of the unit cell, can be characterized by a total phase shift $\theta = kd$ in both the x- and z-directions, along with a characteristic impedance Z_0.

1.4.1 Dispersion Characteristics

For periodic structures, the propagation constant and characteristic impedance are obtained through the periodic analysis of microwave networks, in which an infinite periodic array is characterized using a transmission ($ABCD$) matrix representation of a single unit cell with periodic boundary conditions. The problem is solved by invoking the Floquet–Bloch Theorem, which states that the terminal electric and magnetic fields (voltages and currents) will describe the effective propagation of a "wave" only if they are related through a frequency-dependent phase shift given by an effective propagation constant. The topology of the unit cell and choice of reference plane also specify the effective characteristic impedance, or Bloch impedance, of the structure.

The continuous, two-dimensional dual L–C structure of the previous chapter represented an isotropic medium at all frequencies, hence the circularity of its equifrequency surface (EFS) shown in Fig. 1.13b. However, the use of finite-length transmission lines in the practical NRI-TL structure, although providing the needed dimensionality, results in a spatial anisotropy that makes the propagation and impedance characteristics a function of the propagation angle ϕ, as well as frequency. EFSs for a representative periodically loaded transmission-line metamaterial are shown for the lowest (LH) passband in Fig. 1.15, where the darker curves near the periphery

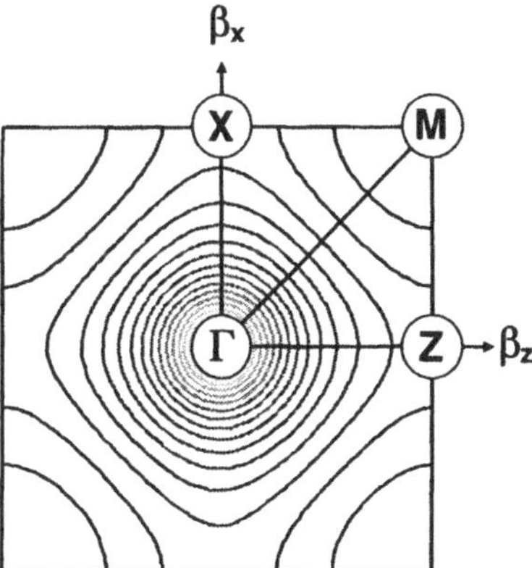

Fig. 1.15 Equifrequency surfaces of the reduced Brillouin zone in the lowest (LH) passband for a representative, periodically loaded transmission-line metamaterial indicating the symmetry points Γ ($\beta_x = \beta_z = 0$), X ($\beta_x = \pi/d, \beta_z = 0$), Z ($\beta_x = 0, \beta_z = \pi/d$), and M ($\beta_x = \pi/d, \beta_z = \pi/d$), where d is the unit cell dimension, or lattice spacing.

represent lower frequencies and the lighter curves near the centre represent higher frequencies. The labels on the figure identify the symmetry points of the reduced Brillouin zone [37], where Γ represents $\beta_x = \beta_z = 0$, X represents $\beta_x = \pi/d, \beta_z = 0$, Z represents $\beta_x = 0, \beta_z = \pi/d$, M represents $\beta_x = \pi/d, \beta_z = \pi/d$, and d is the period of the lattice, or the dimension of the unit cell of Fig. 1.14. The spatial anisotropy evident in Fig. 1.15 is a consequence of the fact that waves scattered by a particular unit cell are restricted to propagation along the axes of the host transmission-line medium, impacting the propagation characteristics along the diagonals ($\phi = \pi/4$) most severely. This can be seen in Fig. 1.16a, which depicts an array of NRI-TL unit cells viewed from above (the x–z plane of Fig. 1.14). Although this anisotropy can be minimized by making the interconnecting lines electrically small, it is clear that the full description of propagation in the NRI-TL structure requires a two-dimensional periodic analysis (see Ref. 38 and Chapter 3). Nevertheless, *axial* propagation along the NRI-TL metamaterial (that is, at $\phi = 0$ or $\phi = \pi/2$) is a highly representative case that provides physical insight into its behaviour and considerably simplifies the analysis. This is unlike propagation along a one-dimensional NRI-TL metamaterial, since we must account for the effect of the transverse loading by the host transmission-line medium. This can be intuitively understood by considering the picture of axial propagation in the two-dimensional grid, as shown in Fig. 1.16b. For axial propagation in z (or, employing the notation of Fig. 1.15, the Γ–Z direction) the impedance in x is infinite, and so no current flows in this direction. Consequently, each of the two transverse transmission-line segments become open-circuited at their centres, and, therefore, contribute capacitively to the unit cell. This is illustrated in Fig. 1.17. Thus, the equivalent NRI-TL unit cell for axial propagation may be depicted as in Fig. 1.18, where the shunt branch now includes the combined admittance

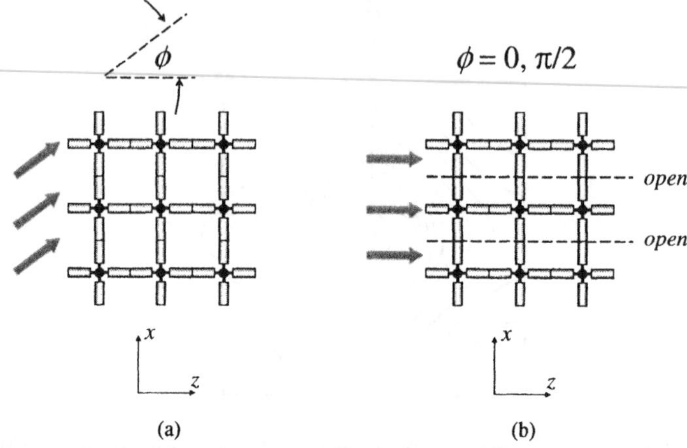

Fig. 1.16 Plane wave incident on an array of NRI-TL unit cells (top view) at (a) angle ϕ and (b) axially ($\phi = 0, \phi = \pi/2$), resulting in the formation of open circuits in the medium transverse to the direction of propagation. After Ref. [27]. Copyright © 2002 IEEE.

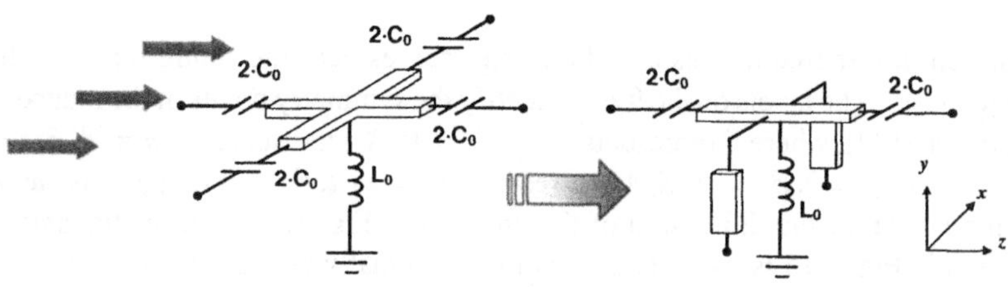

Fig. 1.17 Capacitive loading due to host TL medium transverse to direction of axial propagation.

of the open-circuited stubs $2Y_{OC} = 2(-jZ_0 \cot(\theta/2))^{-1}$ [27]. A periodic analysis applied to this equivalent NRI-TL unit cell reveals the axial propagation constant and terminal Bloch impedance Z_B:

$$\cos \beta d = \left\{ \cos\theta - \frac{1}{2\omega^2 L_0 C_0} \cos^2\frac{\theta}{2} + \frac{1}{2\omega}\left(\frac{1}{C_0 Z_0} + \frac{1}{L_0 Y_0}\right)\sin\theta \right\} - \sin\frac{\theta}{2}\left(2\sin\frac{\theta}{2} - \frac{1}{\omega C_0 Z_0}\cos\frac{\theta}{2}\right) \tag{1.36}$$

$$Z_B^{\pm} = \pm \frac{Z_0 \tan\frac{\theta}{2} - \frac{1}{2\omega C_0}}{\tan\frac{\beta d}{2}} \tag{1.37}$$

The reader may wish to note that the terms enclosed in curly brackets in (1.36) represent the dispersion relation of the one-dimensional NRI-TL line and that the additional

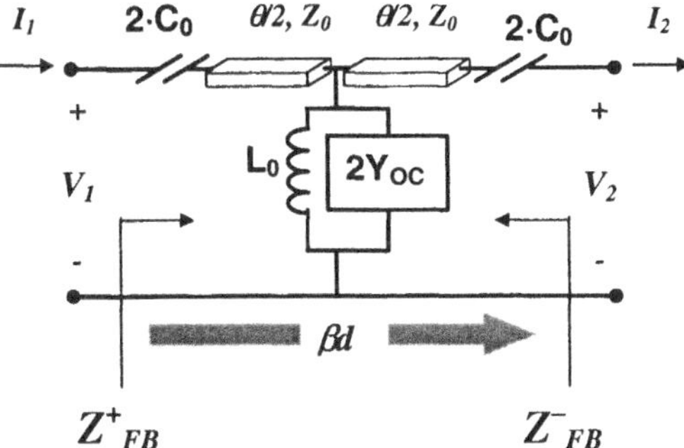

Fig. 1.18 Symmetric NRI-TL unit cell modeling axial propagation in the 2-D NRI-TL grid.

term represents the correction to cos βd required to account for the transverse TL segments in the two-dimensional medium. Thus, the behaviour of electromagnetic waves propagating axially through the infinite two-dimensional NRI-TL periodic structure is thoroughly characterized by (1.36) and (1.37), provided that the properties of the host medium (θ , Z_0) and the loading elements (L_0, C_0) are given.

One suitable host transmission line is the microstrip, which is a class of electromagnetic waveguide consisting of a strip conductor and a backplane, separated by a thin dielectric. Microstrip lines are easily integrated with discrete or printed passive components, as well as active components, and the equipment required to design, construct, and test a microstrip circuit is inexpensive and readily accessible. Their use is further validated by the ease with which they may be fabricated and integrated with existing planar RF/microwave circuits and devices. Microstrip-based NRI-TL structures employing both discrete loading using chip elements [27] and printed elements [39,40] are depicted schematically in Fig. 1.19. Assuming quasi-static conditions, we may replace the inhomogeneous space surrounding the microstrip with a homogeneous dielectric with effective permittivity $\epsilon_P = \epsilon_{eff}\epsilon_0$, as described in Refs. 4 and 41, for which the intrinsic impedance may be represented as η_{eff}, and the per-unit-length capacitance C' and inductance L' are related through the microstrip characteristic impedance Z_0. Although the propagation constant along the microstrip $k = \omega\sqrt{L'C'}$ can be said to be equal to that in the homogeneous surrounding medium (owing to the quasi-TEM nature of the microstrip fields), $\beta = \omega\sqrt{\epsilon_{eff}\epsilon_0\mu_0}$, the characteristic impedance $Z_0 = \sqrt{L'/C'}$ is not, in general, equal to the wave impedance $\eta_{eff} = \sqrt{\mu_0/\epsilon_{eff}\epsilon_0}$ and is, instead, related to it by a factor g determined by the mapping of the quasi-static field quantities **E** and **H** to voltages $\widetilde{V}$ and currents $\widetilde{I}$ through the geometry of the transmission line. Thus, the effect of the geometry is also evident in the relationship between the per-unit-length capacitance and inductance of the transmission line and the permittivity and permeability of the intrinsic medium:

$$k = \omega\sqrt{L'C'} = \omega\sqrt{\epsilon_P\mu_P} \tag{1.38}$$

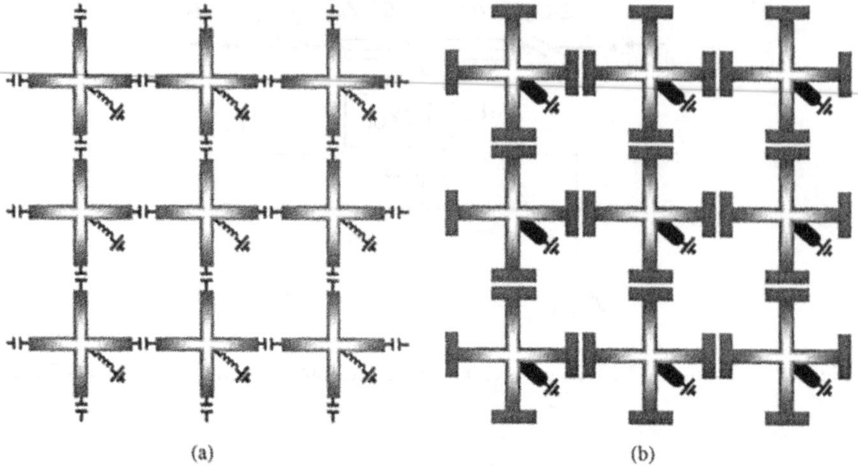

Fig. 1.19 Microstrip-based NRI-TL medium employing loading with (a) discrete inductors and capacitors and (b) vias and printed gaps.

$$Z_0 = \sqrt{\frac{L'}{C'}} = g\sqrt{\frac{\mu_P}{\epsilon_P}} \Rightarrow \begin{cases} \epsilon_P = C'g = \epsilon_{eff}\epsilon_0 \\ \mu_P = L'/g = \mu_0 \end{cases} \tag{1.39}$$

For example, consider the field cell of Fig. 1.8 for propagation in the z-direction. The per-unit-length capacitance and inductance are $C' = \epsilon(\omega_0)\Delta x/\Delta y$ and $L' = \mu(\omega_0)\Delta y/\Delta x$, respectively, and these yield $Z_0 = (\Delta y/\Delta x)\sqrt{\mu(\omega_0)/\epsilon(\omega_0)}$. The factor $g = \Delta y/\Delta x$ is explicitly seen in all three quantities. For microstrip lines of width w and substrate height h, for which the field distributions are not as simple, g may be determined through a mapping of the quasi-static fields and is as follows (see Refs. 4, 41, 42, and 36):

$$g = \begin{cases} \frac{1}{2\pi}\ln(8h/w + w/4h), & w/h \le 1 \\ \left[w/h + 1.393 + 0.667\ln(w/h + 1.444)\right]^{-1}, & w/h > 1 \end{cases} \tag{1.40}$$

In the previous discussion on transmission-line networks representing continuous media, the frequency response of the dual L–C structure permitted propagation at all frequencies, as would be expected for a continuous medium. However, in loading the unit cells periodically across finite lengths of transmission line, the dispersion diagram develops frequency bands in which propagation is forbidden. Indeed, it is evident from (1.36) that the axial propagation constant β can assume imaginary values at frequencies lying within such *stopbands*. Choosing the representative values $L_0 = 5.6$ nH, $C_0 = 1.0$ pF, $\epsilon_r = 2.94$, $h = 1.524$ mm, $w = 0.4$ mm, and $d = 5$ mm, the full dispersion relation is of the form depicted in Fig. Although Figs. 1.20a and 1.20b reassure us that the introduction of finiteness and periodicity to the NRI-TL structure has not affected its left-handed characteristic at low frequencies, the previously infinite extent of the left-handed passband has been truncated and replaced

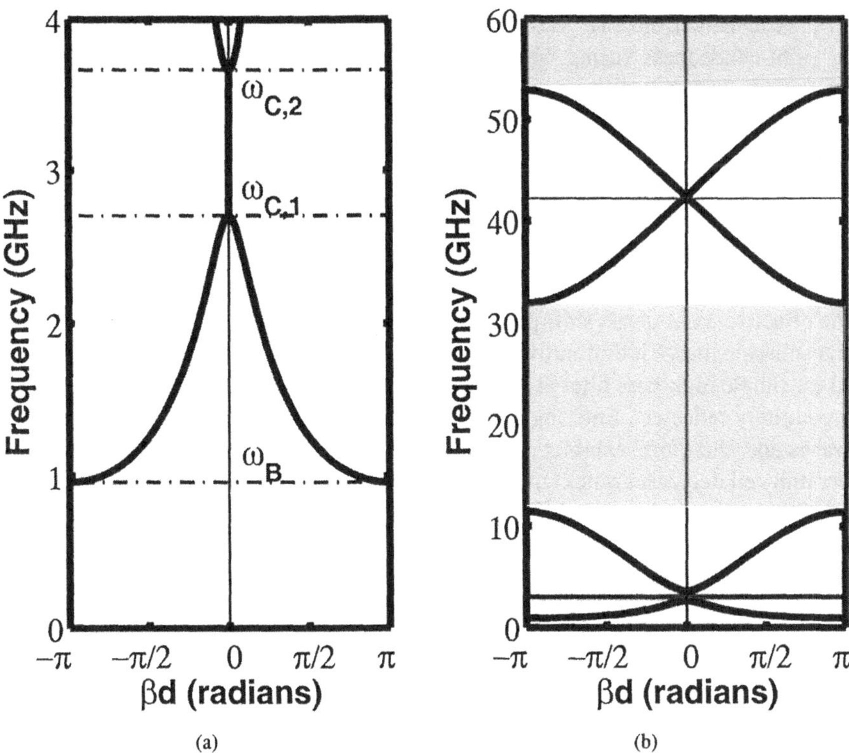

Fig. 1.20 Dispersion relation for a representative NRI-TL medium ($L_0 = 5.6$ nH, $C_0 = 1.0$ pF, $\epsilon_r = 2.94$, $h = 1.524$ mm, $w = 0.4$ mm, and $d = 5$ mm) as obtained through periodic analysis: (a) The lowest left-handed (LH) passband, corresponding to the fundamental spatial harmonic, is enclosed by a Bragg frequency ω_B and a stopband with cutoffs $\omega_{C,1}$ and $\omega_{C,2}$, followed by a right-handed (RH) passband. (b) Reduced Brillouin zone depicted for higher frequencies, showing stopbands at Γ-points (darkly shaded regions near 3 GHz and 42 GHz) and Z-points (lightly shaded).

with a chain of passbands separated by finite stopbands. The darkly shaded regions (near 3 GHz and 42 GHz) and lightly shaded regions of Fig. 1.20b correspond to stopbands in which $\cos\beta d \geq 1$ and $\cos\beta d \leq -1$, respectively, the edges of which may be described as the Γ-point and the Z-point, respectively, of the reduced Brillouin zone for the planar, two-dimensional periodic structure depicted in Fig. 1.14. It should be noted that, although the Γ-point stopbands shown in 1.20b become increasingly narrower with frequency (spanning approximately 900 MHz around 3 GHz and 140 MHz around 42 GHz), they nevertheless represent frequency regions in which waves propagating in the infinite periodic structure experience severe Bragg reflection and exhibit zero group velocity.

It is evident from Fig. 1.20 that successive passbands alternately exhibit left- or right-handedness (using Veselago's terminology), corresponding to particular backward- or forward-wave spatial harmonics, since the concavity of the dispersion surface changes with frequency. Although all periodic structures support an infinite number of forward- and backward-wave spatial harmonics [3] the dispersion diagrams of Fig. 1.20 show that the periodic NRI-TL network supports a backward wave *fundamental spatial harmonic* (extending over the region $-\pi < \beta d < 0$).

We now examine Fig. 1.20a in more detail. Beginning at dc (i.e., where $\theta \to 0$) and proceeding upwards, we first encounter a cutoff frequency near 1 GHz where the effective axial phase shift per unit cell βd is equal to π. Since the effect of the transmission lines θ is diminutive at these frequencies, the structure looks and behaves like a simple high-pass filter at its Bragg resonance, below which the incident waves are entirely reflected. Entering the left-handed passband, which supports backward waves and, therefore, exhibits negative values of β, the magnitude of the phase shift per unit cell decreases quickly, and the corresponding effective wavelength becomes much longer than the period of the lattice. The homogeneous, or effective medium, limit is achieved when approaching the edge of the first stopband, where $\beta d = 0$ and the effective wavelength in the medium is regarded to be infinite. It is true that, as frequency is increased, θ also increases; however, the unit cell may be designed such that it remains electrically small even as $\beta \to 0$.

When θ is small, $\cos\theta \to 1$ and $\sin\theta \to 0$. This is a first-order approximation that expands these functions to a single term in their Taylor expansions and essentially produces the results obtained for the continuous L–C networks considered in the previous section [see (1.32)], which possessed an infinite left-handed band (i.e., $\beta \to 0$ only as $\omega \to \infty$). Hereafter, we employ a second-order analysis that retains a second term in the Taylor expansion of $\cos\theta$ and $\sin\theta$. The resulting form of the dispersion relation is as follows:

$$\cos\beta d = 1 - \frac{\theta^2}{2} - \frac{1}{2\omega^2 L_0 C_0} + \frac{1}{2\omega}\left(\frac{1}{C_0 Z_0} + \frac{1}{L_0 Y_0}\right)\theta - \frac{\theta}{2}\left(\theta - \frac{1}{\omega C_0 Z_0}\right) \tag{1.41}$$

The Bragg frequency encountered when $\beta d = \pi$ is approximately given by

$$\omega_B = \frac{1}{2\sqrt{L_0 C_0}} \tag{1.42}$$

A more accurate expression for ω_B can be found in Ref. 27.

1.4.2 Effective Medium Limit

As frequency is increased, we enter the homogeneous limit, which is of greatest interest to the present development, since, as mentioned previously, this is also the effective medium limit in which we shall find our effective material parameters. In the homogeneous limit, we retain the second-order terms in the expansion of $\cos\beta d$, and arrive at the following expression:

$$(\beta d)^2 \approx \left\{ \left(\theta - \frac{1}{\omega C_0 Z_0} \right) \left(\theta - \frac{1}{\omega L_0 Y_0} \right) \right\} + \theta \left(\theta - \frac{1}{\omega C_0 Z_0} \right)$$
$$= \left(\theta - \frac{1}{\omega C_0 Z_0} \right) \left(2\theta - \frac{1}{\omega L_0 Y_0} \right) \quad (1.43)$$

from which it is evident that the loading due to the transverse TL medium serves to double θ in the factor containing the inductance L_0, the physical meaning of which shall be revealed shortly. By expressing θ and Z_0 in terms of the host transmission-line equivalent permittivity ϵ_P, permeability μ_P, and geometric factor g, the axial propagation constant in the effective medium limit is given to be

$$\beta = \pm\omega \sqrt{\left(2\epsilon_P - \frac{g}{\omega^2 L_0 d} \right) \left(\mu_P - \frac{1/g}{\omega^2 C_0 d} \right)} = \pm\omega\sqrt{\epsilon_N(\omega)\mu_N(\omega)} \quad (1.44)$$

which has also been represented in terms of the effective material parameters $\epsilon_N(\omega)$ and $\mu_N(\omega)$ that we seek. The picture is completed in the determination of the Bloch impedance, which, under the same set of second-order assumptions, reduces to

$$Z_B = \frac{\omega\mu_P g - \frac{1}{\omega C_0 d}}{\beta} = g\sqrt{\frac{\mu_P - \frac{1/g}{\omega^2 C_0 d}}{2\epsilon_P - \frac{g}{\omega^2 L_0 d}}} = g\sqrt{\frac{\mu_N(\omega)}{\epsilon_N(\omega)}} \quad (1.45)$$

where the relationship between the Bloch impedance and effective wave impedance has been defined analogously to that in the host transmission-line medium presented in (1.39). Revealed by the effective axial propagation constant and wave impedance are the effective constitutive parameters of the NRI-TL metamaterial in the effective medium limit:

$$\mu_N(\omega) = \mu_P - \frac{1/g}{\omega^2 C_0 d} \quad (1.46)$$

$$\epsilon_N(\omega) = 2\epsilon_P - \frac{g}{\omega^2 L_0 d} \quad (1.47)$$

from which it can be seen that each consists of a positive contribution due to the host transmission-line medium, and a negative, dispersive contribution due to the loading (corresponding to a negative effective susceptibility). This is unlike the continuous dual *L–C* medium discussed previously, in which the material parameters (1.30) and (1.31) were negative over all frequencies; with the host medium in place, the negation of the material parameters requires that the series and shunt branches be dominated by C_0 and L_0, respectively. Also evident is the fact that the contribution of the open-circuited stubs of the transverse TL medium is, effectively, to double the relative permittivity ϵ_P of the host medium [27]. Consequently, the per-unit-length capacitance C' of a transmission-line segment, given by ϵ_P/g, can be represented as $2\epsilon_P/g$ in the two-dimensional transmission-line *medium*. Accordingly, the characteristic impedance seen by a wave propagating axially through a two-dimensional transmission-line medium is less than that seen in a one-dimensional transmission line by a factor of $\sqrt{2}$. Until now, we have used the symbol Z_0 to denote the characteristic impedance of the one-dimensional transmission line. In what is to follow, we generalize Z_0 to mean, simply, "characteristic impedance," and we request that the reader recognize its application to the one- and two-dimensional cases through the context of the discussion.

The transition frequencies, beyond which the material parameters become positive, are the zeros of the permittivity and permeability functions, or the plasma frequencies of the effective medium. By setting $\beta = 0$, the edges of the stopband are analytically determined to be

$$\omega_{C,1} = \sqrt{\frac{1}{C_0 g \mu_P d}}, \qquad \mu_N(\omega_{C,1}) = 0 \tag{1.48}$$

$$\omega_{C,2} = \sqrt{\frac{1}{L_0 \frac{2\epsilon_P d}{g}}}, \qquad \epsilon_N(\omega_{C,2}) = 0 \tag{1.49}$$

It is evident from these equations that the stopband cutoffs are pushed to infinity as the period d is reduced, arbitrarily widening the bandwidth of the LH passband and essentially restoring the continuous case [22,27]. Although an infinitesimal cell period is impracticable, what is important is that the dispersion characteristics of the NRI-TL metamaterial are inherently controllable and are not restricted by unnecessary resonances [22, 31]. Consequently, negative-refractive-index metamaterials based on the dual L–C concept can be expected to offer dramatically larger operating bandwidths than the wire/SRR composite medium; supporting results are presented in Section 1.6, as well as in Refs. 27 and 36.

1.4.3 Closure of the Stopband: The Impedance-Matched Condition

One particularly interesting feature, independent of the cell period d, is that the stopband shown in Fig. 1.20a may be closed by setting $\omega_{C,1} = \omega_{C,2}$, which adjoins the LH and RH passbands and leads to

$$\sqrt{\frac{1}{C_0 g \mu_P d}} = \sqrt{\frac{1}{L_0 \frac{2\epsilon_P d}{g}}} \Rightarrow \sqrt{\frac{L_0}{C_0}} = g\sqrt{\frac{\mu_P}{2\epsilon_P}} = \sqrt{\frac{L'}{C'}} = Z_0 \tag{1.50}$$

which is an impedance-matched condition that states that the characteristic impedance of the host medium is equal to that of the underlying purely LH distributed medium consisting of the loading elements alone. Condition (1.50) of closing the stopband in a transmission-line NRI metamaterial was originally reported in equation 29 of Ref. 27, and subsequently adopted by Sanada et al. [43].

The closure of the stopband is illustrated in Fig. 1.21a for a NRI-TL design that differs from the design represented in Fig. 1.20 only in the choice of the loading inductance L_0, which has been increased to lower the upper band edge and thus meet the impedance-matched condition. Alternatively, the loading capacitance C_0 could have been decreased while maintaining the original value of L_0. In either case, it is the impedance quantity $\sqrt{L_0/C_0}$ that must change to meet Z_0. It is evident from Fig. 1.21 that this condition permits, at least in theory, the restoration of a nonzero group velocity at the point $\beta = 0$, or the Γ-point, if Brillouin zone notation is employed.

Although we have shown that the impedance-matched condition ensures the closure of the stopband at the Γ-point and restores propagation into the RH passband that follows, we must recall that this result was derived under the assumption of homogeneity, which required both that we be sufficiently close to the Γ-point and that we operate at frequencies low enough to ensure that the effect of the interconnecting host medium is diminutive. Of course, any periodic structure supports an infinite number of spatial harmonics and, consequently, exhibits an infinite number of stopbands at higher frequencies, both at the Γ-points and at the Z-points. It should, therefore, be interesting to see what effect, if any, the impedance-matched condition of (1.50) has on these higher stopbands. To investigate these higher bands, the small-angle approximations employed for the transmission-line host medium parameter θ in the preceding analysis must be abandoned, and we must revisit the full dispersion relation of (1.36). At the Γ-points ($\beta = 0$), it can be shown that the general solutions to (1.36) are in the form of transcendental equations:

$$\cot\frac{\theta}{2} = \frac{2C_0}{\frac{\epsilon_P d}{g}}\theta \tag{1.51}$$

$$\cot\frac{\theta}{2} = \frac{4L_0}{g\mu_P d}\theta \tag{1.52}$$

The corresponding solution 'pairs' of (1.51)–(1.52) cooperatively describe the edges of a single Γ-point stopband. As in (1.48)–(1.49), when the host medium parameters are specified, one edge of each stopband is determined by the series loading capacitance C_0, while the other is determined by the shunt loading inductance L_0. What is immediately evident from the form of these equations is that the solutions can be made to coincide when

$$\frac{2C_0}{\frac{\epsilon_P}{g}d} = \frac{4L_0}{g\mu_P d} \Rightarrow Z_0 = g\sqrt{\frac{\mu_P}{2\epsilon_P}} = \sqrt{\frac{L_0}{C_0}} \tag{1.53}$$

which is the impedance-matched condition of (1.50). The result, therefore, is that the closure of the first stopband through (1.50) guarantees that each one of the infinitely many successive Γ-point stopbands in the reduced Brillouin zone is closed. This is illustrated in Fig. 1.21b, which, when compared to Fig. 1.20b, indicates that the application of (1.50) eliminates the Γ-point stopbands near 3 GHz and 42 GHz, and, indeed, at all other Γ-points.

It can be shown that further manipulations of (1.50) result in

$$\sqrt{\frac{\mu_P - \frac{1/g}{\omega^2 C_0 d}}{2\epsilon_P - \frac{g}{\omega^2 L_0 d}}} = \sqrt{\frac{\mu_P}{2\epsilon_P}} \Rightarrow \eta_{NRI-TL} = \eta_{TL} \tag{1.54}$$

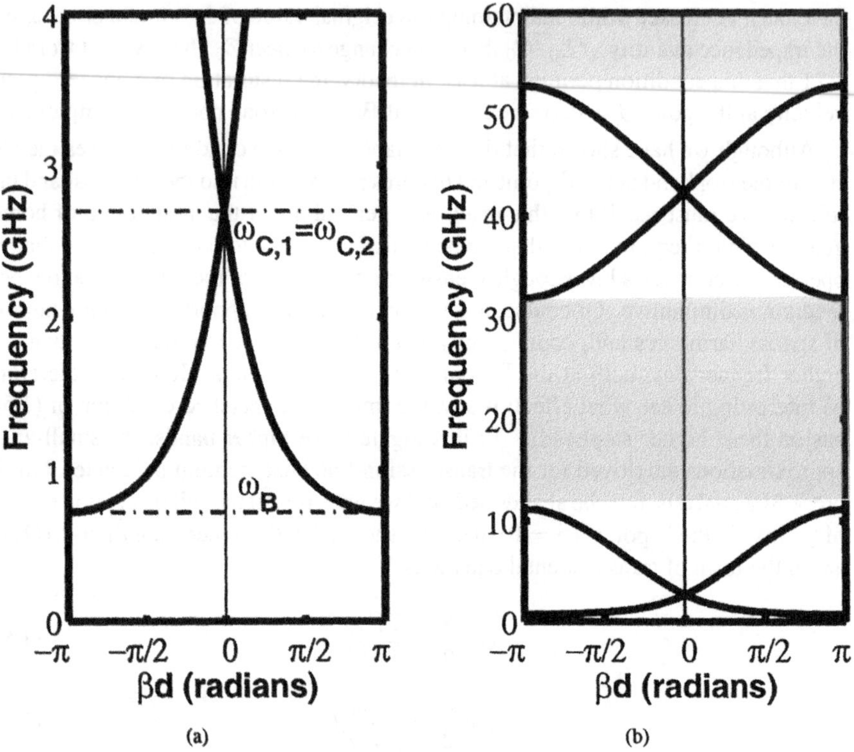

Fig. 1.21 Dispersion relation for an NRI-TL medium designed for a closed stopband ($L_0 = 10\,\text{nH}$, $C_0 = 1.0\,\text{pF}$, $\epsilon_r = 2.94$, $h = 1.524\,\text{mm}$, $w = 0.4\,\text{mm}$, and $d = 5\,\text{mm}$) as obtained through periodic analysis: (a) Closure of the stopband through the application of (1.50) adjoins the LH and RH passbands, indicating the restoration of a nonzero group velocity at the point of closure. (b) Reduced Brillouin zone depicted for higher frequencies, in which all stopbands at Γ-points have been eliminated, but those at Z-points (lightly shaded) remain.

which states that the effective wave impedance of the periodically loaded 2-D NRI-TL metamaterial designed for a closed stopband is matched to the effective wave impedance of an *unloaded* 2-D transmission-line grid. This provides some insight into the design of *interfaces* between positive-refractive-index (PRI) effective media and NRI-TL metamaterials. Interfaces are essential to the demonstration of negative refraction and focusing (see the following sections of this chapter), and impedance-matched interfaces are crucial to the demonstration of perfect lensing (see Chapter 3). Specifically, the above results suggests the use of a 2-D unloaded transmission-line host grid (microstrip in the present discussion, with effective material parameters (μ_P, $2\epsilon_P$) in the homogeneous limit) as our PRI medium. What is also evident from (1.50) and (1.54) is that the impedance-matched condition is frequency insensitive, suggesting that such an interface will be matched over all frequencies where the structure remains isotropic and homogeneous (the regime in which (1.46)–(1.47) retain their meaning) [44].

1.4.4 Equivalent NRI-TL Unit Cell in the Effective Medium Limit

In the previous discussion on continuous media, we developed the following expressions for the propagation constant β and characteristic impedance Z_0:

$$\beta = \pm\sqrt{-Z'Y'} \tag{1.55}$$

$$Z_0 = \sqrt{\frac{Z'}{Y'}} \tag{1.56}$$

where Z' and Y' are per-unit-length quantities. For a unit cell of dimension d and under the assumption of homogeneity, these distributed parameters can be represented by a *total* series impedance per unit cell $Z' = Z/d$ and shunt admittance per unit cell $Y' = Y/d$. When (1.55) and (1.56) are modified to reflect this and are equated to the NRI-TL axial propagation constant (1.44) and Bloch impedance (1.45), the following expressions for Z and Y result:

$$\beta = \pm\sqrt{-Z'Y'} = \pm\omega\sqrt{\mu_N(\omega)\epsilon_N(\omega)} \tag{1.57}$$

$$Z_0 = \sqrt{\frac{Z'}{Y'}} = g\sqrt{\frac{\mu_N(\omega)}{\epsilon_N(\omega)}} \Rightarrow \begin{cases} Z = j\omega g\mu_N(\omega)d \\ Y = j\omega\frac{1}{g}\epsilon_N(\omega)d \end{cases} \tag{1.58}$$

This result suggests that the assumption of homogeneity reduces the practical NRI-TL unit cell, consisting of a reactively loaded transmission-line host medium, into an equivalent circuit consisting of a single series impedance Z and shunt admittance Y. The nature of these elements is revealed through the substitution of (1.46) and (1.47) into (1.58):

$$Z = j\omega g\mu_N(\omega)d = j\omega L'd + \frac{1}{j\omega C_0} \tag{1.59}$$

$$Y = j\omega\frac{1}{g}\epsilon_N(\omega)d = j\omega 2C'd + \frac{1}{j\omega L_0} \tag{1.60}$$

where the quantities $\mu_P g$ and ϵ_P/g, describing only the host medium, have been recognized as the transmission-line distributed inductance L' and capacitance C', respectively. It is immediately evident that the series branch of the equivalent unit cell consists of an inductor, contributed by L' over the length d of the unit cell, in series with a capacitor provided by the loading. Similarly, the shunt branch consists of a capacitor, contributed by $2C'$ over the length d of the unit cell, in parallel with an inductor provided by the loading. This equivalent unit cell is depicted in Fig. 1.22, and its topology is also evident from the effective material parameter expressions (1.46)–(1.47), to which the transmission-line medium contributed positively and the loading contributed negatively [27]. In the absence of loading, the material parameters degenerate to those of the host medium, modeled in Fig. 1.22 by an inductor and

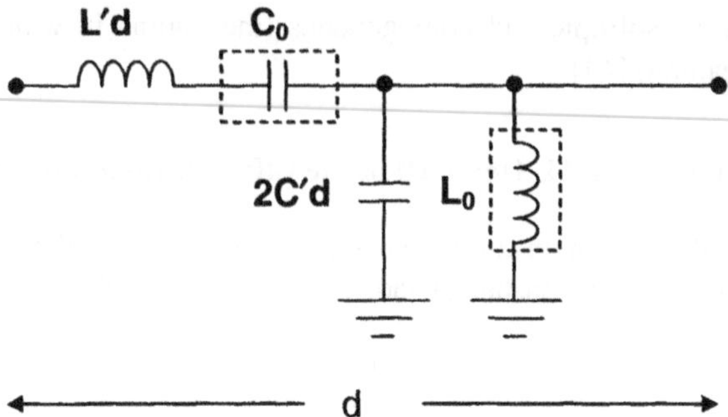

Fig. 1.22 Unit cell for the practical NRI-TL network describing axial propagation in the homogeneous limit. The host transmission-line medium, which appears embedded as an equivalent series inductor and shunt capacitor, is loaded using a lumped (e.g., chip or printed) series capacitor and shunt inductor. After Ref. [27]. Copyright © 2002 IEEE.

capacitor arranged in a low-pass topology. When the loading is dominant, the material parameters become negative, and the equivalent unit cell is restored to the dual topology. Thus, it is seen that the equivalent unit cell of Fig. 1.22, along with the dispersion relation from which it was derived, include the complete RH and LH responses of the periodically loaded NRI-TL metamaterial [27]. This topology also accounts for the magnetic and electric plasma frequencies [equivalently, the edges of the first Γ-point stopband in the NRI-TL dispersion relation, (1.48)–(1.49)], which are, respectively, equal to the series and shunt resonances of the equivalent unit cell. This perspective also reveals that the closure of this stopband by way of the impedance-matched condition of (1.50), in fact, causes the series and shunt branches of Fig. 1.22 to resonate at the same frequency, at which the equivalent circuit represents a direct connection from input to output (see also Ref. 45). Under these conditions, the NRI-TL medium, yet periodic, appears to be completely unperturbed (i.e., homogeneous).

The topology of Fig. 1.22 can also be literally taken as a model for a one-dimensional NRI transmission line. As an example, we suggest here an implementation based on a coaxial host medium, depicted in Fig. 1.23a. The outer conductor has been made semi-transparent in order to see the internal structure. The series capacitors are realized by chopping the central conductor and inserting cylindrical parallel plates, which may be filled with a high-permittivity dielectric or glue. The shunt inductors are realized using thin metallic pins connecting the central and outer conductors. This structure differs from the microstrip implementation in that it is completely shielded and the host medium supports a pure TEM mode; furthermore, although the loading is most certainly lumped, this structure is unlimited by component self-resonances and may consequently be realized at higher frequencies. Finite-element full-wave simulation results for a representative design verify that such a structure supports backward waves and, as shown in Fig. 1.23b, can be designed to have a stopband that is, for all practical purposes, closed (compare with Fig. 1.21a).

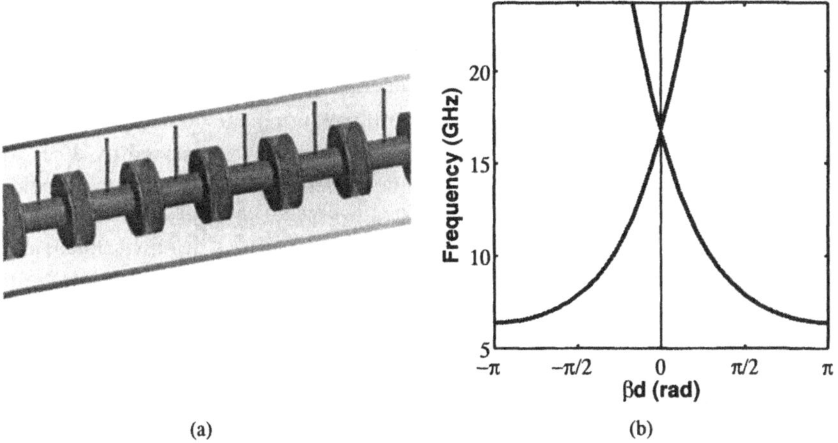

Fig. 1.23 (a) Implementation of one-dimensional NRI-TL structure using coaxial host medium. The outer conductor has been made semitransparent in order to see the internal structure. (b) Dispersion relation for a representative impedance-matched (closed-stopband) coaxial NRI-TL design obtained by finite-element full-wave simulation.

When thus designed and operated at the frequency of closure, the one-dimensional line can be expected to "transmit," with no net phase incurred, electric fields from input to output.

We now pose the following question: What is the nature of the impedance-matched NRI-TL medium at the other (Γ-) points of closure in higher bands? Certainly, the equivalent unit cell of Fig. 1.22 is not helpful in answering this question, since its topology was based on the assumption that the interconnecting transmission lines are electrically short. In higher bands, the electrical length of the interconnecting lines, θ, is large, and the stopband edges described by (1.51) and (1.52) occur when the physical length of the interconnecting lines, d, is near an integer multiple of $\lambda/2$. For the design represented in Fig. 1.21 for which $d = 5$ mm, the first two such instances occur at frequencies of approximately 21 GHz ($d = \lambda/2$) and 42 GHz ($d = \lambda$); while we know from Fig. 1.21 that the latter point, indeed, represents a point of stopband closure, the former seems to lie inside a Z-point stopband, where propagation is forbidden. The discrepancy is resolved when one considers the effect of the loading by the transverse transmission-line medium in the 2-D grid. Referring to Fig. 1.16b, it is clear that the open circuits formed due to axial propagation at 21 GHz occur at a distance of $\lambda/4$ from the node; thus, these quarter-wavelength lines create short circuits to ground at the node site, ensuring no propagation at this frequency. At 42 GHz, the interconnecting lines are one wavelength long; thus, the half-wavelength open-circuited stubs produce an open circuit at the node site, and propagation is once again permitted. It is noteworthy that a one-dimensional impedance-matched NRI-TL line would have permitted propagation under both conditions.

Departing slightly from the above condition (i.e., d is either slightly shorter or slightly longer than λ), we see that the propagation characteristics as measured at each node begin to resemble those in the homogeneous limit, returning us to the question of how impedance matching will affect the propagation characteristics at higher Γ-points. In this regime, the interconnecting lines may be represented by a cascade of continuous, symmetric L–C unit cells (one-dimensional versions of the topology depicted in Fig. 1.10), and the periodic loading may be modeled by a solitary, symmetrical dual L–C unit cell (Fig. 1.12) at each node of the two-dimensional structure. This is depicted in Fig. 1.24a. The dual L–C unit cells at the nodes of the structure appear as a periodic perturbation on the propagation characteristics of the transmission lines, which becomes weaker as the lines become electrically longer, hence the reduction in width of the higher-order Γ-point stopbands observed in Fig. 1.20. It is, therefore, a reasonable guess that the *closure* of the stopbands through the impedance-matched condition of (1.50) would eliminate the perturbation entirely, locally restoring the propagation characteristics of the NRI-TL structure to those of the host transmission lines. This can be verified by noting the interaction of the dual L–C unit cell in Fig. 1.24a with its nearest L–C neighbours, which, it should be noted, represent one-dimensional transmission lines with per-unit-length inductance L' and and capacitance C', crossing at the two-dimensional node. This region of interest is enclosed by the marquee in Fig. 1.24a. The impedance-matched condition (which, the reader will recall, implicates L' and $2C'$ for the two-dimensional case) ensures that, at the frequency of closure, L' (over some appropriately small length) resonates with the capacitive loading to form a series short circuit, which permits, simultaneously, two of the shunt C' (over the same small length) to resonate with the inductive loading to create a shunt open circuit. What remains (depicted in Fig. 1.24b) is a node representing a two-dimensional grid of transmission lines, with the node capacitance doubled to accommodate the two directions of propagation. Thus, the closure of the higher-order Γ-point stopbands through the impedance-matched condition does, indeed, eliminate the perturbation otherwise contributed by the dual L–C loading, so that the propagation characteristics are restored to those of the host transmission-line medium alone.

1.5 MICROWAVE CIRCUIT SIMULATIONS

To verify negative refraction and focusing using the transmission-line metamaterial concept, we must recreate the conditions illustrated in the phase-matching argument of Fig. 1.2—namely, an interface between a positive-refractive-index transmission-line (PRI-TL) medium and a NRI-TL medium. The periodic PRI-TL medium must be excited with the Floquet–Bloch equivalent of a plane wave to observe negative refraction, and of a cylindrical wave to see focusing. The phase progression must then be monitored in both media.

The distributed network approach lends itself to examination using microwave circuit simulations. These were performed using Agilent's Advanced Design System

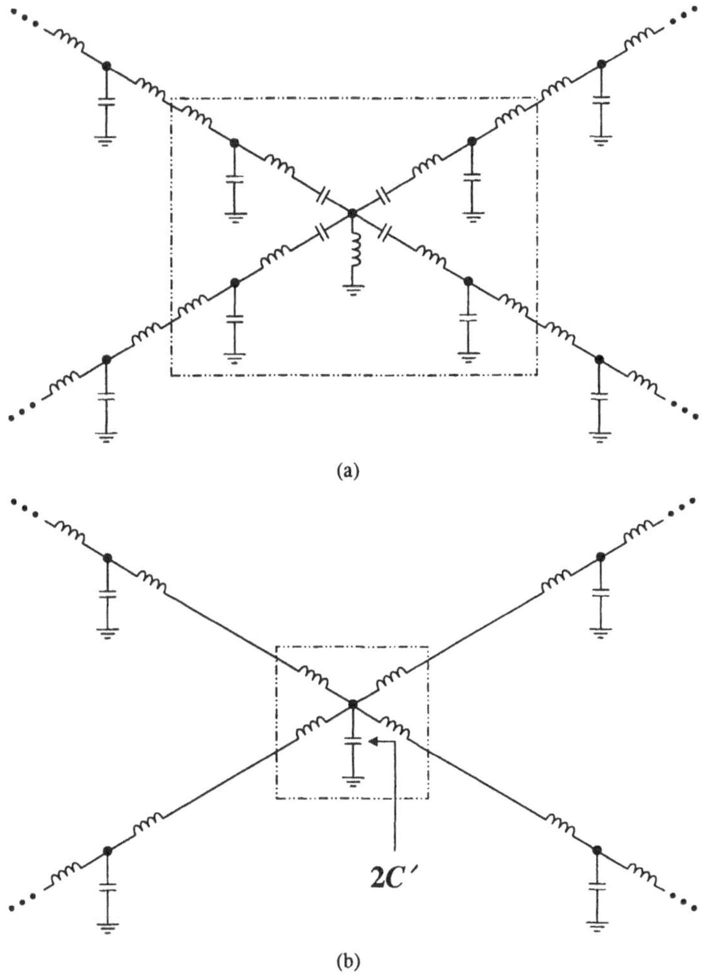

Fig. 1.24 (a) Node of NRI-TL metamaterial at frequencies at which the interconnecting transmission lines are electrically long, employing the continuous L–C and dual L–C unit cells of Figs. 1.10 and 1.12. (b) Reduction of the NRI-TL node in (a) to that of a two-dimensional transmission-line grid under impedance-matched conditions.

(ADS) microwave circuit simulator [46], which represents the unit cell as a network of idealized cascaded transmission-line models. The design of a PRI-TL/NRI-TL interface begins with the selection of a generic host transmission-line medium (characterized by a particular phase shift and characteristic impedance at a selected design frequency), which is periodically loaded using inductors and capacitors. While it is clear that the NRI-TL array must be designed according to the proposed dual L–C topology to realize a particular negative refractive index at the design frequency, this generally requires that the PRI-TL medium be loaded as well, but in a conventional

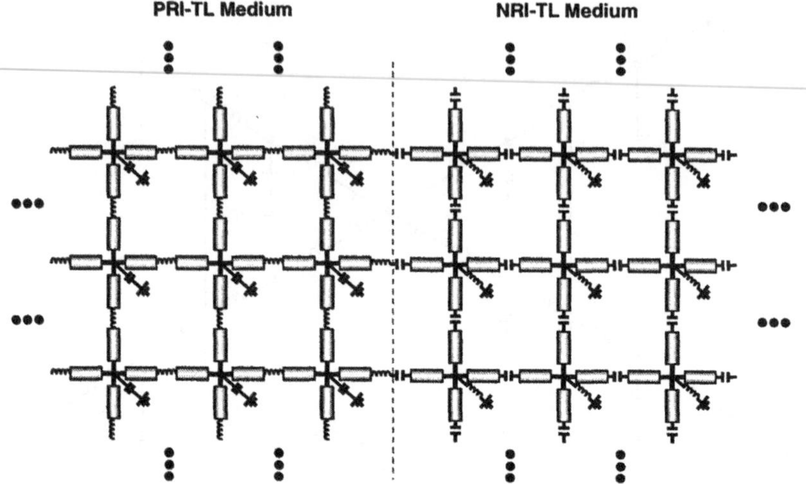

Fig. 1.25 Schematic representation of PRI-TL/NRI-TL interface using 2-D L–C loaded unit cells.

low-pass L–C topology. The final design therefore consists of two large, periodic arrays of 2-D L–C loaded unit cells, a portion of which is schematically represented in Fig. 1.25. We now present negative refraction and focusing simulation results, which originally appeared in Refs. 26 and 27.

1.5.1 Negative Refraction

A suitable PRI-TL/NRI-TL interface was designed for a frequency of 2 GHz using 42×42-cell PRI-TL and NRI-TL metamaterial arrays with effective per-unit-cell phase shifts of $\beta_{PRI-TL}d = +0.25$ and $\beta_{NRI-TL}d = -0.5$, respectively, yielding a relative refractive index, n_{REL}, of -2. The unit cells comprising the arrays are 5 mm square, the characteristic impedances of both media are designed to be equal to 377 Ω (free space) to prevent reflections at the interface, and the arrays are terminated at their boundaries using resistors matched to 377 Ω. The corresponding equivalent, absolute index of refraction of the NRI-TL metamaterial could therefore be determined to be -2.4. To maintain $n_{REL} = -2$, the loading elements of the PRI-TL unit cell were chosen so that the absolute index of refraction of the PRI-TL medium was $+1.2$.

The plane-wave excitation k_1 shown in Fig. 1.2 was synthesized by placing a series of sequentially phased voltage sources along the leftmost boundary of the PRI-TL array. Their progressive phase specifies the tangential wavevector components $k_{1t} = k_{2t}$, hence the effective propagation constants in both media, and the conditions for refraction across the interface. One such excitation is shown in Fig. 1.26, for an incident angle of θ_{PRI}=+29°. Since the interface was designed for $n_{REL} = -2$,

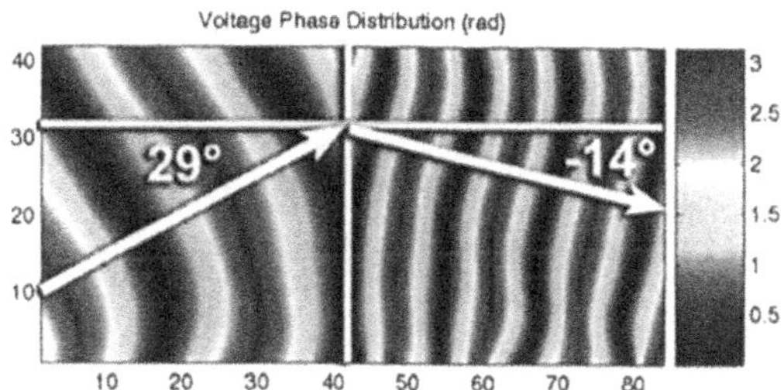

Fig. 1.26 Plane wave illuminating a PRI-TL/NRI-TL interface at 2 GHz (relative refractive index -2, finite air-filled transmission lines with $d = 5$ mm included in each unit cell) at an incident angle of $+29°$. Refraction observed at $-14°$. The axes are labeled according to cell number, and the right-hand-side vertical scale designates radians. After Refs. [26, 27]. Copyright © 2002 IEEE.

Veselago's "negative Snell's Law" predicts refraction at $-14°$. Indeed, the steepest phase descent in the NRI-TL metamaterial is observed at $\theta_{NRI-TL} = -14°$ from the normal, verifying Snell's Law for the given design parameters. The curvature of the wavefronts in the lower portion of the PRI-TL medium and the consequent rippling effect in the lower portion of the NRI-TL medium are due to the fact that voltage sources were not placed on the lower boundary of the PRI-TL region to assert the phase of the incident plane wave at these points (see Ref. 47 for improved results). Furthermore, small reflections are apparent from the use of 377 Ω terminating resistors throughout; this is strictly valid only at normal incidence, and must, in general, be modified by a factor related to the angle of incidence/refraction. Nevertheless, this circuit arrangement unambiguously demonstrates the phenomenon of negative refraction in the NRI-TL metamaterial and reaffirms the idea of a "negative Snell's Law," as predicted by Veselago.

1.5.2 Focusing

The source depicted in the focusing arrangement of Fig. 1.3 represents a point source emanating spherical waves in a three-dimensional medium or an infinite line source emanating cylindrical waves in a two-dimensional medium. Sources in the planar media under study model the latter case, and they may be synthesized by applying a voltage to ground at a single node in the PRI-TL region. Thereafter, it remains only to examine the magnitude and phase of the node voltages to ground in the NRI-TL region to observe the focusing effect. The reversal of power flow predicted by the negative Snell's Law suggests that focusing in the NRI-TL medium should manifest itself as a concentration of the node voltages near the expected focal plane. However, to justify these expectations, it is necessary to examine, with equal attention, the evolution of a cylindrical wave excited in a PRI-TL medium (which we shall denote PRI-TL$_1$) as it enters a second PRI-TL medium (PRI-TL$_2$). In the this case, n_{REL} is

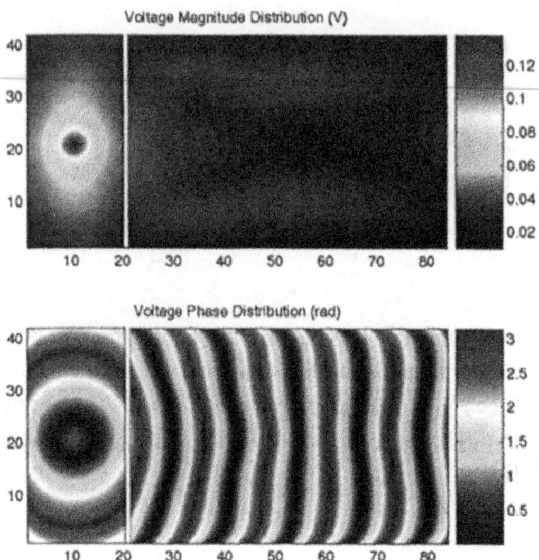

Fig. 1.27 Point source illuminating a PRI-TL/PRI-TL interface at 2 GHz ($n_{REL} = +2$, finite air-filled transmission-lines with $d = 5$ mm included in each unit cell). No focusing is observed in either phase or magnitude. The axis labels refer to cell number. After Refs. [26,27]. Copyright © 2002 IEEE.

positive, and Snell's Law predicts a continued divergence from PRI-TL$_1$ to PRI-TL$_2$. To verify this notion, a 42×21-cell PRI-TL array was interfaced with a 42×63-cell PRI-TL array designed such that $n_{REL} = +2$. The source in PRI-TL$_1$ was placed 11 cells from the interface and the voltage magnitudes and phases at the nodes in both media were observed. Figure 1.27 presents these results, and confirms that the cylindrical wave excitation diverges into PRI-TL$_2$. As was observed in the negative refraction results of Fig. 1.26, small reflections are apparent at the upper and lower edges; this result, although unwanted, is somewhat comforting, since it verifies that power is continually carried outward in all directions from the source, even after entering the second medium.

We now consider focusing across a PRI-TL/NRI-TL interface with $n_{REL} = -2$. The sizes of the two arrays and the location of the source are left unchanged. As shown in Fig. 1.28, the corresponding magnitude and phase results show increased voltage amplitudes in the NRI-TL region along the source axis, as well as a dramatic reversal of the concavity of the wave fronts at both the PRI-TL/NRI-TL boundary and the expected focal point. It is important to note that the focal region resembles an elongated "beam" rather than a "spot," and this can be directly attributed to the choice of a relative refractive index of -2. In this case, as was previously noted, geometric optics predicts spherical aberration. However, as a reference, one can make use of the fact that, in the paraxial limit, the focus in the NRI-TL metamaterial should appear at twice the distance of the source from the interface, or near cell 43 of

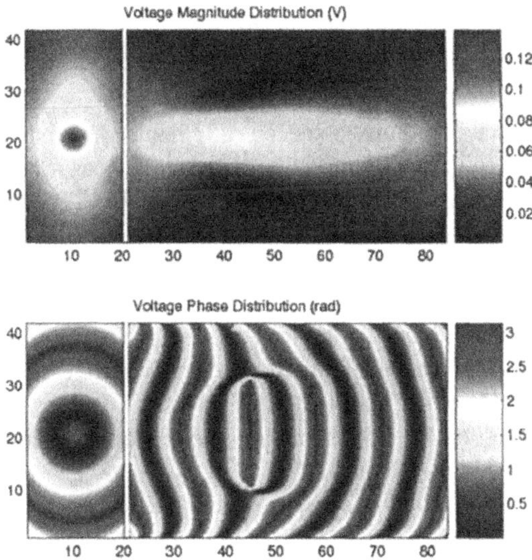

Fig. 1.28 Point source illuminating a PRI-TL/NRI-TL interface at 2 GHz ($n_{REL} = -2$, finite air-filled transmission-lines with $d = 5$ mm included in each unit cell). Focusing is observed in both phase and magnitude. The axis labels refer to cell number. After Refs. [26,27]. Copyright © 2002 IEEE.

the composite array. Indeed, the phase fronts undergo a strong change in concavity in this region, indicating the convergence of the component rays of the cylindrical wave. Furthermore, minimal reflections are observed near the upper and lower edges of the array, indicating that the direction of power flow has truly been reversed. For a detailed discussion on these designs and their associated results, including full-wave simulations of focusing, we refer the reader to Refs. 26 and 27.

1.6 EXPERIMENTAL VERIFICATION OF FOCUSING

The synthesis of the microstrip-based NRI-TL metamaterial prototypes to be described in this section employed standard printed circuit board (PCB) fabrication techniques comprising the following essential steps: A mask is generated using a CAD software package (e.g., Agilent's ADS microwave circuit simulator); photoresist is applied to the sample through spinning or lamination and hardened by exposure to UV light through the mask; the excess photoresist is removed, and the sample is etched in acid to reveal the microstrip host medium. The loading elements arc realized using chip inductors and capacitors, which are soldered manually to the PCB. Another purely printed implementation presented and investigated in Refs. 39 and 40 employs vias and printed gaps, as in Fig. 1.19b. The resulting NRI-TL metamaterial is then interfaced with a PRI-TL medium, which can be synthesized using a parallel-plate waveguide or, alternatively, an unloaded microstrip host grid resembling a parallel-plate waveguide, to ensure good matching between the two media.

The first left-handed medium to successfully demonstrate focusing was a planar, microstrip-based NRI-TL lens comprising 11×6 cells measuring 5 mm on each side, and printed on a 1.524 mm (60 mil) Rexolite substrate [27]. An adjacent parallel-plate waveguide was used as a PRI-TL region. We refer the interested reader to the above reference for a full characterization of this proof-of-concept prototype, which, in addition to experimentally verifying the focusing property predicted by Veselago [1] and the wideband, low-loss properties of the NRI-TL metamaterial, also helped dispel some fundamental doubts raised over the validity of the negative index phenomenon [48–51]. The following discussion, however, shall be restricted to a second, larger prototype employing an unloaded microstrip grid as the PRI-TL instead of a parallel-plate waveguide [36].

The device under consideration consists of a PRI-TL region measuring 21×21 cells (105 mm×105 mm) and an adjacent NRI-TL region measuring 21×40 cells (105 mm×200 mm), as shown in Fig. 1.29. The unloaded microstrip grid constituting the PRI-TL medium and the NRI-TL host medium has a period $d = 5$ mm and comprises $w = 0.4$-mm-wide microstrip lines etched onto a Rogers RT Duroid 6002 ceramic ($\epsilon_r = 2.94$) substrate of height $h = 1.524$ mm (60 mil). In the NRI-TL region, 5.6-nH chip shunt inductors are embedded into rectangular holes punched into the substrate at each cell site, and chip capacitors of 1 pF are surface-mounted between gaps separating the unit cells. To maintain uniformity throughout, 2-pF capacitors are placed at the array edges, followed by matching resistors. The inset of Fig. 1.29 depicts the NRI-TL unit cell, and the corresponding PRI-TL unit cell consists of the microstrip grid alone. The PRI-TL medium was excited in the centre using a shorted SMA (coaxial) connector placed beneath the grid surface (indicated by the arrow in Fig. 1.29), and transmission S-parameter readings were taken at each unit cell using an HP8753D vector network analyzer connected to a near-field detecting probe (also depicted in Fig. 1.29) that was scanned over the grid surfaces using a computer-controlled stepper motor. The detecting probe (essentially a short vertical dipole) was designed to measure the vertical electric fields through capacitive coupling at a small distance above the device plane. The measured fields, in actuality, are fringing fields; nevertheless, these are proportional to the fields within the NRI-TL structure and represent their phase and relative magnitudes faithfully.

The microstrip parameters of this design yield $\epsilon_{eff} = 2.1119$, $g = 0.544186$, $2\epsilon_P = 2\epsilon_{eff}\epsilon_0$, and $\mu_P = \mu_0$, where we remind the reader that ϵ_{eff} represents the effective permittivity surrounding a single microstrip line. These values, along with the loading values, place the first Z-point Bragg frequency at $\omega_B = 2\pi\times$ 960 MHz and place the first Γ-point stopband cutoffs at $\omega_{C,1} = 2\pi\times2.72$ GHz and $\omega_{C,2} = 2\pi\times3.63$ GHz. The dispersion characteristics of this NRI-TL design, as determined through a periodic analysis of the NRI-TL unit cell, were first presented in Fig. 1.20. In Fig. 1.30, the LH passband of the TL theory is presented (dashed

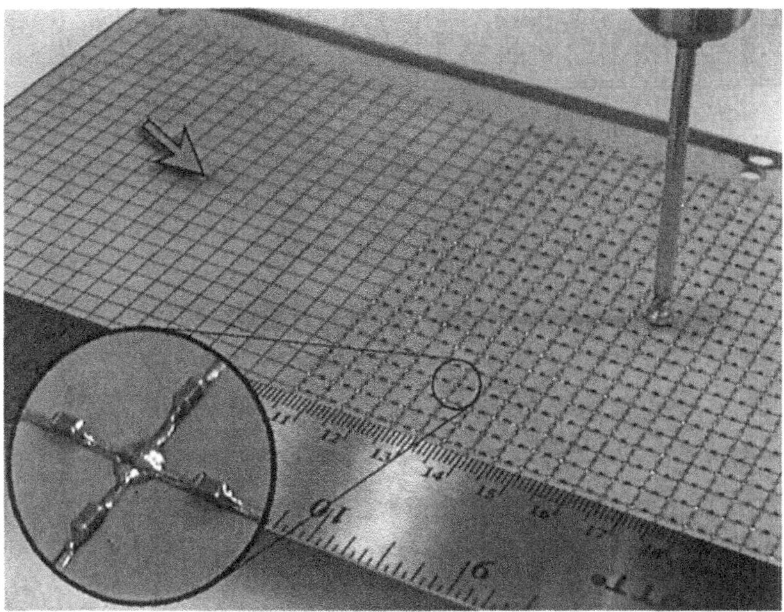

Fig. 1.29 Experimental prototype NRI-TL lens consisting of a 21×40-cell (105 mm×200 mm) NRI-TL metamaterial interfaced with a 21×21-cell (105 mm×105 mm) PRI-TL region. The inset magnifies a single NRI-TL unit cell, consisting of a microstrip grid loaded with surface-mounted chip capacitors and a chip inductor embedded into the substrate at the central node. The near-field detecting probe is also depicted, and the arrow indicates the location of the vertical excitation probe beneath the PRI-TL surface. After Ref. [36]. Copyright © 2003 Optical Society of America, Inc.

curve) and compared with experimental dispersion data (solid curve) extracted from the average observed phase shift per unit cell (βd) for the region $\beta d = \{-\pi, 0\}$ and is reflected in the $\beta d = 0$ axis for the space-reversed solution. At each frequency, this value was obtained from the line-of-best-fit corresponding to the slope of the phase advance profile along the central row (row 11) of the NRI-TL device. Also depicted in Fig. 1.30 is the theoretical dispersion of the PRI-TL medium using the parameters (μ_P, $2\epsilon_P$), and the intersection of the PRI-TL and theoretical NRI-TL dispersion curves indicates $\omega_0 = 2\pi \times 2.18$ GHz, the frequency at which the relative refractive index between the NRI-TL and PRI-TL media, n_{REL}, is equal to -1. The reader will recall that this is the special case in which Veselago envisioned that propagating waves emanating from a source would be perfectly focused without spherical aberration.

The experimentally obtained dispersion relation is largely in excellent agreement with that predicted by periodic analysis of the corresponding infinite structure and exhibits a left-handed characteristic extending from the Bragg frequency at 960 MHz to approximately 2.5 GHz, corresponding to a bandwidth of over 85%. As previously noted, the extremely wideband nature of NRI-TL metamaterials is a direct conse-

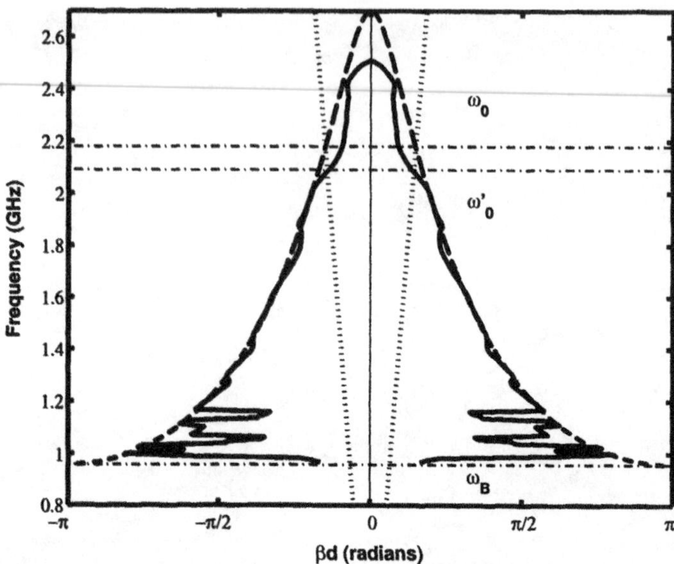

Fig. 1.30 Experimentally obtained NRI-TL dispersion relation (solid curve) indicating the Bragg frequency at 960 MHz and a well-defined NRI region extending to approximately 2.5 GHz (data obtained for region $\beta d = -\pi$ to $\beta d = 0$ and reflected in the $\beta d = 0$ axis for the space-reversed solution). Also depicted are the theoretical NRI-TL (dashed curve) and PRI-TL (dotted curve) dispersion relations. The intersection of the PRI-TL dispersion with the experimental NRI-TL dispersion at ω_0' differs slightly from the predicted intersection at ω_0. After Ref. [36]. Copyright © 2003 Optical Society of America, Inc.

quence of the fact that these media do not rely explicitly on resonances to synthesize the negative index property [22, 31]. The fluctuations at low frequencies result from the fact that measurements taken at any finite interval ($d = 5$ mm in the present case) cannot adequately sample the rapidly varying phase in the approach towards ω_B, where the guided waves begin to experience the coarseness of the transmission-line grid. This phenomenon, which becomes evident near $d = \lambda/4$ [36], may therefore be used to establish a lower frequency limit on the condition of homogeneity in NRI-TL metamaterials required to validate the concept of a negative refractive index.

The effective refractive index can be computed from the dispersion data as $n = \phi c/\omega d$, where c is the speed of light in vacuum and ϕ is the measured average phase shift per unit cell βd. In the most well defined guided-wave region from 1.17 GHz to 2.03 GHz, the absolute refractive index lies between -14.76 and -2.63. Since the PRI-TL grid is expected to possess an absolute refractive index of $\sqrt{2\epsilon_{eff}} = +2.05$, the relative refractive index between the two media, n_{REL}, varies from approximately -7.20 to -1.28. The condition $n_{REL} = -1$, which was predicted to occur near $\omega_0 = 2\pi \times 2.18$ GHz, and indicated in Fig. 1.30 by the intersection between the theoretical NRI-TL and PRI-TL dispersion curves, is achieved

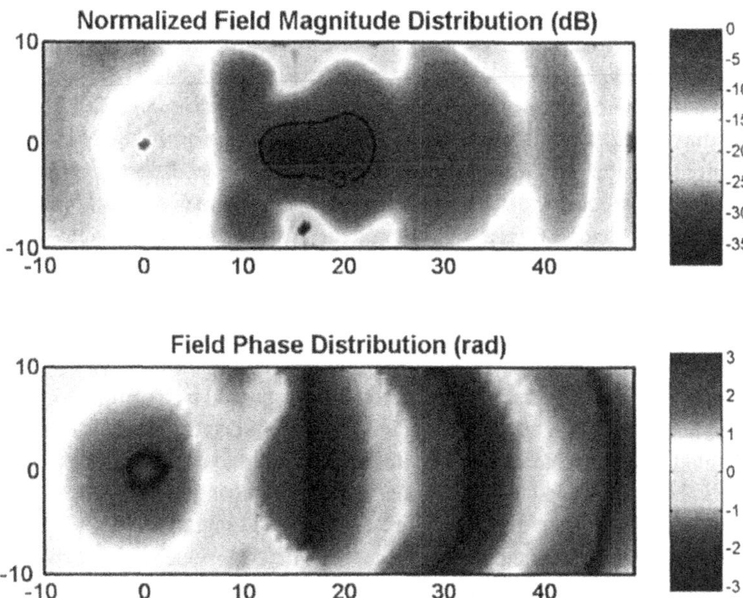

Fig. 1.31 Experimental data showing field magnitude and phase distributions in PRI-TL and NRI-TL regions at 2.09 GHz (PRI-TL: cells −10 to 10, NRI-TL: cells 11 to 50). The −3-dB contour indicates a maximum beam width of 32 mm, corresponding to 0.46 wavelengths in the NRI-TL metamaterial, which is suggestive of subwavelength focusing. After Ref. [36]. Copyright © 2003 Optical Society of America, Inc.

near $\omega_0' = 2\pi \times 2.09$ GHz, approximately where the experimental NRI-TL dispersion curve intersects the theoretical PRI-TL dispersion curve. This deviation is an artifact of the contradirectional coupling between the backward-wave mode and a surface wave mode, weakly bound to the NRI-TL/air boundary, whose dispersion follows the light line (not depicted). This phenomenon is discussed in greater depth in Refs. 39 and 52.

The magnitude and phase of the measured vertical electric fields over the structure at 2.09 GHz are depicted in Fig. 1.31. Cells −10 to 10 on the horizontal axis indicate the extent of the PRI-TL medium, and cells 11 to 50 indicate that of the NRI-TL metamaterial. Once again, the increased transmission at the focal plane and convergent progression of the phase fronts are indicative of focusing. The experimental phase distribution reveals a guide wavelength at 2.09 GHz of λ_{NRI-TL}=70 mm, suggesting, from the measured field magnitudes, a maximum −3-dB beam width of 32 mm, corresponding to $0.46\lambda_{NRI-TL}$. This is suggestive of subwavelength focusing, consistent with the proposal of Pendry [53], although the PRI-TL and NRI-TL regions were not designed to be impedance-matched at this frequency. The behaviour of the field distribution as the frequency is varied also exhibits characteristics typical of NRI-TL focusing, and animations showing the evolution of the measured field magnitude and phase as frequency is varied can be found online at the internet address listed in Ref. 36. For the present purpose, a simple description of this behaviour should suffice. Beginning at ω_B, where it is clear that the effective NRI-TL wavelength is small, the transmission from the PRI-TL region into the NRI-TL region steadily increases, and the first evidence of a focused beam becomes apparent near 1.3 GHz. At this frequency, the PRI-TL and NRI-TL regions exhibit distinct phase lag and phase advance characteristics, respectively, and as the frequency is increased, the structure enters the region of homogeneity and the phase distribution indicates a steadily increasing guide wavelength. Near 1.8 GHz the transmission at the focal plane increases, and at approximately 2 GHz, where, according to the dispersion data, it is expected that the device will encounter the condition $n_{REL} = -1$, the focus appears. As the frequency is increased (accompanied by a diminishing relative refractive index) the focal spot recedes slowly to the interface. Here, the experimentally obtained field distributions also indicate what seem to be surface-wave effects at the PRI-TL/NRI-TL boundary that appear near, and persist well beyond, ω_0'. These phenomena (which are evident at the PRI-TL/NRI-TL interface in Fig. 1.31) are reminiscent of the surface modes described in [12, 54], which vary exponentially away from the PRI-TL/NRI-TL interface and are known to participate in the focusing process. Beyond the stopband edge near 2.5 GHz, the phase exhibits a near-zero progression, and transmission into the NRI-TL region is weak.

1.7 CONCLUSION

In a seminal 1967 work, Veselago proposed that materials with simultaneously negative permittivity and permeability are physically permissible and possess a negative index of refraction. His conceptual exploration of this phenomenon revealed that, through negative refraction, planar slabs of such media would cause light or electromagnetic radiation to focus in on itself. Over thirty-five years later, following a short but intense flurry of research dedicated to the study of these theories, Veselago's principal ideas have been verified experimentally using electromagnetic artificial dielectrics. These structures have been popularly referred to as *metamaterials*, a term that emphasizes their unique transcendent properties.

It is well known that dielectric properties like permittivity and permeability can be modeled using distributed *L–C* networks. For a conventional dielectric, the per-unit-length capacitance and inductance can be directly related to the free-space permittivity and permeability, respectively, implying the use of a low-pass topology. In this chapter, it was shown that reversing the positions of the inductor and capacitor is equivalent to negating the effective material parameters, yielding a high-pass configuration, the dual of the conventional topology. The resulting structures were referred to as negative-refractive-index transmission-line (NRI-TL) metamaterials. Under certain conditions, these structures exhibit simultaneously negative material parameters, and their dispersion characteristics indicate the support of backward waves. A practical, periodic, two-dimensional implementation using a microstrip host medium, reactively loaded in a dual configuration using lumped or printed elements, was then presented. The essential features of the practical NRI-TL dispersion relation were obtained using the method of periodic analysis of microwave networks. These were simplified to identify a regime in which the NRI-TL periodic structure may be perceived as a homogeneous effective medium and reveal expressions for its effective material parameters. Through the appropriate selection of the properties of the host medium and loading, these characteristics could be designed and controlled, and interesting new phenomena were noted, such as the closure of the NRI-TL stopband at $\beta = 0$. These principles were then applied to the design of planar interfaces between positive-refractive-index transmission-line (PRI-TL) media and NRI-TL metamaterials, which were used in microwave circuit simulations to verify negative refraction and focusing. The simulated structures were then fabricated, and experimental results demonstrated focusing of an incident cylindrical wave and dispersion characteristics depicting an extremely broadband region over which the refractive index remains negative, in excellent correspondence with theory and simulations; furthermore, these results were suggestive of the subwavelength focusing phenomenon predicted by Pendry.

The success of NRI-TL metamaterials establishes that the conceptualization and design of LHM may rely not entirely on physical ideas but also on more familiar practical concepts like the design of artificial dielectrics and the backward wave. The NRI-TL concept also offers some advantages, including the fact that these structures need not be lossy and do not rely explicitly on resonances to synthesize the negative material parameters and, thus, offer dramatically increased operating bandwidths. Moreover, their unit cells are connected through a transmission-line network and they may therefore be equipped with lumped printed or discrete elements, which permit them to be compact over a broad range of frequencies. The flexibility thus gained enables NRI-TL metamaterials to be scalable from the MHz to the tens of GHz range. In addition, by employing tunable loading elements, the effective material properties can be dynamically varied and optimized for specific applications. Furthermore, in the planar form described in this chapter, these metamaterials inherently support 2-D wave propagation and are thus well-suited to applications in RF/microwave devices and circuits. Most importantly, as effective media, their effective material parameters are simultaneously and inherently negative.

The ability to bend electromagnetic waves at negative angles has important and numerous implications and great potential for application. One can envision devices that employ the phase-advance characteristic of NRI-TL metamaterials in phase compensation, coupler, and antenna applications; many such novel microwave devices have been realized and are described in Chapter 2. Of course, the most intriguing property of LHM, hence NRI-TL metamaterials, is the subwavelength resolution property suggested by Pendry; based on these notions and preliminary experimental data corroborating Pendry's theories, lenses may be designed to extend the near field and resolve features significantly below the conventional diffraction limit. The first such lens to experimentally demonstrate these ideas is a NRI-TL lens, and it is described in Chapter 3. This has obvious implications to photolithography and electromagnetic probing applications—for example, biomedical imaging or microscopy—which should benefit greatly as the technology develops in the appropriate directions.

Acknowledgments

This work was supported by the Natural Sciences and Engineering Research Council (NSERC) of Canada through Discovery and Strategic grants.

REFERENCES

1. V. G. Veselago, "The electrodynamics of substances with simultaneously negative values of ϵ and μ," *Sov. Phys. Usp*, vol. 10, no. 4, pp. 509–514, January–February 1968 (translation based on the original Russian document, dated 1967, as suggested in Ref. 2.)
2. I. V. Lindell, S. A. Tretyakov, K. I. Nikoskinen, and S. Ilvonen, "BW Media—Media with negative parameters, capable of supporting backward waves," *Microwave Opt. Tech. Lett.*, vol. 31, no. 2, pp. 129–133, October 2001.
3. S. Ramo, J. R. Whinnery, and T. Van Duzer, *Fields and Waves in Communication Electronics*, 3rd ed., John Wiley & Sons, Toronto, 1994.
4. R. E. Collin, *Foundations for Microwave Engineering*, 2nd ed., McGraw-Hill, Singapore, 1992.
5. A. Hessel, "General characteristics of traveling-wave antennas," in *Antenna Theory*, vol. 1, R. E. Collin and F. J. Zucker, eds. McGraw-Hill, New York, 1969, pp. 151–258.
6. W. E. Kock, "Metallic delay lenses," *Bell Syst. Tech. J.*, vol. 27, pp. 58–82, January 1948.
7. W. E. Kock, "Radio lenses," *Bell Lab. Rec.*, vol. 24, pp. 177–216, May 1946.
8. W. E. Kock, "Metal lens antennas," in *Proceedings, IRE and Waves and Electrons*, pp. 828–836, November 1946.
9. R. E. Collin, *Field Theory of Guided Waves*, 2nd ed., Wiley-IEEE Press, Toronto, 1990.
10. R. N. Bracewell, "Analogues of an ionized medium: Applications to the ionosphere," *Wireless Eng.*, vol. 31, pp. 320–326, December 1954.
11. W. Rotman, "Plasma simulation by artificial dielectrics and parallel-plate media," *IRE Trans. Antennas Propag.*, vol. AP-10, no. 1, pp. 82–85, January 1962.

12. J. B. Pendry, A. J. Holden, W. J. Stewart, and I. Youngs, "Extremely low frequency plasmons in metallic mesostructures," *Phys. Rev. Lett.*, vol. 76, no. 25, pp. 4773–4776, June 1996.

13. D. F. Sievenpiper, M. E. Sickmiller, and E. Yablonovitch, "3D wire mesh photonic crystals," *Phys. Rev. Lett.*, vol. 76, no. 14, pp. 2480–2483, April 1996.

14. J. B. Pendry, A. J. Holden, D. J. Robbins, and W. J. Stewart, "Magnetism from conductors and enhanced nonlinear phenomena," *IEEE Trans. Microwave Theory Tech.*, vol. 47, no. 11, pp. 2075–2084, November 1999.

15. D. R. Smith, W. J. Padilla, D. C. Vier, S. C. Nemat-Nasser, and S. Schultz, "Composite medium with simultaneously negative permeability and permittivity," *Phys. Rev. Lett.*, vol. 84, no. 18, pp. 4184–4187, May 2000.

16. D. R. Smith and N. Kroll, "Negative refractive index in left-handed materials," *Phys. Rev. Lett.*, vol. 85, no. 14, pp. 2933–2936, October 2000.

17. D. R. Smith, D. C. Vier, N. Kroll, and S. Schultz, "Direct calculation of permeability and permittivity for a left-handed metamaterial," *Appl. Phys. Lett.*, vol. 77, no. 14, pp. 2246–2248, October 2000.

18. R. Marqués, F. Medina, R. Rafii-El-Idrissi, "Role of bianisotropy in negative permeability and left-handed metamaterials," *Phys. Rev. B*, vol. 65 no. 144440, April 2002.

19. N. Katsarakis, T. Koschny, M. Kafesaki, E. N. Economou, and C. M. Soukoulis, "Electrical coupling to the magnetic resonance of split ring resonators," *Appl. Phys. Lett.*, vol. 84, no. 15, pp. 2943–2945, April 2004.

20. R. A. Shelby, D. R. Smith, S. C. Nemat-Nasser, and S. Schultz, "Microwave transmission through a two-dimensional, isotropic, left-handed metamaterial," *App. Phys. Lett.*, vol. 78, no. 4, pp. 489–491, January 2001.

21. R. A. Shelby, D. R. Smith, and S. Schultz, "Experimental verification of a negative index of refraction," *Science*, vol. 292, pp. 77–79, April 6, 2001.

22. G. V. Eleftheriades, O. Siddiqui, and A. K. Iyer, "Transmission line models for negative refractive index media and associated implementations without excess resonators," *IEEE Microwave Wireless Components Lett.*, vol. 13, no. 2, pp. 51–53, February 2003.

23. R. W. Ziolkowski and E. Heyman, "Wave propagation in media having negative permittivity and permeability," *Phys. Rev. E*, vol. 64 no. 056625, October 2001.

24. G. Kron, "Equivalent circuit of the field equations of Maxwell," *Proc. IRE*, vol. 32, no. 5, pp. 289-299, May 1944.

25. J. R. Whinnery, S. Ramo, "A new approach to the solution of high-frequency field problems," *Proc. IRE*, vol. 32, no. 5, pp. 284-288, May 1944.

26. A. K. Iyer and G. V. Eleftheriades, "Negative refractive index metamaterials supporting 2-D waves," in *IEEE MTT-S International Microwave Symposium Digest*, vol. 2, June 2–7, 2002, Seattle, WA, pp. 1067–1070.

27. G. V. Eleftheriades, A. K. Iyer, and P. C. Kremer, "Planar negative refractive index media using periodically L–C loaded transmission lines," *IEEE Trans. Microwave Theory Tech.*, vol. 50, no. 12, pp. 2702–2712, December 2002.

28. C. Caloz, H. Okabe, H. Iwai, and T. Itoh, "Transmission line approach of left-handed materials," in *USNC/URSI National Radio Science Meeting Digest*, June 16–21, 2002, San Antonio, TX, p. 39.

29. C. Caloz and T. Itoh, "Novel microwave devices and structures based on the transmission line approach of meta-materials," in *IEEE MTT-S International Microwave Symposium Digest*, vol. 1, June 8–13, 2003, Philadelphia, PA, pp. 195–198.

30. A. A. Oliner, "A planar negative-refractive-index medium without resonant elements," in *IEEE MTT-S International Microwave Symposium Digest*, vol. 1, June 8–13, 2003, Philadelphia, PA, pp. 191–194.

31. A. A. Oliner, "A periodic-structure negative-refractive-index medium without resonant elements," in *IEEE APS/URSI International Symposium Digest*, June 16–21, 2002, San Antonio, TX, p. 41.

32. K. G. Balmain, A. A. E. Lüttgen, and P. C. Kremer, "Power flow for resonance cone phenomena in planar anisotropic metamaterials," *IEEE Trans. Antennas Propag.*, Special Issue on Metamaterials, vol. 51, no. 10, pp. 2612–2618, October 2003.

33. K. G. Balmain, A. A. E. Lüttgen, and G. V. Eleftheriades, "Resonance cone radiation from a planar, anisotropic metamaterial," in *2003 URSI Digest*, June 23, 2003, Columbus, OH, p. 24.

34. K. G. Balmain, A. A. E. Lüttgen, and P. C. Kremer, "Using resonance cone refraction for compact RF metamaterial devices," in *Proceedings of the International Conference on Electromagnetics in Advanced Applications (ICEAA'03)*, September 8–12, 2003, Torino, Italy, ISBN 88-8202-008-8 (on CD, ISBN 88-8202-009-6), pp. 419–422.

35. A. K. Iyer, K. G. Balmain, and G. V. Eleftheriades, "Dispersion analysis of resonance cone behaviour in magnetically anisotropic transmission-line metamaterials," in *2004 IEEE Antennas and Propagation Society International Symposium Digest*, June 23, 2004, Monterey, CA, pp. 3147–3150.

36. A. K. Iyer, P. C. Kremer, and G. V. Eleftheriades, "Experimental and theoretical verification of focusing in a large, periodically loaded transmission line negative refractive index metamaterial," *Opt. Express*, vol. 11, pp. 696–708, April 2003, http://www.opticsexpress.org/abstract.cfm?URI=OPEX-11-7-696.

37. L. Brillouin, *Wave Propagation in Periodic Structures: Electric Filters and Crystal Lattices*, Dover, New York, 1946.

38. A. Grbic and G. V. Eleftheriades, "Periodic analysis of a 2-D negative refractive index transmission line structure," *IEEE Trans. Antennas Propag.*, Special Issue on Metamaterials, vol. 51, no. 10, pp. 2604–2611, October 2003.

39. A. Grbic and G. V. Eleftheriades, "Dispersion analysis of a microstrip-based negative refractive index periodic structure," *IEEE Microwave Wireless Components Lett.*, vol. 3, no. 4, pp. 155–157, April 2003.

40. G. V. Eleftheriades, "Planar negative refractive index metamaterials based on periodically L–C Loaded Transmission Lines," in *Workshop on Quantum Optics*, Kavli Institute for Theoretical Physics, University of Santa Barbara, July 11, 2002, http://online.kitp.ucsb.edu/online/qo02/eleftheriades/

41. D. M. Pozar, *Microwave Engineering*, 2nd ed., John Wiley & Sons, Toronto, 1998.

42. J. D. Kraus, *Electromagnetics*, 4th (international) ed., McGraw-Hill, Singapore, 1992.

43. A. Sanada, C. Caloz, and T. Itoh, "Characteristics of the composite right/left-handed transmission lines," *IEEE Microwave Wireless Components Lett.*, vol. 14, no. 2 , pp. 68–70, February 2004.

44. A. K. Iyer, A. Grbic, G. V. Eleftheriades, "Sub-wavelength focusing in loaded transmission line negative refractive index metamaterials," in *IEEE MTT-S International Microwave Symposium Digest*, vol. 1, June 8–13, 2003, Philadelphia, PA, pp. 200–202.

45. A. Alù and N. Engheta, "Pairing an epsilon-negative slab with a mu-negative slab: Resonance, tunneling, and transparency," *IEEE Trans. Antennas Propag.*, Special Issue on Metamaterials, vol. 51, no. 10, pp. 2558–2571, October 2003.

46. Agilent Technologies, http://www.agilent.com/

47. A. Grbic and G. V. Eleftheriades, "Growing evanescent waves in negative-refractive-index transmission-line media," *Appl. Phys. Lett.*, vol. 82, no. 12, pp. 1815–1817, March 2003.

48. G. V. 't Hooft, "Comment on 'negative refraction makes a perfect lens'," *Phys. Rev. Lett.*, vol. 87, no. 24, p. 249701, December 2001.

49. J. M. Williams, "Some problems with negative refraction," *Phys. Rev. Lett.*, vol. 87, no. 24, 249703, December 2001.

50. N. Garcia and M. Nieto-Vesperinas, "Is there an experimental verification of a negative index of refraction yet?" *Opt. Lett.*, vol. 27, no. 11, pp. 885–887, June 2002.

51. P. M. Valanju, R. M. Walser, and A. P. Valanju, "Wave refraction in negative-index media: always positive and very inhomogeneous," *Phys. Rev. Lett.*, vol. 88, no. 18, 187401, May 2002.

52. A. K. Iyer, G. V. Eleftheriades, "Leaky-wave radiation from a two-dimensional negative-refractive-index transmission-line metamaterial," in *Proceedings of 2004 URSI EMTS International Symposium on Electromagnetic Theory*, May 26, 2004, Pisa, Italy, pp. 891–893.

53. J. B. Pendry, "Negative refraction makes a perfect lens," *Phys. Rev. Lett.*, vol. 85, no. 18, pp. 3966–3969, October 2000.

54. R. Ruppin, "Surface polaritons of a left-handed material slab," *J. Phys. Condens. Matter*, vol. 13, pp. 1811–1819, 2001.

2 Microwave Devices and Antennas Using Negative-Refractive-Index Transmission-Line Metamaterials

GEORGE V. ELEFTHERIADES

The Edward S. Rogers Sr. Department of Electrical and Computer Engineering
University of Toronto
Toronto, Ontario, M5S 3G4
Canada

2.1 INTRODUCTION

This chapter describes a number of microwave devices and antenna applications of the transmission-line metamaterials examined in Chapter 1. Such RF/Microwave devices include lenses that can overcome the diffraction limit, small and broadband phase-shifting lines, backward leaky-wave antennas, small and low-profile antennas, antenna feed networks and baluns, novel power-combining architectures, and high-directivity coupled-line couplers. As explained in Chapter 1, the term "metamaterials" is meant here to refer to artificial media with properties that transcend those of natural media ("meta" means "beyond" in Greek). Specifically in this chapter we consider periodic media for which the periodicity is much smaller than the wavelength of the incident electromagnetic wave. Therefore, effective material parameters such as a permittivity, a permeability, and a refractive index can be defined. This definition establishes a direct relationship with artificial dielectrics [1]. In this chapter, we will limit our discussion to isotropic metamaterials in which the permittivity and permeability are simultaneously negative, hence leading to a negative refractive index [2]. Such "left-handed" or "negative-refractive-index" (NRI) media were first implemented using periodic arrays of thin wires to synthesize negative permittivity and split-ring resonators to synthesize negative permeability [3, 4]. This led to the first experimental demonstration of negative refraction of cylindrical waves at microwave frequencies (X band). A different approach for implementing NRI media has been proposed in Refs. 5 and 6 by loading a planar network of printed transmission lines (TL) with series capacitors and shunt inductors in a dual (high-pass) configuration, when compared to a standard transmission line. A two-dimensional (2-D) NRI medium was interfaced with a commensurate conventional dielectric, arguably leading to the first experimental demonstration of focusing from a left-handed metamaterial [6, 7]. This transmission-line methodology for making planar NRI media results in wide operating bandwidths over which the refractive index remains negative; for example, Ref. 7 reports focusing due to negative refraction over an octave bandwidth. Moreover, a three-region planar lens arrangement was used to observe focusing beyond the diffraction limit [8, 9], as was predicted by J. B. Pendry [10]. A similar TL approach was followed by T. Itoh, C. Caloz, and co-workers leading to interesting and useful microwave circuits [11, 12]. This chapter is limited to isotropic NRI metamaterials synthesized using loaded transmission lines. However, intriguing and useful planar anisotropic transmission-line metamaterials have been developed by K. G. Balmain et al. [13, 14] and are described in Chapter 6.

2.2 FUNDAMENTAL PROPERTIES

Veselago was the first to examine in the open literature the feasibility of media characterized by simultaneously negative permittivity and permeability [2]. He concluded that such media are allowed by Maxwell's equations and that plane waves propagating in them would have their electric field, $\mathbf{E}$, magnetic field, $\mathbf{H}$, and propagation constant, $\mathbf{k}$, forming a left-handed triplet. Therefore, he coined the term "left-handed" to describe these hypothetical media. Also, Veselago realized that one has to choose the negative branch of the square root to properly define the corresponding refractive index, that is, $n = -\sqrt{\varepsilon\mu}$. Thus, such left-handed media support negative refraction of electromagnetic waves, something that was demonstrated experimentally more than three decades later by Shelby, Smith, and Schultz [3]. Moreover, due to the fact that $\mathbf{E}$, $\mathbf{H}$, and $\mathbf{k}$ form a left-handed triplet whereas the $\mathbf{E}$ and $\mathbf{H}$ vectors and the Poynting vector $\mathbf{S}$ form a right-handed triplet, Veselago concluded that in left-handed media the propagation constant $\mathbf{k}$ is antiparallel to the Poynting vector $\mathbf{S}$. In retrospect, what Veselago was describing were backward waves. Certainly, one-dimensional backward-wave lines are not new to the microwave and antenna communities, and there is an interesting connection to familiar concepts and structures [15, 16]. However, what is remarkable and surprising in Veselago's work is his realization that isotropic and homogeneous media supporting backward waves should be characterized by a negative refractive index. Consequently, when such two- or three-dimensional media are interfaced with conventional dielectrics, Snell's Law is reversed, leading to the negative refraction of an incident electromagnetic plane wave. One way to understand negative refraction is through the notion of phase matching as is explained in Fig. 1.2. Another way to show this is by invoking the radiation condition, as discussed in the next section.

An issue worth clarifying here is the terminology "negative group velocity," which was used by Veselago in his original paper to characterize left-handed media [2].

As was previously mentioned, what Veselago had in mind were backward waves, in which the phase velocity is antiparallel to the group velocity [and the phase advances under an $\exp(j\omega t)$ time-harmonic variation], although he never mentioned backward waves explicitly in Ref. 2. Hence, if we define the phase velocity to be negative, then in left-handed media the group velocity should be positive to describe power flowing away from the source. In this context, negative group velocity would indicate anomalous dispersion which prevails in conventional dielectric media close to absorption lines. Such left-handed media exhibiting negative phase velocity and anomalous negative group velocity have recently been realized by inserting lossy resonators in transmission-line metamaterials [17]. A detailed description of these negative-group-velocity NRI loaded TL structures can be found in Chapter 10.

Harnessing the phenomenon of negative refraction, entirely new refraction-based devices can be envisioned such as a flat lens without an optical axis, also proposed by Veselago, as shown in Fig. 2.1. Of course this is a peculiar kind of lens since it does not bring to focus incident plane waves. However, lenses that do not focus plane waves are not unusual; one example is the hyperhemispherical oil-immersion lens used in optical microscopy. More interesting lens applications are described in Chapter 5. Other intriguing possibilities predicted by Veselago include the reversal of Čerenkov radiation [18] and of the Doppler shift [19]. Each of these new phenomena can be utilized to make useful new devices. For example, the reversal of Čerenkov radiation inspired the development of backward-wave antennas radiating their fundamental spatial harmonic [18,20], whereas the reversal of the Doppler shift can be exploited for making wideband millimeter-wave sources [19].

2.3 EFFECTIVE MEDIUM THEORY

A practical periodic 2-D transmission-line based NRI metamaterial can be realized using an array of unit cells, each as depicted in Fig. 2.2. A host transmission-line medium (e.g., microstrip) is periodically loaded using discrete series capacitors and shunt inductors [5–7]. From the onset, the key observation is that there is a correspondence between negative permittivity and a shunt inductor (L), as well as between negative permeability and a series capacitor (C). This allows one to synthesize artificial media (metamaterials) with a negative permittivity and a negative permeability and hence a negative refractive index [6,7]. When the unit cell dimension d is much smaller than a guided wavelength, the array can be regarded as a homogeneous effective medium and as such can be described by effective constitutive parameters $\mu_N(\omega)$ and $\varepsilon_N(\omega)$, which are determined through a rigorous periodic analysis to be of the form [7] (assuming 2-D $\mathbf{TM_y}$ wave propagation in Fig. 2.2)

$$\varepsilon_N(\omega) = 2\varepsilon_P - \frac{g}{\omega^2 L_0 d}, \quad \mu_N(\omega) = \mu_P - \frac{1/g}{\omega^2 C_0 d} \tag{2.1}$$

Here ε_P and μ_P are positive constants which are proportional to the per-unit-length capacitance and inductance of the host transmission line medium, respectively. This

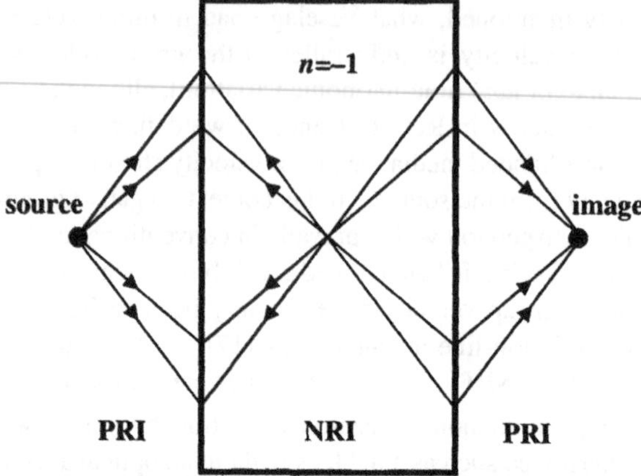

Fig. 2.1 Veselago's lens. As shown, negative refraction is utilized in order to focus a point to a point. This leads to a peculiar lens with flat surfaces and no optical axis. The rays converge to the same point when $n = -1$, thus leading to aberration-free focusing and no reflections from the lens surfaces. The thickness of the lens is half the distance from the source to the image. The arrows represent k vectors; PRI, positive refractive index; NRI, negative refractive index.

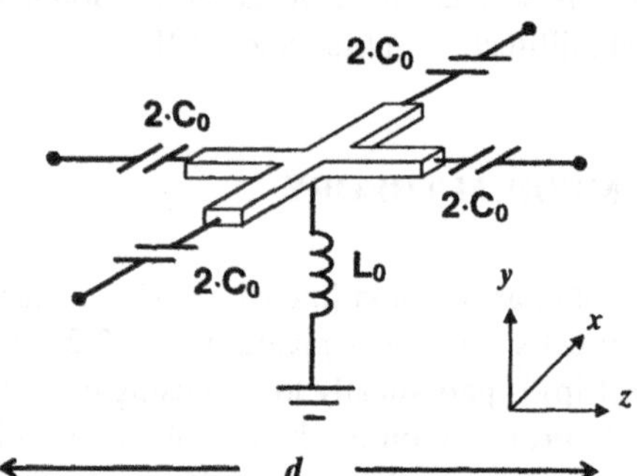

Fig. 2.2 Unit cell for the 2-D transmission-line-based NRI metamaterial. A host transmission line is loaded periodically with series capacitors and shunt inductors in a dual (high-pass) configuration. After Ref. [7]. Copyright © 2003 Optical Society of America, Inc.

particular arrangement of the inclusions L_0 and C_0 provides the desired negative material contribution that diminishes with frequency ω. In fact, the underlying dependence on $1/\omega^2$ is important for satisfying the Poynting Theorem (conservation of energy) in such a dispersive medium [1, 2]. On the other hand, the geometrical factor g relates the characteristic impedance of the transmission-line network to the wave impedance of the effective medium. Moreover, the factor of 2 in front of ε_P is required in order to properly account for the loading from the z-directed (x-directed)

TLs when propagating along the x-axis (z-axis); note that for the 1-D case this factor becomes 1. When the parameters are simultaneously negative, these structures exhibit a negative effective refractive index and have experimentally demonstrated the predicted associated phenomena, including negative refraction, focusing, and focusing with subwavelength resolution [5–9]. A physical interpretation of the negative material parameters of (2.1) in terms of elementary electric and magnetic dipoles is provided in Chapter 3.

Once it is established that a negative permittivity and permeability characterizes these media, then the refractive index is given by $n = \pm\sqrt{\varepsilon_N \mu_N}$ where one has to choose the correct branch of the square root. For this purpose, consider low-loss propagation in which the permeability is slightly complex such that $\mu_N = \mu'_N - j\mu''_N$ with $\mu'_N < 0$ and $\mu''_N > 0$ (passive medium). Under this limit of low-loss propagation, the refractive index can be approximated by

$$n \cong \pm\sqrt{\varepsilon_N \mu'_N}\left(1 - j\frac{\mu''_N}{2\mu'_N}\right) \tag{2.2}$$

Moreover, a plane wave propagating along the positive z-axis $\exp(-jk_0nz)$ should decay with distance. Hence, the imaginary part of the refractive index $n = n' - jn''$ should be positive; that is, one has to choose the negative branch of the square root in (2.2) to satisfy the radiation condition at infinity.

2.4 A SUPER-RESOLVING NEGATIVE-REFRACTIVE-INDEX TRANSMISSION-LINE LENS

Classical electrodynamics imposes a resolution limit when imaging using conventional lenses. This fundamental limit, called the "diffraction limit," in its ultimate form, is attributed to the finite wavelength of electromagnetic waves. The electromagnetic field emanating from a luminous object, lying over the x–y plane, consists of a continuum of plane waves $\exp(-jk_xx - jk_yy)\exp(-jk_zz)$. Each plane wave has a characteristic amplitude and propagates at an angle with the optical z-axis given by the direction cosines $(k_x/k_0,\ k_y/k_0)$, where k_0 is the propagation constant in free space. The plane waves with real-valued direction cosines $(k_x^2 + k_y^2 < k_0^2)$ propagate without attenuation, while the evanescent plane waves with imaginary direction cosines $(k_x^2 + k_y^2 > k_0^2)$ attenuate exponentially along the optical z-axis. A conventional lens focuses only the propagating waves, resulting in an imperfect image of the object, even if the lens diameter were infinite. The finer details of the object, carried by the evanescent waves, are lost due to the strong attenuation these waves experience $(\exp(-z\sqrt{k_x^2 + k_y^2 - k_0^2}))$ when traveling from the object to the image through the lens. The Fourier transform uncertainty relation $k_{t\_max}\Delta\rho \sim 2\pi$, relating the maximum transverse wavenumber k_{t_max} to the smallest transverse spatial detail $\Delta\rho$, implies that spatial details smaller than a wavelength are eliminated from the image,

$$\Delta\rho \sim 2\pi/k_0 = \lambda \tag{2.3}$$

This loss of resolution, which is valid even if the lens diameter were infinite, constitutes the origin of the diffraction limit in its ultimate form. For the typical case of imaging a point source, the diffraction limit manifests itself as an image smeared over an area approximately one wavelength in diameter, in compliance with equation (2.3).

In 2000, John Pendry extended the analysis of Veselago's lens (see Fig. 2.1) to include evanescent waves and observed that such lenses could overcome the diffraction limit [10]. Pendry suggested that Veselago's lens would allow "perfect imaging" if it were completely lossless and its refractive index n were exactly equal to -1 relative to the surrounding medium. The left-handed lens achieves imaging with super-resolution by focusing propagating waves as would a conventional lens (see Fig. 2.1), but in addition it supports growing evanescent waves which restore the decaying evanescent waves emanating from the source. This restoration of evanescent waves at the image plane extends the maximum accessible wavenumbers $k_{t\_max} > k_0$ and allows imaging with super-resolution. The physical mechanism behind the growth of evanescent waves is quite interesting: Within the NRI (left-handed) slab, multiple reflections result in both growing and attenuating evanescent waves; however, $n = -1$ corresponds to a resonant phenomenon in which the attenuating solution is canceled out, thus leaving only the growing wave present. This is achieved because when $n = -1$ the second NRI/PRI interface in Fig. 2.1 corresponds to an infinite reflection coefficient whereas the first PRI/NRI interface is matched. A quantitative description of the latter point-of-view is outlined in the Appendix at the end of this chapter. In a sense, one may think of Veselago's lens as an inverse system that exactly restores propagation in free space (at least the entire region behind the source).

A picture of a planar version of Veselago's lens that was constructed at the University of Toronto is shown in Fig. 2.3 [9]. The NRI lens is a slab consisting of a 5×19 grid of printed microstrip strips, loaded with series capacitors (C_0) and shunt inductors (L_0). This NRI slab is sandwiched between two unloaded printed grids that act as homogeneous media with a positive index of refraction. The first unloaded grid is excited with a monopole (point source) attached to the leftmost grid, which is imaged by the NRI lens to the second unloaded grid. The vertical electric field over the entire structure is measured using a detecting probe (for details, see Ref. 7).

The measured half-power beamwidth of the point-source image at 1.057 GHz is 0.21 effective wavelengths, which is appreciably narrower than that of the diffraction-limited image corresponding to 0.36 wavelengths (see Fig. 2.4a). The enhancement of evanescent waves for the specific structure under consideration was demonstrated in Ref. 22. Figure 2.4b shows the measured vertical electric field above the central row of the lens, which verifies the exponential growth of the fields inside the NRI medium predicted in Ref. 9. Since there is some controversy regarding losses in NRI metamaterials, it could be useful to report that the loss tangent of the NRI medium at 1.05 GHz is estimated to be $\tan\delta = 0.062$, which attests to the low-loss nature of the NRI transmission-line lens. However, even such a slight loss is sufficient to deteriorate the growth of evanescent waves to $k_{t\_max} = 3k_0$ [9]. This implies that a

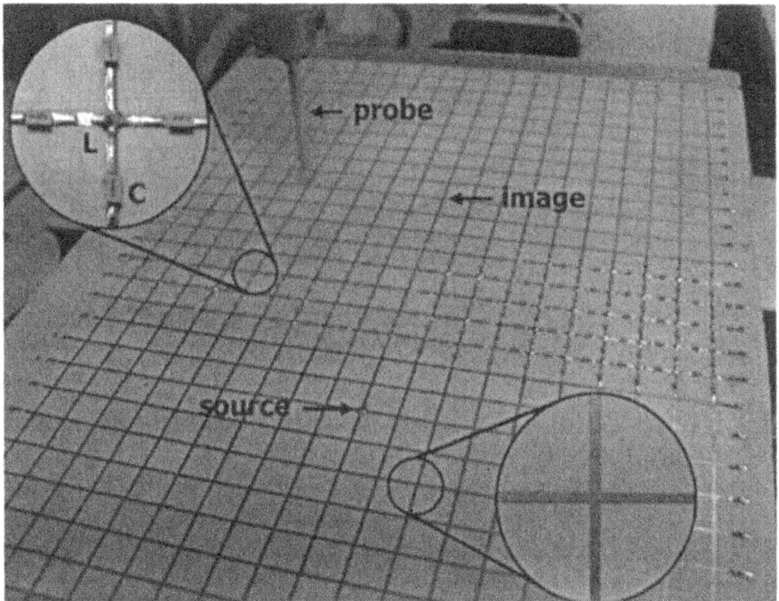

Fig. 2.3 Photograph of a planar super-resolving Veselago lens around 1 GHz. Reprinted figure with permission from Ref. [9]. Copyright © 2004 by the American Physical Society.

resolution equal to $\lambda/6$ is achieved; in comparison, a conventional lens would have produced a diffraction-limited image with a resolution of $\lambda/2$ at best (peak-to-null beamwidth). The periodic Green's function formulation described in Ref. 21 can be extended to quantitatively characterize the resolution limitations of these NRI-TL lenses due to material and mismatch losses. The full details are described in Chapter 3, where it is shown that the maximum transverse wavenumber in the structure of Fig. 2.3 can be approximated by

$$R_{res} = \frac{k_{t\_max}}{k_0} \cong \frac{\ln Q_c}{k_0 h} \tag{2.4}$$

where Q_c is the quality factor of the loading series capacitors and $k_0 h$ is the electrical length of the NRI slab in radians. This result is compatible with the analysis in Ref. 23, which is, however, carried out in terms of the effective lossy material parameters. This is quite an interesting result since it informs one that (*i*) in order to double the resolution, the quality factor of the capacitors has to quadruple, and so on, and (*ii*) electrically thin lenses will lead to higher resolution. Another silent conclusion that can be drawn from (2.4) is the answer to the question: What will happen to an evanescent wave when the thickness of the NRI region h starts increasing to infinity? Will the evanescent wave amplitude diverge to infinity? Clearly, (2.4) suggests that the answer is no, since as the thickness h increases, the maximum transverse wavenumber k_{t_max} decreases inversely proportionally to h, thus eventually clipping

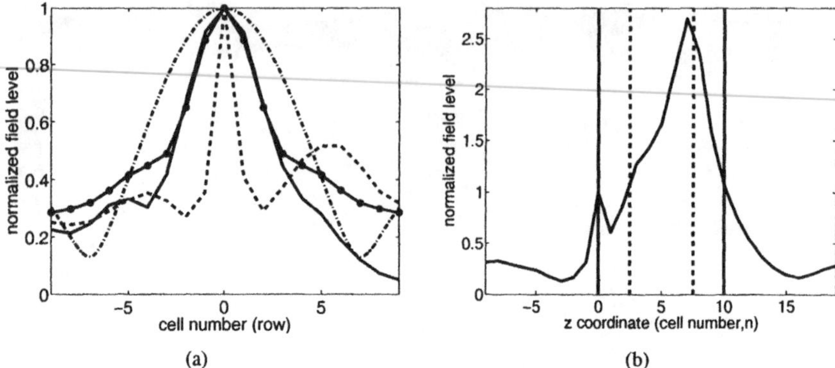

Fig. 2.4 (a) The interior dotted line designates the experimental E-field source pattern. The exterior dotted line designates the diffraction-limited E-field image. The intermediate solid line designates the measured image pattern and the line with the circles indicates the theoretical image pattern with losses included. (b) Experimental verification of growing evanescent waves in a super-resolving NRI-TL lens. The dotted lines indicate the location of the NRI region and the solid lines indicate the location of the source (left) and external image (right).

the growth of that particular evanescent wave. Indeed, for a given Q_c and frequency, there is a characteristic thickness h_{cutoff} after which all evanescent waves will start attenuating instead of growing. This can be determined from (2.4) by insisting that $R_{res} < 1$, that is,

$$h_{cutoff} = \frac{\ln(Q_c)}{k_0} \tag{2.5}$$

As was mentioned before, the super-resolving imaging properties of the structure shown in Fig. 2.3 have been theoretically investigated by means of a rigorous periodic Green's function analysis in Ref. 21. From this analysis, it is quite useful to actually examine the complete 2-D profile of the vertical electric field above the structure of Fig. 2.3. This is shown in Fig. 2.5 for the ideal, lossless case. First, observe the nature of the cylindrical waves formed in the three regions of the PRI/NRI/PRI lens. By inspection, one can verify that the phase center of these cylindrical waves in the three regions indeed coincide with the location of the source, internal and external images (foci) predicted by the ray picture of Fig. 2.1. What is remarkable is that the so identified location of the external (or internal) image *does not* coincide with the location of maximum field intensity. Indeed, the maximum field intensity takes place at the exit interface of the NRI lens and along the line joining the source and the images. This is due to the peculiar function of the PRI/NRI/PRI structure, which enhances the amplitude of the evanescent waves emanating from the source (see the Appendix at the end of this chapter). Another interesting aspect arising from examining Fig. 2.5 is that, ideally, the entire half-space behind the source is perfectly reproduced by the half-space in front of the external image.

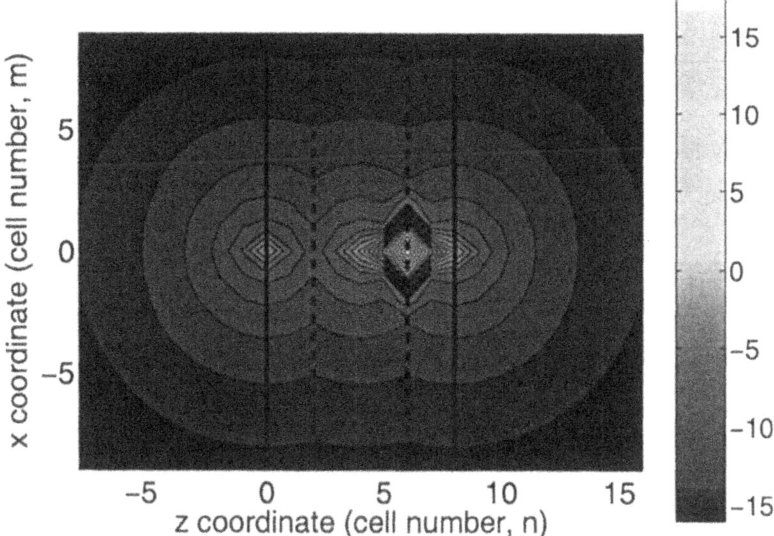

Fig. 2.5 Vertical electric field (magnitude) above the structure of Fig. 2.3; lossless case. The dashed lines indicate the location of the NRI region and the solid lines indicate the location of the source (left) and external image (right). (After Ref. [21].)

The analysis of Ref. 21 also reveals that, due to the periodicity of these realistic structures, the maximum possible transverse wavenumber is limited to

$$k_{t\_cutoff} = \pi/d \tag{2.6}$$

where d is the periodicity. Therefore, even for a lossless lens, the longitudinal wavenumber k_z never assumes infinite values and the solution to the imaging problem avoids the singularities encountered when the NRI lens is assumed homogeneous [10]. This cutoff transverse wavenumber will yield an absolute upper bound to the resolution that can be achieved when a NRI periodic lens is sandwiched between two homogeneous dielectric media.

Another question that is often raised concerns the origin of the energy stored in the growing evanescent waves: From where does this energy come? This question is not difficult to answer if one considers the PRI/NRI/PRI setting as a resonator which stores energy in the form of evanescent waves. Consequently, this energy is stored during the transient period, much like in any passive resonator. In other words, one has to keep in mind that what is shown in Fig. 2.5 is the steady-state solution. For a quantitative treatment of this aspect, the reader is referred to Ref. 24, in which it is actually shown that there is an intrinsic time scale associated with any desired lateral resolution (i.e., one has to wait longer and longer in order to resolve finer and finer spatial details).

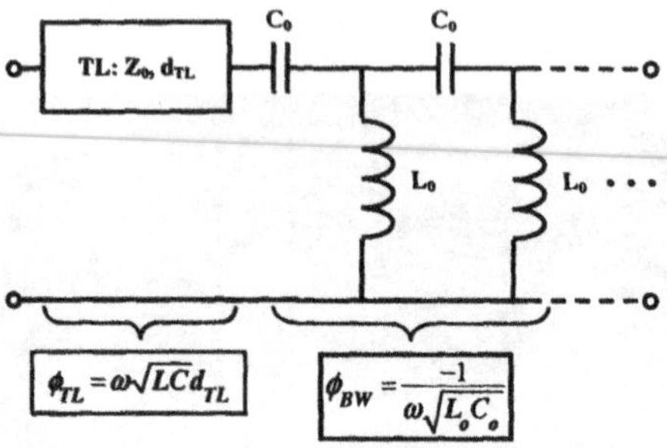

Fig. 2.6 Phase compensating structure based on a conventional TL and a NRI (backward-wave) line.

Finally, the corresponding dispersion characteristics for these distributed structures were derived in Ref. 25 using periodic two-dimensional transmission-line theory. In the case that the loading is achieved using printed instead of chip loading lumped elements—for example, microstrip gaps and vias or coils to implement series capacitors and inductors, respectively [26]—the corresponding dispersion characteristics have been examined using finite-element electromagnetic simulations in Ref. 27.

2.5 COMPACT AND BROADBAND PHASE-SHIFTING LINES

In conventional positive-refractive-index (PRI) transmission lines (TLs), the phase lags in the direction of positive group velocity, thus incurring a negative phase. It therefore follows that phase compensation can be achieved at a given frequency by cascading a section of a NRI line (e.g., backward-wave line) with a section of a PRI line to synthesize positive, negative, or zero transmission phase at a short physical length (see Fig. 2.6) [28].

The structure of Fig. 2.6 can be rearranged to form a series of symmetric metamaterial unit cells as proposed in Refs. 6 and 28. Such a unit cell is shown in Fig. 2.7 and it is nothing but a transmission line of characteristic impedance Z_0, periodically loaded with series capacitors C_0 and shunt inductors L_0 [6]. A representative dispersion diagram for typical host transmission line and loading parameters is shown in Fig. 2.8. The metamaterial phase-shifting lines can then be constructed by cascading these unit cells. The edges of the stopband f_{c1} and f_{c2} in Fig. 2.8 are determined at the series resonance between the inductance of the transmission-line section and the loading capacitor C_0, and the shunt resonance between the capacitance of the transmission-line section and the loading inductance L_0, respectively. Alternatively, these are the frequencies at which the effective permeability $\mu_N(\omega)$ and effective permittivity $\varepsilon_N(\omega)$ vanish:

$$\varepsilon_N(\omega) = 0, \qquad \mu_N(\omega) = 0$$

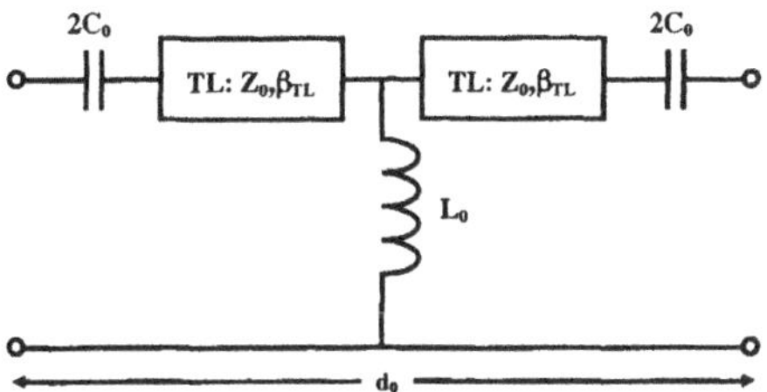

Fig. 2.7 Unit cell of a metamaterial phase-shifting line comprising a host transmission line periodically loaded with series capacitors and shunt inductors. After Ref. [28]. Copyright © 2003 IEEE.

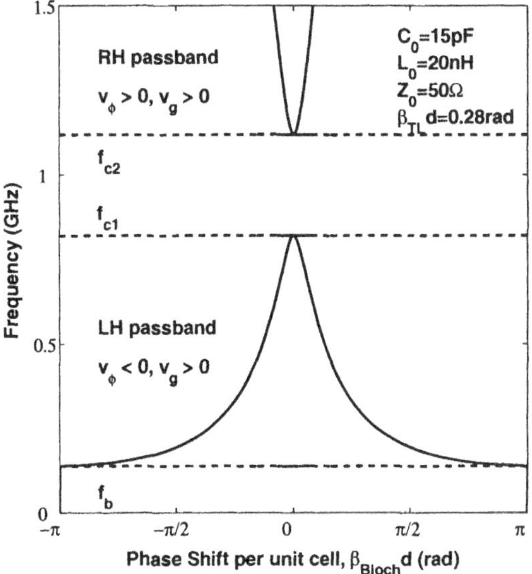

Fig. 2.8 The dispersion diagram for the periodic structure of Fig. 2.7 with typical line and loading parameters. The edges of the stopband are designated by f_{c1} and f_{c2}. After Ref. [28]. Copyright © 2003 IEEE.

Hence by setting the effective material parameters of equation (2.1) to zero, these cutoff frequencies are readily determined to be

$$f_{c1} = \frac{1}{2\pi}\sqrt{\frac{1/g}{\mu_p C_0 d}} \tag{2.7}$$

$$f_{c2} = \frac{1}{2\pi}\sqrt{\frac{g}{\varepsilon_p L_0 d}} \tag{2.8}$$

Fig. 2.9 Top: A 2-stage phase-shifting line (16 mm). Bottom: A 4-stage phase-shifting line (32 mm) at 0.9 GHz. Note: Reference $-360°$ TL line (283.5 mm, not shown).

where the characteristic impedance of the host transmission line is $Z_0 = g\sqrt{\mu_p/\varepsilon_p} = \sqrt{L/C}$. By equating f_{c1} and f_{c2}, the stopband in Fig. 2.8 can be closed, thus allowing one to access phase shifts around the zero mark. The condition for a closed stopband is therefore determined to be

$$Z_0 = \sqrt{\frac{L_0}{C_0}} \tag{2.9}$$

This condition also implies that the transmission line of Fig. 2.6 is matched to the NRI line. The closed stopband condition (2.9) was originally derived in Ref. 6 (equation 29) and later also reported in Ref. 29. Under this condition, it has been shown in Ref. 28 that the total phase shift per unit cell is

$$\beta_{\mathit{eff}} \approx \omega\sqrt{LC} + \frac{-1}{\omega\sqrt{L_0 C_0}} \tag{2.10}$$

This expression can be interpreted as the sum of the phase incurred by the host transmission line and a uniform backward wave *L–C* line as shown in Fig. 2.6.

Various 1-D phase-shifting lines were constructed in coplanar waveguide (CPW) technology at 0.9 GHz, as shown in Fig. 2.9. The simulated and measured phase and magnitude responses for a 2-stage and a 4-stage 0° phase-shifting lines are shown in Fig. 2.10, compared to the phase response of a conventional $-360°$ TL. It can be observed that the experimental results correspond very closely to the simulated results, highlighting the broadband nature of the phase-shifting lines as well as their small losses.

In summary, these metamaterial phase-shifting lines offer some significant advantages when compared to conventional delay lines. They are compact in size, can be easily fabricated using standard etching techniques, and exhibit a linear phase response around the design frequency. They can incur *either* a negative *or* a positive phase, as well as a 0° phase depending on the values of the loading elements, while maintaining a short physical length. In addition, the phase incurred is independent

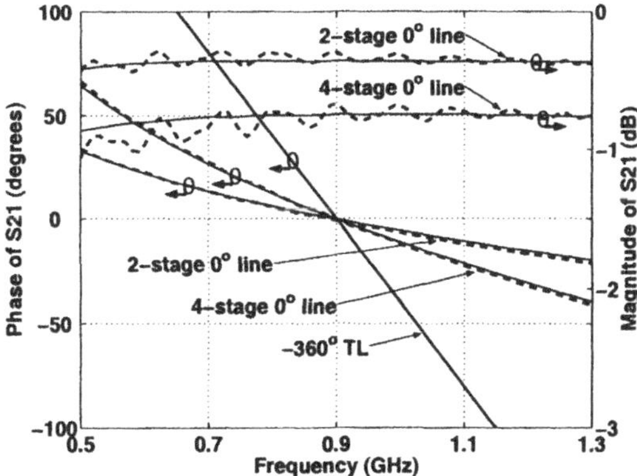

Fig. 2.10 Phase and magnitude responses of 2- and 4-stage 0° phase-shifting lines compared to a conventional $-360°$ TL at 0.9 GHz; (- - -) measured, (—) Agilent-ADS simulation.

of the length of the structure. Due to their compact, planar design, they lend themselves easily toward integration with other microwave components and devices. The metamaterial phase-shifting lines are therefore well suited for broadband applications requiring small, versatile, linear devices.

It should be pointed out that these phase-shifting lines offer an advantage in terms of size and bandwidth when phase shifts about the zero-degree mark are needed. In this scenario, the proposed devices are superior to corresponding delay lines about one wavelength long. This advantage arises from their short electrical length, which implies a broadband response (always when comparing to a one-wavelength delay line). For electrically long PRI/NRI phase-shifting lines, their broadband nature could be retained if the constituent NRI section is designed to also exhibit a negative group velocity as was done in Ref. 17. In this case, not only the signs but also the slopes of the propagation constants (vs. frequency) of the NRI and PRI lines compensate, thus leading to a broadband response. However the difficulty now is how to synthesize a negative group velocity over a broad bandwidth. Moreover, the NRI lines of Ref. 17 are lossy and restoring amplifiers would need to be included for acceptable performance.

2.6 SERIES-FED ANTENNA ARRAYS WITH REDUCED BEAM SQUINTING

The one-dimensional (1-D) metamaterial phase-shifting lines presented in the previous section can be used to develop compact, broadband, nonradiating, metamaterial feed networks for antenna arrays. These can be used to replace conventional TL-based feed networks, which can be bulky and narrowband [30]. For series-fed arrays, the proposed metamaterial feed networks have the advantage of being compact in

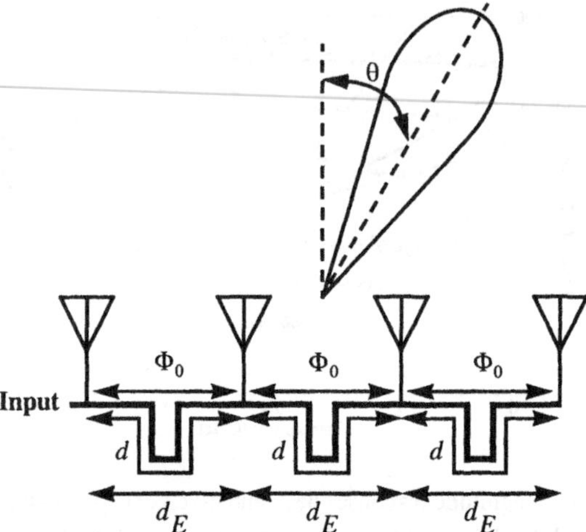

Fig. 2.11 Series-fed linear array using conventional -2π transmission-line meandered feed lines.

size, therefore eliminating the need for conventional TL meander lines. In addition, the metamaterial feed networks are more broadband when compared to conventional TL feed networks, which enables series-fed broadside arrays to experience less beam squint when operated away from the design frequency.

In a typical series-fed linear array designed to radiate at broadside, the antenna elements must be fed in phase. In addition, an interelement spacing d_E of less than half a free-space wavelength ($d_E < \lambda_0/2$) is necessary to avoid capturing grating lobes in the visible region of the array pattern. In order to achieve these design constraints, traditional designs employing TL-based feed networks have resorted to a meander-line approach, as shown in Fig. 2.11. This allows the antenna elements to be physically separated by a distance of $d_E = \lambda_0/2$, while still being fed in phase with a one-guided-wavelength λ_g long meandered line that incurs a phase of -2π radians. Because the phase incurred by the TLs is frequency-dependent, a change in the operating frequency will cause the emerging beam to squint from broadside, which is generally an undesirable phenomenon. In addition, the fact that the lines are meandered causes the radiation pattern to experience high cross-polarization levels, particularly in CPW implementations, as a result of parasitic radiation due to scattering from the corners of the meandered lines [31]. These feed networks employ nonradiating metamaterial phase-shifting lines within a series-fed linear array (see Fig. 2.12) to mitigate some of the problems encountered with conventional TL-based feed networks.

The phase-shifting lines presented in Ref. 28, whose unit cell is shown again in Fig. 2.13a, can incur an arbitrary insertion phase, are compact in size and exhibit a more linear, flatter phase response with frequency compared to conventional TL delay lines. In order to ensure that the phase-shifting lines do not radiate, they can

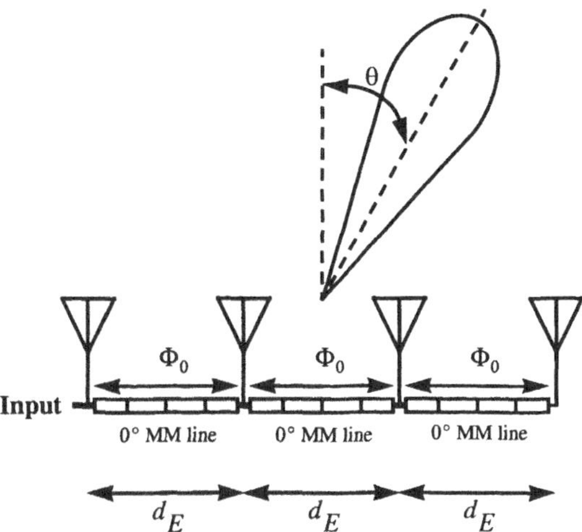

Fig. 2.12 Series-fed linear array using 0° metamaterial (MM) feed lines.

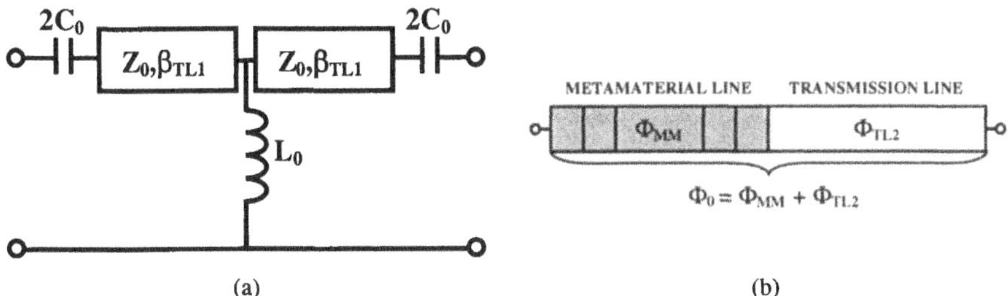

Fig. 2.13 (a) A 1-D metamaterial phase-shifting unit cell. (b) A combined slow-wave metamaterial phase-shifting line.

be operated in the NRI backward-wave region, while simultaneously ensuring that the propagation constant of the line exceeds that of free space. This will effectively produce a slow-wave structure with a positive insertion phase, Φ_{MM}. Cascading this with a conventional TL, which inherently incurs a negative insertion phase, Φ_{TL2}, results in a combined slow-wave metamaterial phase-shifting line, as shown in Fig. 2.13b. If Φ_{MM} and Φ_{TL2} are equal but opposite in value, then the structure will incur a zero insertion phase, given by $\Phi_0 = \Phi_{MM} + \Phi_{TL2} = 0$. The metamaterial phase-shifting lines can incur a positive, negative, or zero insertion phase, by adjusting the values of the loading elements C_0 and L_0. Thus, for a given section of TL with intrinsic phase shift $\Phi_{TL2} = \omega\sqrt{LC}d_{TL1}$ and characteristic impedance Z_0, the phase shift for an n-stage metamaterial line is given by (2.11), subject to the impedance matching condition of (2.12):

$$\Phi_{MM} = n\left(\omega\sqrt{LC}d_{TL1} + \frac{-1}{\omega\sqrt{L_0C_0}}\right) \tag{2.11}$$

$$Z_0 = \sqrt{\frac{L_0}{C_0}} = \sqrt{\frac{L}{C}} \tag{2.12}$$

Assuming that the same type of TL sections are used for $TL1$ and $TL2$, then Z_0, L, and C will be the same for both lines. Therefore, Φ_{TL2} is given by $\Phi_{TL2} = \omega\sqrt{LC}d_{TL2}$. Correspondingly, for a transmission line of length λ_g, the phase as a function of frequency is given by $\Phi_{\lambda_g} = \omega\sqrt{LC}\lambda_g$. The scan angle for each of the metamaterial-based and TL-based linear arrays with an interelement phase shift Φ_0 can therefore be written as

$$\theta_{SCAN,MM} = \sin^{-1}\left(-\frac{\Phi_0}{k_0 d_E}\right) = \sin^{-1}\left(-\frac{\Phi_{MM} + \Phi_{TL2}}{k_0 d_E}\right) \tag{2.13}$$

$$\theta_{SCAN,TL} = \sin^{-1}\left(-\frac{\Phi_0}{k_0 d_E}\right) = \sin^{-1}\left(-\frac{\Phi_{\lambda_g}}{k_0 d_E}\right) \tag{2.14}$$

The metamaterial-based and TL-based feed networks were evaluated in CPW technology at a design frequency of 2 GHz. Two designs were considered: An array with an interelement spacing of $d_E = \lambda_0/2$, and an array with a spacing of $d_E = \lambda_0/4$. The corresponding scan-angle characteristics for the metamaterial-based and TL-based linear arrays with $d_E = \lambda_0/2$ are shown in Fig. 2.14a. It can be observed that the scan angle for the TL-fed array exhibits its full scanning range from $+90°$ to $-90°$ within a bandwidth of 2.67 GHz, while the corresponding scanning bandwidth for the metamaterial-fed array is 4.27 GHz. Thus, the metamaterial-fed array offers a more broadband scan angle characteristic, while simultaneously eliminating the need for meander lines. Also shown in Fig. 2.14a is the scan-angle characteristic for a low-pass loaded slow-wave TL, also of length $\lambda_0/2$. It can be observed that the performance of this line is identical to that of the TL feed line (the two curves are on top of each other in Fig. 2.14a). Thus, although the low-pass loaded line can eliminate the need for meander lines, it does not provide the advantage of an increased scan-angle bandwidth that the metamaterial feed lines offer.

The scan-angle characteristics for the $\lambda_0/4$ feed network are shown in Fig. 2.14b. It can be observed that the bandwidth of the scanning angle for the TL-fed array and the low-pass loaded TL array decreases to 1.07 GHz, while the corresponding scanning bandwidth for the metamaterial-fed array remains at 4.27 GHz. Thus, as the spacing between the antenna elements decreases, the scan-angle characteristic for a metamaterial-fed array remains constant, while the corresponding scan-angle characteristic for the TL-fed array becomes more narrowband.

Once more it should be pointed out that an even more broadband response could be achieved (especially for longer interelement spacings) if the NRI-TL sections in Fig. 2.13b are designed to also exhibit a negative group velocity [17]. However such lines are inherently lossy and restoring amplifiers would need to be inserted in them.

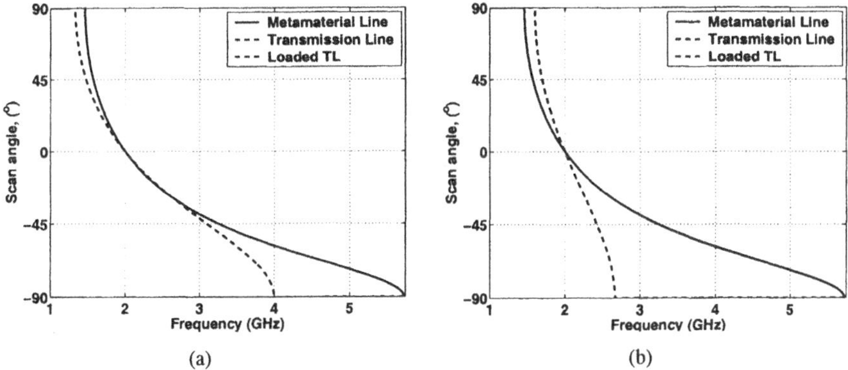

Fig. 2.14 (a) Scan-angle performance of a series-fed linear array with $d_E = \lambda_0/2$ using different feeding techniques. (b) Scan-angle performance of a series-fed linear array with $d_E = \lambda_0/4$ using different feeding techniques.

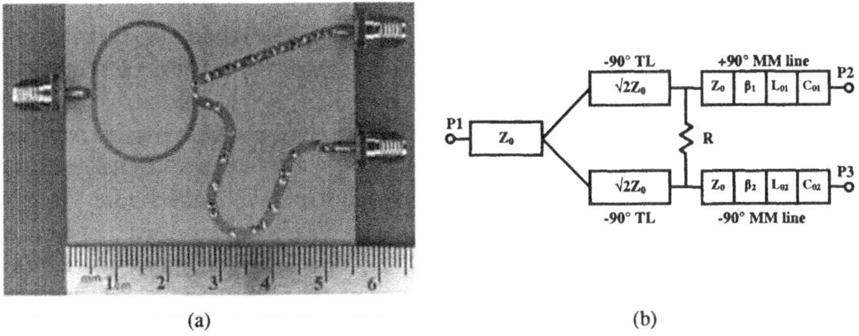

Fig. 2.15 (a) Metamaterial balun; An example design in microstrip at 1.56 GHz on a RO3003 substrate ($\epsilon_r = 3.0$, height $= 0.762$ mil). (b) Block diagram for the balun's architecture. After Ref. [32]. Copyright © 2005 IEEE.

2.7 A BROADBAND METAMATERIAL BALUN IN MICROSTRIP

Baluns are particularly useful for feeding two-wire antennas, where balanced currents on each branch are necessary to maintain symmetrical radiation patterns with a given polarization. Two-wire antennas have input ports that are closely spaced, therefore their feeding structures should be chosen to accommodate for this requirement. A broadband balun with such a quality can be synthesized using the phase-shifting lines discussed previously in Section 2.5 [28].

The proposed metamaterial (MM) balun is shown in Figs. 2.15a and 2.15b and consists of a Wilkinson power divider, followed by a $+90°$ MM phase-shifting line

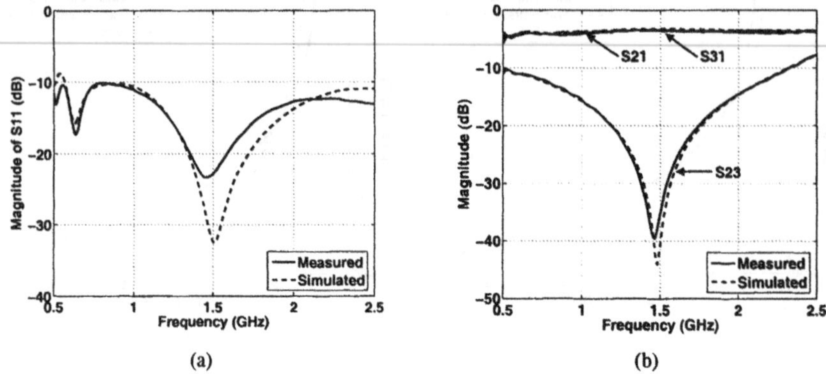

Fig. 2.16 (a) Measured and simulated return loss for port 1 of the metamaterial balun. (b) Measured and simulated isolation S_{23} and through (S_{21} and S_{31}) magnitude responses of the metamaterial balun. After Ref. [32]. Copyright © 2005 IEEE.

along the top branch and a $-90°$ MM phase-shifting line along the bottom branch [33]. In order to match the phase response of the $-90°$ MM line with that of the $+90°$ MM line, and therefore create a broadband differential output phase, the slopes of their phase characteristics are designed to be equal at the design frequency. Moreover, to ensure that the MM phase-shifting lines do not radiate, each unit cell is operated in the region outside the light cone on the Brillouin diagram. Thus, the $+90°$ MM phase-shifting line is operated in the NRI backward-wave region, while simultaneously ensuring that its phase velocity does not exceed that of free space, resulting in a slow-wave structure with a positive insertion phase. Correspondingly, the $-90°$ MM phase-shifting line is operated in the positive-refractive-index (PRI) forward-wave region, while simultaneously ensuring that its phase velocity also does not exceed that of free space, resulting in a slow-wave structure with a negative insertion phase.

The MM Wilkinson balun was implemented in microstrip technology on a Rogers RO3003 substrate at a design frequency of $f_o = 1.5$ GHz. A five-stage design was chosen for the $+90°$ MM phase-shifting line as well as for the $-90°$ MM phase-shifting line. The experimental results were compared with the simulated ones obtained using Agilent's ADS. Figure 2.16a shows the measured versus the simulated return loss magnitude response for port 1, demonstrating good agreement between the two, indicating that the device is well matched, especially around $f_o = 1.5$ GHz. The measured and simulated return losses for ports 2 and 3 exhibit similar responses. Figure 2.16b shows excellent isolation for the device, as well as equal power split between the two output ports. Moreover, the insertion loss on each output port is better than 0.5 dB.

Figure 2.17a shows the measured versus the simulated phase responses of the two balun branches. The experimental results agree very closely with the simulated ones.

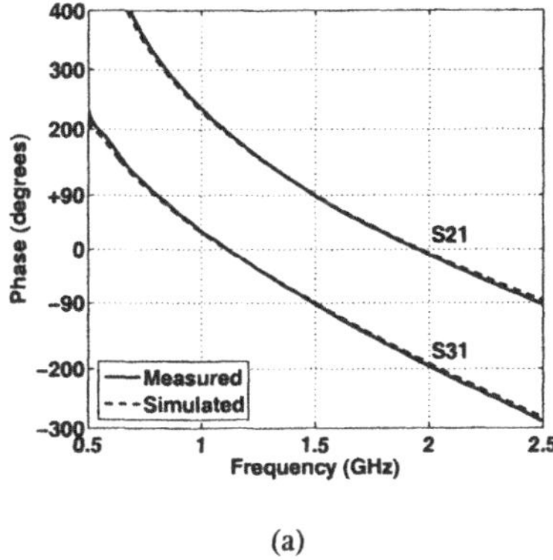

(a)

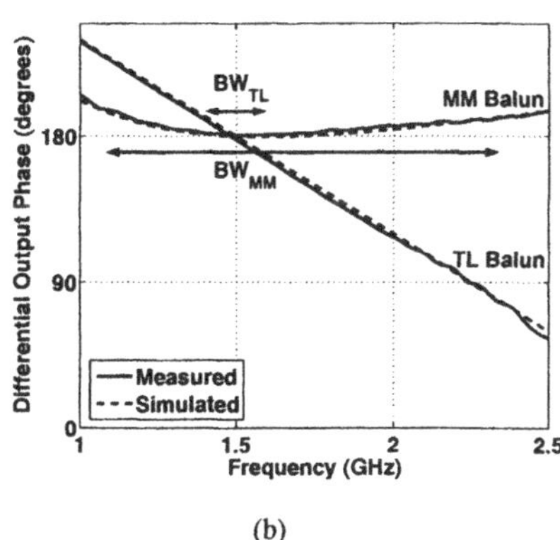

(b)

Fig. 2.17 (a) Measured and simulated phase responses of S_{21} (+90° MM line) and S_{31} (−90° MM line) of the MM balun. (b) Measured and simulated differential phase comparison between the MM balun and the TL balun. After Ref. [32]. Copyright © 2005 IEEE.

It can be observed that the phase of S_{21} is exactly equal to $+90°$ at $f_o = 1.5$ GHz, while the phase of S_{31} is exactly equal to $-90°$ at $f_o = 1.5$ GHz, and that the phase responses of the two branches are quite similar.

Figure 2.17b shows the measured and simulated differential output phase of the MM balun, with excellent agreement between the two. It can be observed that the differential output phase remains flat over a large frequency band, which follows directly form the fact that the phase characteristics of the $+90°$ and $-90°$ lines shown in Fig. 2.17a correspond very closely. The flat differential output phase has a $180° \pm 10°$ bandwidth of 1.16 GHz, from 1.17 to 2.33 GHz. Since the device exhibits excellent return loss, isolation and through characteristics over this frequency range, it can be concluded that the MM balun can be used as a broadband single-ended to differential converter in the frequency range from 1.17 to 2.33 GHz.

For comparison, a distributed TL Wilkinson balun employing $-270°$ and $-90°$ TLs instead of the $+90°$ and $-90°$ MM lines was also simulated, fabricated and measured at $f_o = 1.5$ GHz, and the differential output phase of the TL balun is also shown in Fig. 2.17b. It can be observed that the phase response of the TL balun is linear with frequency, with a slope equal to the difference between the phase slopes of the $-270°$ and $-90°$ TLs. Since the gradient of the resulting phase characteristic is quite steep, this renders the output differential phase response of the TL balun narrowband. Thus, the TL balun exhibits a measured differential phase bandwidth of only 11%, from 1.42 to 1.58 GHz, compared to 77% exhibited by the MM balun. In addition, the TL balun occupies an area of 33.5 cm^2, compared to 18.5 cm^2 for the MM balun. Thus, the MM balun is more compact, occupying only 55% of the area that the conventional TL balun occupies. Furthermore, the MM balun exhibits more than double the bandwidth compared to a lumped-element implementation using low-pass/high-pass lines, which typically exhibits a bandwidth around 30%.

This can be attributed to the fact that the low-pass line has a linear phase response, while the response of the high-pass line has a varying slope with frequency. Thus, the shapes of the phase responses of the two lines do not match, resulting in a more narrowband differential output phase.

It should be pointed out that, although we have considered here the case of producing a broadband 180° response, the same method can be utilized to produce other desired differential phase responses over a broad bandwidth (i.e., a +45° line and a −45° line could be utilized in Fig. 2.15b to produce a broadband 90° differential phase-shift, useful in I/Q RF receiver front ends).

2.8 BROADBAND POWER COMBINERS USING ZERO-DEGREE PHASE-SHIFTING LINES

The zero-degree phase shifting lines described previously can be used to implement novel broadband power combiners. The basic idea can be understood by considering Fig. 2.18a. As shown, a number of three-terminal gain devices (in this case field-effect microwave transistors) are combined in parallel using one wavelength lines to connect the input (gate) and output (drain) terminals of the devices (FETs). Since the devices are connected in phase, a power-combining effect takes place. This method leads to a reduced real-estate usage when compared to standard power-combining schemes by means of corporate power-dividing/combining trees. However, the architecture of

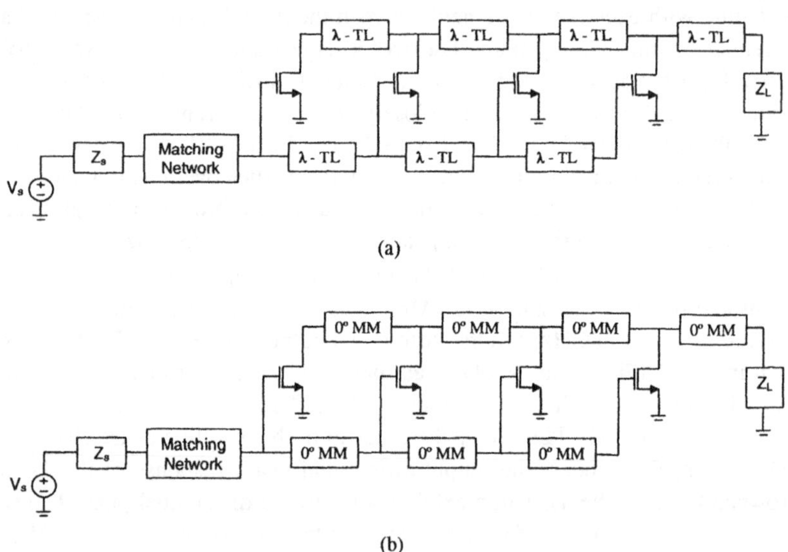

Fig. 2.18 (a) Conventional power-combining scheme using one-wavelength interconnecting lines. (b) Metamaterial power-combining scheme using zero-degree phase-shifting interconnecting lines.

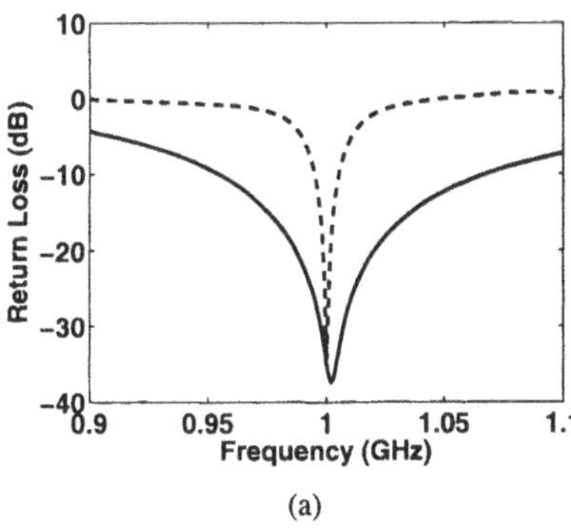

(a)

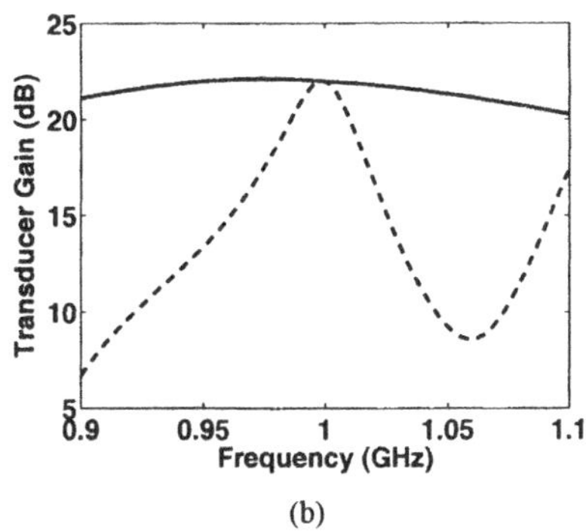

(b)

Fig. 2.19 (a) Return loss of conventional (dotted line) vs. metamaterial (solid line) power-combining scheme; 4-stages of ATF-34143 PHEMTs. (b) Transducer gain of conventional (dotted line) vs. metamaterial (solid line) power-combining scheme; 4-stages of ATF-34143 PHEMTs.

Fig. 2.18a is inherently narrowband since the devices are combined in phase only at the frequency where the interconnecting lines are exactly one wavelength long. To this end, the zero-degree phase-shifting lines described earlier in Section 2.5 could be harnessed to replace these one-wavelength lines. Since the corresponding zero-degree phase-shifting lines could be made significantly shorter than the one-wavelength lines they replace, the resulting scheme shown in Fig. 2.18b is not only physically smaller, but also significantly more broadband than that shown in Fig. 2.18a. The architecture of Fig. 2.18b is similar to the scheme of series-fed antennas described earlier in Section 2.6 [30].

To evaluate the proposed power-combining architecture, microwave circuit simulations have been carried out using realistic devices. Specifically, Figs. 2.19a and 2.19b show the simulated return loss and transducer power gain as a function of frequency when combining four Agilent ATF-34143 PHEMT transistors using the scheme of Fig. 2.18b. The same figures also shows the corresponding quantities when one-wavelength interconnecting lines are used according to the conventional scheme of Fig. 2.18a. Clearly, the metamaterial-based power-combining scheme is significantly more broadband compared to its traditional counterpart.

2.9 ELECTRICALLY SMALL RING ANTENNA WITH VERTICAL POLARIZATION

Another example of harnessing the phase-shifting lines of Ref. [28] is to wrap around a zero-degree phase-shifting line to implement a small printed antenna. This is shown in Fig. 2.20 for a realization at 1.5 GHz. As shown, there are four metamaterial phase-shifting sections arranged in a square ring. Each constituent section comprises

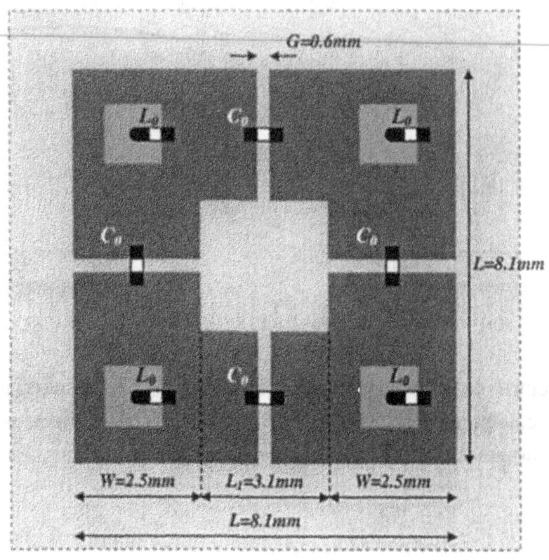

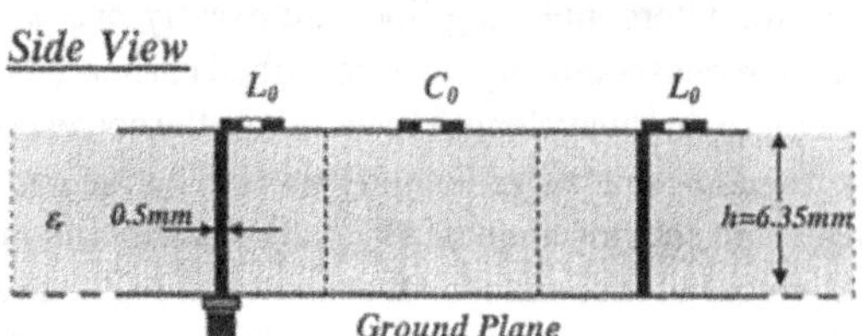

Fig. 2.20 Diagram of the metamaterial ring antenna at 1.5 GHz. The loading capacitance and inductance required to feed the vias in phase are $C_0 = 3.70$ pF and $L_0 = 71.08$ nH.

a negative-refractive-index (NRI) microstrip transmission line (TL), designed to incur a zero insertion phase at the antenna operating frequency. This allows the inductive posts to ground, which act as the main radiating elements, to be fed in phase. Hence, the antenna operates as a 2-D array of closely spaced monopoles that are fed in phase through a compact feed network. This leads to a ring antenna with a small footprint (diameter of $\lambda/25$) and a low-profile (height $\lambda/31$) capable of radiating vertical polarization. Figure 2.21 shows the measured versus the simulated return loss obtained from an equivalent circuit model based on Fig. 2.7. It can be observed that the antenna is well-matched at 1.51 GHz, with a measured return loss bandwidth below -10 dB of approximately 1.5%. This bandwidth can be increased to 3–4% if the dielectric substrate is reduced to about the size of the ring or if the via height is increased. Moreover, the ring antenna could be implemented entirely in air, in which case also the bandwidth increases to $> 5\%$. In fact, such low-cost air-filled ring antennas could be attractive for emerging multiple-input–multiple-output (MIMO)

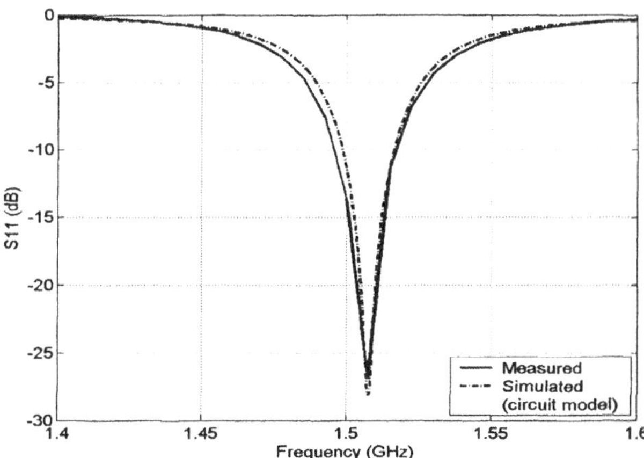

Fig. 2.21 Measured and simulated (HFSS and circuit models) return loss for the metamaterial ring antenna.

wireless telecommunication systems. Specifically, two or three such low-profile antennas could be mounted on hand-held or laptop units in orthogonal directions for creating independent channels based on polarization diversity. Figure 2.22 shows the measured versus the simulated E- and H-plane patterns obtained from Agilent's HFSS, which demonstrates good agreement. It can be observed that the antenna exhibits a radiation pattern with a vertical linear electric field polarization, similar to a short monopole on a finite ground plane. The radiation in the back direction is reduced compared to the forward direction due to the effect of the finite ground plane used; however, it is not completely eliminated. Moreover, there is good cross-polarization purity in the E-plane, with a maximum measured electric field cross-polarization level of -17.2 dB. In the H-plane, the maximum electric field cross-polarization level is only -6.6 dB.

The loading with chip passive lumped elements is effective at RF and low microwave frequencies. At higher frequencies, these can be replaced by printed lumped elements. For example, a fully printed version of this antenna at 30 GHz, in which the loading lumped-element chip capacitors and inductors were replaced by gaps and vias respectively, was reported in Ref. 34.

2.10 A LEAKY-WAVE BACKWARD ANTENNA RADIATING ITS FUNDAMENTAL SPATIAL HARMONIC

The transmission-line (TL) approach to synthesizing NRI metamaterials has led to the development of a new kind of leaky-wave antenna (LWA). By appropriately choosing the circuit parameters of the dual TL model, a fast-wave structure can be designed

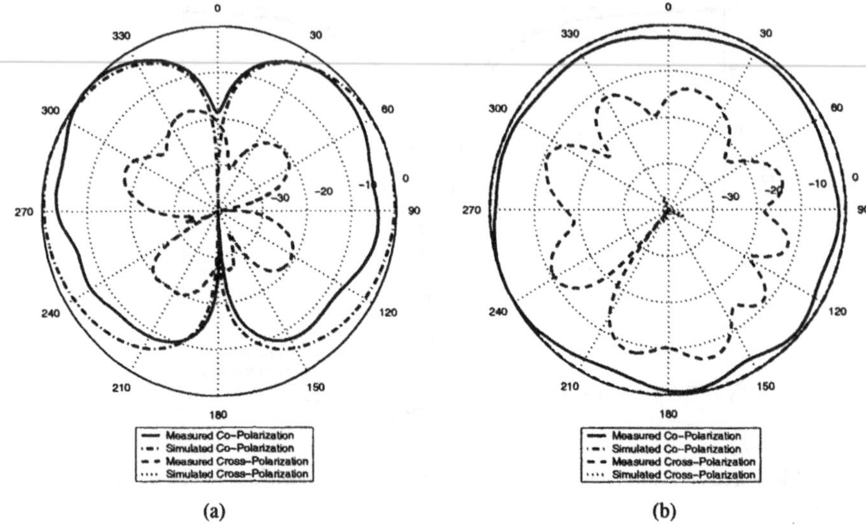

Fig. 2.22 (a) E-plane and (b) H-plane measured and simulated patterns.

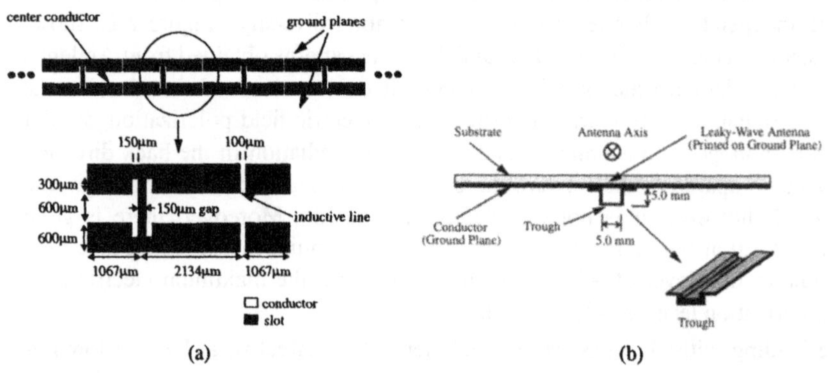

Fig. 2.23 (a) Backward leaky-wave antenna based on the dual TL model at 15 GHz [18,20]. (b) Unidirectional backward leaky-wave antenna design at 15 GHz. Reprinted with permission from Ref. [18]. Copyright © 2002 American Institute of Physics.

that supports a fundamental spatial harmonic which radiates toward the backward direction [18,20].

The CPW implementation of this leaky-wave antenna is shown in Fig. 2.23a. The gaps in the CPW feedline serve as the series capacitors of the dual TL model, while the narrow lines connecting the center conductor to the coplanar ground planes serve as the shunt inductors. The capacitive gaps are the radiating elements in this leaky-wave antenna, and excite a radiating transverse magnetic (TM) wave. Due to the antiparallel currents flowing on each pair of narrow inductive lines, the shunt inductors remain nonradiating. Simulated and experimental results for this bidirectional leaky-wave antenna were reported in Ref. 18. Simulation results for a unidirectional LWA design were also presented in Ref. 20. The unidirectional design is simply the leaky-wave antenna described in Ref. 20 backed by a long metallic trough as shown in Fig. 2.23b. Since the LWA's transverse dimension is electrically small, the backing trough can be narrow (below resonance). The trough used is one quarter wavelength in height and in width and covers the entire length of the antenna on the conductor side of the substrate. It acts as a waveguide below cutoff and recovers the back radiation, resulting in unidirectional far-field patterns.

Here, we present experimental results for the unidirectional design proposed in Ref. 20. As noted in Ref. 20, a frequency shift of 3%, or 400 MHz, was observed in the experiments compared to the method of moments simulations of the LWA using Agilent's Advanced Design System. As a result, the experimental unidirectional radiation patterns are shown at 14.6 GHz while the simulation patterns are shown at 15 GHz. The E-plane and H-plane patterns are shown in Figs. 2.24a and 2.24b, respectively. A gain improvement of 2.8 dB was observed for the unidirectional design over the bidirectional design, indicating that effectively all of the back radiation is recovered with the trough.

It should be pointed out that a complementary forward unidirectional leaky-wave antenna, also radiating the fundamental spatial harmonic, has been reported in Ref. 35. As was first noted in Ref. 35, both the forward and the backward LWAs offer the advantage of a simple feed, unlike the conventional microstrip LWA which operates on a higher-order mode and thus requires a special feed mechanism to guard against the excitation of the fundamental microstrip mode [36]. Moreover, the same approach can be extended to 2-D leaky metamaterial surfaces that can form pencil, instead of fan, beams which can be frequency scanned about the broadside direction [37].

2.11 A HIGH-DIRECTIVITY BACKWARD NRI/MICROSTRIP COUPLER

A peculiar coupled-line coupler (see Fig. 2.25) can be realized using a regular microstrip (MS) line that is edge-coupled to a negative-refractive-index (NRI) line [38,39]. Such a coupler exhibits co-directional phase but contradirectional Poynting vectors on the lines, thus leading to backward power coupling.

Using coupled-mode theory, it can be shown that coupled modes with *complex-conjugate propagation constants* are excited in this coupler at the frequency where the propagation constants of the two isolated lines become equal [39]. For a sufficiently long coupler operated at this frequency, the exponentially increasing modes can be discarded, in which case line voltage/current expressions take the following form:

$$\begin{pmatrix} V_1 \\ V_2 \\ I_1 \\ I_2 \end{pmatrix} = \begin{pmatrix} 1 & 1 \\ j & -j \\ 1/Z_c & 1/(-Z_c^*) \\ j/(-Z_c) & -j/(Z_c^*) \end{pmatrix} \begin{pmatrix} V_c^+ e^{-\gamma_c z} \\ V_\pi^+ e^{-\gamma_\pi z} \end{pmatrix} \tag{2.15}$$

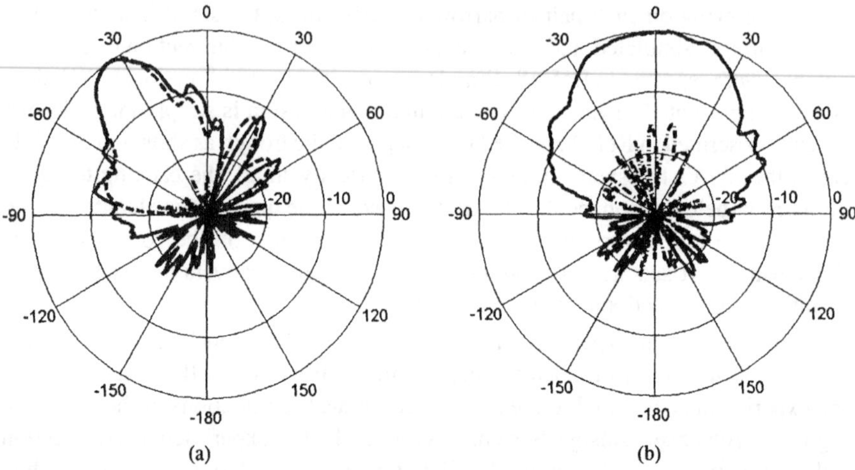

Fig. 2.24 (a) E-plane pattern for the unidirectional leaky-wave antenna. (—) Experimental co-polarization; (- · -) experimental cross-polarization; (- - -) simulated co-polarization using Agilent ADS. $f = 15$ GHz. (b) H-plane pattern for the unidirectional leaky-wave antenna. (—) Experimental co-polarization; (- · -) experimental cross-polarization; (- - -) simulated co-polarization using Agilent ADS. $f = 15$ GHz.

Here, γ_c and γ_π are complex conjugate eigenvalues, $\gamma_c = \alpha + j\beta$ and Z_c is the impedance of the symmetric c-mode on the MS line. If port 1 (see Fig. 2.25) is excited, from (2.15) it can be shown that

$$\frac{1}{2}\text{Re}(V_1 I_1^*|_{z=0}) = \frac{\text{Re}(Z_c)}{2|Z_c|^2}\left(|V_c|^2 - |V_\pi|^2\right) = -\frac{1}{2}\text{Re}(V_2 I_2^*|_{z=0}) \tag{2.16}$$

Equation (2.16) demonstrates that there is complete backward transfer of power from port 1 to port 2 (see Fig. 2.25). In order to compare the performance of a MS/NRI coupler to its regular MS/MS counterpart of equal length, line spacing, and propagation constant, a benchmark microstrip coupler was designed and is shown in Fig. 2.25. This benchmark MS/NRI coupler was constructed with unit cells 5 mm long and loading elements of 2.7-nH shunt inductors and 0.9-pF series capacitors for the NRI line. The line widths are 2.45 mm (MS) and 2 mm (NRI), and the transverse line separation is 0.4 mm. The corresponding performance of this 3-unit cell MS/NRI coupler is compared to that of a quarter-wavelength regular MS/MS coupler at 2.8 GHz, when the isolated propagation constants of the two lines are similar. This comparison is shown in Figs. 2.26a and 2.26b, from which it is evident that, compared to its conventional counterpart of the same length and line spacing, this new MS/NRI coupler exhibits better performance in terms of coupled power (higher coupling), lower return loss and isolation, without any bandwidth degradation or significant change in insertion loss. This results in improved directivity (20 dB) when compared to the ordinary MS/MS quarter-wavelength coupler (7 dB).

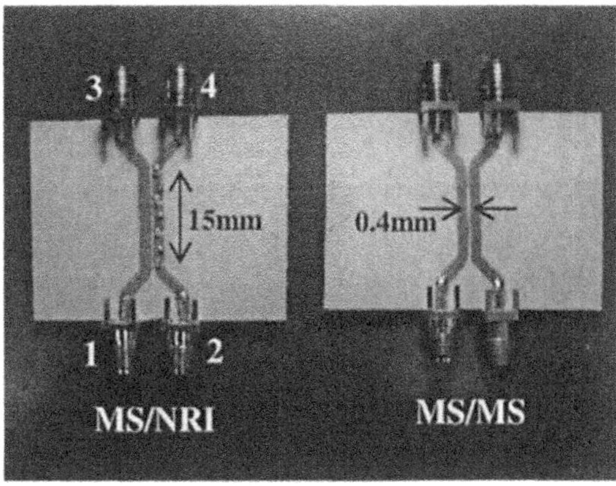

Fig. 2.25 MS/NRI and MS/MS ($\lambda/4$) couplers of equal length, line spacing, and propagation constants designed for operation at 2.8 GHz. (Port 1: input; port 2: coupled; port 3: through; port 4: isolated.)

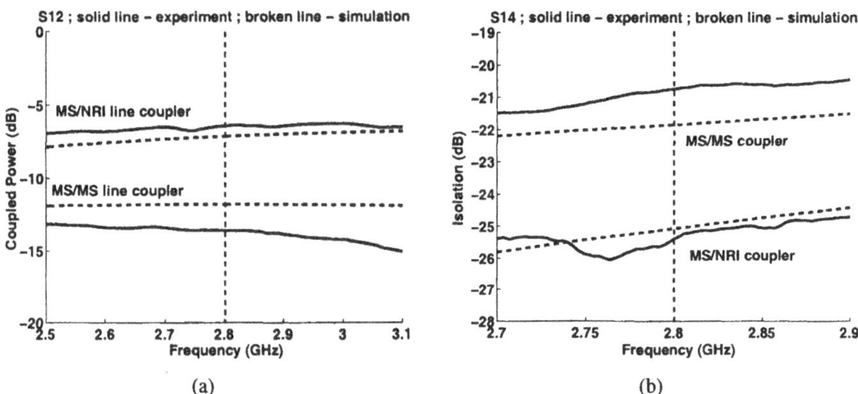

(a) (b)

Fig. 2.26 (a) Comparison of coupled power levels for the MS/NRI and the MS/MS couplers. (b) Comparison of isolation for the MS/NRI and the MS/MS couplers.

A 3-dB coupler has also been constructed and tested as shown in Fig. 2.27 [39]. Corresponding simulation and experimental results for this coupler are presented in Figs. 2.28a and 2.28b. As shown, the measured power splits between the through and coupled ports at the level of −3.03 dB and −3.68 dB respectively, using this 6-cell-long coupler (24 mm) at the design frequency of 3 GHz. Moreover, the return loss and isolation are found to be below −20 dB, whereas the directivity is better than 20 dB. For low coupling applications (e.g., reflectometry, VSWR signal monitoring, etc.), the directivity can be optimized in practice to values in the range of 35 dB, which is very difficult to achieve with conventional microstrip couplers [41].

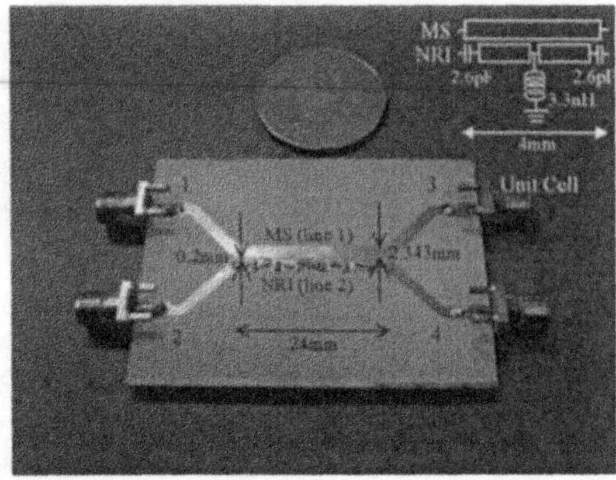

Fig. 2.27 A 3-dB MS/NRI coupled line coupler (6 unit cells long) constructed on a 50-mil Rogers TMM4 ($\varepsilon_r = 4.6$) substrate. After Ref. [39]. Copyright © 2004 IEEE.

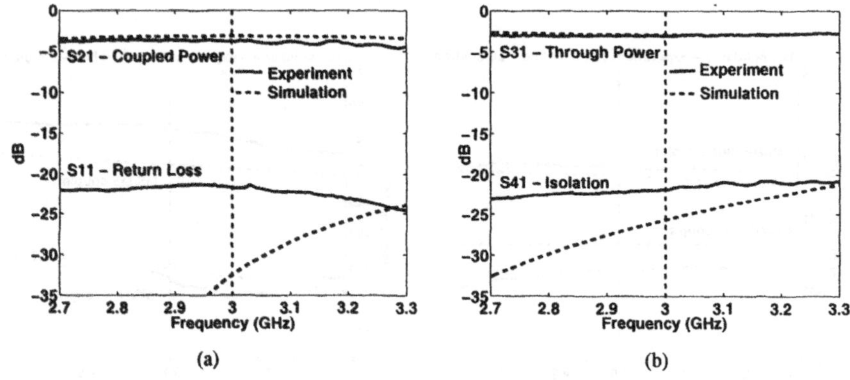

Fig. 2.28 (a) Simulation and experimental results for 3-dB MS/NRI coupled-line coupler: Return loss and coupled power. (b) Simulation and experimental results for 3-dB MS/NRI coupled line coupler: Through power and isolation.

Figure 2.29 shows the dispersion diagram of the coupled modes for the 3-dB coupler. In the same diagram, the dispersion curves of the isolated microstrip and NRI (backward-wave) transmission lines that make up the coupler are also depicted. As shown, both Ansoft HFSS finite-element simulations and coupled-mode theory results verify the formation of a stopband in the dispersion diagram of the coupler at the location where the dispersion curves of the isolated lines meet. Therefore, the two eigenmodes of the coupler become complex, thus leading to exponential field decay along the lines. This enables enhanced coupling and high isolation with only moderate line lengths and interline spacings. Pierce [16] pointed out the existence of the coupled mode stopband and suggested that it occurs when a mode in a periodic

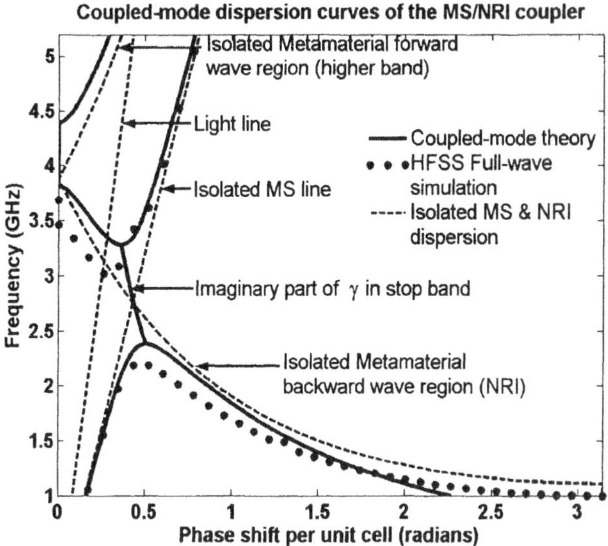

Fig. 2.29 Coupled-mode dispersion diagram for the 3-dB MS/NRI coupled-line coupler of Fig. 2.27. After Ref. [39]. Copyright © 2004 IEEE.

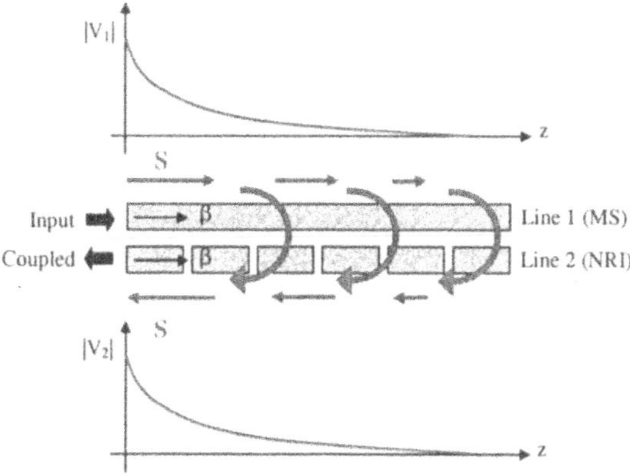

Fig. 2.30 Physical mechanism by which power continuously leaks from the microstrip line to the NRI line.

structure couples to a higher-order backward-wave spatial harmonic. The unique feature presented here is that this coupling takes place between two lines involving their fundamental spatial harmonics.

At the point where the dispersion curves of the two isolated lines comprising the coupler meet, the corresponding attenuation constant is maximum [40]. Around this point and within the "contradirectional" stopband of Fig. 2.29, one could think that the miscrostrip mode continuously "leaks" power backwards to the NRI line as suggested in Fig. 2.30. This leakage effect has to be described in terms of a complex propagation constant, which is consistent with the previously presented analysis. At this frequency, the corresponding coupled-line system supports leaky complex conjugate modes ($\alpha \pm j\beta$) which are associated with complex conjugate impedances ($Z_0 \pm j\Delta Z$). The attenuation of these modes is accounted for by the backward power leakage from the excited line to the coupled line. It is assumed that the imaginary parts of the mode impedances are small, which is valid for moderate spacing between the two lines. If port 1 of the coupler is excited with a matched source and the remaining ports are terminated with an impedance of Z_0, then the scattering parameters for a coupler of length D can be derived by applying the boundary conditions at the four ports. Ignoring second- and higher-order terms in $\Delta Z/Z_0$, these parameters are listed below in equations (2.17) to (2.20), with port 2, port 3 and port 4 referring to the coupled, through and isolated ports, respectively [41]. It can be seen from (2.17) that the return loss is minimized when the line spacing of the coupler is increased, as it causes ΔZ to become smaller. From (2.18), the coupled power decreases with line spacing (causing the attenuation factor α to decrease), and it increases with coupler length D. Hence a coupler with a length on the order of a few $1/\alpha$'s will direct almost all the input power to the coupled port. Finally, (2.20) reveals that it is possible to obtain perfect isolation when the length of the coupler is made to be an integer multiple of half its guide wavelength:

$$S_{11} = j\frac{\Delta Z}{Z_0}\tanh(\alpha D) \tag{2.17}$$

$$S_{21} = j\tanh(\alpha D) \tag{2.18}$$

$$S_{31} = e^{-j\beta D}\text{sech}(\alpha D) \tag{2.19}$$

$$S_{41} = j\frac{\Delta Z}{Z_0}\sin(\beta D)\text{sech}(\alpha D) \tag{2.20}$$

It should be pointed out that a structure related to this backward-wave coupler has been reported in Ref. 42. Specifically, Ref. 42 deals with a rectangular metallic waveguide which is partially filled with a NRI metamaterial slab along the longitudinal direction. This problem and the corresponding theory are presented in detail in Chapter 9. Moreover, a different backward-wave coupler comprising two negative-refractive-index lines, operating about their radiating $\beta = 0$ region, has been described in Ref. 43. On the other hand, the same topology of two identical coupled NRI lines has been used to report a forward coupler in Ref. 11, although it is unclear under which conditions this same topology can lead to either a backward or a forward coupler.

2.12 PHASE-AGILE BRANCH-LINE MICROSTRIP COUPLERS

Two types of branch-line couplers are described here that utilize a combination of regular microstrip (MS) and negative-refractive-index lines [45]. Interesting and

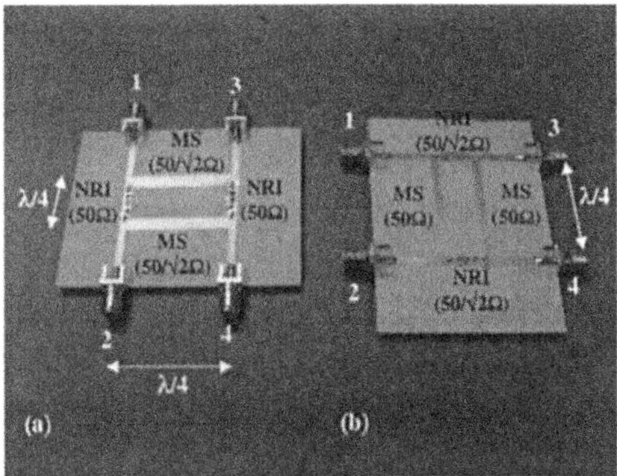

Fig. 2.31 MS/NRI Branch-line couplers: (a) Type 1. (b) Type 2. After Ref. [45]. Copyright © 2004 IEEE.

useful phase compensation (0° phase shift) at the output port and a choice of 90° phase shift at the through port, with respect to the input, can be synthesized. Moreover, one of the two orthogonal dimensions of the coupler is significantly reduced, compared to a corresponding conventional branch-line coupler, but without any bandwidth degradation (see Fig. 2.31) .

The two different coupler designs are denoted as type 1 and type 2. The former uses regular microstrip lines (MS) for the low-impedance branches and NRI lines for the high-impedance ones (see Fig. 2.31a). The latter (type 2) is the dual of the former (type 1) and utilizes NRI lines for the low-impedance branches and microstrip lines for the high-impedance ones (see Fig. 2.31b). In both couplers, the power splits equally between the two output ports with a 0° phase shift (with respect to the input) at the coupled port. Furthermore, the type 1 MS/NRI branch-line coupler offers a negative phase quadrature (−90°) while type 2 provides a positive phase quadrature (+90°) at their through ports with respect to the input port.

Figures 2.32a and 2.32b show the measured and simulated phase response at the through and coupled ports of the two types of MS/NRI branch-line couplers. These results verify the previously described phase relationships among the input and output ports. These kinds of couplers could find several applications. For example, Fig. 2.33 shows a combination of a type 1 MS/NRI metamaterial coupler with two conventional branch-line couplers in a 1:4 power-divider network. In this arrangement, the input power is divided equally among the four output ports but with a linear phase taper having a step of 90°. Such a feed network can be utilized for antenna beamforming.

2.13 CONCLUSION

The emerging field of negative-refractive-index metamaterials is very exciting for two interrelated reasons. On the one hand, there is new science being developed as-

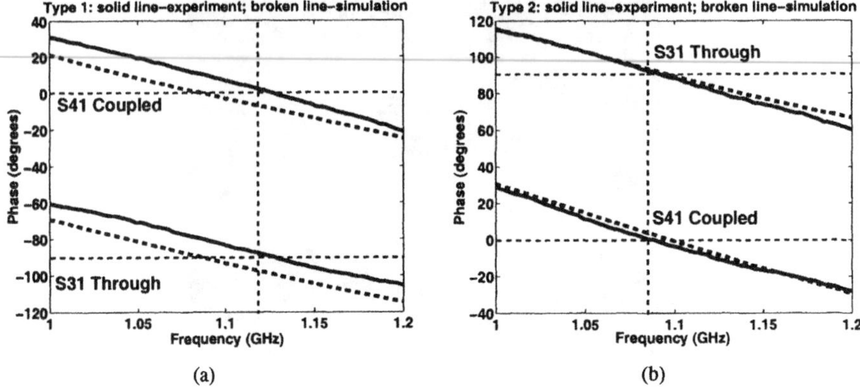

Fig. 2.32 (a) Output phase response for branch-line coupler type 1. (b) Output phase response for branch-line coupler type 2. After Ref. [45]. Copyright © 2004 IEEE.

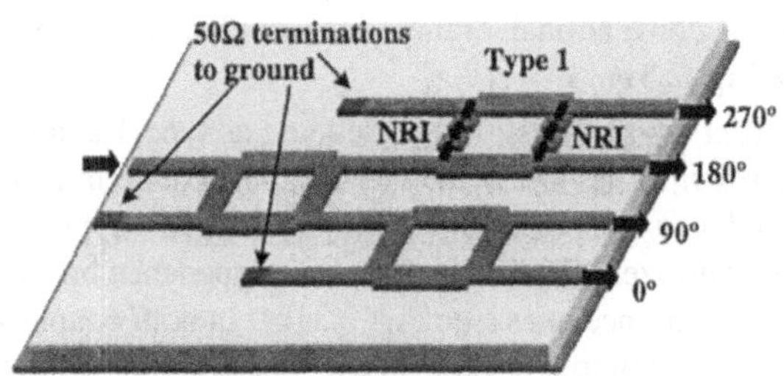

Fig. 2.33 A 1:4 power-dividing feed network with a 90° phase taper.

sociated with these fascinating phenomena and their interpretation. These intriguing phenomena include negative refraction, growing evanescent waves and surpassing the classical diffraction limit, reversal of the Čerenkov and Doppler effects, and others yet to be discovered! On the other hand, these new phenomena can be harnessed to make RF/microwave passive devices including antennas and their feed networks offering unique properties in terms of functionality, size reduction, and performance. In this chapter a number of these devices based on negative-refractive-index transmission-line metamaterials have been demonstrated. Specifically, we have presented lenses that can overcome the diffraction limit, compact and broadband phase-shifting lines, baluns, electrically small antennas, series-fed antenna arrays with reduced beam squinting, novel broadband power-combining schemes, backward leaky-wave antennas, small antennas and high-directivity coupled-line couplers. For more applications and properties of metamaterials the reader is referred to Refs. 46 and 47.

Acknowledgments

The author would like to thank his graduate students at the University of Toronto who carried out all the hard work and for their dedication and enthusiasm. For the material presented in this chapter the author specifically thanks Ashwin Iyer, Tony Grbic, Marco Antoniades, Rubaiyat Islam, and Francis Elek.

Financial support from the Natural Sciences and Engineering Research Council of Canada (NSERC) through Discovery and Strategic grants, as well as through a Steacie Fellowship, is gratefully acknowledged. Moreover, the author thanks Dr. Louis Varga and Dr. Satish Kashyap at the Department of National Defence (DRDC/Ottawa) for financial support and for a fruitful collaboration.

APPENDIX

Consider the two-dimensional lens of Fig. 2.3 and let us assume that the origin of the coordinate system is located at the second interface. A simple transmission-line equivalent circuit for TM_y wave propagation along the longitudinal z-direction is shown in Fig. A.1 (e.g., see Ref. 15, pp. 306–309). We will now carry out the wave propagation analysis based on the homogeneous limit. The reader is referred to Chapter 3 for the more general discrete periodic case.

The longitudinal wavenumbers in the PRI and NRI regions for evanescent waves are given by

$$k_z^{PRI} = \sqrt{k_x^2 - k_0^2}, \quad k_z^{NRI} = \sqrt{k_x^2 - k_0^2 n^2} \tag{A.1}$$

Since the refractive index is $n = -1$, these are equal, that is,

$$k_z^{PRI} = k_z^{NRI} = k_z > 0 \tag{A.2}$$

On the other hand, the transverse characteristic impedances are

$$Z_z^{PRI} = \frac{j\omega\mu_{PRI}}{k_z}, \quad Z_z^{NRI} = \frac{j\omega\mu_{NRI}}{k_z} \tag{A.3}$$

Since $-\mu_{NRI} = \mu_{PRI} > 0$, this implies that the PRI region is inductive, $\text{Im}(Z_z^{PRI}) > 0$, whereas the NRI region is capacitive, $\text{Im}(Z_z^{NRI}) < 0$. Furthermore, these are conjugately matched, that is, $Z_z^{PRI} + Z_z^{NRI} = 0$ [10,21,44].

Let us now write down the general form of the evanescent voltage waves in the three regions of Fig. A.1.

$$\text{First PRI medium:} \quad V_1 = e^{-k_z(z+h)} + Re^{+k_z(z+h)} \tag{A.4}$$

$$\text{NRI medium:} \quad V_2 = V_{inc} + V_{ref} = V_0^+ e^{-k_z z} + V_0^+ \Gamma_2 e^{+k_z z} \tag{A.5}$$

$$\text{Second PRI medium:} \quad V_3 = V_t e^{-k_z z} \tag{A.6}$$

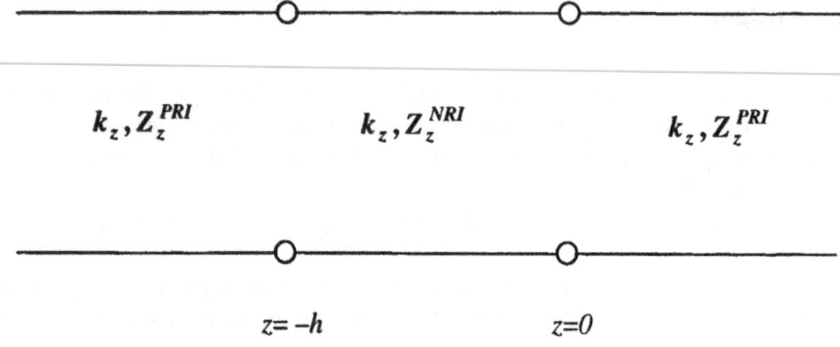

Fig. A.1 Equivalent transmission-line network for a negative-refractive-index slab sandwiched between two semi-infinite positive-refractive-index (PRI) regions.

The corresponding current waves will be given by

$$\text{First PRI medium:} \quad I_1 = e^{-k_z(z+h)}/Z_z^{PRI} - Re^{+k_z(z+h)}/Z_z^{PRI} \tag{A.7}$$

$$\text{NRI medium:} \quad I_2 = V_0^+ e^{-k_z z}/Z_z^{NRI} - V_0^+\Gamma_2 e^{+k_z z}/Z_z^{NRI} \tag{A.8}$$

$$\text{Second PRI medium:} \quad I_3 = V_t e^{-k_z z}/Z_z^{PRI} \tag{A.9}$$

Now let us impose the boundary conditions. At the first interface, $z = -h$:

$$V_1 = V_2 \Rightarrow 1 + R = V_0^+ e^{+k_z h} + V_0^+\Gamma_2 e^{-k_z h} \tag{A.10}$$

Also,

$$I_1 = I_2 \Rightarrow 1/Z_z^{PRI} - R/Z_z^{PRI} = V_0^+ e^{+k_z h}/Z_z^{NRI} - V_0^+\Gamma_2 e^{-k_z h}/Z_z^{NRI} \tag{A.11}$$

At the second interface, $z = 0$:

$$V_2 = V_3 \Rightarrow V_0^+ + V_0^+\Gamma_2 = V_t \tag{A.12}$$

Also,

$$I_2 = I_3 \Rightarrow V_0^+/Z_z^{NRI} - V_0^+\Gamma_2/Z_z^{NRI} = V_t/Z_z^{PRI} \tag{A.13}$$

By dividing (A.12) and (A.13) one obtains the familiar expression

$$\Gamma_2 = \frac{Z_z^{PRI} - Z_z^{NRI}}{Z_z^{PRI} + Z_z^{NRI}} \tag{A.14}$$

Also, from (A.10) and (A.11),

$$Z_z^{PRI}\frac{1+R}{1-R} = Z_z^{NRI}\frac{1+\Gamma_2 e^{-2k_z h}}{1-\Gamma_2 e^{-2k_z h}} \tag{A.15}$$

From (A.10) it follows that

$$V_0^+ = \frac{1+R}{e^{+k_z h} + \Gamma_2 e^{-k_z h}} \tag{A.16}$$

On the other hand, from (A.12) and (A.16) the transmitted voltage wave is

$$V_t = \frac{V^{interf2}}{V^{interf1}} = V_0^+(1+\Gamma_2) = \frac{(1+\Gamma_2)(1+R)}{e^{+k_z h} + \Gamma_2 e^{-k_z h}} \tag{A.17}$$

Now let us apply the limit that at resonance $Z_z^{PRI} + Z_z^{NRI} = 0$ and hence, from (A.14), $\Gamma_2 \to \infty$. Under this limit, (A.15) implies that the first interface is matched, that is,

$$R = 0 \tag{A.18}$$

Moreover, from (A.16) and (A.18), V_0^+ vanishes as well, that is, $V_0^+ \to 0$ (no incident wave in the NRI region). On the other hand, the product $V_0^+\Gamma_2$ remains finite—that is, $V_0^+\Gamma_2 \to e^{+k_z h}$—and hence, from (A.17),

$$V_t \to e^{+k_z h} \tag{A.19}$$

In summary, the final voltage waves in the three regions are

$$\text{First PRI medium:} \quad V_1 = e^{-k_z(z+h)} \tag{A.20}$$

$$\text{NRI medium:} \quad V_2 = 0 + V_{ref} = e^{+k_z h} e^{+k_z z} \tag{A.21}$$

$$\text{Second PRI medium:} \quad V_3 = e^{+k_z d} e^{-k_z z} \tag{A.22}$$

Observe the growing exponential wave in the NRI region and the annihilation of the incident wave. Moreover, there are no voltages that blow up to infinity anywhere within the lens. On the other hand, the ratio of the voltages between the second (NRI/PRI) and first (PRI/NRI) interface is given by

$$V_t = \frac{V^{interf2}}{V^{interf1}} = \frac{V_3(z=0)}{V_1(z=-h)} = \frac{e^{+k_z h}}{1} = e^{+k_z h} \tag{A.23}$$

The reader is reminded that the distance from the source to the image is twice the thickness of the lens, h (see Fig. 2.1). Therefore, (A.23) implies that the exponential growth within the lens exactly counters the exponential decay in the two PRI media, thus perfectly restoring the amplitude of the evanescent wave. A key aspect for this restoration process is that it happens for *any transverse wavenumber* k_x (of course, up to the cutoff k_x determined by (2.5) in the lossy case and (2.6) in the lossless case). This latter observation is essential for imaging where ideally *all* transverse wavenumbers should be restored for "perfect" imaging. Incidentally, it should be pointed out that the analysis presented in this Appendix can be extended in a trivial manner to show that restoration is also achieved for propagating waves in compliance with the ray picture of Fig. 2.1 (the details are presented in Chapter 3).

REFERENCES

1. R. E. Collin, *Field Theory of Guided Waves*, 2nd ed., Wiley–IEEE Press, Toronto, 1990, Chapter 12.

2. V. G. Veselago, "The electrodynamics of substances with simultaneously negative values of ε and μ," *Sov. Phys. Usp.*, vol. 10, no. 4, pp. 509–514, January 1967.

3. R. A. Shelby, D. R. Smith, and S. Schultz, "Experimental verification of a negative index of refraction," *Science*, vol. 292, pp. 77–79, April 6, 2001.

4. J. B. Pendry, A. J. Holden, D. J. Robbins, and W. J. Stewart, "Magnetism from conductors and enhanced nonlinear phenomena," *IEEE Trans. Microwave Theory Tech.*, vol. 47, no. 11, pp. 2075–2084, November 1999.

5. A. K. Iyer and G. V. Eleftheriades, "Negative refractive index metamaterials supporting 2-D waves," in *IEEE MTT-S International Microwave Symposium Digest*, vol. 2, June 2–7, 2002, Seattle, WA, pp. 1067–1070.

6. G. V. Eleftheriades, A. K. Iyer, and P. C. Kremer, "Planar negative refractive index media using periodically L–C loaded transmission lines," *IEEE Trans. Microwave Theory Tech.*, vol. 50, no. 12, pp. 2702–2712, December 2002.

7. A. K. Iyer, P. C. Kremer, and G. V. Eleftheriades, "Experimental and theoretical verification of focusing in a large, periodically loaded transmission line negative refractive index metamaterial," *Opt. Express*, vol. 11, pp. 696–708, April 2003, http://www.opticsexpress.org/abstract.cfm?URI=OPEX-11-7-696.

8. A. K. Iyer, A. Grbic, and G. V. Eleftheriades, "Sub-wavelength focusing in loaded transmission line negative refractive index metamaterials," in *IEEE MTT-S International Microwave Symposium Digest*, June 8–13, 2003, Philadelphia, PA, pp. 199–202.

9. A. Grbic and G. V. Eleftheriades, "Overcoming the diffraction limit with a planar left-handed transmission-line lens," *Phys. Rev. Lett.*, vol. 92, no. 11, 117403, March 19, 2004.

10. J. B. Pendry, "Negative refraction makes a perfect lens," *Phys. Rev. Lett.*, vol. 85, no. 18, pp. 3966–3969, October 2000.

11. L. Liu, C. Caloz, C. Chang, and T. Itoh, "Forward coupling phenomenon between artificial left-handed transmission lines," *J. Appl. Phys.*, vol. 92, no. 9 , pp. 5560–5565, November 2002.

12. C. Caloz and T. Itoh, "Novel microwave devices and structures based on the transmission line approach of meta-materials," in *2003 IEEE International Microwave Symposium Digest*, June 2003, pp. 195–198.

13. K. G. Balmain, A. A. E. Lüttgen, and P. C. Kremer, "Power flow for resonance cone phenomena in planar anisotropic metamaterials," *IEEE Trans. Antennas Propagat.*, Special Issue on Metamaterials, vol. 51, no. 10, pp. 2612–2618, October 2003.

14. K. G. Balmain, A. A. E. Lüttgen, and P. C. Kremer, "Resonance cone formation, reflection, refraction, and focusing in a planar anisotropic metamaterial," *IEEE Antennas Wireless Propag. Lett.*, vol. 1, no. 7, pp. 146–149, 2002.

15. S. Ramo, J. R. Whinnery, and T. Van Duzer, *Fields and Waves in Communication Electronics*, 3rd ed., John Wiley & Sons, Toronto, 1994.

16. J. R. Pierce, *Almost All About Waves*, The MIT Press, Cambridge, MA, 1974.

17. O. Siddiqui, M. Mojahedi, and G. V. Eleftheriades, "Periodically loaded transmission line with effective negative refractive index and negative group velocity," *IEEE Trans. Antennas Propag.*, Special Issue on Metamaterials, vol. 51, no. 10, pp. 2619–2625, October 2003.

18. A. Grbic and G. V. Eleftheriades, "Experimental verification of backward-wave radiation from a negative refractive index metamaterial," *J. Appl. Phys.*, vol. 92, no. 10, pp. 5930–5935, November 2002.

19. N. Seddon and T. Bearpark. "Observation of the inverse Doppler effect," *Science*, vol. 302, pp. 537–1540, November 28, 2003.

20. A. Grbic and G. V. Eleftheriades, "A backward-wave antenna based on negative refractive index L–C networks," in *Proceedings of the IEEE International Symposium on Antennas and Propagation*, vol. IV, San Antonio, TX, June 16–21, 2002, pp. 340–343.

21. A. Grbic and G. V. Eleftheriades, "Negative refraction, growing evanescent waves and sub-diffraction imaging in loaded-transmission-line metamaterials," *IEEE Trans. Microwave Theory Tech.*, vol. 51, no. 12., pp. 2297–2305, December 2003. (see also Erratum in *IEEE T-MTT*, vol. 52, no. 5, page 1580, May 2004.)

22. A. Grbic and G. V. Eleftheriades, "Growing evanescent waves in negative-refractive-index transmission-line media," *Appl. Phys. Lett.*, vol. 82, no. 12, pp. 1815–1817, March 24, 2003.

23. D. R. Smith, D. Schurig, M. Rosenbluth, S. Schultz, S. A. Ramakrishna, and J. B. Pendry, "Limitations on subdiffraction imaging with a negative refractive index slab," *Appl. Phys. Lett.*, vol. 82, no. 10, pp. 1506–1508, March 10, 2003.

24. G. Gomez-Santos, "Universal features of the time evolution of evanescent moes in a left-handed perfect lens," *Phys. Rev. Lett.*, vol. 90, no. 7, 077401, February 2003.

25. A. Grbic and G. V. Eleftheriades, "Periodic analysis of a 2-D negative refractive index transmission line structure," *IEEE Trans. Antennas Propag.*, Special Issue on Metamaterials, vol. 51, no. 10, pp. 2604–2611, October 2003.

26. G. V. Eleftheriades, "Planar negative refractive index metamaterials based on periodically L–C loaded transmission lines," Workshop on Quantum Optics, Kavli Institute of Theoretical Physics, University of Santa Barbara, July 2002. http://online.kitp.ucsb.edu/online/qo02/eleftheriades/ (slide #12).

27. A. Grbic and G. V. Eleftheriades, "Dispersion analysis of a microstrip based negative refractive index periodic structure," *IEEE Microwave Wireless Components Lett.*, vol. 13, no. 4, pp. 155–157, April 2003.

28. M. A. Antoniades and G. V. Eleftheriades, "Compact, linear, lead/lag metamaterial phase shifters for broadband applications," *IEEE Antennas Wireless Propag. Lett.*, vol. 2, no. 7, pp. 103–106, July 2003.

29. A. Sanada, C. Caloz, and T. Itoh, "Characteristics of the composite right/left-handed transmission lines," *IEEE Microwave Wireless Components Lett.*, vol. 14, pp. 68–70, February 2004.

30. G. V. Eleftheriades, M. A. Antoniades, A. Grbic, and R. Islam, "Electromagnetic applications of negative-refractive-index transmission-line metamaterial," in *27th ESA Antenna Technology Workshop on Innovative Periodic Antennas*, Santiago, Spain, March 2004, pp. 21–28.

31. M. Qiu, M. Simcoe, and G. V. Eleftheriades, "High-gain meander-less slot arrays on electrically thick substrates at mm-wave frequencies," *IEEE Trans. Microwave Theory Tech.*, vol. MTT-50, no. 2, pp. 517–528, February 2002.

32. M. A. Antoniades and G. V. Eleftheriades, "A broadband Wilkinson balun using microstrip metamaterial lines," *IEEE Antennas Wireless Propag. Lett.*, 2005.

33. M. Antoniades, and G. V. Eleftheriades, "A broadband balun using metamaterial phase-shifting lines," *2005 IEEE Intl. Symposium on Antennas and Propagation Digest*, Washington D.C., July 2005.

34. G. V. Eleftheriades, A. Grbic, and M. Antoniades, "Negative-refractive-index metamaterials and enabling electromagnetic applications," in *2004 IEEE International Symposium on Antennas and Propagation Digest*, Monterey, CA, June 2004, pp. 1399–1402.

35. A. Grbic and G. V. Eleftheriades, "Leaky CPW-based slot antenna arrays for millimeter-wave applications," *IEEE Trans. Antennas Propag.*, vol. 50, pp. 1494–1504, November 2002.

36. W. Menzel, "A new travelling-wave antenna in microstrip," *Arch. Elek. Ubertragung. (AEU)*, pp. 137–140, April 1979.

37. A. K. Iyer and G. V. Eleftheriades, "Leaky-wave radiation from a two-Dimensional negative-refractive-index transmission-line metamaterials," in *2004 URSI International Symposium on Electromagnetic Theory*, vol. 2, Pisa, Italy, May 2004, pp. 891–893.

38. R. Islam and G. V. Eleftheriades, "A planar metamaterial co-directional coupler that couples power backwards," in *2003 IEEE International Microwave Symposium Digest*, Philadelphia, PA, June 8–13, 2003, pp. 321–324.

39. R. Islam, F. Elek, and G. V. Eleftheriades, "Coupled-line metamaterial coupler having co-directional phase but contra-directional power flow," *Electron. Lett.*, vol. 40, no. 5, March 4, 2004.

40. F. Elek and G. V. Eleftheriades, "Dispersion analysis of Sievenpiper's shielded structure using multi-conductor transmission-line theory," *IEEE Microwave Wireless Components Lett.*, vol. 14, no. 9, pp. 434–436, September 2004.

41. R. Islam and G. V. Eleftheriades, "Analysis of a finite length microstrip/negative-refractive-index coupled-line coupler," in *2005 IEEE International Symposium on Antennas and Propagation Digest*, Washington D.C., July 2005.

42. A. Alù and N. Engheta, "Guided modes in a waveguide filled with a pair of single-negative (SNG), double-negative (DNG), and/or double-positive (DPS) layers," *IEEE Trans. Microwave Theory Tech.*, vol. 52, pp. 199–210, January 2004.

43. C. Caloz, A. Sanada, and T. Itoh, "A novel composite right-/left-handed coupled-line directional coupler with arbitrary coupling level and broad bandwidth," *IEEE Trans. Microwave Theory Tech.*, vol. 52, pp. 980–992, March 2004.

44. A. Alù and N. Engheta, "Pairing an epsilon-negative slab with a mu-negative slab: resonance, tunneling and transparency," *IEEE Trans. Antennas Propag.*, Special Issue on Metamaterials, vol. 51, no. 10, pp. 2558–2571, October 2003.

45. R. Islam and G. V. Eleftheriades, "Phase-agile branch-line couplers using metamaterial lines," *IEEE Microwave Wireless Components Lett.*, vol. 14, no. 7, pp. 340–342, July 2004.

46. J. B. Pendry, "Negative refraction," *Contemp. Phys.*, vol. 45, no. 3, pp. 191–202, May–June 2004.

47. D. R. Smith, J. B. Pendry, and M. C. K. Wiltshire, "Metamaterials and negative refractive index," *Science*, vol. 305, pp. 788–792, August 6, 2004.

3 Super-Resolving Negative-Refractive-Index Transmission-Line Lenses

ANTHONY GRBIC and GEORGE V. ELEFTHERIADES

The Edward S. Rogers Sr. Department of Electrical and Computer Engineering
University of Toronto
Toronto, Ontario, M5S 3G4
Canada

This chapter focuses on planar microwave lenses that are based on the transmission-line (TL) approach to implementing negative-refractive-index (NRI) media. The transmission-line approach involves loading a planar network of transmission lines with series capacitors and shunt inductors in order to synthesize an isotropic and homogenous NRI medium. This planar high-pass network will be referred to as the dual TL since it is the dual of a conventional low-pass TL. The chapter starts with the theory behind 1-D and 2-D dual transmission lines. Microwave network theory and an effective medium approach are used to describe the propagation characteristics of the dual TL. A physical interpretation of how a series capacitor leads to negative permeability and a shunt inductor to negative permittivity is also given.

A dual TL implementation of the NRI or "left-handed" lens described by V. G. Veselago is presented and characterized in this chapter [1]. This lens is referred to as the NRI-TL lens. The ability of a NRI-TL lens to form images that overcome the diffraction limit is demonstrated. This entails showing that the NRI-TL lens focuses propagating waves and restores the amplitude of evanescent waves at its foci. Super-resolution (imaging beyond the diffraction limit) is shown analytically for the NRI-TL lens, and supporting experimental evidence is presented. The resolution limitations of practical NRI-TL lenses are discussed in detail.

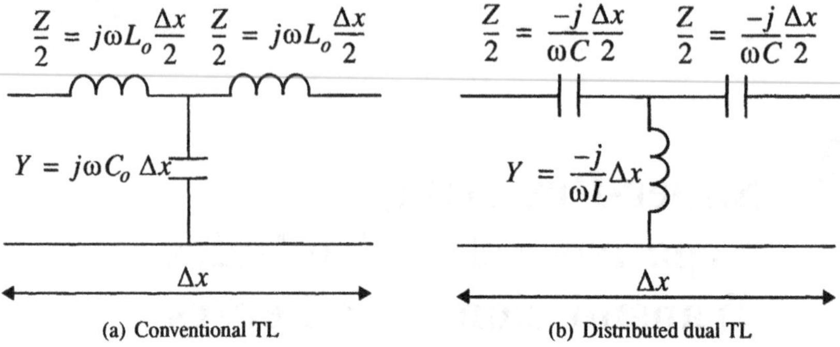

(a) Conventional TL (b) Distributed dual TL

Fig. 3.1 Distributed transmission line (TL) representations.

3.1 THE DISTRIBUTED DUAL TRANSMISSION LINE

The synthesis of the original split-ring resonator/wire negative-refractive-index (NRI) medium was based on separately considering an array of straight conducting wires to attain negative permittivity and an array of split-ring resonators to attain negative permeability [2,3]. Here a different approach is taken, originating from the analogy that is readily drawn between transverse electromagnetic (TEM) propagation on transmission lines (TLs) and plane-wave propagation in a homogeneous isotropic medium with positive material parameters, ϵ and μ. This equivalence forms the basis of a numerical technique for solving Maxwell's equations called the transmission-line modelling method [4,5]. Comparing the differential equations governing transmission-line and plane-wave propagation, the distributed inductance L_o and capacitance C_o of the transmission line depicted in Fig. 3.1a become equivalent to the permeability μ and permittivity ϵ of the medium supporting plane-wave propagation:

$$\epsilon \equiv C_o \qquad \mu \equiv L_o \tag{3.1}$$

This transmission-line model of a medium with positive material parameters, ϵ and μ, also offers insight into devising materials with negative ϵ and μ. Intuition suggests that in order to synthesize a negative-refractive-index medium ($\epsilon < 0$ and $\mu < 0$), the series reactance and shunt susceptance shown in Fig. 3.1a should become negative, given that the material parameters are directly proportional to these circuit quantities [6–9]. This change in sign implies the distributed system with series capacitors and shunt inductors shown in Fig. 3.1b [10]. It is the dual of the conventional transmission line and as a result will be referred to as the distributed dual TL.

As in the case of a conventional transmission line, the propagation constant and characteristic impedance of the distributed dual TL can be found from the distributed (per-unit-length) impedance Z and distributed admittance Y (assuming a $e^{j\omega t}$ time-harmonic progression) [10]:

$$k = -j\sqrt{ZY} = \frac{-1}{\omega\sqrt{LC}} \tag{3.2}$$

$$Z_o = \sqrt{\frac{Z}{Y}} = \sqrt{\frac{j\omega L}{j\omega C}} = \sqrt{\frac{L}{C}} \tag{3.3}$$

The positive square root is chosen in (3.3) because we are dealing with power flow in the positive z-direction of a passive structure and therefore Z_o must be a positive quantity. The sign ($\pm$) of the square root in (3.2), which determines the sign of the propagation constant k, is not as obvious. For the time being, we will assume that k is a negative quantity as in (3.2). The phase (v_p) and group (v_g) velocities of the waves guided by the distributed dual TL are

$$v_p = \frac{\omega}{k} = -\omega^2\sqrt{LC} \tag{3.4}$$

$$v_g = \frac{d\omega}{dk} = \frac{1}{\omega^2\sqrt{LC}} \tag{3.5}$$

The expressions above indicate that the distributed dual TL supports backward waves—that is, waves that have phase and group velocities of opposite sign. Since the group velocity v_g represents energy flow in this lossless system, a positive group velocity identifies power flow in the positive direction. The phase velocity is negative for such power flow, reassuring us that the propagation constant given by $k = \omega/v_p$ is in fact negative, as initially assumed in 3.2. Since the refractive index n is the ratio of v_p to c (the speed of light in a vacuum), one may also say that the refractive index is negative even for this 1-D distributed dual TL.

3.2 THE PERIODIC DUAL TRANSMISSION LINE

The distributed dual TL shown in Fig. 3.1b is a hypothetical structure which cannot be realized, but can be approximated by periodically loading a conventional TL with series capacitors ($2C$) and shunt inductors (L) as depicted in Fig. 3.2. For short interconnecting transmission-line lengths ($\beta d \ll 1$, where β is the propagation constant of the interconnecting transmission lines and d is the unit cell dimension), this periodic dual TL approaches the distributed dual TL shown in Fig. 3.1b. This periodic dual TL will be referred to as simply the dual TL from this point on.

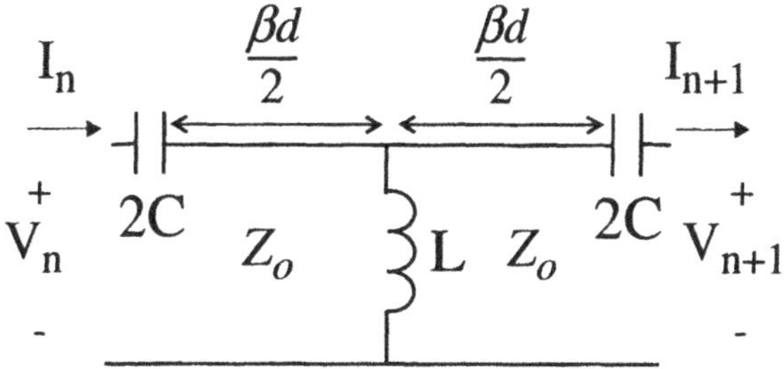

Fig. 3.2 A 1-D periodic dual transmission line. It is realized by periodically loading a host TL with series capacitors and shunt inductors.

The propagation characteristics of an infinite dual TL can be derived from a single unit cell using the Floquet Theorem. This theorem serves as a basis for the study of periodic structures. The Floquet Theorem may be stated as follows [11]: "*For a given mode of propagation at a given steady-state frequency, the fields at one cross section differ from those one period away only by a complex constant.*" A unit cell of the dual TL consists of a length d of transmission line with a shunt inductor L at its center and $2C$ capacitors at its ends. The Bloch voltages and currents on either side of the nth unit cell in an infinite cascade can be related using the transmission ($ABCD$) matrix of the unit cell (see Fig. 3.2) [12]:

$$\begin{pmatrix} V_n \\ I_n \end{pmatrix} = \begin{pmatrix} A & B \\ C & D \end{pmatrix} \begin{pmatrix} V_{n+1} \\ I_{n+1} \end{pmatrix} \tag{3.6}$$

where A, B, C, D are the elements of the transmission matrix for unit cell. The transmission matrix of the unit cell is the product of the transmission matrices of the constitutive circuit elements: the capacitor $2C$, the $\beta d/2$ section of transmission line, the inductor L, the second $\beta d/2$ section of transmission line and second $2C$ capacitor (see Fig. 3.2). According to the Floquet Theorem, the guided voltage and current waves in one cell differ by a complex constant from those one period away. In other words, the voltage and current at the $(n+1)$th terminal differs from the voltage and current at the nth terminal by a complex constant e^{-jkd}:

$$\begin{pmatrix} V_n \\ I_n \end{pmatrix} = \begin{pmatrix} A & B \\ C & D \end{pmatrix} \begin{pmatrix} V_{n+1} \\ I_{n+1} \end{pmatrix} = \begin{pmatrix} V_{n+1}e^{-jkd} \\ I_{n+1}e^{-jkd} \end{pmatrix} \tag{3.7}$$

where k is a complex number commonly referred to as the Bloch wavenumber. Bringing all the terms to one side of the equation yields the following homogeneous matrix equation:

$$\begin{pmatrix} A - e^{-jkd} & B \\ C & D - e^{-jkd} \end{pmatrix} \begin{pmatrix} V_{n+1} \\ I_{n+1} \end{pmatrix} = 0 \tag{3.8}$$

Setting the determinant equal to zero for this homogeneous matrix equation, gives us the dispersion equation for the dual TL:

$$\cos(kd) = \frac{A + D}{2} \tag{3.9}$$

The dispersion equation can be further simplified when the unit cell is symmetric ($A = D$) about its center, as is the case for the dual TL [12]:

$$\cos(kd) = A \tag{3.10}$$

Using (3.8) and (3.10), a periodic impedance known as the Bloch impedance ($Z_B = \frac{V_n}{I_n} = \frac{V_{n+1}}{I_{n+1}}$) can be defined at the terminals of the unit cell:

$$Z_B^{\pm} = \frac{\pm B}{\sqrt{A^2 - 1}} = \frac{\mp jB}{D\tan(kd)} \tag{3.11}$$

The $\pm$ superscripts denote the Bloch impedance solutions for forward (+) and reverse (−) traveling waves. Forward traveling waves carry power in the positive z-direction and reverse traveling waves carry power in the negative z-direction. Expressing the matrix elements (A, B, C, D) in terms of the transmission-line parameters (βd, Z_o) and inductor L and capacitor $2C$ values, the dispersion relation (3.10) and Bloch impedance expression (3.11) simplify to

$$\sin^2(kd/2) = \left[\sin(\beta d/2) - \frac{\cos(\beta d/2)}{2\omega C Z_o}\right]\left[\sin(\beta d/2) - \frac{Z_o \cos(\beta d/2)}{2\omega L}\right] \tag{3.12}$$

$$Z_B^{\pm} = \pm \frac{Z_o \tan(\beta d/2) - \frac{1}{2\omega C}}{\tan(kd/2)} \tag{3.13}$$

Just as a conventional TL is described by its characteristic impedance and propagation constant, the dual TL is characterized by its Bloch impedance Z_B and Bloch wavenumber k. The dispersion equation (3.12) defines the passbands and stopbands of the dual TL. Yet, it is not obvious from the dispersion relation that the dual TL can support a backward-wave propagation band. One may implicitly differentiate (3.13) and check whether the group and phase velocities are of opposite sign for a given set of transmission-line parameters and L, C values. A simpler approach is to take a closer look at the Bloch impedance under certain assumptions. First, the dual TL is assumed to have short interconnecting TL sections ($\beta d \ll 1$) so that it resembles the hypothetical distributed dual TL from the previous section. Next, a small per-unit-cell phase shift $kd \ll 1$ is assumed so that the Bragg condition is avoided. These two conditions restrict the operation to the long wavelength regime. In the frequency range where $Z_o \tan(\beta d/2) < \frac{1}{2\omega C}$, the numerator of the Bloch impedance becomes negative. Power flow in the positive z-direction requires that the Bloch impedance remains positive. Therefore, the denominator of the Bloch impedance must also be negative, implying that the wavenumber k is a negative quantity. A negative wavenumber given positive power flow implies backward-wave propagation. Note that the only way to get a propagating solution (a real-valued k) is to further assume that $Z_o \tan(\beta d/2) < \frac{1}{2\omega L}$.

This argument can be made mathematically as well. With the above-mentioned assumptions ($\beta d \ll 1, kd \ll 1$), the dispersion relation and Bloch impedance simplify to

$$k^2 = \left[\beta - \frac{1}{\omega C d Z_o}\right]\left[\beta - \frac{Z_o}{\omega L d}\right] = \left[\omega L_o - \frac{1}{\omega C d}\right]\left[\omega C_o - \frac{1}{\omega L d}\right] \tag{3.14}$$

$$Z_B = Z_o \frac{\sqrt{\beta d - \frac{1}{\omega C Z_o}}}{\sqrt{\beta d - \frac{Z_o}{\omega L}}} = \frac{\sqrt{\omega L_o - \frac{1}{\omega C d}}}{\sqrt{\omega C_o - \frac{1}{\omega L d}}} \tag{3.15}$$

where L_o and C_o are the per-unit-length inductance and capacitance of the unloaded transmission line: $\beta = \omega\sqrt{L_o C_o}$ and $Z_o = \sqrt{L_o/C_o}$. When $\omega L_o < \frac{1}{\omega C d}$, the series reactance of the dual TL becomes negative making the effective permeability

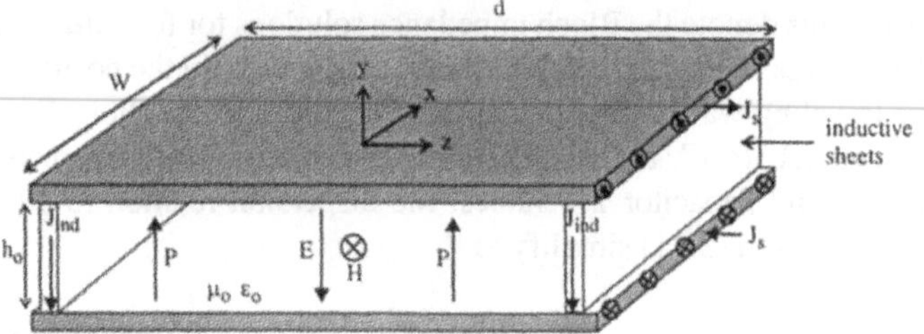

Fig. 3.3 Parallel-plate waveguide with periodic shunt inductive sheets. The periodic spacing d of the sheets is much smaller than the wavelength of operation.

negative. Similarly when $\omega C_o < \frac{1}{\omega L d}$, the shunt susceptance becomes negative making the permittivity negative. When both series reactance and shunt susceptance are negative, the dual TL supports backward waves and acts as a negative-refractive-index medium.

3.3 INTERPRETING NEGATIVE PERMITTIVITY AND PERMEABILITY

It has been shown that the periodic dual TL depicted in Fig. 3.2 supports backward waves. Circuit analysis brings us to this conclusion, but it does not provide us with a physical understanding of how a series capacitor leads to negative permeability and a shunt inductor to negative permittivity. In this section, the simplest transmission-line geometry is studied in order to develop this physical understanding. A parallel-plate waveguide operating in its TEM mode of operation is considered [13]. A parallel-plate waveguide is chosen for its convenient geometry that lends itself to simple mathematical expressions. The waveguide is infinite in both the x- and z-directions and is sufficiently thin (thickness = h_o in the y -direction) such that higher-order TE and TM modes are cut off. The waveguide serves as the host TL in the dual TL model. First, the parallel-plate waveguide is loaded with shunt inductive sheets (see Fig. 3.3) in order to understand how shunt inductors give rise to a negative effective permittivity. Next, the same waveguide is loaded with capacitive transverse slots (see Fig. 3.4) in order to understand how periodically spaced series capacitors give rise to a negative effective permeability. Finally, a parallel-plate waveguide simultaneously loaded with capacitive transverse slots and shunt inductive sheets is studied.

3.3.1 Negative Permittivity

A parallel-plate waveguide that is loaded with vertical (y-directed) inductive sheets is depicted in Fig. 3.3. The inductive sheets serve as the shunt inductors in the dual TL model. They are infinitely thin and periodically spaced at a distance d from each other that is much smaller than the wavelength of operation ($d \ll \lambda$), so that quasi-static

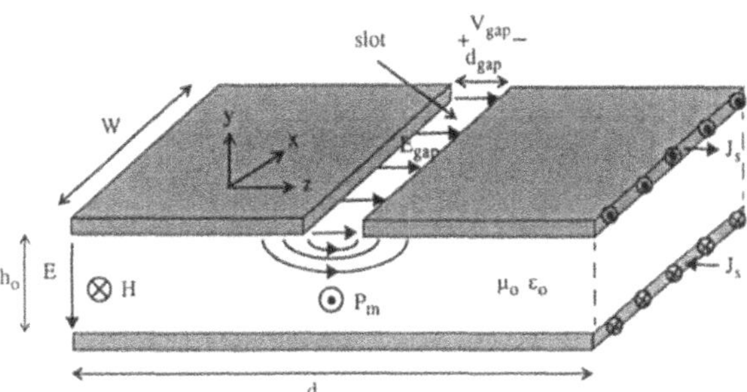

Fig. 3.4 Parallel-plate waveguide with capacitive transverse slots. The periodic spacing of the transverse slots is much smaller than the wavelength of operation.

approximations can be made. An inductive sheet can be implemented in practice as an array of closely spaced thin vias between the two plates of the waveguide.

First, a qualitative description is given of how inductive sheets give rise to a negative permittivity. In short, the impinging voltage wave (or equivalently the vertical electric field intensity **E**) between the top and bottom plates of the parallel-plate waveguide produces an electric flux density $\mathbf{D}_o = \epsilon_o \mathbf{E}$ in the y-direction. The voltage also excites an electric surface current density $\mathbf{J}_{ind}$ along the inductive sheets. This surface current density $\mathbf{J}_{ind}$ produces an electric polarization **P** that is antiparallel to the electric flux density $\mathbf{D}_o$ that is present in an unperturbed parallel-plate waveguide [14, 15]. When the electric polarization **P** caused by the inductive sheets overcomes $\mathbf{D}_o$, the effective permittivity of the medium becomes negative.

Now let us proceed with the detailed mathematical formulation. A surface impedance Z_{ind} can be defined for the inductive sheets. This surface impedance is the ratio of the tangential electric field to the tangential magnetic field, over the inductive sheet's surface. The surface impedance has units of Ohms per square. A section of the inductive sheet W wide in the x-direction has an impedance equal to $Z_s h_o / W\ \Omega$. The surface current density $\mathbf{J}_{ind}$ excited on the inductive sheet can be related to the impinging electric field intensity **E** by the surface impedance Z_{ind}:

$$\mathbf{J}_{ind} = \frac{-E\hat{y}}{Z_{ind}} \tag{3.16}$$

where E is simply the potential difference V between the parallel-plates divided by the waveguide height h_o. Since the current sheet is inductive, it has a surface impedance of the form $Z_{ind} = j\omega L$, where L has units of Henry per square. As such, the surface current density on the inductive sheets can be rewritten as

$$\mathbf{J}_{ind} = \frac{\mathbf{E}}{j\omega L} = \frac{-E\hat{y}}{j\omega L} \tag{3.17}$$

The inductive sheets can now be replaced by periodically spaced sheets of surface current $\mathbf{J}_{ind}$, as shown in Fig. 3.5. Since the spacing d is much smaller than λ, the array of current sheets can be treated as a continuous current density **J** (amps / m^2) over the entire parallel-plate waveguide, where **J** is given by

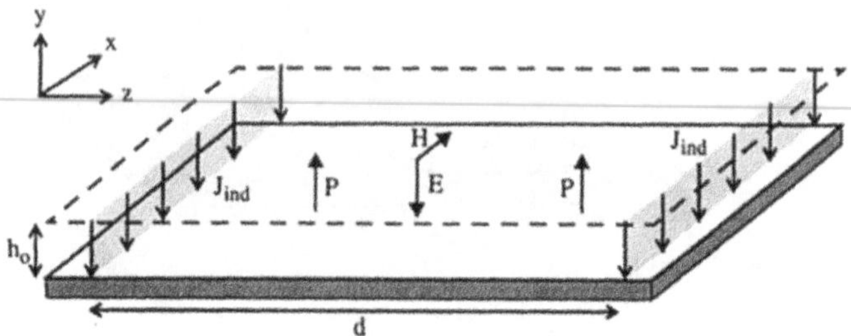

Fig. 3.5 Replacing the inductive sheets with periodically spaced sheets of electric surface current J_{ind}. The top plate of the parallel-plate waveguide is shown with dashed lines.

$$\mathbf{J} = \mathbf{J}_{ind}/d \tag{3.18}$$

This continuous current density **J** is depicted in Fig. 3.6. Ampere's Law can now be written for the inductively loaded parallel plate waveguide as follows:

$$\nabla \times \mathbf{H} = j\omega\epsilon_o\mathbf{E} + \mathbf{J} \tag{3.19}$$

where $\mathbf{E} = -E\hat{y}$. Substituting in the expression for **J** yields

$$\nabla \times \mathbf{H} = j\omega\left(\epsilon_o\mathbf{E} - \frac{\mathbf{E}}{\omega^2 Ld}\right) \tag{3.20}$$

From the above equation, the total electric flux density in the inductively loaded parallel-plate waveguide is

$$\mathbf{D}_{total} = \left(\epsilon_o - \frac{1}{\omega^2 Ld}\right)\mathbf{E} \tag{3.21}$$

The effective permittivity of the inductively loaded parallel-plate waveguide can be obtained by dividing $\mathbf{D}_{total}$ by **E**:

$$\epsilon = \epsilon_o - \frac{1}{\omega^2 Ld} \tag{3.22}$$

The electric polarization **P** (the electric moment per unit volume) resulting from the inductive sheets is therefore

$$\mathbf{P} = -\frac{\mathbf{E}}{\omega^2 Ld} = \frac{E\hat{y}}{\omega^2 Ld} \tag{3.23}$$

The electric polarization **P** is antiparallel to the electric flux density $\mathbf{D_o} = \epsilon_o\mathbf{E}$ that would be present in an unperturbed parallel-plate waveguide. Equation (3.21) has the form $\mathbf{D} = \epsilon\mathbf{E} = \epsilon_o\mathbf{E} + \mathbf{P} = (1 + \chi_e)\epsilon_o\mathbf{E}$, where χ_e is the electric susceptibility. Therefore, the electric susceptibility of the inductively loaded parallel-plate waveguide is $\chi_e = \frac{-1}{\epsilon_o\omega^2 Ld}$. The electric susceptibility is negative, since the electric polarization produced by the inclusions (the inductive sheets) is antiparallel

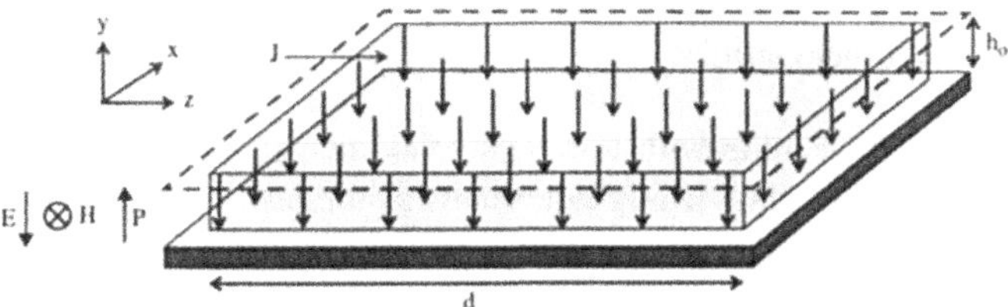

Fig. 3.6 Replacing the periodically spaced electric current sheets J_{ind} (A/m) with a continuous electric current density J (A/m^2).

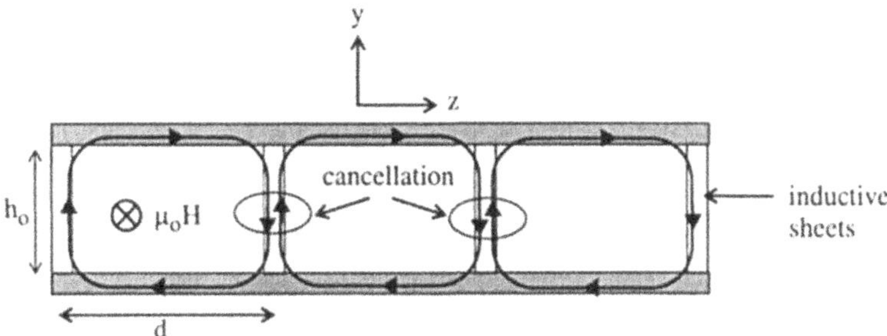

Fig. 3.7 Cancellation of the electric currents induced on the inductive sheets by the impinging magnetic field intensity H. The electric currents are shown with arrows.

to the electric flux density $\mathbf{D}_o = \epsilon_o \mathbf{E}$. At frequencies where $\chi_e < -1$, the effective permittivity $\epsilon = \epsilon_o(1 + \chi_e)$ of the inductively loaded waveguide becomes negative. The situation where the electric polarization $\mathbf{P}$ overcomes the impinging flux density $\mathbf{D}_o$ ($\chi_e < -1$) may seem unexpected or impossible at first. It implies that the response, $\mathbf{P}$, is greater than the initial excitation $\mathbf{D}_o$. This, however, is not uncommon. Consider the situation where the field strength inside a passive resonator is much greater than its initial excitation. The large response simply builds up with time. From the above analysis, we see that effective permittivity of this system can be tuned by varying the sheet inductance L.

In deriving the effective permittivity of the inductively loaded parallel-plate waveguide, it was assumed that the impinging magnetic field intensity $\mathbf{H}$ does not affect the electric polarization. Using Faraday's Law, it can be shown that this assumption is true for the inductively loaded parallel-plate waveguide with $d \ll \lambda$. According to Faraday's Law, the time-varying $\mathbf{H}$ field induces circulating electric currents in the loaded waveguide as depicted in Fig. 3.7, causing the electric currents induced in the vertical inductive sheets to cancel. From this simple argument, we see that the impinging $\mathbf{H}$ does not induce electric dipoles and the electric polarization remains unaffected.

3.3.2 Negative Permeability

A parallel-plate waveguide with transverse slots of gap width d_{gap} is depicted in Fig. 3.4. The parallel-plate waveguide serves as the host transmission line and the slots as the series capacitors in the dual TL model. The slots are periodically spaced at a distance d that is much smaller than the wavelength of operation ($d \ll \lambda$). As in the previous section, a qualitative description of how capacitive transverse slots give rise to negative permeability is first given. The impinging electric current density J_s in the top and bottom plates of the waveguide (or equivalently the magnetic field intensity H) produces a time-varying magnetic flux density between the plates $B_o = \mu_o H$. This same electric current density J_s also excites potential drops across the periodically spaced transverse slots. These potential drops generate a magnetic polarization $\mathbf{P}_m$ that is antiparallel to the magnetic flux density $\mathbf{B}_o$ that is present in an unperturbed parallel-plate waveguide. When the magnetic polarization $\mathbf{P}_m$ due to the slots overcomes the magnetic flux density $\mathbf{B}_o$, the effective permeability of the capacitively loaded parallel-plate waveguide becomes negative [16].

Now let us proceed with the mathematical formulation. The voltage across a slot can be related to the electric current density (J_s) in the plates of the parallel-plate waveguide by the per-unit-width capacitance (C) of the slot:

$$J_s = j\omega C V_{gap} \tag{3.24}$$

where C has units of pF/m in the x-direction. We have simply equated the displacement current density across the slot to the electric current density in the plates of the parallel-plate waveguide. Using conformal mapping, the value of C can be derived [17]:

$$C = \frac{2\epsilon_o}{\pi} \cosh^{-1}\left(\frac{d}{d_{gap}}\right) \tag{3.25}$$

From Ampere's Law, we find that the electric current density J_s is equal to the average magnetic field intensity H between the plates of the parallel-plate waveguide. Therefore, (3.24) can be rewritten as

$$V_{gap} = \frac{H}{j\omega C} \tag{3.26}$$

Next we apply Faraday's Law to the cross section of the capacitively loaded parallel-plate waveguide depicted in Fig. 3.8:

$$\int_{C_1} \mathbf{E}\cdot d\mathbf{l} = \int_{l_1+l_2} \mathbf{E}\cdot d\mathbf{l} + \int_{l_3} \mathbf{E}\cdot d\mathbf{l} = -j\omega\mu_o \int_{A_1} \mathbf{H}\cdot d\mathbf{A} \tag{3.27}$$

The line integral of the electric field over l_3 is simply V_{gap} given by (3.26). Therefore, (3.27) can be rewritten as

$$\int_{l_1+l_2} \mathbf{E}\cdot d\mathbf{l} = -j\omega\mu_o \int_{A_1} \mathbf{H}\cdot d\mathbf{A} - \frac{\mathbf{H}}{j\omega C} \tag{3.28}$$

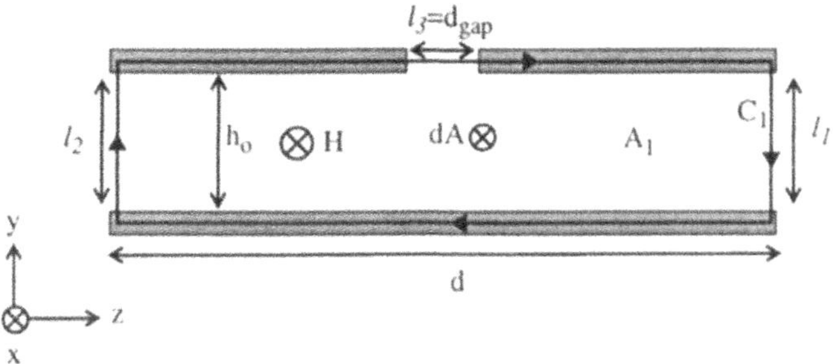

Fig. 3.8 Applying Faraday's Law to the cross section of the parallel-plate waveguide with capacitive slots.

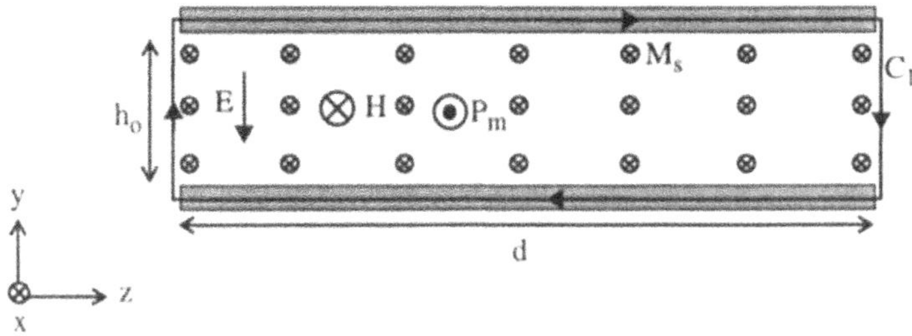

Fig. 3.9 A parallel-plate waveguide possessing a permeability μ_o and a magnetic current density (M_s).

Since both the height h_o and the spacing d of the gaps is assumed to be much smaller than λ, V_{gap} can be averaged over the cross-sectional area of the unit cell ($A_1 = h_o d$) and brought into the integrand, yielding

$$\int_{l_1+l_2} \mathbf{E} \cdot \mathbf{dl} = \int_{A_1} \left(-j\omega\mu_o \mathbf{H} - \frac{\mathbf{H}}{j\omega C h_o d} \right) \cdot \mathbf{dA} \tag{3.29}$$

This equation has the same form as Faraday's Law when applied to a parallel-plate waveguide possessing a permeability μ_o and magnetic current density $\mathbf{M}_s$ (see Fig. 3.9):

$$\int_{l_1+l_2} \mathbf{E} \cdot \mathbf{dl} = \int_{A_1} (-j\omega\mu_o \mathbf{H} - \mathbf{M}_s) \cdot \mathbf{dA} \tag{3.30}$$

This suggests that the capacitor acts as a magnetic current density $\mathbf{M}_s$ given by the following expression:

$$\mathbf{M_s} = \frac{\mathbf{H}}{j\omega C h_o d} = \frac{H\hat{x}}{j\omega C h_o d} \tag{3.31}$$

Expressed in differential form, (3.29) and (3.30) become

$$\nabla \times \mathbf{E} = -j\omega\mu_o \mathbf{H} - \frac{\mathbf{H}}{j\omega C h_o d} \tag{3.32}$$

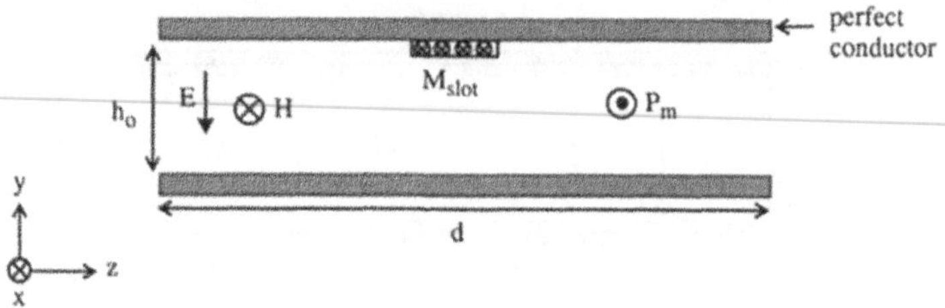

Fig. 3.10 Replacing the capacitive transverse slots with periodically spaced magnetic current densities M_{slot}.

$$\nabla \times \mathbf{E} = -j\omega\mu_o\mathbf{H} - \mathbf{M}_s \tag{3.33}$$

The link between the capacitive slot and the magnetic current $\mathbf{M}_s$ can be established in a more intuitive manner using Love's Equivalence Principle. The electric field intensity across the slot is equal to $E_{gap} = V_{gap}/d_{gap}$. Substituting this result into (3.26) yields

$$\mathbf{E}_{gap} = \frac{H\hat{z}}{j\omega C d_{gap}} \tag{3.34}$$

By Love's Equivalence Principle, the slot can be replaced by an equivalent magnetic surface current density $\mathbf{M}_{slot}$ that is covered by a perfect electric conductor [18], as shown in Fig. 3.10. The value of $\mathbf{M}_{slot}$ is given by

$$\mathbf{M}_{slot} = -\hat{n} \times E_{gap}\hat{z} = \hat{y} \times E_{gap}\hat{z} = E_{gap}\hat{x} \tag{3.35}$$

where the normal unit vector $\hat{n}$ is taken in the negative y-direction, since the region of interest is within the parallel-plate waveguide. Therefore, we have a parallel-plate waveguide with periodically spaced magnetic current densities. Substituting (3.34) into (3.35) yields an expression for $\mathbf{M}_{slot}$:

$$\mathbf{M}_{slot} = \frac{H\hat{x}}{j\omega C d_{gap}} \tag{3.36}$$

Assuming the slot is narrow, the total magnetic current ($\mathbf{I}_m$) per unit cell length d is

$$\mathbf{I}_m = \mathbf{M}_{slot} d_{gap} = \frac{H\hat{x}}{j\omega C} \tag{3.37}$$

The closely spaced ($d \ll \lambda$) magnetic currents $\mathbf{I}_m$ can be averaged over the cross-sectional area of the unit cell in order to find the continuous magnetic current density $\mathbf{M}_s$ (see Fig. 3.9):

$$\mathbf{M}_s = \frac{\mathbf{I}_m}{h_o d} = \frac{H\hat{x}}{j\omega C h_o d} \tag{3.38}$$

The value of $\mathbf{M}_s$ given by (3.31) has now been recovered using the notion that the slot acts as a magnetic current, giving us further physical insight.

Now let us get back to deriving the effective permeability of the capacitively loaded parallel-plate waveguide. Equation (3.32) can be rewritten in the following form:

$$\nabla \times \mathbf{E} = -j\omega \left(\mu_o - \frac{1}{\omega^2 C h_o d} \right) \mathbf{H} \tag{3.39}$$

From the above equation, we see that the total magnetic flux density $\mathbf{B}_{total}$ in the capacitively loaded parallel-plate waveguide is

$$\mathbf{B}_{total} = \left(\mu_o - \frac{1}{\omega^2 C h_o d} \right) \mathbf{H} \tag{3.40}$$

The effective permeability of the slotted parallel-plate waveguide can be obtained by dividing $\mathbf{B}_{total}$ by $\mathbf{H}$:

$$\mu = \mu_o - \frac{1}{\omega^2 C h_o d} \tag{3.41}$$

The magnetic polarization $\mathbf{P}_m$ (the magnetic moment per unit volume) resulting from the capacitive slots is therefore

$$\mathbf{P}_m = \frac{-\mathbf{H}}{\mu_o C \omega^2 d h_o} \tag{3.42}$$

Equation (3.40) has the form $\mathbf{B}_{total} = \mu\mathbf{H} = \mu_o(\mathbf{H}+\mathbf{P}_m) = (1+\chi_m)\mu_o\mathbf{H}$, where χ_m is the magnetic susceptibility. Therefore, the effective magnetic susceptibility of this slotted parallel-plate waveguide is $\chi_m = \frac{-1}{\mu_o C \omega^2 d h_o}$. The susceptibility is negative since the slots generate a magnetic polarization $\mathbf{P}_m$ that is antiparallel to the magnetic flux density $\mathbf{B}_o = \mu_o\mathbf{H}$ of an unperturbed parallel-plate waveguide. At frequencies where $\chi_m < -1$, the effective permeability $\mu = \mu_o(1 + \chi_m)$ becomes negative. From the expression above, we see that the effective permeability of this system can be tuned by varying the slot capacitance.

To derive the effective permeability of the slotted parallel-plate waveguide, the total magnetic flux $\mathbf{B}_{total}$ between the plates was divided by the impinging $\mathbf{H}$ field. It was, however, assumed that the impinging electric field $\mathbf{E}$ does not affect the magnetic polarization. It will be shown through symmetry arguments that this assumption is in fact true for the slotted parallel-plate waveguide considered. The y components of the circulating electric field around each slot cancel due to their close spacing ($d \ll \lambda$) and only the longitudinal (z-directed) electric field is excited, as shown in Fig. 3.11. For this reason, the impinging y-directed electric field $\mathbf{E}$ cannot couple to the slots resulting in an unchanged magnetic polarization.

3.3.3 Combining Negative ϵ and Negative μ

In the previous two sections, it was shown that shunt inductors give rise to negative electric polarization while series capacitors give rise to negative magnetic polarization. Specifically, the impinging time-varying $\mathbf{E}$ field excites electric currents along the inductive sheets which produce an electric polarization $\mathbf{P}$ that is antiparallel to

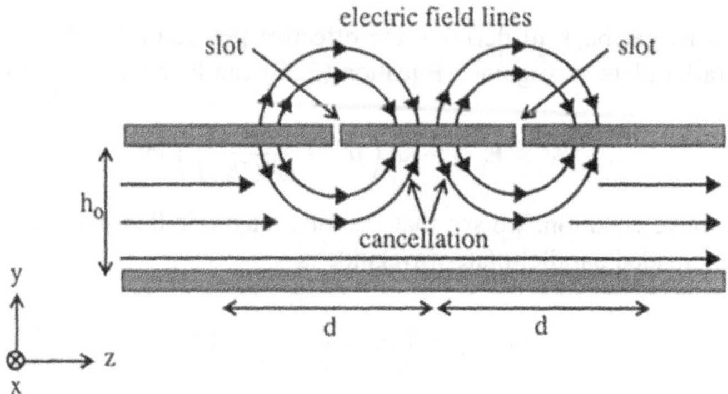

Fig. 3.11 Cancellation of the vertical electric field components produced by adjacent slots in the parallel-plate waveguide with capacitive transverse slots

the electric flux density $\mathbf{D}_o$ that would be present in an unperturbed parallel-plate waveguide. Similarly, the impinging time-varying $\mathbf{H}$ field excites magnetic currents in the capacitive slots which produce a magnetic polarization $\mathbf{P}_m$ that is antiparallel to the magnetic flux density $\mathbf{B}_o$ that would be present in an unperturbed parallel-plate waveguide. Ensuring that $\chi_m < -1$ for the slotted waveguide leads to a structure with negative μ, while $\chi_e < -1$ for the inductively loaded waveguide leads to a structure with negative ϵ. Consequently, one may argue that combining both the inductive sheets and capacitive slots into a single parallel-plate waveguide will lead to simultaneously negative μ and ϵ. In order to design the permittivity and permeability independently as in the previous two sections, the induced electric and magnetic currents in the combined structure (with inductive sheets and capacitive slots) should not interact with each other. Again, it will be shown through symmetry arguments that this is in fact the case in the combined dual TL parallel-plate waveguide.

The magnetic field intensity produced by adjacent inductive sheets cancels due to their close spacing as shown in Fig. 3.12. Therefore, the electric current density (J_{ind}) along the inductive sheets does not produce a time-varying magnetic flux between the plates of the loaded waveguide and, as a result, does not induce an electromotive force across the slots. Consequently, the electric currents along the inductive sheets do not magnetically couple to the induced magnetic currents of the capacitive slots. Recall that the electric field radiated by the inductive sheets is vertical (y-directed) while the electric field radiated by the closely spaced slots is horizontal (z-directed); therefore, not only are the induced electric and magnetic dipoles magnetically decoupled, they are also electrically decoupled.

3.4 THE 2-D DUAL TRANSMISSION LINE

Transmission-line analysis, together with the Floquet Theorem, was used to show that the 1-D dual TL supports backward waves and acts as a 1-D negative-refractive-index

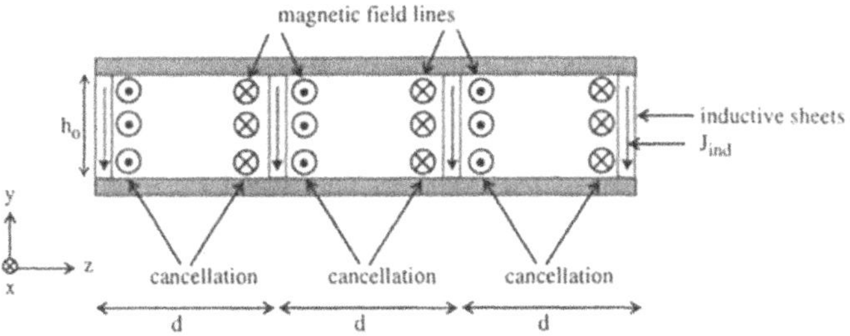

Fig. 3.12 Cancellation of magnetic field produced by adjacent inductive sheets.

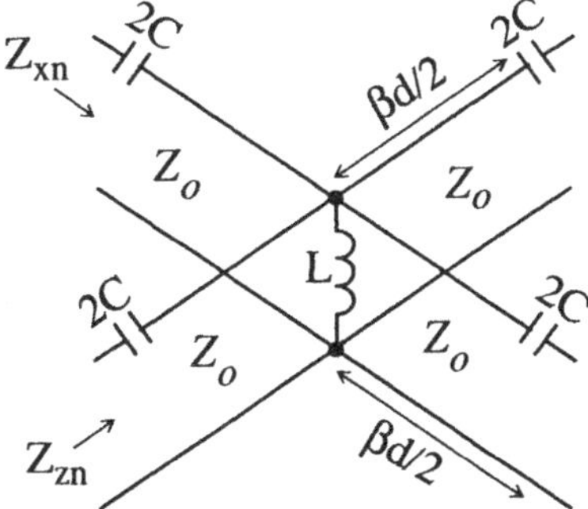

Fig. 3.13 The unit cell of the 2-D dual transmission line. After Ref. [19]. Copyright © 2003 IEEE.

medium. The same approach can be taken to show that a 2-D network of transmission lines [19] loaded with series capacitors and shunt inductors acts as a two-dimensional negative-refractive-index medium. The 2-D network of loaded transmission lines will be referred to as the 2-D dual TL. A schematic of the 2-D dual TL is shown in Fig. 3.13. The analysis of the 2-D dual TL presented in the previous two chapters was restricted to propagation along the principal axes (x- and z- in Fig. 3.13). In this section, the Floquet Theorem is applied in two dimensions and the propagation characteristics of the 2-D dual TL are found for all directions of propagation. Voltage and current relationships, Bloch impedance expressions and dispersion equations are first developed for a generalized 2-D periodic electrical network. These basic relations are then applied to the 2-D dual TL. Various resonances are examined which define the passbands and stopbands of the 2-D dual TL . Expressions for effective material parameters at frequencies of backward-wave homogeneous and isotropic operation are also derived, which provide a simplified understanding of the underlying band structure.

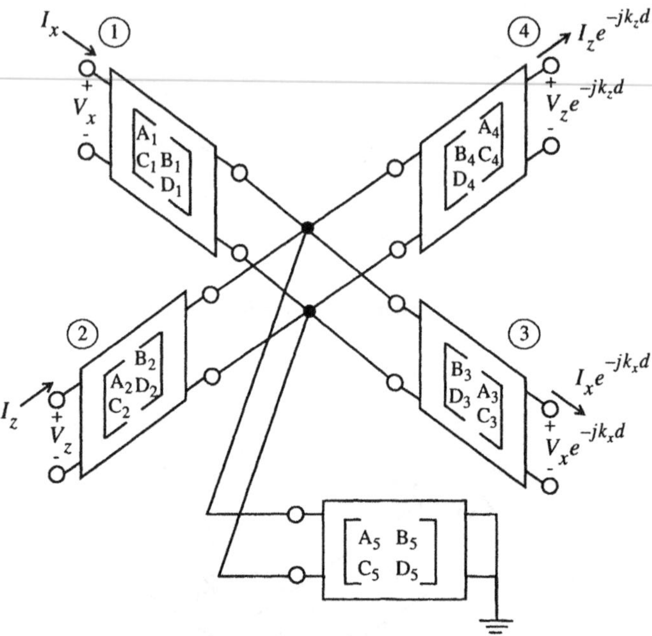

Fig. 3.14 The unit cell of a general 2-D periodic electrical T-network. After Ref. [19]. Copyright © 2003 IEEE.

3.4.1 The Generalized 2-D Periodic Electrical Network

The propagation characteristics of the generalized 2-D periodic electrical network shown in Fig. 3.14 will be investigated in this section and the results subsequently applied to the 2-D dual TL. This generic four-port electrical network is studied since it applies to a wide range of circuits and electromagnetic structures. The network can provide insight into recent electromagnetic bandgap structures and planar anisotropic metamaterials [20–23]. The general formulation applies to non-reciprocal periodic structures as well. The periodic electrical network depicted in Fig. 3.14 is general in the sense that the unit cell is represented by a generic T-network of transmission ($ABCD$) matrices. As illustrated, the four series branches of the T-network are characterized by four separate transmission matrices while a fifth transmission matrix represents the shunt branch. Four different series branches are considered to account for anisotropies that may exist in a structure. An infinite structure consisting of a 2-D array of such unit cells can be analyzed by applying the Floquet Theorem to the Bloch voltages and currents at the ports of the unit cell. As shown in Fig. 3.14, the voltage and current at one port can be related to those at the opposite port by a wavenumber k_x or k_z in the x- and z-directions, respectively:

$$V_1 = V_x, \qquad V_3 = V_x e^{-jk_x d}, \qquad I_1 = I_x, \qquad I_3 = I_x e^{-jk_x d} \tag{3.43}$$

$$V_2 = V_z, \qquad V_4 = V_z e^{-jk_z d}, \qquad I_2 = I_z, \qquad I_4 = I_z e^{-jk_z d} \tag{3.44}$$

where V_1 to V_4 and I_1 to I_4 represent the Bloch voltages and currents at the four ports of the unit cell . Two equations can be obtained by applying Kirchhoff's voltage law (KVL) in the x- and z-directions. One equation relates V_x to I_x, while the other relates V_z to I_z. Applying KVL from port 3 to port 4 yields a third equation relating V_x, I_x, V_z and I_z. Finally, applying Kirchhoff's current law to the central node of the T-network provides a fourth equation relating V_x, I_x, V_z, and I_z. These four equations form the following system of linear homogeneous equations:

$$(\,F\,)\begin{pmatrix} V_x \\ I_x \\ V_z \\ I_z \end{pmatrix} = \begin{pmatrix} f_{11} & f_{12} & 0 & 0 \\ 0 & 0 & f_{23} & f_{24} \\ f_{31} & f_{32} & f_{33} & f_{34} \\ f_{41} & f_{42} & f_{43} & f_{44} \end{pmatrix} \begin{pmatrix} V_x \\ I_x \\ V_z \\ I_z \end{pmatrix} = \begin{pmatrix} 0 \\ 0 \\ 0 \\ 0 \end{pmatrix} \tag{3.45}$$

$$\begin{aligned}
f_{11} &= 1 - \frac{R_1 D_3}{R_3 D_1} e^{-jk_x d} & f_{12} &= -\frac{R_1 B_3}{R_3 D_1} e^{-jk_x d} - \frac{B_1}{D_1} \\
f_{23} &= 1 - \frac{R_2 D_4}{R_4 D_2} e^{-jk_z d} & f_{24} &= -\frac{R_2 B_4}{R_4 D_2} e^{-jk_z d} - \frac{B_2}{D_2} \\
f_{31} &= \frac{D_3}{R_3} e^{-jk_x d} & f_{32} &= \frac{B_3}{R_3} e^{-jk_x d} \\
f_{33} &= -\frac{D_4}{R_4} e^{-jk_z d} & f_{34} &= -\frac{B_4}{R_4} e^{-jk_z d} \\
f_{41} &= \frac{-C_1}{R_1} - \frac{C_3}{R_3} e^{-jk_x d} - \frac{D_5 D_3}{B_5 R_3} e^{-jk_x d} & f_{42} &= \frac{A_1}{R_1} - \frac{A_3}{R_3} e^{-jk_x d} - \frac{D_5 B_3}{R_3 B_5} e^{-jk_x d} \\
f_{43} &= \frac{-C_2}{R_2} - \frac{C_4}{R_4} e^{-jk_z d} & f_{44} &= \frac{A_2}{R_2} - \frac{A_4}{R_4} e^{-jk_z d}
\end{aligned} \tag{3.46}$$

where

$$\begin{aligned}
R_1 &= A_1 D_1 - B_1 C_1 \\
R_2 &= A_2 D_2 - B_2 C_2 \\
R_3 &= A_3 D_3 - B_3 C_3 \\
R_4 &= A_4 D_4 - B_4 C_4
\end{aligned} \tag{3.47}$$

Substituting the first two rows of the coefficient matrix F into the last two rows and assuming the series branches of the T-network are reciprocal ($R_1 = R_2 = R_3 = R_4 = 1$) yields the following simplified system of linear homogeneous equations:

$$(\,G\,)\begin{pmatrix} V_x \\ V_z \end{pmatrix} = \begin{pmatrix} g_{11} & g_{12} \\ g_{21} & g_{22} \end{pmatrix} \begin{pmatrix} V_x \\ V_z \end{pmatrix} = \begin{pmatrix} 0 \\ 0 \end{pmatrix} \tag{3.48}$$

$$g_{11} = \frac{(B_1D_3 + B_3D_1)e^{-jk_xd}}{B_3e^{-jk_xd} + B_1}$$

$$g_{12} = \frac{-(B_2D_4 + B_4D_2)e^{-jk_zd}}{B_4e^{-jk_zd} + B_2}$$

$$g_{21} = \frac{1 + e^{-2jk_xd} - e^{-jk_xd}\left(C_1B_3 + B_1C_3 + A_1D_3 + D_1A_3 + \frac{B_1D_3D_5 + D_1B_3D_5}{B_5}\right)}{B_1 + B_3e^{-jk_xd}}$$

$$g_{22} = \frac{1 + e^{-2jk_zd} - e^{-jk_zd}(C_2B_4 + B_2C_4 + A_2D_4 + D_2A_4)}{B_2 + B_4e^{-jk_zd}} \tag{3.49}$$

For a nontrivial solution, the determinant of the coefficient matrix G must vanish. This yields the following dispersion equation for the generalized 2-D periodic electrical network that is reciprocal:

$$\begin{aligned} 0 &= (B_1D_3 + B_3D_1)\left[2\cos(k_zd) - (C_2B_4 + B_2C_4 + A_2D_4 + D_2A_4)\right] \\ &\quad + (B_2D_4 + B_4D_2)\left[2\cos(k_xd) - \left(C_1B_3 + B_1C_3 + A_1D_3 + D_1A_3 \right.\right. \\ &\qquad \left.\left. + \frac{B_1D_3D_5}{B_5} + \frac{B_3D_1D_5}{B_5}\right)\right] \end{aligned} \tag{3.50}$$

The periodic TL networks that will be analyzed in this chapter act as isotropic media. They consist of T-networks with identical series branches so the following simplifications apply:

$$\begin{aligned} A &= A_1 = A_2 = A_3 = A_4 \\ B &= B_1 = B_2 = B_3 = B_4 \\ C &= C_1 = C_2 = C_3 = C_4 \\ D &= D_1 = D_2 = D_3 = D_4 \end{aligned} \tag{3.51}$$

Given these simplifications, the dispersion equation (3.50) reduces to

$$0 = BD\left[\cos(k_xd) + \cos(k_zd) + 2 - 4AD - BD\frac{D_5}{B_5}\right] \tag{3.52}$$

This dispersion relation defines the various passbands and stopbands of a periodic electrical T-network with identical series branches.

Bloch impedances as in a standard 1-D periodic structure [12] can be defined using the first two rows of matrix F in (3.45). The Bloch impedances, which will be denoted as Z_x and Z_z, are the ratio of the voltage and current at ports 1 and 2 of the unit cell. The Bloch impedances for power flow in the positive x- and z-directions are therefore

$$Z_x = \frac{V_x}{I_x} = \frac{-f_{12}}{f_{11}} = \frac{B_1 + B_3e^{-jk_xd}}{D_1 - D_3e^{-jk_xd}} \tag{3.53}$$

$$Z_z = \frac{V_z}{I_z} = \frac{-f_{24}}{f_{23}} = \frac{B_2 + B_4e^{-jk_zd}}{D_2 - D_4e^{-jk_zd}} \tag{3.54}$$

If the periodic structure consists of unit cells with identical series branches defined by (3.51), the Bloch impedance expressions simplify to

$$Z_x = \frac{V_x}{I_x} = \frac{-jB}{D\tan(k_x d/2)} \tag{3.55}$$

$$Z_z = \frac{V_z}{I_z} = \frac{-jB}{D\tan(k_z d/2)} \tag{3.56}$$

Finally, an expression that relates V_x and V_z can be found, given a direction of propagation defined by k_x and k_z. It can be derived using the first row of matrix G in (3.48):

$$V_x = V_z \frac{g_{12}}{g_{11}} = V_z \left[\frac{B_2 D_4 + B_4 D_2}{B_4 + B_2 e^{jk_z d}}\right] \left[\frac{B_3 + B_1 e^{jk_x d}}{B_1 D_3 + B_3 D_1}\right] \tag{3.57}$$

Once more, assuming that the series branches consist of identical networks, (3.57) reduces to

$$V_x = V_z \frac{1 + e^{jk_x d}}{1 + e^{jk_z d}} \tag{3.58}$$

In summary, (3.43) defines the relationship between the voltages and currents at ports 1 and 3, while (3.44) specifies the relationship between the voltages and currents at ports 2 and 4. The Bloch impedances (3.53)–(3.56), on the other hand, define the dependence between the voltage and current at the same port. Equations (3.57) and (3.58) complete the picture by specifying the dependence between the voltages at ports 1 and 2. All four port voltages and currents can now be related given k_x and k_z.

3.4.2 Periodic Analysis of the 2-D Dual Transmission Line

The expressions derived in the previous section for a generalized 2-D periodic electrical network will now be applied to the 2-D dual TL shown in Fig. 3.13, in order to study its propagation characteristics. The 2-D dual TL has identical reciprocal series branches which can be represented by the transmission matrix of a $2C$ capacitor connected in series with a $\beta d/2$ section of transmission line [12]:

$$\begin{pmatrix} A & B \\ C & D \end{pmatrix} = \begin{pmatrix} \cos(\frac{\beta d}{2}) + \frac{1}{Z_o \omega 2C}\sin(\frac{\beta d}{2}) & jZ_o \sin(\frac{\beta d}{2}) - \frac{j}{\omega 2C}\cos(\frac{\beta d}{2}) \\ \frac{j}{Z_o}\sin(\frac{\beta d}{2}) & \cos(\frac{\beta d}{2}) \end{pmatrix} \tag{3.59}$$

where Z_o is the characteristic impedance, β the propagation constant, and d the length of the interconnecting transmission-line sections. The shunt branch is represented by the transmission matrix of a series inductance L:

$$\begin{pmatrix} A_5 & B_5 \\ C_5 & D_5 \end{pmatrix} = \begin{pmatrix} 1 & j\omega L \\ 0 & 1 \end{pmatrix} \tag{3.60}$$

Substituting the matrix elements of the 2-D dual TL given by (3.59) and (3.60) into the general dispersion equation (3.52) yields

$$\sin^2\left(\frac{k_{xn}d}{2}\right) + \sin^2\left(\frac{k_{zn}d}{2}\right) = \frac{1}{2}\left[2\sin\left(\frac{\beta d}{2}\right) - \frac{1}{Z_o\omega C}\cos\left(\frac{\beta d}{2}\right)\right] \cdot \left[2\sin\left(\frac{\beta d}{2}\right) - \frac{Z_o}{2\omega L}\cos\left(\frac{\beta d}{2}\right)\right] \tag{3.61}$$

where $\beta = \omega/v_\phi$, v_ϕ is the phase velocity of the interconnecting transmission lines and k_{xn} and k_{zn} are the wavenumbers in the x- and z-directions, respectively. This dispersion equation defines the underlying band structure of the 2-D dual TL.

Similarly, substituting the matrix elements of the dual TL into the general Bloch impedance equations (3.55) and (3.56), yields the Bloch impedances of the 2-D dual TL:

$$Z_{xn} = \frac{Z_o\tan(\frac{\beta d}{2}) - \frac{1}{2\omega C}}{\tan(\frac{k_x d}{2})}, \qquad Z_{zn} = \frac{Z_o\tan(\frac{\beta d}{2}) - \frac{1}{2\omega C}}{\tan(\frac{k_z d}{2})} \tag{3.62}$$

These expressions for Z_{xn} and Z_{zn} resemble the Bloch impedance expression Z_B for the 1-D dual TL given by (3.13). The only difference is that the wavenumber k in the 1-D dual TL is replaced by k_{xn} and k_{zn} for the respective x- and z-directions in the 2-D dual TL. The unit cells of the 2-D dual TL are symmetrical, therefore the Bloch impedances looking into the positive and negative x- and z-directions are negatives of each other. Just as a conventional TL is described by its characteristic impedance and propagation constant, the 2-D dual TL is characterized by its dispersion equation and by the x- and z-directed Bloch impedances.

A dispersion characteristic for a representative 2-D dual TL is plotted in Fig. 3.15 in the form of a Brillouin diagram. A physical understanding of this band structure can be gained through examining the various resonances which identify the location and nature of the 2-D dual TL's passbands and stopbands. It is clear from the high-pass configuration (series C and shunt L) of the 2-D dual TL that a stopband exists at low frequencies of operation. However, as the frequency is increased, the unit cells begin to resonate marking the onset of the structure's first passband of operation. The frequency at which the first passband begins (point a in Fig. 3.15) can be solved for by setting $k_x d = k_z d = \pi$ in dispersion equation (3.61). Rearranging this dispersion equation yields the following expression:

$$0 = \cos\left(\frac{\beta d}{2}\right)\left[\left(4 - \frac{1}{2\omega^2 LC}\right)\cos\left(\frac{\beta d}{2}\right) + 2\sin\left(\frac{\beta d}{2}\right)\left(\frac{1}{Z_o\omega C} + \frac{Z_o}{2\omega L}\right)\right] \tag{3.63}$$

The start of the passband (a) is found by setting the second product term in (3.63) equal to zero. Setting this term to zero is equivalent to computing the resonant frequency, ω_1, of the unit cell with all four terminals short-circuited to ground as shown in Fig. 3.16. This is evident from the fact that the Bloch impedances Z_x and Z_z vanish when $k_{xn}d = k_{zn}d = \pi$. At ω_1, a resonance occurs between the

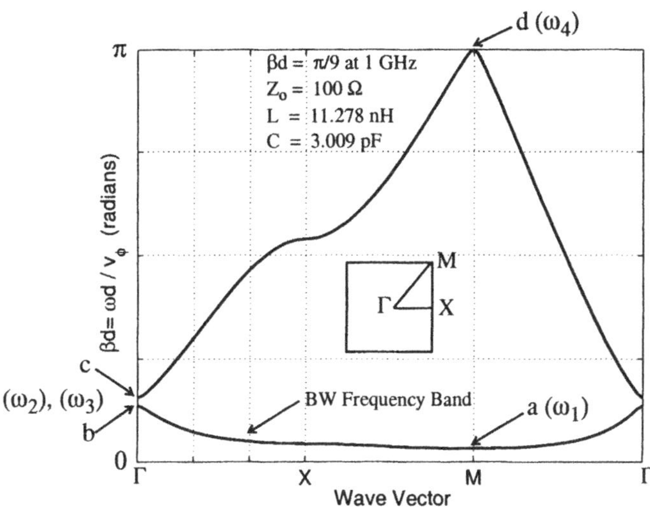

Fig. 3.15 Brillouin diagram for a representative 2-D dual TL. After Ref. [19]. Copyright © 2003 IEEE.

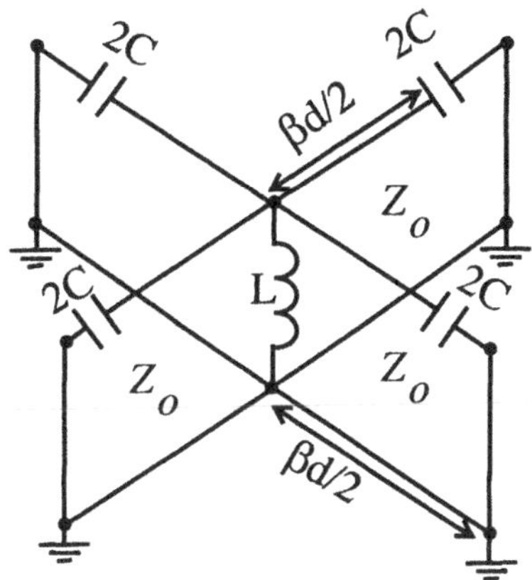

Fig. 3.16 Resonance identifying the onset of the backward-wave passband in the 2-D dual TL (ω_1). After Ref. [19]. Copyright © 2003 IEEE.

four parallel sections of $\beta d/2$ transmission lines grounded by $2C$ capacitors and the loading inductor L:

$$\frac{Z_o\left[\frac{-1}{\omega_1 2C}+Z_o\tan\left(\frac{\beta_1 d}{2}\right)\right]}{\left[Z_o+\frac{\tan\left(\frac{\beta_1 d}{2}\right)}{\omega_1 2C}\right]}=-4\omega_1 L \tag{3.64}$$

where $\beta_1 = \omega_1/v_\phi$. The left-hand side of (3.64) represents the impedance of the series branch in the 2-D dual TL: a $\beta d/2$ section of TL terminated in a $2C$

capacitor. Within this first passband of operation (between points a and b), the structure supports backward-wave (BW) propagation because the wavevector $\mathbf{k}_n = [k_{xn}, k_{zn}]$ and direction of power flow (given by the gradient to the dispersion surface) are antiparallel. A more intuitive explanation as to why this band supports backward wave propagation is provided in the next section. As the frequency is increased within the backward-wave passband, the magnitude of the wavenumbers decreases from $k_{xn} = k_{zn} = \pi/d$ at the initial resonance (point a) to $k_{xn} = k_{zn} = 0$ at point b. At point b, the 2-D dual TL enters its second stopband region which extends to point c. The frequencies ω_2 and ω_3 of the stopband edges (points b and c in Fig. 3.15) can be solved for by setting $k_{xn}d = k_{zn}d = 0$ in the dispersion relation. The stopband edge, ω_2, is found by setting the first term in (3.61) to zero:

$$\frac{1}{\omega_2(2C)} = Z_o \tan\left(\frac{\beta_2 d}{2}\right) \tag{3.65}$$

where $\beta_2 = \omega_2/v_\phi$. The frequency, ω_2, represents either point b or c depending on the relative values of L and C in the 2-D dual structure. Equation (3.65) indicates that at ω_2, the resonance shown in Fig. 3.17a occurs. The ports and central node of the unit cell act as if they are shorted to ground, thereby short-circuiting the shunt inductors. At this resonance, each shorted $\beta d/2$ transmission-line section resonates with a $2C$ capacitor. In other words, the series inductance $L_o d/2$ (where $Z_o = \sqrt{L_o/C_o}$) of each $\beta d/2$ TL section resonates with the $2C$ capacitor.

The other stopband edge, ω_3 (either point b or c), can be solved for by setting the second term in (3.61) equal to zero:

$$\frac{1}{\omega_3(4L)} = Y_o \tan\left(\frac{\beta_3 d}{2}\right) \tag{3.66}$$

where $Y_o = 1/Z_o$ is the characteristic admittance of interconnecting transmission-line sections and $\beta_3 = \omega_3/v_\phi$. This expression suggests that the resonance depicted in Fig. 3.17b occurs at ω_3. The central node and ports of the unit cell act as if they are open-circuited, thereby open-circuiting the series capacitors. Accordingly, each open-circuited $\beta d/2$ section of transmission line resonates with a $4L$ inductor. In other words, the shunt capacitance $C_o d/2$ (where $Z_o = \sqrt{L_o/C_o}$) of each $\beta d/2$ TL section resonates with a $4L$ inductor.

Beyond the second stopband, there exists another passband supporting forward-wave propagation that extends from points c to d. Its upper cutoff frequency ω_4 (labelled d in Fig. 3.15) can be found by setting the first term in (3.63) equal to zero:

$$\cos\left(\frac{\beta_4 d}{2}\right) = 0 \tag{3.67}$$

where $\beta_4 = \omega_4/v_\phi$. This occurs when the interconnecting transmission-line sections become an odd multiple of half a wavelength. It is important to note that as the electrical length of the interconnecting transmission lines vanishes, the stopband edges ω_2 and ω_3 are pushed to infinity. As a result, large bandwidths of NRI operation

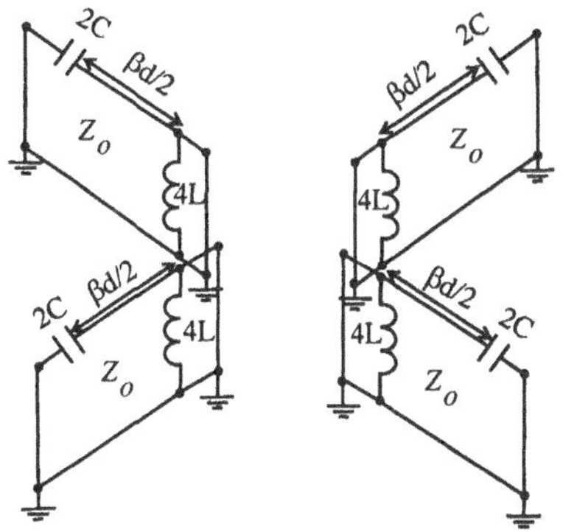

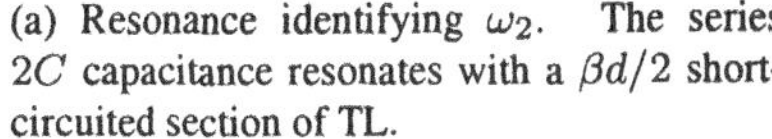
(a) Resonance identifying ω_2. The series $2C$ capacitance resonates with a $\beta d/2$ short-circuited section of TL.

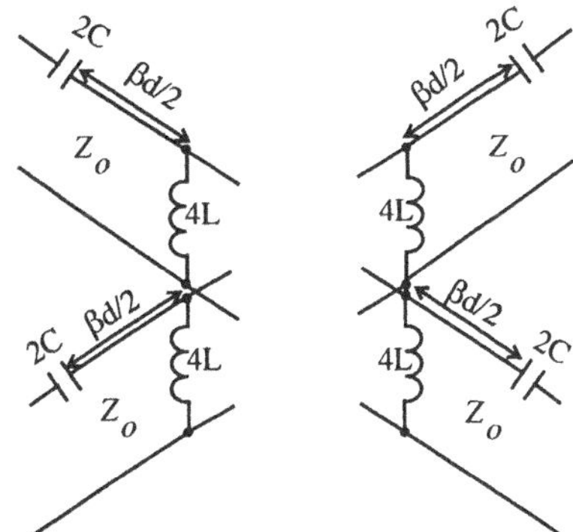

(b) Resonance identifying ω_3. The shunt $4L$ inductance resonates with a $\beta d/2$ section of open-circuited TL.

Fig. 3.17 Resonances identifying the band edges of the stopband just above the backward-wave passband. After Ref. [19]. Copyright © 2003 IEEE.

are achievable with such TL structures. This was confirmed by the experimental focusing results of Ref. 24, which reported a backward wave propagation band over an octave bandwidth.

In order to eliminate the stopband that extends from point b to c in Fig. 3.15 and allow a continuous transition between the backward-wave propagation band (between points a and b) and the forward-wave propagation band (between points c and d), one can make the resonances at ω_2 and ω_3 coincide. The required condition for closing the stopband can be derived by setting $\omega_2 = \omega_3$ and combining (3.65) and (3.66). The condition for closing the stopband simplifies to [25]

$$\frac{Z_o}{\sqrt{2}} = \sqrt{\frac{L}{C}} \tag{3.68}$$

recalling that Z_o is the characteristic impedance of the interconnecting transmission lines and L and C are the loading inductance and capacitance, respectively.

3.4.3 The 2-D Dual TL as an Effective Medium

Within a given frequency range of the backward-wave passband, the 2-D dual TL shown in Fig. 3.13 appears isotropic and homogeneous. This frequency range occurs near the top of the backward-wave passband, just below point b in Fig. 3.15. The 2-D dispersion surface of the backward-wave passband has been plotted in Fig. 3.18 with the points a and b labelled as in Fig. 3.15. This dispersion surface has been plotted for the same L–C loading elements and transmission-line parameters identified in Fig. 3.15. As can be seen in Fig. 3.18, the fixed-frequency contours of the 2-D dispersion surface are circular just below point b, indicating that the propagation is isotropic: the phase velocity does not vary with direction. The frequency range of

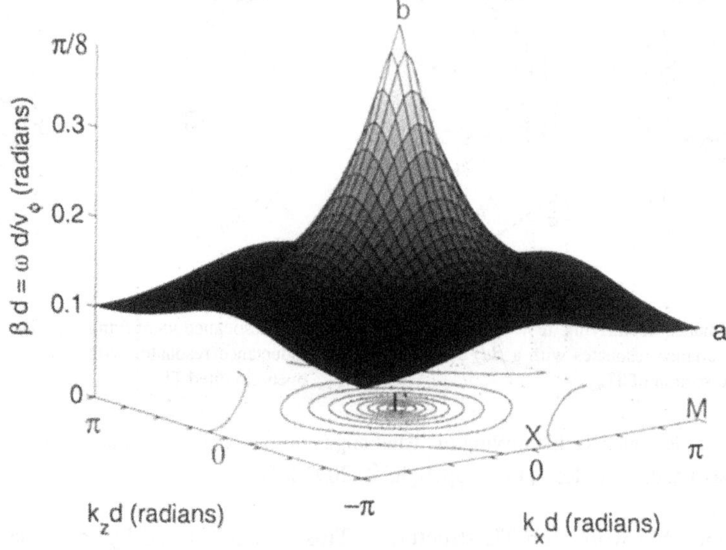

Fig. 3.18 Dispersion surface of the backward-wave (BW) passband in the 2-D dual TL.

isotropic propagation is also sufficiently far from any Bragg conditions so that the 2-D dual TL appears homogeneous. Since the 2-D dual TL is both isotropic and homogeneous, it can be considered an effective medium. As a result, effective material parameters such as permittivity and permeability can be assigned to it. An effective medium perspective provides additional insight and a simplified understanding of the 2-D dual TL's propagation characteristics.

The frequency range where the 2-D dual TL acts as an effective medium occurs when the interconnecting transmission-line sections are electrically short ($\beta d \ll 1$) and the per-unit-cell phase delay is small. The per-unit-phase delays remain small when the right-hand side of (3.61) is much less than 1. Under these two conditions, the dispersion equation (3.61) simplifies to

$$k_n^2 = k_{xn}^2 + k_{zn}^2 = \left(\beta - \frac{1}{Z_o \omega C d}\right)\left(2\beta - \frac{Z_o}{\omega L d}\right) \tag{3.69}$$

The wavenumber k_n is simply the intrinsic wavenumber of the medium; that is,

$$k_{xn} = k_n \sin\phi, \qquad k_{zn} = k_n \cos\phi \tag{3.70}$$

where ϕ is angle between $\mathbf{k_n}$ and the z-axis. This dispersion equation can also be rewritten in terms of the distributed inductance L_o and capacitance C_o of the interconnecting TL sections:

$$k_n^2 = k_{xn}^2 + k_{zn}^2 = \left(\omega L_o - \frac{1}{\omega C d}\right)\left(\omega 2 C_o - \frac{1}{\omega L d}\right) \tag{3.71}$$

where $Z_o = \sqrt{L_o/C_o}$ and $\beta = \omega\sqrt{L_o C_o}$. The dispersion equation above (3.71) closely resembles the dispersion equation (3.14) of the 1-D dual TL. The only difference is that the distributed shunt capacitance appears as $2C_o$ in the 2-D dual TL as opposed to simply C_o in the 1-D dual TL. The reason for this becomes apparent if we consider on-axis propagation in the 2-D dual TL. For propagation along the z-axis, the x-directed Bloch impedance Z_{xn} becomes infinite and the $\beta d/2$ TL sections that run in the x-direction appear as open-circuited stubs. For short TL sections ($\beta d \ll 1$), the open-circuited TL stubs appear as a shunt capacitance $C_o d$ along the z-directed TL sections. Therefore, the distributed capacitance of the 2-D TL network becomes effectively twice that of a 1-D TL. The open-circuited stubs contribute a distributed capacitance C_o and the z-directed TL section contributes C_o, thereby making the effective distributed shunt capacitance $2C_o$ [4].

For frequency bands of homogeneous and nearly isotropic propagation, the Bloch impedance expressions reduce to the following:

$$Z_{xn} = \frac{Z_n}{\cos\phi}, \quad Z_{zn} = \frac{Z_n}{\sin\phi}, \qquad \text{where } Z_n = Z_o\sqrt{\frac{\beta - \frac{1}{\omega C d Z_o}}{2\beta - \frac{Z_o}{\omega L d}}} \tag{3.72}$$

where Z_n is the intrinsic impedance of the 2-D dual TL. The Bloch impedances of (3.72) can also be rewritten in terms of L_o, C_o and ω:

$$Z_{xn} = \frac{Z_n}{\cos\phi}, \quad Z_{zn} = \frac{Z_n}{\sin\phi}, \qquad \text{where } Z_n = Z_o\sqrt{\frac{\omega L_o - \frac{1}{\omega C d}}{\omega 2C_o - \frac{1}{\omega L d}}} \tag{3.73}$$

In short, the intrinsic wavenumber k_n and intrinsic impedance Z_n can be found by considering propagation along one of the principal (x- or z-) axes, at a frequencies of isotropic and homogeneous propagation. For example, if propagation along the z-axis is used, we obtain

$$k_n = k_{zn} \quad \text{when} \quad k_x = 0, \qquad Z_n = Z_{zn} \quad \text{when} \quad k_x = 0 \tag{3.74}$$

The effective material parameters μ_n and ϵ_n of the 2-D dual TL can now be found using the intrinsic wavenumber k_n and intrinsic impedance Z_n. Having the effective material parameters allows one to make comparisons with idealized isotropic and uniform negative-refractive-index materials considered in most theoretical studies [26–43]. In addition, it allows one to directly compare the 2-D dual TL and the wire/split-ring resonator medium which is readily described in terms of effective material parameters [44–47]. Just as the material parameters filling a parallel-plate waveguide can be related to the waveguide's TL parameters for a TEM mode of operation, the effective material parameters of the 2-D dual TL can be related to k_n and Z_n. The effective permeability μ_n, as in all isotropic and homogeneous media, is given by the product of the wave impedance η_n and wavenumber k_n, divided by the angular frequency ω. Similarly, the effective permittivity ϵ_n is given by k_n divided by the product of η_n and ω:

$$\mu_n = \frac{k_n \eta_n}{\omega}, \quad \epsilon_n = \frac{k_n}{\eta_n \omega} \tag{3.75}$$

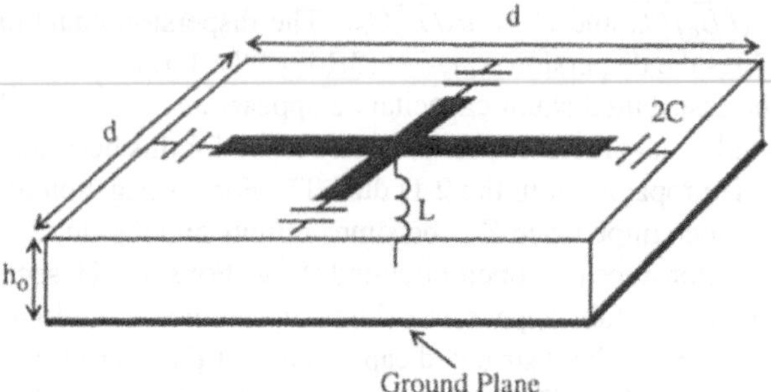

Fig. 3.19 Microstrip implementation of the 2-D dual TL. The capacitors can be implemented as packaged components or printed elements.

The wave impedance η_n can be related to the intrinsic impedance Z_n by a geometrical factor g_2, where $\eta = Z_n/g_2$. Therefore, the effective material parameters become

$$\mu_n = \frac{k_n Z_n}{g_2 \omega} \simeq \frac{Z_o \beta - \frac{1}{\omega C d}}{\omega g_2} = \frac{L_o - \frac{1}{\omega^2 C d}}{g_2} \tag{3.76}$$

$$\epsilon_n = \frac{k_n g_2}{Z_n \omega} \simeq \frac{g_2 \left[\frac{2\beta}{Z_o} - \frac{1}{\omega L d}\right]}{\omega} = \left(2C_o - \frac{1}{\omega^2 L d}\right) g_2 \tag{3.77}$$

For unit cell geometries that are rectangular boxes of dimension $d \times d \times h_o$, such as the microstrip implementation shown in Fig. 3.19, the geometrical factor $g_2 = h_o/d$. Finally, the distributed capacitance C_o and inductance L_o of a transmission line can be expressed in terms of the material parameters of the host medium and another geometrical factor g in the following manner:

$$\mu = L_o/g \tag{3.78}$$

$$\epsilon = C_o g \tag{3.79}$$

In other words, the geometrical factor g relates the characteristic impedance Z_o of the interconnecting transmission-line sections to the wave impedance of the host medium:

$$\sqrt{\frac{\mu}{\epsilon}} = \frac{Z_o}{g}$$

This leads to the following expressions for ϵ_n and μ_n in terms of the host material's ϵ and μ:

$$\mu_n = \frac{\mu g - \frac{1}{\omega^2 C d}}{g_2} \tag{3.80}$$

$$\epsilon_n = \left(\frac{2\epsilon}{g} - \frac{1}{\omega^2 L d}\right) g_2 \tag{3.81}$$

Planar unit cell geometries such as the microstrip implementation depicted in Fig. 3.19 can be approximated as a parallel-plate waveguide of thickness h_o filled with a dielectric having material parameters μ_n and ϵ_n.

Equation (3.80) indicates that the loading series capacitor provides a negative magnetic susceptibility $\chi_m = -\frac{1}{g_2\omega^2 Cd}$, since it reduces the effective permeability of the host medium. On the other hand, (3.81) indicates that the loading shunt inductance provides a negative electric susceptibility $\chi_e = -\frac{g_2}{\omega^2 Ld}$, since it reduces the host medium's permittivity. Within the backward-wave propagation band, the negative electric and negative magnetic susceptibilities dominate and the effective material parameters (ϵ_n, μ_n) become negative. Therefore, the structure acts as a NRI medium at these frequencies. The negative material parameters explain the backward-wave nature of the propagation within the first passband of operation. Equations (3.80) and (3.81) also suggest that at high frequencies the effective material parameters become positive, due to a decrease of the electric and magnetic susceptibilities. The positive material parameters give rise to a high-frequency passband supporting forward-wave propagation. This propagation band corresponds to the passband in Fig. 3.15 that extends from c to d. At frequencies between the backward-wave and forward-wave propagation bands, one of the effective material parameters (ϵ_n or μ_n) is positive while the other is negative. This causes the wavenumber k_n to be imaginary, indicating the existence of a stopband. This corresponds to the stopband stretching from point b to c in Fig. 3.15.

3.5 THE NEGATIVE-REFRACTIVE-INDEX (NRI) TL LENS

In the 1960s, Victor G. Veselago theoretically investigated the electrodynamics of materials possessing negative permittivity ϵ and negative permeability μ. Veselago showed that electromagnetic waves undergo negative refraction as they pass from a regular medium with positive material parameters (ϵ_p and μ_p), such as air, to a medium with negative material parameters (ϵ_n and μ_n). Therefore, these materials with negative ϵ and μ possess a negative refractive index (NRI). Veselago explained how negative refraction allows a flat slab of material with negative material parameters to act as an unusual lens which focuses rays of light emanating from a source to an image on the opposite side of the slab. Figure 3.20 depicts the imaging of a monochromatic source by the NRI flat lens system (slab) envisioned by Veselago. The flat lens shown has $\epsilon_n(\omega) = -\epsilon_p$ and $\mu_n(\omega) = -\mu_p$ at the frequency of operation. The rays in Fig. 3.20 denote the negative refraction of the propagating Fourier components. This simple ray diagram suggests that an internal focus exists within the slab at $z = 2d_1$ and an external focus exists beyond the second slab interface at $z = 2d_2 - 2d_1$.The evanescent Fourier components, not shown in Fig. 3.20, decay away from the source in the positive refractive index (PRI) media on either side of the lens. The PRI medium would typically be air with $\mu_p = \mu_o$ and $\epsilon_p = \epsilon_o$. In 2000, John B. Pendry discovered that evanescent waves grow inside the NRI lens such that they are restored to their original source amplitudes at the foci [26]. Therefore, the decay the evanescent waves experience on either side of the lens is compensated by the growth within the lens. On the contrary, evanescent waves decay within a conventional curved lens and are lost at the focal plane, thereby creating an imperfect image which is limited in resolution to approximately one wavelength. Therefore, it can be said that the flat NRI lens described by Veselago achieves "perfect imaging" since it focuses the source's/object's propagating waves and restores the amplitude of the evanescent waves at the focal plane. The NRI lens completely recreates the source plane at the internal and external focal planes. The "perfect imaging" predicted by Pendry only occurs when the lens and the surrounding PRI medium are lossless and two specific conditions are met. The NRI and PRI media

1. must be impedance matched

2. must have a relative refractive index of -1 with respect to each other.

These two conditions ensure that the material parameters of the lens (μ_n, ϵ_n) are the negative of those for the surrounding medium: $\mu_n{=}-\mu_p$ and $\epsilon_n = -\epsilon_p$. The impedance match eliminates any reflections at the interfaces of the NRI lens and the relative refractive index of -1 ensures there are no aberrations; all Fourier components focus to the same point. Another important constraint to "perfect" or "near perfect" imaging is that the source's/object's evanescent spectrum must reach the NRI lens. In other words, the first interface of the lens must be in the source's/object's near field, which extends to approximately $\lambda/2$ from an elementary source. Therefore, the distance d_1 should be less than $\lambda/2$ in Fig. 3.20. For distances longer than this, the evanescent waves typically fall below the noise floor of the system and are lost.

It was shown in the previous section that for a given frequency range, the 2-D dual TL acts as an isotropic and homogeneous medium with simultaneously negative values of permittivity and permeability. In this section, a transmission-line implementation of Veselago's flat lens using the 2-D dual TL will be introduced. The continuous NRI slab depicted in Fig. 3.20 will be replaced by a network of 2-D dual TL unit cells which will be referred to as the negative-refractive-index transmission-line (NRI-TL) lens, and the PRI media on either side of the lens will be replaced by a 2-D network of unloaded transmission lines. In addition, the conditions for "perfect imaging" using the NRI-TL lens will be derived. The peculiar electromagnetic phenomena associated with negative ϵ and μ materials such as negative refraction, the growth of evanescent waves and "perfect imaging" will be demonstrated using the NRI-TL lens in the subsequent sections.

3.5.1 The Transmission-Line Implementation of Veselago's NRI Lens

Veselago's flat lens depicted in Fig. 3.20 consists of a NRI lens with PRI media on either side of the lens. A practical implementation of Veselago's lens using periodic TL-based networks is depicted in Fig. 3.21. The NRI lens is implemented as a network of 2-D dual TL cells which is finite in the z-direction and infinite in the x-direction. The PRI media on either side of the lens are semi-infinite 2-D networks of unloaded TL unit cells shown in Fig. 3.22, which will be referred to as TL meshes. A TL mesh acts more or less as a parallel-plate waveguide at frequencies well below the Bragg

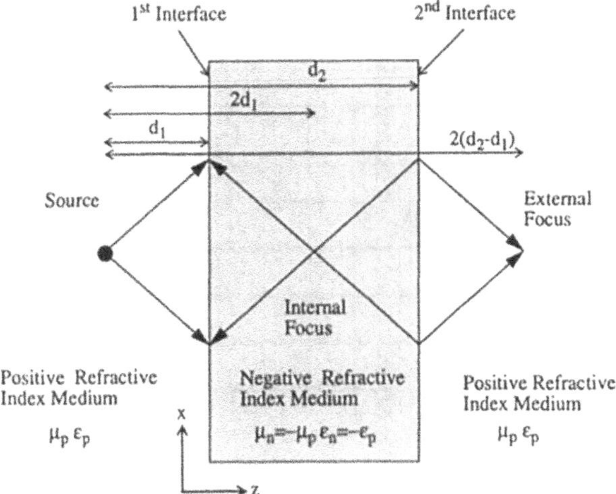

Fig. 3.20 Veselago's flat negative-refractive-index lens (the rays represent wavevectors).

condition of $\beta d = \pi$. Both the TL mesh and NRI-TL lens shown in Fig. 3.21 are operated at frequencies of homogeneous and isotropic propagation and therefore, can be justifiably called effective media. At these frequencies, the 2-D dual TL exhibits backward-wave propagation (antiparallel phase and group velocities) characteristics and the TL mesh forward-wave propagation (parallel phase and group velocities) characteristics, as would a NRI and PRI medium respectively [19]. For clarity, the 2-D space shown in Fig. 3.21 is divided into four regions: A, B, C, and D. Region A extends from $-\infty < z \leq 0$ and region B extends from $0 \leq z \leq hd$, where h is a positive integer. The $z = 0$ line will be referred to as the source plane. Region C encompasses the NRI-TL lens. It extends from $hd \leq z \leq ld$, where l is a positive integer. The boundary between regions B and C will be referred to as the first interface because it corresponds to the distance $z = d_1$ in Fig. 3.20. The TL mesh on the opposite side of the NRI-TL lens ($ld \leq z < \infty$) will be labelled region D. Likewise, the boundary between regions C and D will be referred to as the second interface since it corresponds to the distance $z = d_2$ in Fig. 3.20.

Since Region C encompasses the NRI-TL lens, the Bloch impedance expressions and wavenumbers within this region are given by the Bloch impedance expressions (3.62) and dispersion equation (3.61) for a 2-D dual TL. Regions A, B, and D are TL meshes, therefore the Bloch impedances and wavenumbers in these regions are given by the Bloch impedances and dispersion equation for a TL mesh. To study the NRI-TL lens, we must first investigate the propagation characteristics of a TL mesh.

3.5.2 Propagation Characteristics of the TL Mesh

The TL mesh and 2-D dual TL are electrical networks, therefore it is natural to describe them in terms of propagation constants and impedances rather than effective

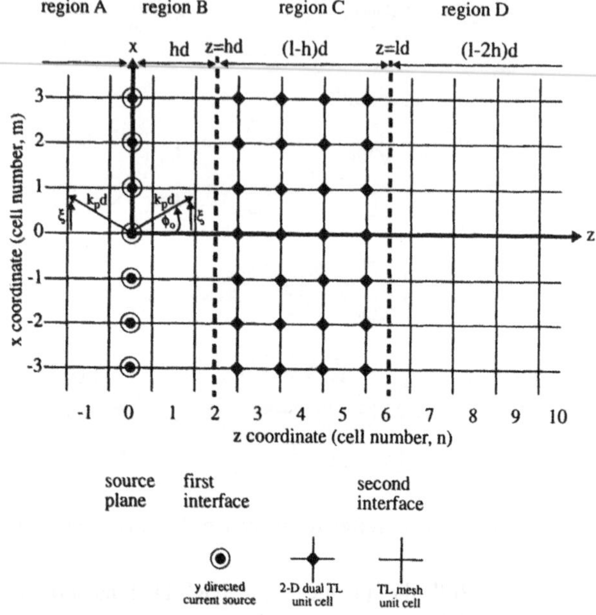

Fig. 3.21 The negative-refractive-index transmission-line (NRI-TL) lens: a practical implementation of Veselago's flat lens. The array of currents provides a Bloch wave excitation (see text). After Ref. [50]. Copyright © 2003 IEEE.

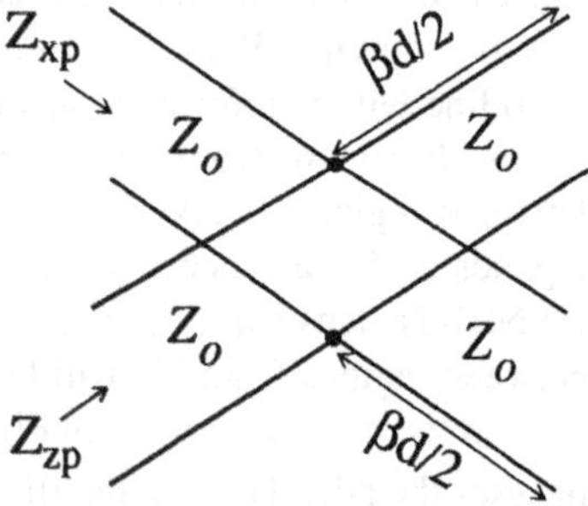

Fig. 3.22 The unit cell of the transmission-line mesh (PRI medium). The unit cell consists of two crossed transmission lines.

material parameters, which are commonly used to describe NRI media. Both TL networks can be completely characterized in terms of their dispersion equations which define the propagation constant, and Bloch impedance expressions that establish the relationship between voltage and current. The dispersion equation (3.61) and Bloch impedances Z_{xn} and Z_{zn} (3.62) have already been derived for the 2-D dual TL network. The dispersion equation for the TL mesh can be found by simply setting $C \to \infty$ and $L \to \infty$ in (3.61). The dispersion equation for TL mesh is therefore

[5, 19]

$$\sin^2\left(\frac{k_{xp}d}{2}\right) + \sin^2\left(\frac{k_{zp}d}{2}\right) = 2\sin^2\left(\frac{\beta d}{2}\right) \tag{3.82}$$

where ω is the angular frequency, k_{xp} and k_{zp} are the wavenumbers in the x- and z-directions, respectively, d is the unit cell dimension, β is the propagation constant of the interconnecting TL sections, and Z_o is their characteristic impedance. The wavenumbers k_{xp} and k_{zp} can be related to the intrinsic wavenumber of the TL mesh by the following equations:

$$k_{xp} = k_p \sin(\phi), \qquad k_{zp} = k_p \cos(\phi) \tag{3.83}$$

where ϕ is the angle between wavevector $\mathbf{k_p}$ and the z-axis. Similarly, the Bloch impedances of the TL mesh can be found by setting $L \to \infty$ and $C \to \infty$ in (3.62) [19]:

$$Z_{xp} = Z_o \frac{\tan\left(\frac{\beta d}{2}\right)}{\tan\left(\frac{k_{xp}d}{2}\right)}, \qquad Z_{zp} = Z_o \frac{\tan\left(\frac{\beta d}{2}\right)}{\tan\left(\frac{k_{zp}d}{2}\right)} \tag{3.84}$$

Under conditions of isotropic and homogenous propagation ($\beta d \ll 1$), the intrinsic wavenumber and intrinsic impedance of the TL mesh are found by considering propagation along one of the main axes:

$$k_p = k_{zp} \quad \text{when} \quad k_{xp} = 0, \qquad Z_p = Z_{zp} \quad \text{when} \quad k_{xp} = 0 \tag{3.85}$$

This leads to the following approximate expressions for the dispersion equation and Bloch impedances of the TL mesh:

$$k_p^2 = 2\beta^2 = \omega^2 L_o 2C_o \tag{3.86}$$

$$Z_{xp} = \frac{Z_p}{\cos\phi}, \qquad Z_{zp} = \frac{Z_p}{\sin\phi}, \qquad Z_p = \frac{Z_o}{\sqrt{2}} = \sqrt{\frac{L_o}{2C_o}} \tag{3.87}$$

Similarly, the effective material parameters of the TL mesh can be derived:

$$\mu_p = \frac{k_p \eta_p}{\omega} = \frac{k_p Z_p}{g_2 \omega} \simeq \frac{Z_o \beta}{\omega g_2} = \frac{L_o}{g_2} = \frac{\mu g}{g_2} \tag{3.88}$$

$$\epsilon_p = \frac{k_p}{\eta_p \omega} = \frac{k_p g_2}{Z_p \omega} \simeq \frac{2\beta g_2}{Z_o \omega} = 2C_o g_2 = \frac{2\epsilon g_2}{g} \tag{3.89}$$

where the η_p is the effective wave impedance of the TL mesh and the geometrical factors g, g_2 are as previously defined for the 2-D dual TL. For planar unit cell geometries such as the microstrip implementation depicted in Fig. 3.23 the geometrical factor $g_2 = h_o/d$ and the TL mesh can be approximated as a parallel-plate waveguide of thickness h_o filled with a dielectric having material parameters μ_p and ϵ_p, under isotropic and homogeneous conditions.

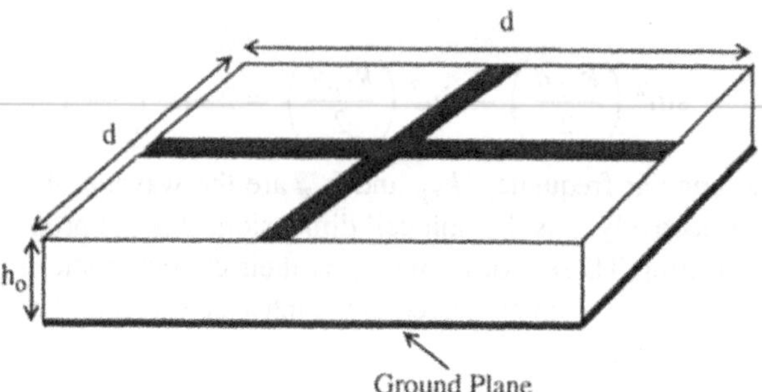

Fig. 3.23 A microstrip implementation of the TL mesh.

3.5.3 Conditions for "Perfect" Imaging in the NRI-TL Lens

Now that the Bloch impedances and dispersion equations have been defined for the TL mesh and 2-D dual TL, the conditions needed to achieve "perfect" imaging can be expressed in terms of the L,C components and the TL parameters βd, Z_o. The 2-D dual TL and TL mesh are ideal, therefore they satisfy the lossless requirement of "perfect" imaging. Both TL networks must also be operated at a frequency where they behave as effective media with homogeneous and nearly isotropic propagation characteristics. It is clear from the dispersion equations (3.61) and (3.82) that such propagation characteristics exist when $k_n d \ll 1$, $k_p d \ll 1$, and $\beta d \ll 1$. The first condition for "perfect" imaging" requires the relative refractive index between the TL mesh and 2-D dual TL to be -1. This involves making the intrinsic wavenumber k_p in the TL mesh and k_n in the 2-D dual TL (NRI-TL lens) equal in magnitude but opposite in sign: $k_p = -k_n$. This condition can be met by equating the right-hand sides of (3.61) and (3.82). The second condition for "perfect imaging" requires that there be an impedance match between the TL mesh and NRI-TL lens. This condition can be met by equating the Bloch impedance expression for the TL mesh and 2-D dual TL. This involves setting the numerator of (3.62) equal to the negative of the numerator in (3.84). Assuming that the same TL parameters (βd, Z_o) are used for the 2-D dual TL and TL mesh, the two conditions of "perfect imaging" reduce to the following two simple expressions:

$$\frac{Z_o}{\sqrt{2}} = \sqrt{\frac{L}{C}} \tag{3.90}$$

$$\tan\left(\frac{\beta d}{2}\right) = \frac{1}{4\omega C Z_o} \tag{3.91}$$

It is worth mentioning that (3.90) is identical to (3.68), which is the condition for closing the stopband between the backward-wave propagation band and the next higher forward-wave band in the 2-D dual TL network.

Table 3.1 lists the electrical parameters of a representative TL mesh and 2-D dual TL that satisfy the conditions of "perfect imaging" at a frequency of 1 GHz. The two complementary networks are impedance-matched and have a relative refractive

Table 3.1 Electrical parameters for the TL mesh and dual TL structure at 1 GHz

$k_p d = -k_n d$	$Z_p = Z_n$	Z_o	βd	L	C
0.33859 rad	90.00000 Ω	128.20565 Ω	0.23885 rad	21.25569 nH	2.58637 pF

index of -1 with respect to each other at a frequency of 1 GHz. For simplicity, the same transmission-line parameters (Z_o and βd) are utilized in both structures; as a result they only differ by the shunt L and series C components in the 2-D dual TL. Figure 3.24 shows the magnitude of the wavenumbers (k_p and k_n) for all propagation directions at 1 GHz for the TL mesh and 2-D dual TL defined in Table 3.1. The two plots are fixed-frequency contours of their respective dispersion surfaces at a frequency of 1 GHz. It is evident from the plot that the magnitude of the refractive index is the same for both structures, since the two curves overlap. The structures are also isotropic since the plots are circular, indicating that there is no spatial dispersion: the magnitudes of $\mathbf{k_p}$ and $\mathbf{k_n}$ do not change with direction of propagation ϕ. It is important to note that for any direction of power flow, the wavevectors in the two complementary networks are antiparallel. The z-directed Bloch impedances (Z_{zp}, Z_{zn}) have been plotted in Fig. 3.25a as a function of propagation direction ϕ. Due to the symmetry of the cell, the Bloch impedances are plotted only for $\phi = 0°$ to $90°$. The impedances exhibit a $1/\cos\phi$ relationship as would a perpendicularly polarized electromagnetic wave. They asymptotically approach infinity at 90° given that the z-directed current is zero for a Bloch wave propagating along the x-axis. The x-directed Bloch impedances (Z_{xp}, Z_{xn}) for both networks are shown in Fig. 3.25b. In this case, the impedances exhibit a $1/\sin\phi$ relationship. They asymptotically approach infinity for $\phi = 0$, since the x-directed current is zero for Bloch wave propagation along the z-axis. Most importantly, Figs. 3.25a and 3.25b show that the two complementary networks are matched for all angles of propagation.

Now that the intrinsic impedances Z_n and Z_p and the intrinsic wavenumbers k_n and k_p of the TL mesh and 2-D dual TL defined in Table 3.1 have been derived, we can compute their respective effective material parameters. Knowing the effective material parameters allows one to directly compare the NRI-TL lens, shown in Fig. 3.21, to Veselago's idealized lens shown in Fig. 3.20. In order to calculate the effective material parameters, we must first assume some unit cell dimensions. A planar geometry is assumed for both the 2-D dual TL and TL mesh (see Figs. 3.19 and 3.23, respectively) with unit cell dimensions $d \times d \times t = 8.4\,\text{mm} \times 8.4\,\text{mm} \times 1.524\,\text{mm}$. These dimensions are chosen since they are the unit cell dimensions of a NRI-TL lens that was designed at the University of Toronto [48]. Due to the fact that the unit cells are planar, the geometrical factor $g_2 = t/d = 5.5118$. Using (3.76) and (3.77) for 2-D dual TL and using (3.88) and (3.89) for the TL mesh defined in Table 3.1, the effective material parameters work out to be

$$\mu_n = -2.5392, \qquad \epsilon_n = -1.4575, \qquad \mu_p = 2.5392, \qquad \epsilon_p = 1.4575 \quad (3.92)$$

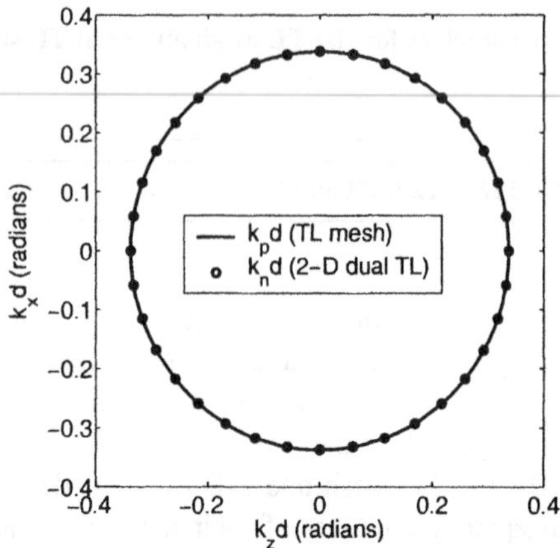

Fig. 3.24 Fixed-frequency contours of the dispersion equations for the 2-D dual TL and TL mesh defined in Table 3.1. After Ref. [50]. Copyright © 2003 IEEE.

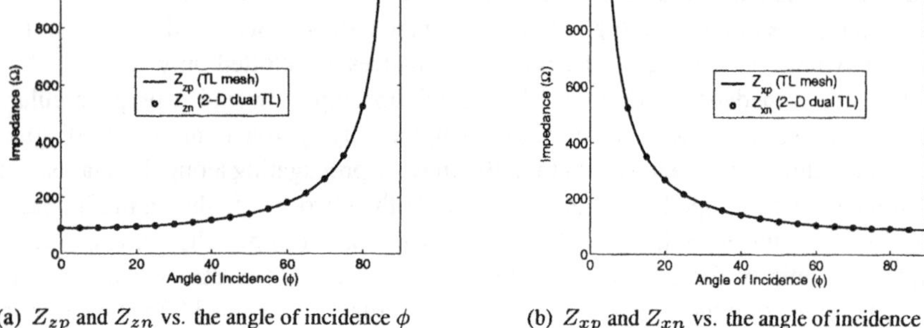

(a) Z_{zp} and Z_{zn} vs. the angle of incidence ϕ

(b) Z_{xp} and Z_{xn} vs. the angle of incidence ϕ

Fig. 3.25 Bloch impedances as a function of incident angle for the 2-D dual TL and TL mesh defined in Table 3.1. After Ref. [50]. Copyright © 2003 IEEE.

The material parameters of the 2-D dual TL are simply the negative of those in the TL mesh. Thus, both TL networks are impedance matched and their relative refractive index is -1. In effect, we have shown that the TL mesh and 2-D dual TL are suitable PRI and NRI media for realizing Pendry's "perfect" lens in the microwave regime.

3.6 REFLECTION AND TRANSMISSION THROUGH THE LOSSLESS NRI-TL LENS

Now that the TL meshes and the NRI-TL lens depicted in Fig. 3.21 have been characterized as both electrical networks and effective media, and the conditions for "perfect imaging" have been defined, we can proceed to investigate the reflection and transmission of Bloch waves through the NRI-TL lens. In this section, the voltage and current solutions are derived in regions A to D (see Fig. /refgrbic-GRBIFG22)for both propagating and evanescent Bloch wave excitations. This is done in order to demonstrate analytically negative refraction and the growth of evanescent waves within the NRI-TL lens. These results are then combined with an "array scanning" approach in the next section to analytically demonstrate "perfect imaging" using the NRI-TL lens.

A Bloch wave excitation is achieved at the source plane ($z = 0$) of the NRI-TL lens by placing an infinite array of y-directed current sources along the x-axis, as shown in Fig. 3.21. The current sources have equal amplitude I_o and possess a progressive phase shift of $-\xi$. The current source at coordinate $(x, z) = (rd, 0)$ has a phase shift of $-r\xi$, where r is an integer and d is the unit cell dimension. The infinite array of current sources actually excites two Bloch waves in the homogeneous and isotropic TL mesh. One Bloch wave in region A and one in region B. The Bloch wave in region B has a propagation direction ϕ dictated by the phase-matching condition along the source plane: $\xi = k_{xp}d$. This leads to the following expression for ϕ_o familiar from antenna array analysis:

$$\phi_o = \sin^{-1}\left(\frac{\xi}{k_p d}\right) \tag{3.93}$$

The direction of propagation of the second Bloch wave excited in region A is $\pi - \phi_o$. Propagating waves are excited by the infinite current array as long as $|\xi| < |k_p d|$. On the other hand, if the magnitude of the progressive phase shift ξ exceeds $|k_p d|$ ($|\xi| > |k_p d|$), then evanescent waves are excited by the infinite current array which decay in amplitude away from the source plane in the z-direction.

Recall that the first interface is located at $z = hd$ ($h = 2$ in Fig. 3.21) while the second interface of the lens is located at $z = ld$ ($l = 6$ in Fig. 3.21). The phase matching condition along these two interfaces further dictates that the transverse wavenumbers must all be equal: $k_x d = k_{xp} d = k_{xn} d = \xi$. Given the current array excitation, the voltage at any point $(x, z) = (md, nd)$ (where m and n are integer values) can be represented as a sum of an incident and a reflected Bloch voltage wave within each region. Matching the Bloch currents and voltages at the interfaces results in the following voltage solution:

$$\begin{aligned}
V(md, nd) &= V_1 e^{jk_{zp}nd} e^{-jm\xi} && z \le 0, \text{ (reg. A)}\\
V(md, nd) &= \left[V_2 e^{-jk_{zp}nd} + V_3 e^{jk_{zp}nd}\right] e^{-jm\xi}, && 0 \le z \le hd \text{ (reg. B)}\\
V(md, nd) &= \left[V_4 e^{-jk_{zn}(n-h)d} + V_5 e^{jk_{zn}(n-h)d}\right] e^{-jm\xi}, && hd \le z \le ld \text{ (reg. C)}\\
V(md, nd) &= V_6 e^{-jk_{zp}(n-l)d} e^{-jm\xi}, && z \ge ld \text{ (reg. D)}
\end{aligned} \tag{3.94}$$

The voltage coefficients of the incident Bloch waves (V_1, V_2, V_4, V_6) and the reflected waves (V_3, V_5) are given by the following expressions:

$$\begin{aligned}
V_1 &= V_2 + V_3 \\
V_2 &= \frac{I_o Z_{zp}}{2} \\
V_3 &= \frac{V_2 e^{-2jk_{zp}hd}\left[\Gamma_1 + \Gamma_2 e^{-2jk_{zn}(l-h)d}\right]}{\left[1 + \Gamma_1\Gamma_2 e^{-2jk_{zn}(l-h)d}\right]} \\
V_4 &= \frac{V_2 T_1 e^{-jk_{zp}hd}}{1 + \Gamma_1\Gamma_2 e^{-2jk_{zn}(l-h)d}} \\
V_5 &= \frac{V_2 \Gamma_2 T_1 e^{-jk_{zp}hd}}{\Gamma_1\Gamma_2 + e^{2jk_{zn}(l-h)d}} \\
V_6 &= \frac{V_2 T_1 T_2 e^{-jk_{zp}hd}}{e^{jk_{zn}(l-h)d} + \Gamma_1\Gamma_2 e^{-jk_{zn}(l-h)d}}
\end{aligned} \tag{3.95}$$

The expressions Γ_1 and Γ_2 are the Fresnel reflection coefficients initially seen by an incident Bloch voltage wave at the first and second interfaces respectively, while T_1 and T_2 are the corresponding Fresnel transmission coefficients. They are given by the following expressions:

$$\begin{aligned}
\Gamma_1 &= \frac{Z_{zc} - Z_{za}}{Z_{zc} + Z_{za}}, & \Gamma_2 &= \frac{Z_{zd} - Z_{zc}}{Z_{zd} + Z_{zc}} \\
T_1 &= \frac{2Z_{zc}}{Z_{za} + Z_{zc}}, & T_2 &= \frac{2Z_{zd}}{Z_{zc} + Z_{zd}}
\end{aligned} \tag{3.96}$$

The z-directed terminal currents $I_z(md, nd)$ generated by the current array excitation can also be evaluated in the four regions (A to D) by dividing the voltage expressions of (3.125) by the z-directed Bloch impedances of the corresponding region:

$$\begin{aligned}
I_z(md,nd) &= \tfrac{V_1}{Z_{zp}} e^{jk_{zp}nd} e^{-jm\xi} & z &\le 0 \text{ (reg. A)} \\
I_z(md,nd) &= \left[\tfrac{V_2}{Z_{zp}} e^{-jk_{zp}nd} - \tfrac{V_3}{Z_{zp}} e^{jk_{zp}nd}\right] e^{-jm\xi} & 0 &\le z \le hd \text{ (reg. B)} \\
I_z(md,nd) &= \Big[\tfrac{V_4}{Z_{zn}} e^{-jk_{zn}(n-h)d} \\
&\qquad - \tfrac{V_5}{Z_{zn}} e^{jk_{zn}(n-h)d}\Big] e^{-jm\xi} & hd &\le z \le ld \text{ (reg. C)} \\
I_z(md,nd) &= \tfrac{V_6}{Z_{zp}} e^{-jk_{zp}(n-l)d} e^{-jm\xi} & z &\ge ld \text{ (reg. D)}
\end{aligned} \tag{3.97}$$

Since the progressive phase shift is equal to the transverse wavenumbers ($\xi = k_x d$), the wavenumbers k_{zp} and k_{zn} can be found using dispersion equations (3.82) and (3.61), respectively:

$$k_{zp}d = 2\sin^{-1}\left\{\left[2\sin^2\left(\frac{\beta d}{2}\right) - \sin^2\left(\frac{\xi}{2}\right)\right]^{\frac{1}{2}}\right\} \tag{3.98}$$

$$k_{zn}d = 2\sin^{-1}\left(\left\{\frac{1}{2}\left[2\sin\left(\frac{\beta d}{2}\right) - \frac{1}{Z_o\omega C}\cos\left(\frac{\beta d}{2}\right)\right]\right.\right.$$
$$\left.\left.\cdot\left[2\sin\left(\frac{\beta d}{2}\right) - \frac{Z_o}{2\omega L}\cos\left(\frac{\beta d}{2}\right)\right] - \sin^2\left(\frac{\xi}{2}\right)\right\}^{\frac{1}{2}}\right) \quad (3.99)$$

The signs (+ or −) of these wavenumbers are determined by the radiation condition. The radiation condition stipulates that propagating waves carry power away from their source toward infinity and evanescent waves decay in amplitude away from their source toward infinity. For propagating waves, it is therefore assumed that the z-directed Bloch impedances are positive quantities. This restricts k_{zn} to have a negative real part and k_{zp} to have a positive real part [according to (3.62) and (3.84)] for frequencies of isotropic propagation. In other words, the TL mesh supports forward-wave propagation and the dual TL network backward-wave propagation. Evanescent waves are assumed to decay, therefore k_{zn} and k_{zp} are required to have negative imaginary parts: $-j\alpha_{zn}, -j\alpha_{zp}$. As a result, Z_{zp} has an inductive reactance (X_{zp}) and Z_{zn} a capacitive reactance (X_{zn}) for evanescent waves [see (3.62) and (3.84)]. In summary, the following conditions apply to lossless "perfect imaging" [49]:

$$k_{zp} = -k_{zn}, \qquad Z_{zn} = Z_{zp} \qquad \text{for propagating waves} \quad (3.100)$$
$$k_{zp} = k_{zn}, \qquad Z_{zn} = -Z_{zp} \qquad \text{for evanescent waves} \quad (3.101)$$

For the particular case of propagating waves incident on the "perfect" NRI-TL lens, the Fresnel coefficients become $\Gamma_1 = \Gamma_2 = 0$ and $T_1 = T_2 = 1$. This implies that there are no reflections at the two interfaces of the NRI-TL lens, causing the voltage coefficients of the reflected waves to vanish; that is, $V_3 = V_5 = 0$. The situation is quite different for evanescent waves. For evanescent waves incident on the "perfect" NRI-TL lens, the Fresnel coefficients $\Gamma_1, \Gamma_2, T_1, T_2$ become infinite and therefore $V_3 = V_4 = 0$. The fact that the voltage coefficient $V_3 = 0$ implies that the first interface of the NRI-TL lens is impedance matched to the TL mesh. On the other hand, $V_4 = 0$ suggests that the incident evanescent wave within the NRI-TL lens (which decays in the positive z-direction) vanishes. The only wave present inside the NRI-TL lens is the reflected evanescent wave with voltage coefficient V_5. This reflected evanescent wave decays in the negative z-direction. A wave that decays in the negative z-direction is in fact a wave that grows in the positive z-direction. Therefore, what we have is in fact a growing evanescent wave within the lens!

Under the conditions of "perfect imaging," the voltage solution [given by (3.125)] for both propagating and evanescent waves simplifies to the following:

$$\begin{aligned} V(md,nd) &= \frac{I_o Z_{zp}}{2} e^{jk_{zp}nd} e^{-jm\xi}, & z \leq 0 \text{ (reg. A)} \\ V(md,nd) &= \frac{I_o Z_{zp}}{2} e^{-jk_{zp}nd} e^{-jm\xi}, & 0 \leq z \leq hd \text{ (reg. B)} \\ V(md,nd) &= \frac{I_o Z_{zp}}{2} e^{jk_{zp}(n-2h)d} e^{-jm\xi}, & hd \leq z \leq ld \text{ (reg. C)} \\ V(md,nd) &= \frac{I_o Z_{zp}}{2} e^{-jk_{zp}(n-2(l-h))d} e^{-jm\xi}, & z \geq ld \text{ (reg. D)} \end{aligned} \quad (3.102)$$

where $k_{zp} = -j\alpha_{zp}$ for evanescent waves. The voltage solution for region C (3.102) shows that the source plane voltage (voltage along $z = 0$) is recovered at $z = 2hd$ ($n = 2h$). This corresponds to the location of the internal focal plane of the NRI-TL lens. Equivalently, the voltage expression for region D (3.102) indicates that the source plane voltage is again recovered at $z = 2(l-h)d$ ($n = 2l-2h$), corresponding to the location of the external focal plane of the NRI-TL lens.

3.6.1 Phase Compensation of Propagating Waves

In Fig. 3.26, the phase of the terminal voltages in all four regions (A to D) are shown for the NRI-TL lens depicted in Fig. 3.21, under a propagating Bloch wave excitation by the current array. The electrical parameters of the TL mesh and 2-D dual TL used are those defined in Table 3.1. In these calculations, the amplitude of the current sources was set to $I_o = 19.1517$ mA so that the voltage amplitude $|V_2| = 1$ V at $z = 0$. The progressive phase shift of the current sources is set to $\xi = 0.169295$ radians. According to (3.93), this progressive phase shift ξ excites a Bloch wave in Region B that is incident at an angle of $\phi = \pi/6$ radians normal to the first interface. As in Fig. 3.21, $h = 2$ and $l = 6$ are used in the computation. The voltage phase progression in Fig. 3.26 clearly shows the negative refraction of a propagating Bloch wave incident at $\pi/6$ radians. As anticipated, the phase of the incident plane wave ($z = 0$) is restored along the internal ($z = 2hd = 4d$) and external focal plane [$z = 2(l-h)d = 8d$]. Therefore, the NRI-TL lens acts as a phase compensator for propagating plane waves, much like a conventional lens [19,50].

3.6.2 Growth and Restoration of Evanescent Waves

Next, a progressive phase shift ξ exceeding $k_p d$ is considered. The transverse wavenumber $k_{xp} = \xi/d$ exceeds the intrinsic wavenumber k_p in regions A and B, forcing the array of current sources to excite an evanescent Bloch wave. In other words, the z-directed wavenumber takes the form $k_{zp} = -j\alpha_{zp}$. The voltage

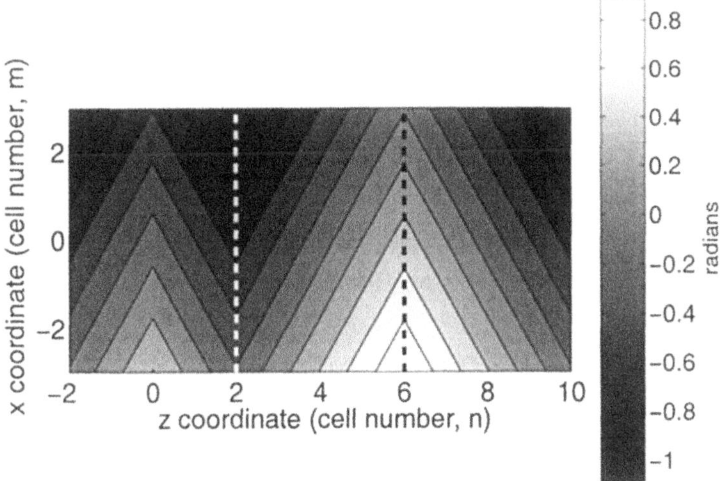

Fig. 3.26 Contour plot of the voltage phase in regions A, B, C, and D for a propagating Bloch wave excitation along the source plane ($n = 0$). The dashed lines identify the interfaces of the NRI-TL lens. After Ref. [50]. Copyright © 2003 IEEE.

expression for region C in (3.102) becomes

$$V(md, nd) = \frac{I_o Z_{za}}{2} e^{\alpha_{zp}(n-2h)d} e^{-jm\xi}, \qquad hd \leq z \leq ld \quad \text{(reg. C)} \tag{3.103}$$

Equation (3.103) indicates that there is a growing evanescent wave within the NRI-TL lens, as predicted by Pendry [26]. In the same manner, it can also be shown that the voltages in regions A, B and D are decaying waves. Fig. 3.27 shows the magnitude plot of the terminal voltages for a current array excitation with progressive phase shift $\xi = 0.5698$. The TL mesh and 2-D dual TL structure defined in Table 3.1 are used once again. According to (3.98), this ξ corresponds to $\alpha_{zp} d = 0.4463$. The current amplitude is set to $I_o = 28.5328$ mA in order to yield a voltage amplitude of $|V_2| = 1$ V at $z = 0$. As anticipated, the voltage plot indicates a growing evanescent wave within the NRI-TL lens (region C) and a decaying evanescent wave in the TL mesh regions (A, B, and D). The evanescent wave reaches a maximum at the second interface of the lens. Along the internal ($z = 2hd = 4d$) and external focus ($z = 2(l - h)d = 8d$), the amplitude of the source ($z = 0$) is restored. Unlike a conventional curved lens, the NRI-TL lens restores the amplitude of evanescent waves and truly acts as a "perfect lens" [50, 51]. The Bloch impedances for this evanescent wave excitation in the positive x- and z-directions are

$$\begin{aligned} Z_{zp} &= 70.095j \ \Omega, & Z_{xp} &= 52.289 \ \Omega \\ Z_{zn} &= -70.095j \ \Omega, & Z_{xn} &= -52.289 \ \Omega \end{aligned} \tag{3.104}$$

Within the TL meshes (regions A, B, and D), the x-directed Bloch impedance is positive and real, indicating that there is net power flow in the positive x-direction.

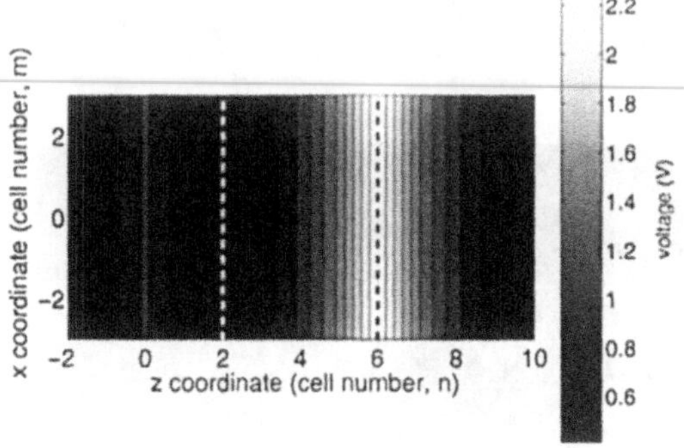

(a) Contour plot of voltage magnitudes

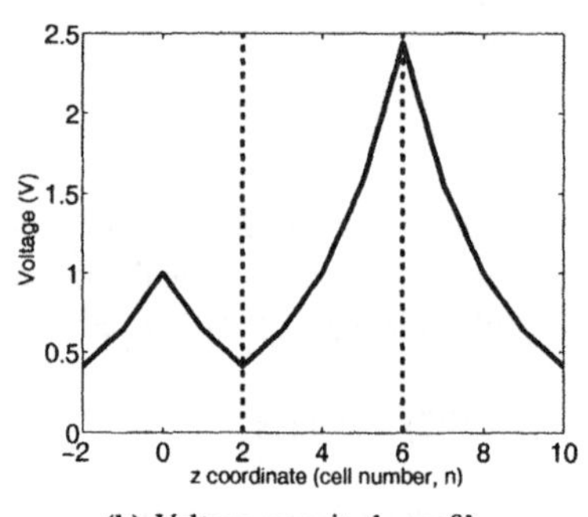

(b) Voltage magnitude profile

Fig. 3.27 Contour plot of the voltage magnitudes in regions A, B, C, and D for an evanescent Bloch wave excitation along the source plane $(n = 0)$. The dotted lines identify the interfaces of the NRI-TL lens. After Ref. [50]. Copyright © 2003 IEEE.

Conversely, the x-directed Bloch impedance is negative within the NRI-TL (region C), indicating that there is net power flow in the negative x-direction. This suggests that for growing evanescent waves, power flow circulates in the x-direction, forming an "S" pattern as one moves from region A to D. The z-directed Bloch impedances are both reactive due to the fact that the evanescent wave does not transport energy in the z-direction. The fact that the z-directed Bloch impedances are complex conjugates of each other under the conditions of "perfect imaging" reveals that the growth of evanescent waves is in fact a resonant phenomenon. The condition $Z_{zn} = -Z_{zp}$ is equivalent to the condition for the existence of a surface plasmon (s-polarized/perpendicularly polarized) at the interface between uniform NRI and PRI semi-infinite half-spaces [52]:

$$\frac{\omega\mu_p}{k_{zp}} = -\frac{\omega\mu_n}{k_{zn}} \tag{3.105}$$

This can be shown by substituting the definitions of μ_p, μ_n for the TL mesh and dual TL network into (3.105):

$$\frac{Z_p k_p}{k_{zp}} = -\frac{Z_n k_n}{k_{zn}} \tag{3.106}$$

which is equivalent to $Z_{zn} = -Z_{zp}$ for frequency bands of isotropic and homogeneous propagation in the dual TL network and TL mesh. A surface plasmon is an electromagnetic mode that decays into both media but propagates along an interface. The evanescent waves decaying from the source couple to the surface plasmon resonances at the two interfaces of the NRI-TL lens. This interaction between the surface plasmons and the decaying incident wave is in fact what produces the growing evanescent wave within the lens. For an in-depth discussion of surface plasmons the reader is referred to Refs. 29 and 52–56.

3.7 THE SUPER-RESOLVING NRI TRANSMISSION-LINE LENS

In the previous section, the incidence of propagating and evanescent Bloch waves on the NRI-TL lens was studied by deriving the voltage and current solution for a y-directed current array excitation. Here, the voltage and current solution resulting from a single y-directed current source is derived, in order to study the super resolving ability of the NRI-TL lens. A transformation known as "analytical array scanning" [57,58] is applied to the previous solutions for a current array excitation in order to find the Green's functions or voltage and current solutions due to a single current source [50]. The technique is essentially a Bloch wave expansion technique analogous to the spectral domain (plane-wave expansion) technique for continuous media [59]. Finally, the resolution limits imposed by periodicity, loss, and impedance mismatches are considered for the NRI-TL lens.

A schematic of the single current source and NRI-TL lens setup being examined is shown in Fig. 3.28. The terminal voltages in all four regions (A to D) caused by a single current source excitation at the origin can be found by integrating the voltage solutions of (3.125) from $-\pi \leq \xi \leq \pi$ and dividing the result by 2π:

$$\begin{aligned}
V'(md, nd) &= \frac{1}{2\pi}\int_{-\pi}^{\pi} V_1 e^{jk_{zp}nd} e^{-jm\xi}\, d\xi, && z \leq 0 \\
V'(md, nd) &= \frac{1}{2\pi}\int_{-\pi}^{\pi} [V_2 e^{-jk_{zp}nd} + V_3 e^{jk_{zp}nd}] e^{-jm\xi}\, d\xi, && 0 \leq z \leq hd \\
V'(md, nd) &= \frac{1}{2\pi}\int_{-\pi}^{\pi} [V_4 e^{-jk_{zn}(n-h)d} + V_5 e^{jk_{zn}(n-h)d}] e^{-jm\xi}\, d\xi, && hd \leq z \leq ld \\
V'(md, nd) &= \frac{1}{2\pi}\int_{-\pi}^{\pi} V_6 e^{-jk_{zp}(n-l)d} e^{-jm\xi}\, d\xi, && z \geq ld
\end{aligned} \tag{3.107}$$

The integration in (3.107) represents a superposition of the phased current array solutions (analyzed in the previous section) over the entire phase space: $-\pi \leq \xi \leq \pi$.

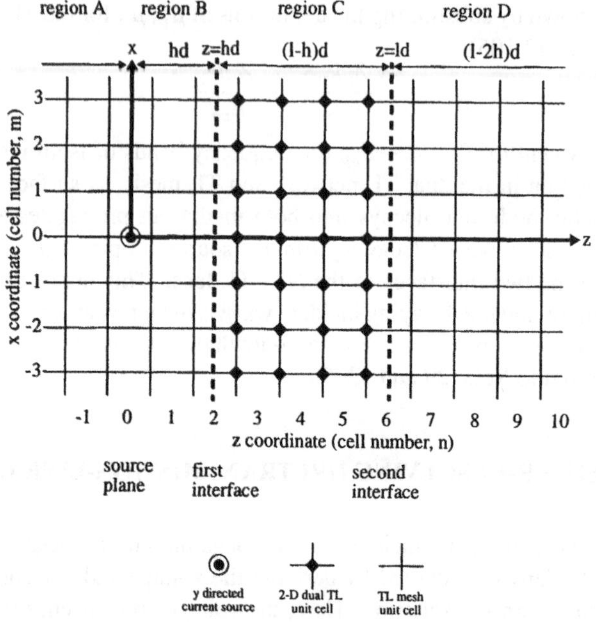

Fig. 3.28 Imaging of an elementary current source using a NRI-TL lens.

This integration cancels out all the current sources in the infinite array along the source plane $z = 0$ except for the one located at $x = 0$, which has a zero phase for each phased current array solution. The cancellation of current sources occurs due to the simple fact that [58]

$$\frac{1}{2\pi}\int_{-\pi}^{\pi} e^{jq\psi}d\psi = \begin{cases} 1, & q = 0 \\ 0, & q \neq 0 \end{cases} \tag{3.108}$$

Similarly, the z-directed terminal currents can be found by integrating the current solutions of (3.97):

$$\begin{aligned}
I_z'(md,nd) &= \tfrac{1}{2\pi}\int_{-\pi}^{\pi} \tfrac{V_1}{Zzp} e^{jk_{zp}nd} e^{-jm\xi}\,d\xi, && z \le 0 \\
I_z'(md,nd) &= \tfrac{1}{2\pi}\int_{-\pi}^{\pi} \left[\tfrac{V_2}{Z_{zp}} e^{-jk_{zp}nd} - \tfrac{V_3}{Z_{zp}} e^{jk_{zp}nd}\right] e^{-jm\xi}\,d\xi, && 0 \le z \le hd \\
I_z'(md,nd) &= \tfrac{1}{2\pi}\int_{-\pi}^{\pi} \left[\tfrac{V_4}{Z_{zn}} e^{-jk_{zn}(n-h)d}\right. \\
&\qquad \left. - \tfrac{V_5}{Z_{zn}} e^{jk_{zn}(n-h)d}\right] e^{-jm\xi}\,d\xi, && hd \le z \le ld \\
I_z'(md,nd) &= \tfrac{1}{2\pi}\int_{-\pi}^{\pi} \tfrac{V_6}{Z_{zp}} e^{-jk_{zp}(n-l)d} e^{-jm\xi}\,d\xi, && z \ge ld
\end{aligned} \tag{3.109}$$

Given the conditions of "perfect" imaging [(3.100) and (3.101)], the terminal voltages at the source plane ($z = 0$) are completely recovered at the internal ($z = 2hd$) and external focus $[z = 2(l - h)d]$. Under these conditions, the terminal voltages at all three planes are

$$V'(md, nd = 0) = \frac{I_o}{4\pi} \int_{-\pi}^{\pi} Z_{zp}(\xi) e^{-jm\xi} \, d\xi \tag{3.110}$$

Under isotropic and homogeneous propagation characteristics, the terminal voltages simplify to

$$V'(md, nd = 0) = \frac{I_o k_p Z_p d}{4\pi} \int_{-\pi/d}^{\pi/d} \frac{e^{-jk_x md}}{k_{zp}} \, dk_x \tag{3.111}$$

where $k_x d = \xi$. The expression above is a zeroth-order Hankel function of the second kind, $\frac{I_o k_p Z_p d}{4} H_o^{(2)}(k_p md)$, with a truncated spatial spectrum $-\pi/d \leq k_x \leq \pi/d$. The periodicity is what truncates the spatial spectrum. The terminal voltage magnitudes at the source and external focal plane are shown in Fig. 3.29 for the NRI-TL lens depicted in Fig. 3.28 with $(l-h)d = 4d$. These voltages are computed using (3.107) for the TL mesh and 2-D dual TL defined in Table 3.1. In the computation, the current source is set to $I_o = 52.17e^{-j1.1604}$ mA in order to yield a voltage of 1 V at the source. As shown in Fig. 3.29, the terminal voltages along the external focal plane are identical to those at the source plane.

Under the conditions of "perfect imaging," the terminal currents at the source plane ($z = 0$) are also completely recovered at the internal and external focal planes. Under these conditions, the z-directed terminal currents at all three planes are given by the following expression:

$$I'(md, nd = 0) = \frac{I_o}{4\pi} \int_{-\pi}^{\pi} e^{-jm\xi} \, d\xi = \frac{I_o \text{sinc}(m\pi)}{2} \tag{3.112}$$

Hence, there is a current null at the terminal of each unit cell ($x = md$) except for $md = 0$. The terminal current magnitudes at the source and external focal plane are shown in Fig. 3.30 for the NRI-TL lens depicted in Fig. 3.28 with $(l - h)d = 4d$. These currents are computed using (3.109) for the TL mesh and 2-D dual TL defined in Table 3.1. The current source is again set to $I_o = 52.17e^{-j1.1604}$ mA for this computation.

In addition to the source and image, the diffraction-limited voltage and current patterns are also shown in Figs 3.29 and 3.30. The two diffraction-limited patterns for voltage, V_{diff1} and V_{diff2}, are plotted in Fig. 3.29. The diffraction-limited pattern V_{diff1} is obtained by inverse Fourier transforming only the propagating Bloch voltage spectrum (Bloch waves) of the source [60]:

$$V_{diff1}(md) = \frac{I_o}{4\pi} \int_{-k_p d}^{k_p d} Z_{zp} e^{-jm\xi} \, d\xi \tag{3.113}$$

This diffraction-limited pattern assumes that the evanescent spectrum of the source is lost in the imaging process, and the image comprises only of propagating waves. This is generally the case with conventional imaging systems such as curved dielectric

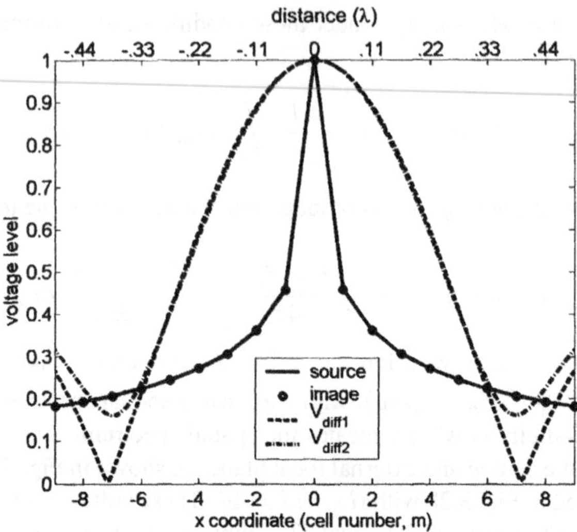

Fig. 3.29 Magnitude of the terminal voltages at the source plane ($n = 0$) and external focal plane ($n = 8$) at 1 GHz when imaging a current source using the NRI-TL lens. The normalized diffraction-limited patterns are also shown. After Ref. [50]. Copyright © 2003 IEEE.

lenses [26]. Under isotropic and homogeneous conditions, V_{diff1} reduces to the following (recall $\xi = k_x d$):

$$V_{diff1}(md) \simeq \frac{I_o Z_p k_p d}{4\pi} \int_{-k_p}^{k_p} \frac{e^{-jk_x md}}{k_{zp}} dk_x = \frac{I_o Z_p k_p d}{4} J_o(k_p md)$$
$$= \frac{I_o \omega \mu_p g_2 d}{4} J_o(k_p md) \qquad (3.114)$$

where k_{zp} is approximated by the isotropic dispersion equation:

$$k_{zp} = \sqrt{{k_p}^2 - {k_x}^2} \qquad (3.115)$$

The diffraction-limited pattern V_{diff1} is a Bessel function of the first kind, of order zero. This Bessel function has its first zeros at $k_p md = \pm 2.4048$. Therefore, the null-to-null beamwidth of V_{diff1} is $\Delta x = 4.809/k_p = 0.77\lambda$, where $\lambda = 2\pi/k_p$ is the wavelength of operation. The beamwidth corresponds to $4.809/(k_p d)$ unit cells in the TL mesh. For the NRI-TL lens defined in Table 3.1 $k_p d = 0.33859$, therefore the beamwidth of V_{diff1} shown in Fig. 3.29 is 14.2 unit cells.

Due to the close proximity of the source and image when imaging using a NRI lens, a second diffraction-limited pattern V_{diff2} has also been plotted in Fig. 3.29. This diffraction-limited pattern takes into account the attenuated evanescent waves that reach the external focal plane. It represents the best image one can hope to achieve with a conventional lens in the near field. To obtain V_{diff2}, the propagating

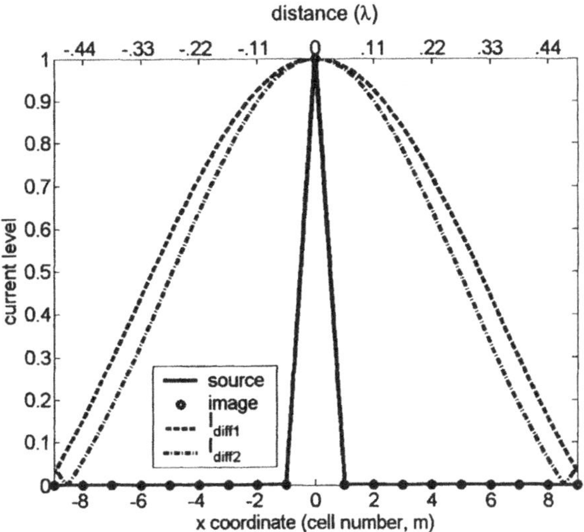

Fig. 3.30 Normalized magnitude of the terminal currents at the source plane $(n = 0)$ and external focal plane $(n = 8)$ at 1 GHz. The normalized diffraction-limited patterns are also shown.

Bloch waves emanating from the source are focused, whereas the evanescent Bloch waves are assumed to exponentially decay from the source to image with attenuation factors corresponding to that in the TL mesh:

$$\begin{aligned} V_{diff2}(md) &= \frac{I_o Z_p k_p d}{4\pi} \int_{-\pi/d}^{\pi/d} \frac{e^{-j(k_{zp}+k_{zn})D} e^{-jk_x md}}{k_{zp}} dk_x \\ &= \frac{I_o \omega \mu_p g_2 d}{4\pi} \int_{-\pi/d}^{\pi/d} \frac{e^{-j(k_{zp}+k_{zn})D} e^{-jk_x md}}{k_{zp}} dk_x \end{aligned} \tag{3.116}$$

where k_{zp} is given by the following isotropic dispersion equations:

$$\begin{aligned} k_{zp} &= -k_{zn} = \sqrt{k_p{}^2 - k_x{}^2} && \text{for } k_x < k_p \\ k_{zp} &= k_{zn} = -j\sqrt{k_x{}^2 - k_p{}^2} && \text{for } k_x > k_p \end{aligned} \tag{3.117}$$

and $D = (l - h)d$ is the thickness of the NRI-TL lens in the z-direction. Both diffraction-limited patterns, V_{diff1} and V_{diff2} are quite similar in Fig. 3.29 since the source-to-image separation for the NRI-TL lens defined in Table 3.1 is sufficiently large such that most of the evanescent wave contribution is removed from V_{diff2}. The source-image separation is $2k_pD = 8k_pd = 8(0.33859) = 2.71$ rad or equivalently 0.43λ, which is very close to the approximate limit of 0.5λ to which the near-field extends.

As with voltage, diffraction-limited patterns for current can also be defined. The two diffraction-limited patterns for current, I_{diff1} and I_{diff2}, are plotted in Fig. 3.30.

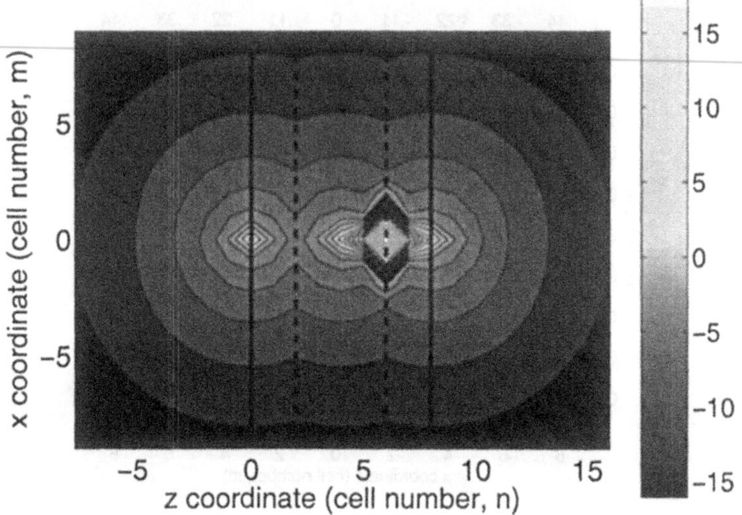

Fig. 3.31 Contour plot of the voltage magnitude when imaging a current source using the NRI-TL lens. After Ref. [50]. Copyright © 2003 IEEE.

Following the same convention, I_{diff1} is the diffraction-limited pattern that neglects the evanescent wave contribution and I_{diff2} is the diffraction-limited current pattern that takes into account the attenuated evanescent waves that reach the external focal plane:

$$I_{diff1}(md) = \frac{I_o}{4\pi} \int_{-k_p}^{k_p} e^{-jk_x md}\, dk_x = \frac{I_o k_p d}{2\pi} \text{sinc}(mk_p d) \tag{3.118}$$

where k_{zp} is approximated using (3.115);

$$I_{diff2}(m) = \frac{I_o}{4\pi} \int_{-\pi/d}^{\pi/d} e^{-j(k_{zp}+k_{zn})D} e^{-jk_x md}\, dk_x \tag{3.119}$$

where k_{zp} is given by (3.117). The diffraction-limited current pattern I_{diff1} is a sinc function which has its first zeros at $k_p md = \pm\pi$. Therefore, the null-to-null beamwidth of I_{diff1} is $\Delta x = 2\pi/(k_p) = \lambda$. This beamwidth corresponds to $2\pi/(k_p d)$ unit cells in the TL mesh. For the NRI-TL lens defined in Table 3.1 $k_p d = 0.33859$, therefore the beamwidth of V_{diff1} shown in Fig. 3.29 is 18.6 unit cells.

The current and voltage images plotted in Figs. 3.29 and 3.30, respectively, clearly show finer resolution than the diffraction-limited patterns. Therefore, the practical microwave lens depicted in Fig. 3.28 acts as a "superlens" allowing imaging beyond

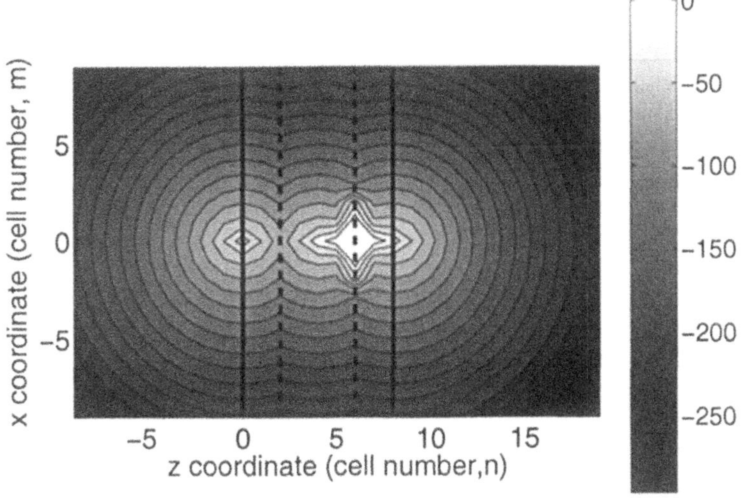

Fig. 3.32 Contour plot of the voltage phase when imaging a current source with the NRI-TL lens.

the diffraction limit and the usual restrictions of the wavelength of operation. This super-resolving ability is due to the lens' ability to restore evanescent waves.

To gain a better understanding of the "perfect imaging" taking place, the entire voltage solution (for regions A, B, C, D) is shown in Figs. 3.31 and 3.32. Figure 3.31 is a contour plot of the voltage magnitudes at the terminals of the unit cells in both TL mesh regions and the NRI-TL lens. These results were computed by numerically solving the integrals in (3.107). High-voltage amplitudes are evident near the second interface ($n = 6$) due to the growing evanescent waves in the NRI-TL lens. Figure 3.33 explicitly plots the voltage magnitude along the central row ($m = 0$). The plot reveals that the source amplitude is recovered at $[n, m] = [4, 0]$ and $[n, m] = [6, 0]$, the internal and external focus, respectively. The voltage magnitude is the highest along the central row ($m = 0$) since the evanescent waves all add in phase. In addition, the voltage magnitudes further from the central row ($m = 0$) form three distinct cylindrical waves with centers that identify the source $[n, m] = [0, 0]$, the internal focus $[n, m] = [4, 0]$ in the NRI-TL lens, and the external focus $[n, m] = [8, 0]$ (refer to Fig. 3.28). To further clarify the nature of the "perfect imaging," the phase distribution of the terminal voltages is shown in Fig. 3.32. Three cylindrical waves are again revealed by the phase plot which identify the source and the internal and external foci. Furthermore, a region of constant phase is formed along the central row at the second interface $[n, m] = [6, 0]$. The fact that the phase is constant is consistent with the notion that this region is dominated by the evanescent Bloch waves.

The magnitude and phase plots of the voltage solution show that the region to the left of the source ($n \leq 0$) is exactly recreated in the region to the right of the external

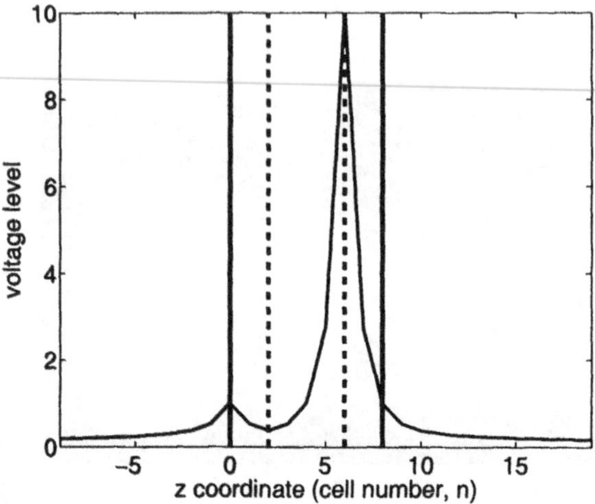

Fig. 3.33 Voltage magnitudes along the central row ($m = 0$) at 1.0 GHz when imaging an elementary current source using the NRI-TL lens. The vertical solid lines identify the source ($n = 0$) and external focal plane ($n = 8$) while the vertical dashed lines identify the interfaces of the NRI-TL lens.

focus ($n \geq 8$). The two half-spaces are completely identical. To an observer located in the region $n \geq 8$, it appears that the source is located at the external focus. The "perfect imaging" observed is peculiar in the sense that the phase center does not coincide with the location of maximum signal amplitude, as in conventional optics. The maximum signal amplitude occurs at the lens' second interface $[n, m] = [6, 0]$ while the phase center (the external focus) occurs at $[n, m] = [8, 0]$.

3.7.1 The Effect of Periodicity on Image Resolution and the Growth of Evanescent Waves

According to Fourier optics, the minimum feature Δx that can be resolved by a lens is related to the maximum transverse wavenumber $k_x = k_{xmax}$ that contributes to the image, by the Fourier transform relationship: $k_{xmax}\Delta x \sim 2\pi$. When imaging using conventional lenses, evanescent waves are lost due to the attenuation they experience from the source/object to the focal plane. Therefore, k_{xmax} is bounded by the wavenumber of the surrounding medium k_p, and the minimum resolvable feature remains on the order of a wavelength: $\Delta x \sim 2\pi/k_p = \lambda$. Lenses with numerical apertures less than 1 are further limited by the fact that not all the propagating wavenumbers k_x are captured by the lens. As was shown, the NRI-TL lens (or NRI lenses in general) supports growing evanescent waves which restore some of the object's evanescent spectrum at the focal plane [26]. This results in $k_{xmax} > k_p$ and the minimum resolvable feature $\Delta x \sim 2\pi/k_{xmax}$ becomes smaller than a wavelength ($\Delta x \leq \lambda$). Hence, the ratio $R_e = k_{xmax}/k_p$, which has been referred to as the resolution enhancement, can be used as a figure of merit for lenses [34, 61]. For example, a conventional lens cannot surpass a resolution enhancement $R_e = 1$ and a perfect lens has a $R_e \to \infty$, while R_e for a practical/imperfect NRI lens lies somewhere in between $1 < R_e < \infty$.

The integration limits of the voltage and current solutions of (3.107) and (3.109) reveal that the periodicity of the PRI (TL mesh) and NRI (2-D dual TL) media impose a resolution limit. According to the integration limits, the maximum x-directed Bloch wavenumber is $k_x = k_{xmax} = \pi/d$. This yields a resolution enhancement of

$$R_e = \frac{k_{xmax}}{k_p} = \frac{\pi}{k_p d} \tag{3.120}$$

Accordingly, the periodicity required for a given R_e is given by the following expression:

$$d = \frac{\pi}{k_p R_e} = \frac{\lambda}{2R_e} \tag{3.121}$$

The lossless NRI-TL lens depicted in Fig. 3.28 and defined by the parameters in Table 3.1 would therefore have a resolution enhancement of $R_e = \pi/.33859 = 9.3$. In other words, the image would only include Bloch waves with transverse wavenumbers in the range $0 \leq |k_{xp}| \leq 9.3k_p$. For the NRI-TL lens system shown in Fig. 3.28, k_{max} is limited right at the source. This occurs because, in addition to the NRI-TL lens being periodic, the PRI medium (TL mesh) in which the current source is embedded is periodic as well. This explains why there is no apparent loss in resolution at the focal planes given by (3.110). A "perfect image" of a spatially filtered source is obtained. Conversely, if the source were in a continuous medium, its spatial spectrum would include wavenumbers in the range $-\infty < k_x < \infty$ and the resolution would be limited solely by the periodicity of the NRI lens. In this case, "perfect imaging" could not be achieved but super resolution ($k_{xmax} > k_p$) would still be possible.

It is important to realize that k_{xmax} places a limit on the maximum α_{zn}, the "amplification" factor of the Bloch evanescent waves [see (3.99)]. This limitation on α_{zn} prevents the voltages and currents from growing to unphysically large values at the second interface of a practical lens having finite thickness $D = (l - h)d$. In fact, the periodicity provides a natural mechanism by which the amplitudes of the evanescent waves are limited. The finer the periodicity, the larger the maximum amplitude of the growing evanescent waves and the higher the resolution [50].

3.7.2 The Optical Transfer Function of the NRI-TL Lens

As with other imaging systems, linear systems theory and Fourier analysis can be applied to the NRI-TL lens. Namely, the lens can be characterized mathematically using the techniques of frequency analysis by defining an optical transfer function that maps the spatial spectrum of the source to that of the image [62]. Before we proceed, however, a few terms need to be defined. The spatial spectrum of the terminal voltages along the external focal plane $[n = 2(l - h)]$ will be labelled the Bloch voltage spectrum of the image $F_V(\xi)$. Similarly, the spatial spectrum of the terminal voltages incident along the source plane ($n = 0$) will be labelled the Bloch voltage spectrum of the source $S_V(\xi)$. The two spectra are given by the following expressions derived from (3.94):

$$S_V(\xi) = \frac{V_2}{2\pi} = \frac{I_o Z_{Zp}}{4\pi} \tag{3.122}$$

$$F_V(\xi) = \frac{V_6}{2\pi} e^{-jk_{zp}(l-2h)d} = \frac{I_o Z_{zp}}{4\pi} \cdot \frac{T_1 T_2 e^{-jk_{zp}D}}{e^{jk_{zn}D} - {\Gamma_1}^2 e^{-jk_{zn}D}} \tag{3.123}$$

The Bloch voltage spectrum of the source extends from $-\pi/(k_p d) \leq k_x/(k_p d) \leq \pi/(k_p d)$. The Bloch voltage spectrum of the source is plotted in Fig. 3.34 for the NRI-TL lens defined in Table 3.1. The current source is set to $I_o = 52.17e^{-j1.1604}$ mA so that the source voltage is 1 V. Since $k_p d = 0.33859$ radians for this particular lens, its Bloch voltage spectrum spans $-9.3 \leq k_x/k_p \leq 9.3$.

Now that the Bloch voltage spectra of the source and image have been defined, an optical transfer function (OTF) can be defined that describes the lens' spatial frequency response. The optical transfer function for the NRI-TL lens is simply the ratio of the Bloch voltage spectrum of the image $F_V(\xi)$ to the Bloch voltage spectrum of the source $S_V(\xi)$:

$$T(\xi) = \frac{F_V(\xi)}{S_V(\xi)} = \frac{T_1 T_2 e^{-jk_{zp}D}}{e^{jk_{zn}D} - {\Gamma_1}^2 e^{-jk_{zn}D}} V_2 \tag{3.124}$$

where $D = (l - h)d$ is the thickness of the NRI-TL lens in the z-direction. The terminal voltages along the external focal plane of the NRI-TL lens depicted in Fig. 3.28 can be recovered by an inverse discrete-space Fourier transform of $F_v(\xi)$:

$$V(m, n = 2l - 2h) = \int_{-\pi}^{\pi} F_V(\xi) e^{-jm\xi}\, d\xi = \int_{-\pi}^{\pi} S_V(\xi) T e^{-jm\xi}\, d\xi \tag{3.125}$$

It is a discrete-space Fourier transform because we are only dealing with the unit cell terminals; a discrete set of points. Note that under the conditions of "perfect imaging" the OTF is unity ($T = 1$), therefore $S_V(\xi) = F_V(\xi)$ and the terminal voltages along the source and external focal plane become identical.

Just as the voltage spectra of the source and image have been defined, the current spectra can also be defined. The spatial spectrum of the terminal currents along the external focal plane $I_z(n = 2l - 2h)$ will be labelled the Bloch current spectrum of the image $F_I(\xi)$, while the spatial spectrum of the incident terminal currents $I_z(n = 0)$ along the source plane, the Bloch current spectrum of the source $S_I(\xi)$. The Bloch current spectra of the source and image are given by

$$S_I(\xi) = \frac{S_V(\xi)}{Z_{zp}} = \frac{I_o}{4\pi} \tag{3.126}$$

$$F_I(\xi) = \frac{F_V(\xi)}{Z_{zp}} = \frac{I_o}{4\pi} \cdot \frac{T_1 T_2 e^{-jk_{zp}D}}{e^{jk_{zn}D} - {\Gamma_1}^2 e^{-jk_{zn}D}} \tag{3.127}$$

The Bloch current spectrum of the source is uniform and spans $-\pi/(k_p d) \leq$

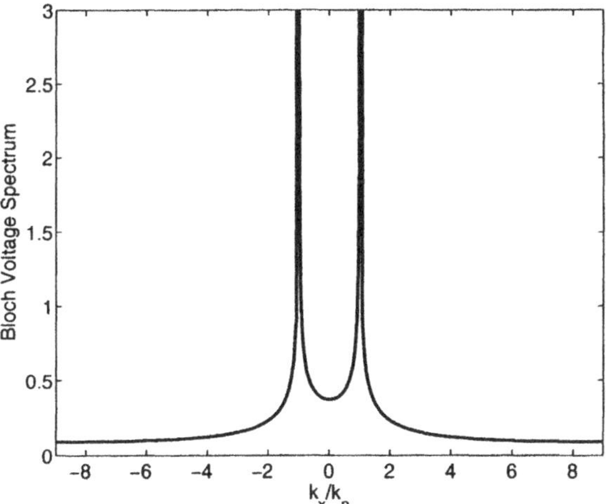

Fig. 3.34 Magnitude of the Bloch voltage spectrum of the source $S_V(\xi)$.

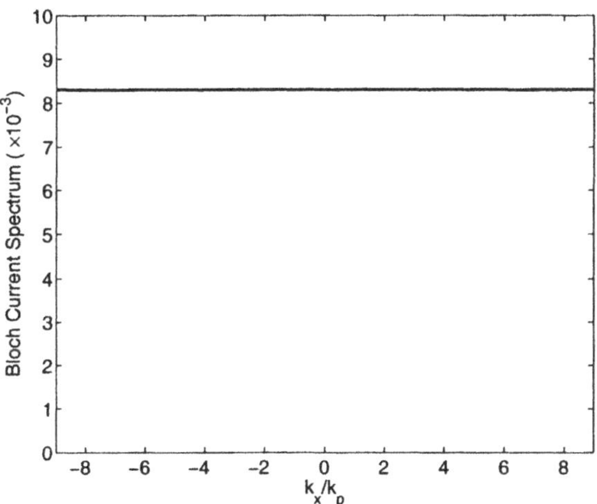

Fig. 3.35 Magnitude of the Bloch current spectrum of the source $S_I(\xi)$.

$k_x/(k_p d) \leq \pi/(k_p d)$. The current spectrum of the source for the NRI-TL lens defined in Table 3.1 is plotted in Fig. 3.35 for $I_o = 52.17e^{-j1.1604}$ mA. The ratio of $F_I(\xi)$ to $S_I(\xi)$ is simply the OTF of the NRI-TL lens. As before, the terminal currents along the external focal plane $[I_z(n = 2l - 2h)]$ can be recovered by a discrete-space inverse Fourier transform:

$$I_z(m, n = 2l - 2h) = \int_{-\pi}^{\pi} F_I(\xi) e^{-jm\xi}\, d\xi = \int_{-\pi}^{\pi} S_I(\xi) T e^{-jm\xi}\, d\xi \qquad (3.128)$$

For the particular case of "perfect imaging," $T = 1$ and the terminal currents at the external focal plane are identical to those at the source plane.

Thus far, it was assumed that the current source is placed at the terminals of a unit cell. However, if the current source is placed at the central node of a TL mesh unit cell, the Bloch voltage spectrum of the source changes slightly to the following:

$$S'_V(\xi) = \frac{I_o Z_{zp} \cos^2\left(\frac{\beta d}{2}\right)}{4\pi \cos^2\left(\frac{k_{zp}}{2}\right)} \tag{3.129}$$

The Bloch voltage spectrum of the image then becomes $F'_V(\xi) = TS'_V(\xi)$.

3.7.3 The Resolving Capability of a Lossy NRI-TL Lens

Using the concept of the optical transfer function, the limitations on image resolution imposed by impedance mismatches and losses are explored in this section [13]. The analytical formulation of imaging using the lossless NRI-TL lens presented earlier is expanded in order to study these effects. The models of the 2-D dual TL and TL mesh are extended to include losses. A conductance $2G$ is added in parallel with the series capacitor $2C$ and a resistance R is added in series with the shunt inductance L, in the 2-D dual TL unit cell depicted in Fig. 3.36. These dissipative elements account for the losses inherent to practical capacitors and inductors, and are commonly expressed in terms of quality factors, Q_L for the inductor and Q_C for the capacitor:

$$Q_L = \frac{\omega L}{R}, \qquad Q_C = \frac{\omega C}{G} \tag{3.130}$$

The loss in the interconnecting TLs can be accounted for by a complex propagation constant: $\beta = -j\alpha_{TL} + \beta_{TL}$. This loss can be expressed in terms of a TL quality factor:

$$Q_{TL} = \frac{\beta_{TL}}{2\alpha_{TL}} \tag{3.131}$$

Accounting for the lossy components simply means replacing $2C$ with $2C(1-j/Q_C)$ and L with $L(1-j/Q_L)$ in the dispersion relation (3.61) and Bloch impedance expressions (3.62) for the lossless 2-D dual TL:

$$\begin{aligned}\sin^2\left(\frac{k_{xn}d}{2}\right) + \sin^2\left(\frac{k_{zn}d}{2}\right) = \frac{1}{2}&\left[2\sin\left(\frac{\beta d}{2}\right) - \frac{1}{Z_o\omega C\,(1-j/Q_C)}\cos\left(\frac{\beta d}{2}\right)\right] \\ &\cdot\left[2\sin\left(\frac{\beta d}{2}\right) - \frac{Z_o}{2\omega L\,(1-j/Q_L)}\cos\left(\frac{\beta d}{2}\right)\right]\end{aligned} \tag{3.132}$$

$$Z_{xn} = \frac{Z_o \tan\left(\frac{\beta d}{2}\right) - \frac{1}{2\omega C(1-j/Q_C)}}{\tan\left(\frac{k_x d}{2}\right)}, \qquad Z_{zn} = \frac{Z_o \tan\left(\frac{\beta d}{2}\right) - \frac{1}{2\omega C(1-j/Q_C)}}{\tan\left(\frac{k_z d}{2}\right)} \tag{3.133}$$

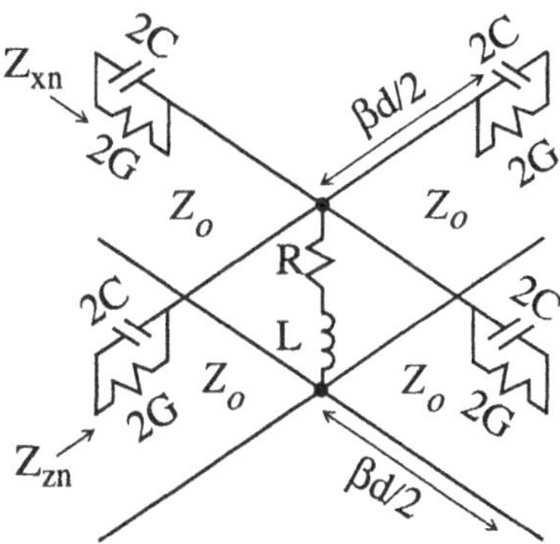

Fig. 3.36 The unit cell of the lossy 2-D dual TL. The conductance $2G$ and resistance R account for the losses inherent to the series capacitor $2C$ and shunt inductor L. The loss in the interconnecting TLs is accounted for by the complex propagation constant $\beta = -j\alpha_{TL} + \beta_{TL}$.

where $\beta = -j\alpha_{TL} + \beta_{TL} = \beta_{TL}(1 - j/(2Q_{TL}))$. The dispersion equation and Bloch impedance expressions for the TL mesh remain the same, other than the fact that the propagation constant β is now a complex number. Correspondingly, the lossy effective material parameters for the 2-D dual TL become

$$\mu_n = \frac{Z_n k_n}{g_2 \omega} \simeq \frac{Z_o\beta - \frac{1}{\omega C d(1-j/Q_c)}}{\omega g_2} = \frac{L_o - \frac{1}{\omega^2 C d(1-j/Q_c)}}{g_2} \tag{3.134}$$

$$\epsilon_n = \frac{k_n g_2}{Z_n \omega} \simeq \frac{g_2\left[\frac{2\beta}{Z_o} - \frac{1}{\omega L d(1-j/Q_L)}\right]}{\omega} = g_2\left[2C_o - \frac{1}{\omega^2 L d(1-j/Q_L)}\right] \tag{3.135}$$

Recall that to achieve "perfect imaging," the TL mesh and 2-D dual TL must be lossless, that is the quality factors must be infinite: $Q_L = Q_C = Q_{TL} \to \infty$.

Under the conditions of "perfect imaging" [see (3.100) and (3.101)], there is perfect transmission $T = 1$ for both propagating and evanescent Bloch waves and the resolution enhancement of the NRI-TL lens is limited only by the lens' periodicity given by (3.120). However, the loss and impedance mismatch of practical NRI-TL lenses further limit the resolution enhancement beyond that imposed by periodicity. The effect of losses and impedance mismatches can be studied by plotting the OTF (T vs. k_x/k_p) for variations in TL parameters ($Z_o, \beta d$), quality factors (Q_C, Q_L, Q_{TL}), and L–C component values. In general, losses and impedance mismatches remove some of the higher transverse wavenumbers k_x (evanescent Bloch waves) from the image that capture the subwavelength features of the source/object. Essentially, the image becomes a spatially lowpass filtered version of the source. As was shown in Refs. 29 and 34, the resolution enhancement of an imperfect lens can be estimated by examining the denominator of the OTF (3.124).

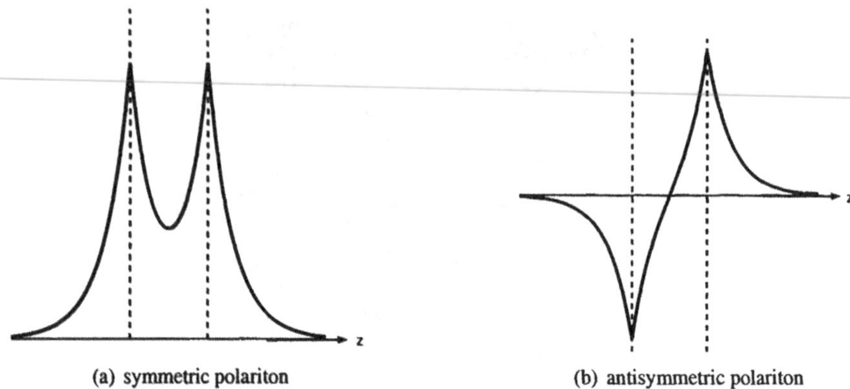

Fig. 3.37 Electric field profile of the polaritons of the NRI-TL lens. The vertical dashed lines identify the interfaces of the NRI-TL lens.

First, let us examine a lossless impedance mismatch between the NRI-TL lens and the TL meshes on either side. A lossless impedance mismatch refers to the situation when the Fresnel reflection coefficient Γ_1 is a real number. At the design frequency, this type of mismatch introduces a pole in the OTF (3.124) since the Γ_1^2 term in the denominator of the OTF is a positive real number. For small Bloch impedance mismatches (good NRI-TL lens designs), this pole occurs for an evanescent Bloch wave having a large transverse wavenumber $k_x > k_p$ and therefore a decaying longitudinal wavenumber $k_{zn} = -j\alpha_{zn}$. Setting the denominator of (3.124) equal to zero yields

$$\Gamma_1{}^2 = e^{2\alpha_{zn} D} \tag{3.136}$$

where $k_{zn} = -j\alpha_{zn}$ is related to k_x by dispersion equation (3.61). This pole disproportionately augments the evanescent Bloch waves in its vicinity and as a result distorts the image. Additionally, as k_x increases beyond the vicinity of the pole, the first term in the denominator of the OTF (3.124) begins to dominate and T exponentially drops in magnitude. Therefore, the transverse wavenumber (k_x) identifying the pole can be used as a conservative estimate of k_{xmax}, the maximum transverse wavenumber contributing to the image. However, the losses inherent to practical lenses dampen this pole resonance in T. Pole resonances such as this represent the excitation of polaritons of the NRI-TL lens [29,53,55,56]. Equation (3.136) is the transverse resonance condition for guided modes that have an evanescent profile (polaritons), that is imaginary z-directed wavenumbers (α_{zn}, α_{zp}) both inside and outside the NRI-TL lens as depicted in Fig. 3.37. The surface plasmons at the two interfaces of the NRI-TL lens [given by (3.106)] couple and give rise to these polaritons. The incident evanescent Bloch wave with transverse wavenumber $k_x = k_{xmax}$ pumps the guided polariton and eventually drives the fields of the mode to infinity under lossless conditions, effectively drowning out the source and its image. By taking the square roots of (3.136), the dispersion relation for the symmetric and antisymmetric polaritons of the NRI-TL lens shown in Fig. 3.37 can be derived:

$$-Z_{zn} \coth\left(\frac{\alpha_{zn} D}{2}\right) = Z_{zp} \tag{3.137}$$

$$-Z_{zn} \tanh\left(\frac{\alpha_{zn} D}{2}\right) = Z_{zp} \tag{3.138}$$

where $D = (l - h)d$ is the thickness of the NRI-TL lens. The dispersion relations of the symmetric and antisymmetric polaritons are plotted in Fig. 3.38 for the NRI-TL lens shown in Fig. 3.28 with parameters defined in Table 3.1. This NRI-TL lens design satisfies the conditions of "perfect imaging"; therefore neither polariton is explicitly excited at the design frequency of 1 GHz. The dispersion curves asymptotically approach 1 GHz as the transverse wavenumber k_x increases. If the lens was mismatched, however, a polariton dispersion relation would have crossed the 1-GHz horizontal line (see Fig. 3.38) at a finite k_x, identifying the wavenumber k_{xmax}. By expressing Z_{zn} and Z_{zp} in the above dispersion equations (3.137) and (3.138) in terms of μ_n and μ_p, the dispersion relations become the same as those for polaritons of a uniform NRI slab for perpendicular / s-polarization [29,53,55,56,63–67]:

$$-\frac{\mu_n}{\alpha_{zn}} \coth\left(\frac{\alpha_{zn} D}{2}\right) = \frac{\mu_p}{\alpha_{zp}} \tag{3.139}$$

$$-\frac{\mu_n}{\alpha_{zn}} \tanh\left(\frac{\alpha_{zn} D}{2}\right) = \frac{\mu_p}{\alpha_{zp}} \tag{3.140}$$

As the electrical thickness of the lens is increased, the polariton cutoff wavenumber (the pole in the OTF) moves to a lower attenuation constant (α_{zp}) corresponding to a lower k_{xmax}. This results in decreasing resolution enhancement $R_e = k_{xmax}/k_p$ with increasing lens thickness ($D = l - h$), for a fixed impedance mismatch defined by Γ_1.

The effect of losses on the OTF is now considered in detail. In this case, Γ_1 is primarily due to the added losses in the NRI-TL lens so that Γ_1 is an imaginary number. This implies that ${\Gamma_1}^2$ is a negative real number, therefore this mismatch due to loss does not introduce a pole in the OTF (3.124). Nevertheless, it still attenuates the higher transverse wavenumbers. Equating the two terms in the denominator of the OTF (3.124) provides a conservative estimate of the transverse wavenumber k_{xmax} at which T exponentially drops in magnitude:

$$-{\Gamma_1}^2 = e^{2\alpha_{zn} D} \tag{3.141}$$

This transverse wavenumber is approximately the -6-dB point in the OTF's low pass response when T is plotted versus k_x. As before, it can be used as an estimate for k_{xmax}. A general formula for estimating k_{xmax}, whether the resolution enhancement R_e is limited by loss or by a lossless impedance mismatch, is

$$|\Gamma_1| = e^{\alpha_{zn} D} \tag{3.142}$$

Assuming the NRI-TL lens is well-designed (Γ_1 is small), the situation given by (3.142) occurs for a large transverse wavenumber ($k_x \gg k_n$) such that $|k_x| \sim |\alpha_{zn}|$.

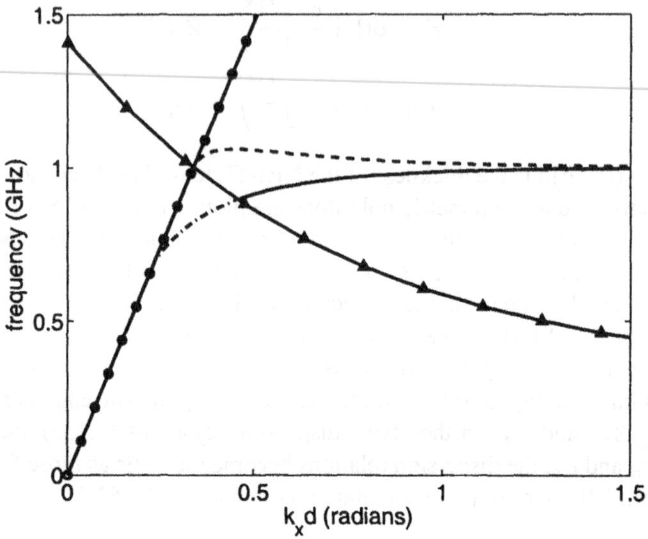

Fig. 3.38 Dispersion relations for the polaritons of a NRI-TL lens. The dispersion relation for the symmetric polariton is shown with a dotted line and the antisymmetric polariton with a solid line. The triangles identify $k_x = k_n$ (the intrinsic wavenumber in the NRI-TL lens) and the circles identify $k_x = k_p$ (the intrinsic wavenumber in the TL mesh). The dispersion relation for a bulk mode (k_{zn} is real and k_{zp} is imaginary) of the NRI-TL lens is shown with the dash-dot line.

The resolution enhancement R_e of the NRI-TL lens then simplifies to

$$R_e = \frac{k_{xmax}}{k_p} \simeq \frac{\alpha_{zn}}{k_p} = \frac{\ln|\Gamma_1|}{k_p D} \simeq \frac{1}{k_p D} \ln\left|\frac{k_n Z_n - k_p Z_p}{k_n Z_n + k_p Z_p}\right| \tag{3.143}$$

Substituting in the effective permittivities and permeabilities of the TL mesh and 2-D dual TL into the above equation, the expression for the R_e of a uniform NRI lens (perpendicular polarization) is recovered [34]:

$$R_e \simeq -\frac{1}{2\pi} \ln|\delta\mu| \frac{\lambda}{D} \tag{3.144}$$

where $\delta\mu = (\mu_n + \mu_p)/(\mu_p - \mu_n)$ and $\lambda = 2\pi/k_p$ is the wavelength of radiation.

For small arguments ($\beta d \ll 1$, $k_p d \ll 1$, $k_n d \ll 1$), namely conditions under which the dimensions considered are much smaller than a wavelength, the resolution enhancement (3.143) further reduces to

$$\begin{aligned} R_e = \frac{k_{xmax}}{k_p} &\simeq \frac{1}{k_p D} \ln\left|\frac{1}{-4Z_o \tan(\beta d/2)\omega C(1 - j/Q_C) + 1}\right| \\ &\simeq \frac{1}{k_p D} \ln\left|\frac{1}{-2Z_o \beta d \omega C(1 - j/Q_C) + 1}\right| \end{aligned} \tag{3.145}$$

Both (3.144) and (3.145) show that the approximation $|k_x| \sim |\alpha_{zn}|$ is in fact a magnetostatic approximation [29]. It neglects the smaller effect of mismatches caused by the shunt inductor L, which represents the permittivity of the NRI-TL lens. Only the larger effect of mismatches due to C, which represents the permeability of the NRI-TL lens, is considered.

The last four equations—(3.142) to (3.145)—apply to both lossless impedance mismatches and mismatches due to loss. For the latter, the TL mesh and dual TL network are assumed to be mismatched only due to component losses (Q_L, Q_C, Q_{TL}), such that (3.90) and (3.91) are still satisfied for $\beta = \beta_{TL}$. Substituting (3.91) $[2Z_o\beta_{TL}d \simeq 1/(\omega C)]$ into (3.145) reduces R_e to the following simple expression for the loss-limited case:

$$R_e \simeq \frac{-\ln[1/Q_C + 1/(2Q_{TL})]}{k_p D} \tag{3.146}$$

In this magnetostatic limit (near field for perpendicular polarization), the expression $Q = 1/Q_C + 1/(2Q_{TL})$ can be thought of as the resolution quality factor of this fascinating resonator—the NRI-TL lens. Typically, the capacitor quality factor Q_C is smaller than $2Q_{TL}$, therefore the expression for the loss-limited resolution enhancement can be further approximated as

$$R_e \simeq \frac{\ln Q_C}{k_p D} \tag{3.147}$$

The resolution enhancement R_e of the NRI-TL lens is approximately equal to the natural logarithm of the capacitor quality factor Q_C divided by the electrical thickness of the lens in radians. The logarithmic dependence on the quality factor places severe restrictions on losses. For instance, in order to double the resolution enhancement R_e of a NRI-TL lens with a given electrical thickness, one must square its resolution quality factor. In addition, the inverse relationship with $k_p d$ indicates that electrically thin lenses perform better than thicker ones. At a certain lens thickness, the losses inherent to a NRI-TL lens overcome the growth the evanescent wave experiences and the corresponding contribution to the image is lost. Losses also prevent the amplitude of an evanescent wave from diverging as the lens thickness $D = (l - h)d$ approaches infinity, as is the case in lossless NRI-TL lenses and lossless uniform NRI lenses. In terms of the uniform left-handed lens, Q_C represents the imaginary part of the lens' permeability μ_n [see (3.76)] which gives rise to an imaginary Fresnel reflection coefficient Γ_1 that limits resolution. The effect of Q_L, which represents the imaginary part of the lens' permittivity, on image resolution is much smaller than that associated with Q_C and therefore it is neglected [34]. This same dependence of resolution enhancement R_e on deviations in permeability has been observed for a uniform left-handed lens for S-polarized/perpendicularly polarized waves in Refs. 29, 34, and 39.

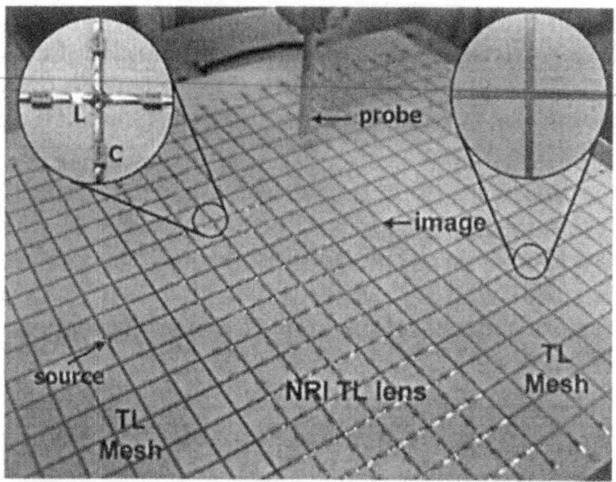

Fig. 3.39 A photograph of an experimental NRI transmission-line lens designed at the University of Toronto. The 2-D dual TL and TL mesh unit cells are shown in the left and right inset, respectively. Reprinted figure with permission from Ref. [48]. Copyright © 2004 by the American Physical Society.

3.8 AN EXPERIMENTAL NRI-TL LENS

A photograph of an experimental NRI-TL lens that was designed at the University of Toronto is shown in Fig. 3.39 [48]. A schematic of this same lens is shown in Fig. 3.40 for clarity. The experimental set up consists of a lens made of a 2-D dual TL network sandwiched between two TL meshes. The interfaces of the experimental lens are located at $h = 2.5$ and $l = 7.5$. The source is attached to the central node of a TL mesh unit cell; therefore its voltage spectrum is defined by (3.129). It is located 2.5 cells away from the first interface of the lens at the location $[n, m] = [0, 0]$, and the image is located at $[n, m] = [10, 0]$ (2.5 cells from the second interface of the lens). The experimental NRI-TL lens shown in Fig. 3.39 extends 5 cells in the z-direction $[D = (l - h)d = 5d]$ and 19 cells in the x-direction ($-9 \leq m \leq 9$). The TL meshes on either side of the lens extend 12 cells in the z-direction and 19 cells in the x-direction. Therefore, the overall experimental setup depicted in Fig. 3.39 is 29×12 cells.

Microstrip implementations of the TL mesh and 2-D dual TL were chosen for the experimental NRI-TL lens, as shown in Fig. 3.39. The interconnecting TLs are microstrip lines with metal thickness of 17 microns and width of 750 microns (see insets of Fig. 3.39). These microstrip lines were etched on a 60 mil ($t = 1.52$ mm) thick grounded microwave substrate with a dielectric constant $\epsilon_r = 3.00$ and a loss tangent $\tan\delta = 0.0013$. The unit cells of the TL mesh and 2-D dual TL, which are $d \times d = 8.4 \times 8.4$ mm in dimension, are identical other than the L, and C loading elements. Off-the-shelf packaged capacitors and inductors were used for L and C loading elements. The quality factors of the inductors and capacitors that were used in

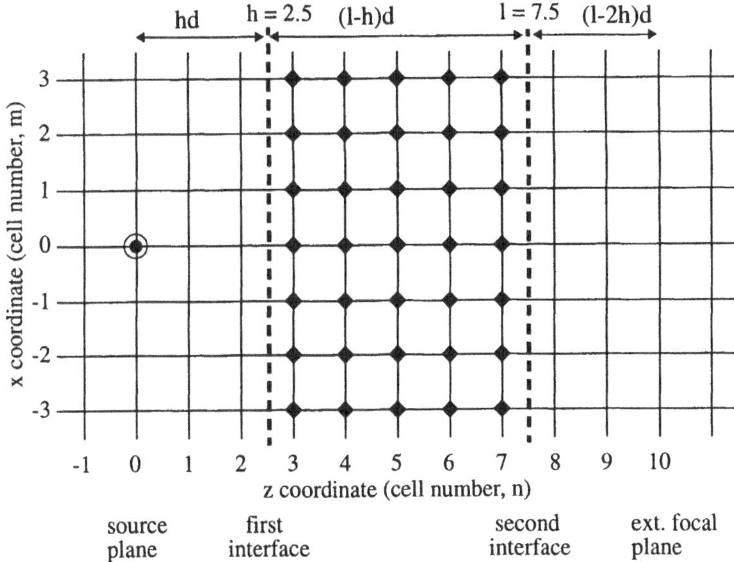

Fig. 3.40 A schematic of the experimental NRI transmission-line lens shown in Fig. 3.39. After Ref. [68]. Copyright © 2005 IEEE.

the experimental NRI-TL lens are $Q_L = 44$ and $Q_C = 150$ at 1.0 GHz, respectively. In addition, the quality factor of the interconnecting microstrip transmission lines was estimated to be $Q_{TL} = 221$ at 1 GHz, using a commercial transmission-line calculator.

A few differences exist between the theoretical set up depicted in Fig. 3.40 and the experimental one shown in Fig. 3.39. In the theoretical calculations, a y-directed current source is the elemental source being imaged. In the experiment, the elemental source is realized by a y-directed monopole that is fed by a coaxial cable through the TL mesh ground plane. The monopole attaches the center conductor of the coaxial cable to the central node of a TL mesh unit cell, while the outer conductor of the coaxial cable attaches to the TL mesh ground plane. Instead of measuring the node voltages in the entire structure (as in the theoretical calculations presented earlier), the vertical electric field is detected above the surface of the entire experimental structure. The electric field is detected through proximity coupling using a coaxial probe that is scanned approximately 0.8 mm above the surface of the entire structure. A Hewlett-Packard Vector Network Analyzer model 8753D is connected to measure the transmission coefficient between the monopole and the probe. This normalized transmission coefficient is proportional to the experimental node voltages. These experimental node voltages can be directly compared to the theoretically predicted ones in order to analyze image resolution and lens performance. Another difference between the theoretical and experimental setup is that the theory assumes that the NRI-TL lens and the TL meshes on either side are infinite in the transverse x-direction while the experimental lens is finite (19 cells). This is a valid approximation since

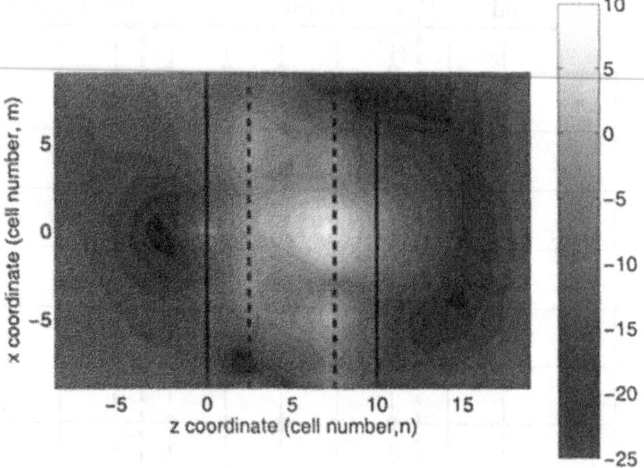

Fig. 3.41 The magnitude of the measured vertical electric field detected 0.8 mm above the surface of the entire experimental structure at 1.057 GHz. The plot has been normalized with respect to the source amplitude (dB scale). Strong fields are evident at the second interface $(n = 7.5)$ of the NRI-TL lens.

the source is placed quite close to the experimental lens and the numerical aperture of the NRI-TL lens is $NA = 0.96$. The edges of the experimental setup was terminated in matching impedances. These matching impedances can be found by taking the ratio of voltage to current at the node of interest [see (3.107) and (3.109)].

The best focusing results were observed at 1.057 GHz for this experimental NRI-TL lens, a frequency slightly higher than the design frequency of 1.00 GHz. The frequency offset was primarily due to the variation in chip inductors and capacitors from their nominal values, as well as fabrication tolerances in printing the grid lines. The magnitude and phase of the measured vertical electric field above each unit cell in the entire experimental structure is shown in Figs. 3.41 and 3.42, respectively, at a frequency of 1.057 GHz. As shown, the enhancement of evanescent waves is quite evident along the central row $(m = 0)$, near the second interface of the NRI-TL lens. The high voltage magnitude and lack of phase progression indicate that evanescent Bloch waves dominate this region. Figure 3.43 explicitly plots the magnitude of the measured electric field along the central row $(m = 0)$ to highlight the growing evanescent fields within the NRI-TL lens.

The measured vertical electric field along the external focal plane $(n = 10)$ is shown in Fig. 3.44. Plotted in the same figure is the measured vertical electric field along the source plane $(n = 0)$, as well as the two diffraction-limited patterns. Both of the diffraction-limited patterns, one that takes into account the attenuated evanescent waves at the external focal plane (V_{diff2}) and the other that neglects them completely (V_{diff1}), are quite similar. This suggests that the distance between source and image (0.54λ) is sufficiently long such that evanescent waves passing

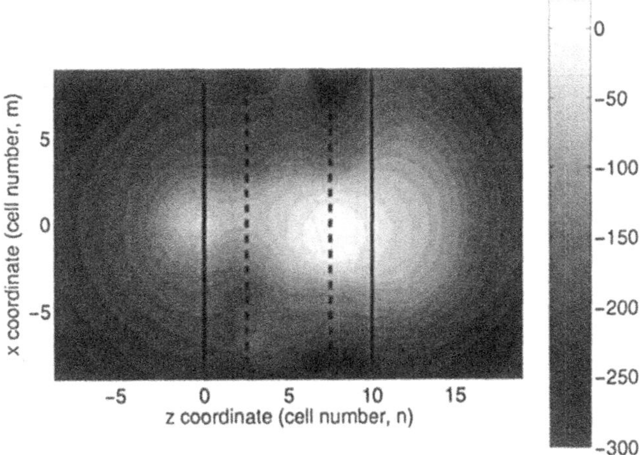

Fig. 3.42 The phase of the measured vertical electric field detected 0.8 mm above the surface of the entire experimental structure at 1.057 GHz. The phase has been normalized with respect to the source (degrees). A lack of phase progression is evident near the second interface of the NRI-TL lens where the evanescent waves dominate.

through a conventional lens would not reach the external focal plane. Therefore, any enhancement in image resolution can be attributed to the experimental NRI-TL lens' ability to restore evanescent waves at the external focal plane.

The measured electric field (image) above the external focal plane is evidently narrower than the theoretical diffraction-limited patterns. The measured half-power beamwidth is 0.21 λ compared to 0.36 λ for the diffraction-limited patterns. This narrowing of the beamwidth beyond the diffraction limit is due to the growing evanescent waves evident in Figs. 3.41 and 3.43. Nevertheless, the image is still imperfect since the source beamwidth is narrower than that of the image. This is not surprising since we have shown earlier that slight mismatches at the lens interfaces significantly degrade the resolution enhancement R_e. More will be said about this in the next section.

3.9 CHARACTERIZATION OF AN EXPERIMENTAL NRI-TL LENS

Both the TL mesh and 2-D dual TL unit cells used in the experimental NRI-TL lens were simulated using a commercial Finite Element Method (FEM) electromagnetic solver. In the FEM simulations, ideal capacitances and inductances were used to model the packaged components. The ideal capacitance and inductance values were extracted at 1 GHz from the S-parameter files provided by the respective manufacturers. Using the FEM solver, the intrinsic Bloch impedances and propagation constants of the TL mesh and 2-D dual TL were matched at a frequency of 1 GHz:

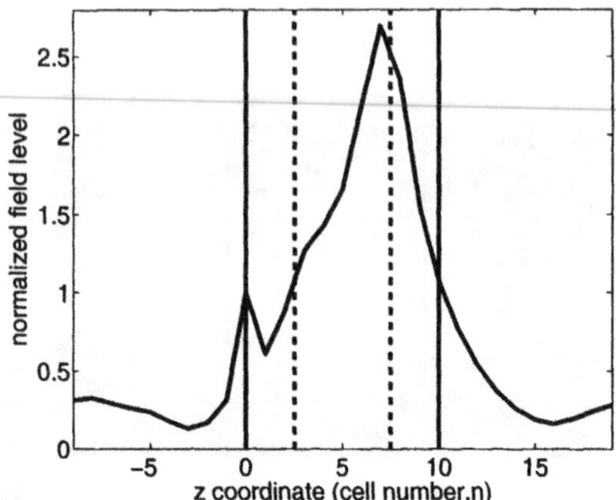

Fig. 3.43 The measured vertical electric field above the central row ($m = 0$) at 1.057 GHz (linear scale). The vertical solid lines identify the source ($n = 0$) and external focal ($n = 10$) plane while the vertical dashed lines identify the interfaces of the NRI-TL lens. The growth of the evanescent waves within the NRI-TL lens is clear. Reprinted figure with permission from Ref. [48]. Copyright © 2004 by the American Physical Society.

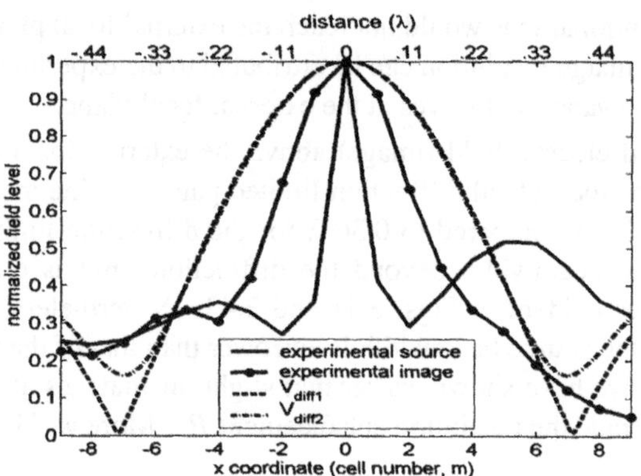

Fig. 3.44 The normalized measured vertical electric field at the source ($n = 0$) and the external focal plane ($n = 10$) at 1.057 GHz along with the theoretical diffraction-limited images (linear scale). Reprinted figure with permission from Ref. [48]. Copyright © 2004 by the American Physical Society.

$Z_n = Z_p = 90\ \Omega$, $k_p d = -k_n d = 0.33859$ radians. Since the same TL parameters (β, Z_o) were used in the TL mesh and dual TL network, the parameters β, Z_o, L, C can be extracted from Z_n, Z_p, k_p, k_n by employing (3.61), (3.82), (3.62), and (3.84). These extracted values are those listed in Table 3.1. They are based on the intrinsic Bloch impedances (Z_n, Z_p) and wavenumbers (k_n, k_p) obtained through full-wave

Table 3.2 Intrinsic wavenumbers and impedances

$k_n d$	Z_n	$k_p d$	Z_p
$-0.338 - 0.011j$	$89.986 - 1.531j\Omega$	$0.339 - 0.001j$	$90.000 + 0.003j$

analysis (with periodic boundary conditions) and therefore take into account mutual coupling interactions within and between the unit cells as well as parasitics introduced by the junctions in the microstrip TLs.

The quality factors ($Q_L = 44$, $Q_C = 150$, $Q_{TL} = 221$ at 1 GHz) and the parameters of the experimental TL lens listed in Table 3.1 can be utilized to compute the lossy Z_n, Z_p, k_p, k_n parameters using (3.82), (3.132), (3.84), and (3.133). The lossy parameters are listed in Table 3.2. From these lossy parameters, the effective material parameters of the experimental lens can be estimated using (3.88), (3.89), (3.134), and (3.135):

$$\mu_n = -2.539 - 0.038j, \qquad \epsilon_n = -1.456 - 0.071j \tag{3.148}$$

$$\mu_p = 2.539 - 0.006j, \qquad \epsilon_p = 1.458 - 0.003j \tag{3.149}$$

If we use the effective material parameters, the refractive indexes n_n and n_p of the NRI-TL lens and TL mesh, respectively, can be found:

$$n_n = k_n/k_o - 1.924 - 0.062i, \qquad n_p = k_p/k_o = 1.925 - 0.004i \tag{3.150}$$

where $k_o = \omega/c$ is the wavenumber in free space and c is the speed of light in vacuum. It is evident from the imaginary parts of the refractive indexes, that losses are higher in the NRI-TL lens than the TL mesh. This is due to the lossy L,C chip components used in the experimental NRI-TL lens . Nevertheless, the 2-D dual TL is still a relatively low-loss composite medium with a loss tangent of only $\tan\delta = 0.064$.

Having theoretically characterized the experimental NRI-TL lens, its performance and resolution enhancement can be predicted and directly compared to the experimental results. The resolution enhancement of the experimental lens can be estimated using either (3.144) or (3.146), depending on whether one views the NRI-TL lens arrangement shown in Fig. 3.39 as an electrical network or an effective medium. The R_e derived for uniform isotropic NRI slabs (3.144) can be used given that the slab thickness of the dual TL network is $D = (l - h)d = (5)0.84$ cm $= 4.2$ cm, and the operating wavelength is $\lambda = 2\pi d/(k_p d) = 2\pi(0.84)/0.339 = 15.59$ cm. Substituting these two values and the effective material parameters μ_p and μ_n into (3.144) results in $R_e = 2.8$. Alternately, the resolution enhancement R_e of the experimental lens can be estimated from the quality factors ($Q_C = 150, Q_{TL} = 221$) and the electrical thickness of the lens [$k_p D = 5(0.339)$ radians] using (3.146). The estimated resolution enhancement is therefore $R_e = \ln[1/150 + 1/(442)]/[5(0.339)] = 2.8$ which is the same as the previous estimate based on material parameters. These estimates can be verified by finding the bandwidth of the theoretically predicted OTF shown in Fig. 3.45 at a 1-GHz frequency of operation. The OTF is computed using

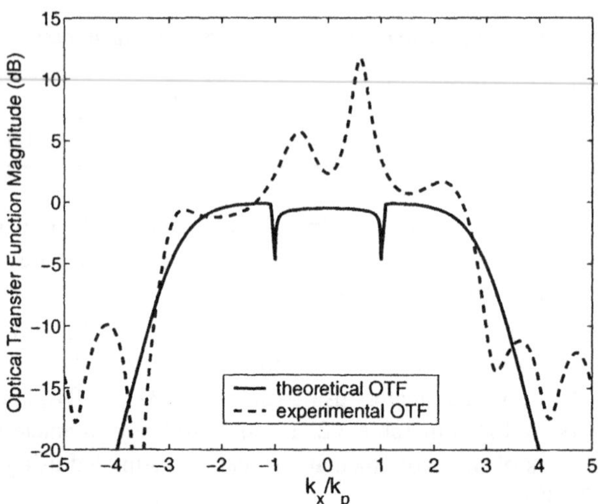

Fig. 3.45 Theoretical and approximate experimental optical transfer functions for the NRI-TL lens.

(3.124) given the quality factors (Q_L, Q_C, Q_{TL}) and parameters listed in Table 3.1 that model the experimental lens. Figure 3.45 indicates that the OTF magnitude drops to 0.5 (-6 dB) at approximately $k_x/k_p = 3$, justifying both of the preliminary R_e estimates, one based on the effective material parameters (3.144) and the other solely based on quality factors (3.146). The approximate OTF of the experimental NRI-TL lens at 1.057 GHz is also shown in Fig. 3.45. The best experimental focusing was observed at 1.057 GHz; therefore, experimental results at this frequency will be used when making comparisons to the theoretical results at 1.0 GHz. The experimental OTF was obtained by dividing the transverse spectrum at the external focal plane ($n = 10$) by the transverse spectrum at the source plane ($n = 0$). The experimental OTF is only approximate since it includes both incident and reflected waves present at the source plane, whereas the theoretical OTF only takes into account the incident wave. The experimental OTF shows a sharp cutoff at approximately $R_e = 3$ corroborating our theoretical prediction of the lens' resolution enhancement. The spikes in the passband of the transfer function are due to the asymmetries in the experimental source due to reflections at the edges of the experimental setup.

Next, the theoretically predicted voltage along the external focal plane ($n = 10$) is compared to the normalized vertical electric field detected above the external focal plane in the experimental setup. Both the experimental and theoretical images are plotted in Fig. 3.46 and show close agreement. The experimental image beamwidth is captured accurately by the expanded analytical formulation that includes losses. The theoretical current (I_z) image is also shown in Fig. 3.46 for the experimental NRI-TL lens. The current image is representative of the x component of the magnetic field intensity H_x that would be detected along the external focal plane of the experimental NRI-TL lens. The current image has its first nulls at $x = \pm 3d$ or approximately

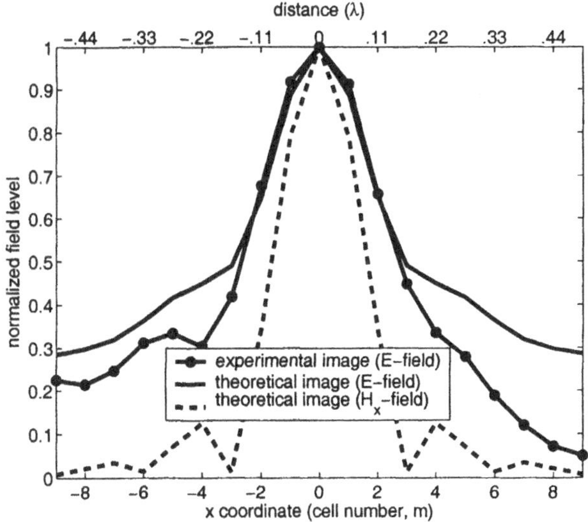

Fig. 3.46 Measured and theoretical electric field image as well as the theoretical H_x field image at the external focal plane ($n = 10$). The current (I'_z) image represents the H_x field image just as the voltage (V') image represents the E-field image.

$\pm\lambda/6$. Since the Bloch current spectrum of the image is uniform (see Fig. 3.35), a null-to-null beamwidth of $\lambda/3$ for the current image again suggests that $R_e = 3$ for the experimental NRI-TL lens. The beamwidth of the source (shown in Fig. 3.44) is limited predominantly by the periodicity to $R_e = \pi/k_p d$. The broadening of the electric field image beyond that of the source arises from the mismatch between the TL mesh and dual TL network, due to added losses in the dual TL network. The added losses are introduced by the quality factors of the components used and further limit the resolution enhancement of the experimental lens to $R_e = 2.8$ beyond that imposed by periodicity $R_e = 9.3$. Even small losses in the NRI-TL lens degrade its resolution enhancement R_e as is evident from (3.146).

3.10 AN ISOTROPIC 3-D TRANSMISSION-LINE METAMATERIAL WITH A NEGATIVE REFRACTIVE INDEX

In this section we explain how to extend the transmission-line approach to synthesizing isotropic 3-dimensional (3-D) NRI metamaterials. The 3-D NRI transmission-line metamaterial presented here draws on Gabriel Kron's work with 3-D electrical networks [69]. In 1943, Kron conceived a 3-D network representation of Maxwell's equations in a charge-free medium with positive ϵ and μ as shown in Fig. 3.47a. This pioneering work by Kron laid the foundation for the transmission-line matrix method (TLM) developed by Johns, for the numerical solution of Maxwell's equations [4,70]. As depicted in Fig. 3.47a, there are two orthogonally oriented transmission lines with

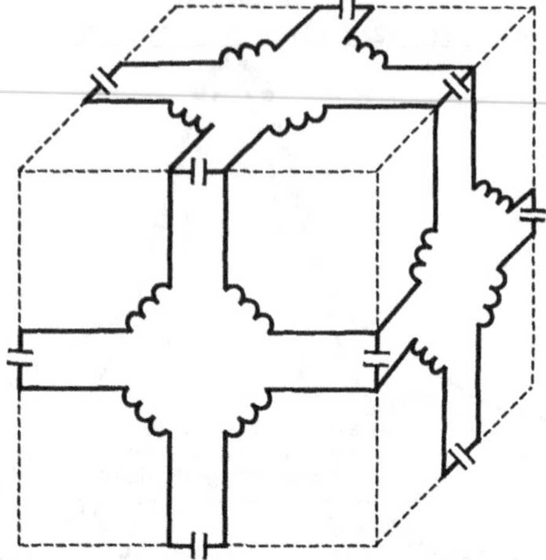

(a) Kron's unit cell (3-D transmission line unit cell)

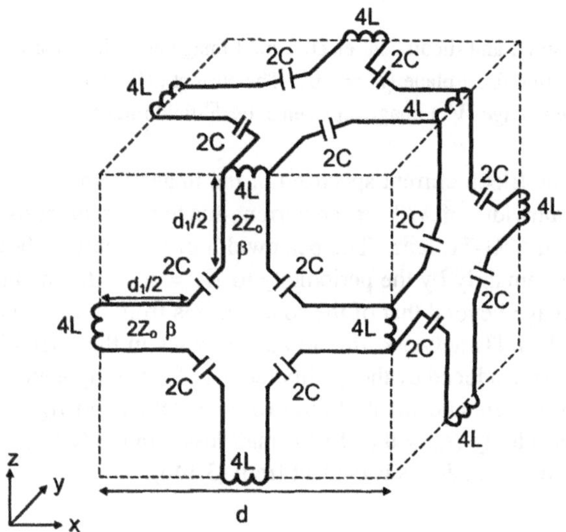

(b) Unit Cell of a 3-D negative-refractive-index metamaterial (3-D dual transmission line unit cell)

Fig. 3.47 3-D transmission-line unit cells.

per-unit-length inductance L_o and capacitance C_o on each face of Kron's cubic unit cell. The symmetry of the unit cell naturally suggests isotropic wave propagation. Each component of the electric field E lies along the corresponding shunt capacitors while each component of the magnetic field H is supported by an orthogonally-oriented inductive loop (there are three orthogonal loops that can be identified when examining any corner of the cube). Kron established a formal analogy between L_o and C_o, and the corresponding permeability and permittivity of the space being represented. Earlier in the chapter it was recognized that in order to synthesize negative permittivity and permeability one should switch the positions of the inductors and capacitors as they appear in a conventional (forward-wave) transmission-line system. When this is done, the series capacitors produce a negative permeability, and the shunt inductors produce a negative permittivity. Applying this same prescription to Kron's unit cell leads to the structure shown in Fig. 3.47b, which we will refer to as the 3-D dual transmission line [71]. The fact that the two-wire transmission lines on each side of the cube are now backward-wave lines [10] further supports the view that the emerging medium has an isotropic negative refractive index. The 3-D dual TL can be implemented by loading a 3-D network of transmission lines (with characteristic impedance Z_o and intrinsic propagation constant β) with lumped L,C elements as shown in Fig. 3.47b. The characteristic impedance Z_o and propagation constant β of the transmission lines can also be expressed in terms of the transmission line's per-unit-length capacitance C_o and inductance L_o:

$$Z_o = \sqrt{\frac{L_o}{C_o}}, \qquad \beta = \omega\sqrt{L_o C_o} \tag{3.151}$$

where ω is the angular frequency.

In order to understand the wave propagation within the 3-D dual TL depicted in Fig. 3.47b, consider the on-axis propagation of a plane wave having an E-field polarized in the z-direction and propagating along the x-axis. In this case, virtual electric walls (tangential E field vanishes) are formed at the top and bottom faces of the cubic unit cell, whereas virtual magnetic walls (tangential H field vanishes) are formed at the midpoint of the right and left faces of the cube shown in Fig. 3.47b. The resulting equivalent circuit for on-axis propagation is shown in Fig. 3.48. The short-circuited and open-circuited stubs periodically load the axial transmission line with series $L_o d_1$ inductors and shunt $C_o d_1$ capacitors, where d_1 is the length of the interconnecting transmission lines. A dispersion relation for on-axis propagation can be derived from the schematic shown in Fig. 3.48 [19]:

$$\begin{aligned}\sin^2\left(\frac{kd}{2}\right) = &\left[2\sin\left(\frac{\beta d_1}{2}\right) - \frac{1}{2Z_o\omega C}\cos\left(\frac{\beta d_1}{2}\right)\right] \\ &\cdot\left[2\sin\left(\frac{\beta d_1}{2}\right) - \frac{Z_o}{2\omega L}\cos\left(\frac{\beta d_1}{2}\right)\right]\end{aligned} \tag{3.152}$$

where k is the wavenumber in the 3-D dual TL and d is the dimension of the unit cell. Using equation (3.151) and assuming that kd and the phase shift βd_1 along the interconnecting lines is small (homogeneous limit), equation (3.152) simplifies to the following:

$$\begin{aligned}k &= \pm\omega\sqrt{\mu_e\epsilon_e} \\ &= \pm\omega\sqrt{\left(\frac{2C_o d_1}{d} - \frac{1}{\omega^2 L d}\right)\left(\frac{2L_o d_1}{d} - \frac{1}{\omega^2 C d}\right)}\end{aligned} \tag{3.153}$$

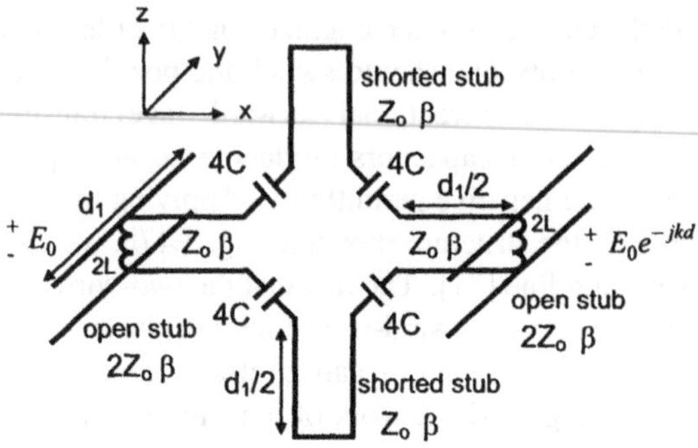

Fig. 3.48 1-D transmission-line model for propagation along the x-axis (E-field is polarized along the z-axis).

where

$$C_o = \frac{\epsilon_o}{g}, \qquad L_o = \mu_o g \tag{3.154}$$

and ϵ_o and μ_o are the permittivity and permeability of free space, since we are assuming that the transmission lines are embedded in free space. The first factor under the square root is equal to the effective permittivity (ϵ_e) of the 3-D dual TL, whereas the second term is equal to the effective permeability (μ_e). The proportionality constant g depends on the geometry of the interconnecting transmission lines. It relates the characteristic impedance of the transmission lines to the wave impedance of free space in the following manner:

$$Z_o = g\sqrt{\frac{\mu_o}{\epsilon_o}} \tag{3.155}$$

Observe that the material parameters (ϵ_e, μ_e) become negative when the loading terms (the second term in each factor) overcome the corresponding per-unit-length capacitance $2C_o$ and inductance $2L_o$ of the 3-D transmission-line system. The wavenumber k becomes negative and the propagation becomes backward-wave (antiparallel phase and group velocities) when

$$2\omega C_o < \frac{1}{\omega L d_1} \quad \text{and} \quad 2\omega L_o < \frac{1}{\omega C d_1} \tag{3.156}$$

To physically implement the metamaterial depicted in Fig. 3.47b, one needs to select an appropriate host transmission-line system. A natural choice is to use parallel broadside strips printed on the two faces of a microwave substrate. The resulting unit cell is shown in Fig. 3.49. Indeed, the topology of the unit cell suggests a method for constructing the 3-D dual TL. One can print each pair of parallel strips on three separate microwave printed circuit boards (PCB). By interleaving the three PCBs that lie along the three Cartesian planes, one can construct each unit cell (shown in Fig. 3.49). In this way, the 3-D metamaterial will be composed of three sets of orthogonally interleaved PCBs. The required series capacitors and shunt inductors

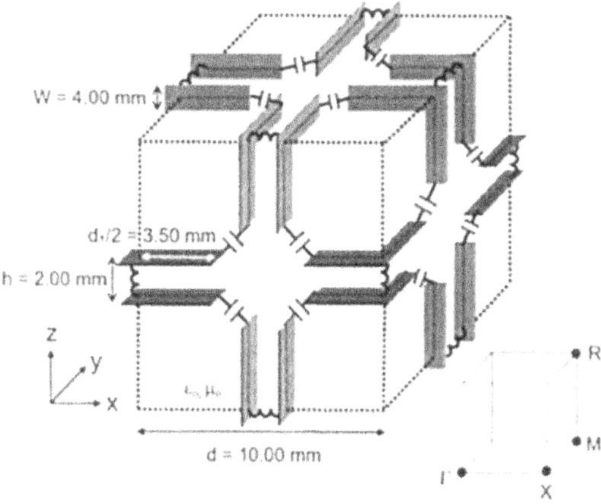

Fig. 3.49 Physical implementation of the 3-D negative-refractive-index metamaterial (3-D dual TL) using L_o, C_o loaded parallel strip transmission lines.

can be implemented in printed form using gap or interdigital capacitors and vias through the PCBs, respectively.

For the specific case of parallel strips having width W and separation h, well known quasi-static formulas for microstrip lines can be utilized to estimate the geometrical factor g [12]:

$$g = \frac{2}{2W/L + 1.393 + 0.667\ln(2W/h + 1.444)} \tag{3.157}$$

For simplicity we have assumed that the parallel strips are printed in free space. Equations (3.151)–(3.157) will yield only approximate expressions for the implementation of the 3-D dual TL of Fig. 3.49 since they do not take into account the parasitics associated with the bends in the parallel strip transmission lines. In order to verify that the 3-D dual TL shown in Fig. 3.49 does in fact behave as a 3-D isotropic negative-refractive-index medium, the structure was simulated using a commercially available finite-element electromagnetic solver. The dimensions of the simulated 3-D dual TL are shown in Fig. 3.49. The L, C components were modelled as lumped elements having the following values: $L = 40.0$ nH and $C = 2.5$ pF.

The simulated 3-D Brillouin diagram is shown in Fig. 3.50. The two fundamental bands exhibit backward-wave propagation and overlap for on-axis propagation (from Γ to X). These two bands correspond to the two orthogonal polarizations that are possible for on-axis propagation. Since the two bands overlap, the structure behaves identically for both TE and TM polarizations. From X to R, however, the two bands split. One band slopes downwards between X and R, while the other slopes upwards from X, peaks at a frequency of 1.15 GHz at M and then slopes back down to R. Due to this splitting, the two bands exhibit identical backward-wave propagation only between 1.15 GHz and 1.37 GHz, which can be considered the useful backward-wave bandwidth. In addition to these two bands there is also another mode present, which

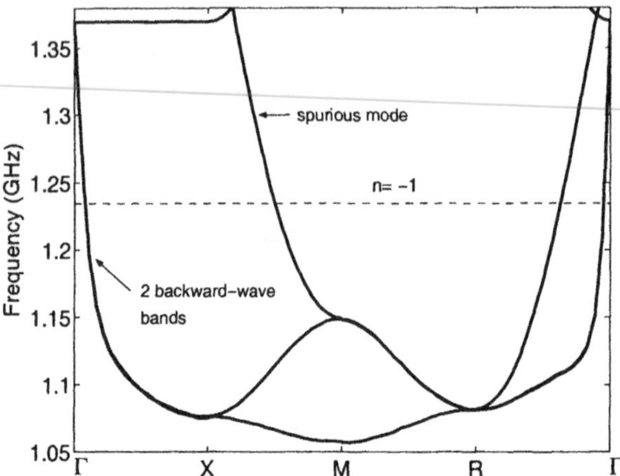

Fig. 3.50 Brillouin diagram for the 3-D negative refractive index medium (fullwave eigenmode solution).

we will call the spurious mode (see Fig. 3.50). This spurious mode coexists with the two backward-wave bands. A similar spurious mode has been observed in the numerical dispersion analysis of 3-D formulations of the TLM numerical method [72].

At a frequency of 1.23 GHz, the backward bands have a wavenumber k equal to $-k_o$, where $k_o = 25.8$ rad/m is the wavenumber in free space at this frequency. Therefore, the 3-D dual TL possesses a relative refractive index equal to $n = -1$. At 1.23 GHz, the cell size is $1/24$ the wavelength of free space. In addition to the backward-wave bands, the spurious mode is also present at this frequency (see Fig. 3.50). To see what adverse effects the spurious mode may have on the performance of the 3-D NRI medium when it is interfaced to free space, as in the case of Veselago's lens (a flat slab of NRI material with $n = -1$ in free space as described earlier), one can examine different equifrequency contours of the dispersion surface at 1.23 GHz. In Fig. 3.51a, the equifrequency contour along the (100) plane is shown. This contour depicts propagation along one of the three Cartesian planes (the wavevector **k** lies within a Cartesian plane). The two backward-wave bands appear as two concentric and overlapping circles. This indicates that propagation is isotropic and identical for both polarizations.

The spurious mode manifests itself at the four corners of Fig. 3.51a (around the M points), at much larger wavenumbers. From Fig. 3.51a it becomes clear that highly attenuating evanescent waves incident from free space on the 3-D dual TL could in fact couple to this (propagating) spurious mode. For this specific geometry, the 3-D dual TL behaves as a NRI material with $n = -1$ for all propagating waves and evanescent waves with transverse wavenumbers up to approximately $5k_o$ (see Fig. 3.51a). Finally, the equifrequency contour along the (110) plane is shown in Fig. 3.51b. The two backward-wave bands for the two orthogonal polarizations appear again as two concentric and overlapping circles with wavenumbers k equal to

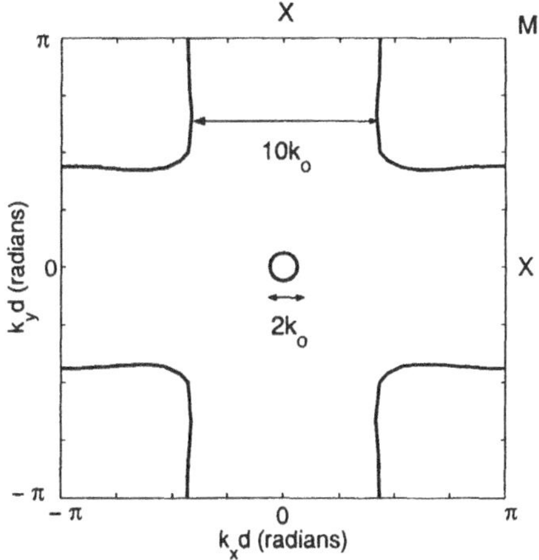

(a) Equifrequency contour along the (100) plane at 1.23 GHz

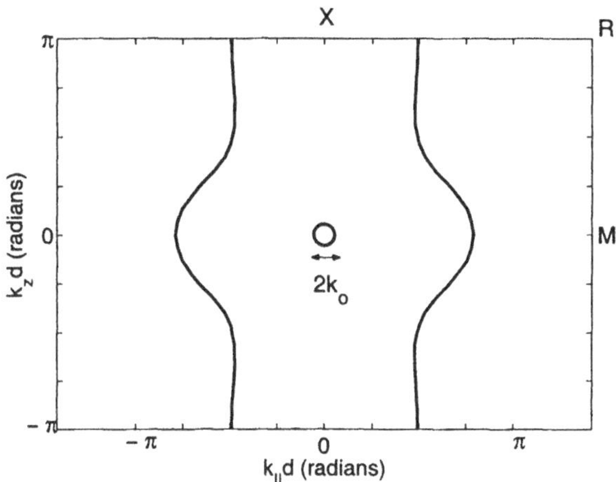

(b) Equifrequency contour along the (110) plane at 1.23 GHz

Fig. 3.51 Equifrequency contours at 1.23 GHz.

$-k_o$. The spurious band now appears on either side of the backward-wave circles, near the edges of the cubic first Brillouin zone.

From the equifrequency contours at 1.23 GHz, two general observations can be made. First, that the dispersion surfaces of the backward-wave bands are two concentric and overlapping spheres, as in an isotropic homogeneous medium. Second, that the spurious band is confined to regions of the reciprocal lattice near the edges of the first Brillouin zone. From the 3-D Brillouin diagram shown in Fig. 3.50, one also observes that as the frequency of operation is increased from 1.15 GHz to 1.37

GHz, the refractive index of the 3-D dual TL decreases, and the dispersion surface of the spurious mode approaches those of the backward-wave bands.

To summarize, a 3-D negative refractive index medium has been outlined that exhibits homogeneous and isotropic backward-wave propagation at microwave frequencies. This is valid for all propagating plane-waves impinging on the medium from free space regardless of polarization. However, an additional spurious mode has also been shown to coexist with the backward-wave propagation bands. This spurious mode can be excited by highly attenuating evanescent waves ($5k_o$) incident on the structure from free space. Nevertheless, the resolution of a practical NRI lens made out of this medium would most likely be limited by loss rather than the excitation of this spurious mode as was explained earlier in this chapter [48].

Furthermore, it should be pointed out that the method outlined in this section can form the basis for synthesizing isotropic and anisotropic metamaterials with prescribed material parameters (other than simultaneously negative permittivity and permeability). For example, by appropriately loading the host 3-D network of parallel strips only with series capacitors, a broadband band-gap magnetic medium can be synthesized.

REFERENCES

1. V. G. Veselago, "The electrodynamics of substances with simultaneously negative values of ϵ and μ," *Sov. Phys. Usp.*, vol. 10, pp. 509–514, January–February 1968.

2. D. R. Smith, W. J. Padilla, D. C. Vier, S. C. Nemat-Nasser, and S. Schultz, "Composite medium with simultaneously negative permeability and permittivity," *Phys. Rev. Lett.*, vol. 84, pp. 4184–4187, May 2000.

3. R. A. Shelby, D. R. Smith, and S. Schultz, "Experimental verification of a negative index of refraction," *Science*, vol. 292, pp. 77–79, April 2001.

4. P. B. Johns and R. L. Beurle, "Numerical solution of 2-dimensional scattering problem using a transmission-line matrix," *Proc. IEE*, vol. 118, pp. 1203–1208, September 1971.

5. W. J. R. Hoefer, "The transmission-line matrix method: Theory and applications," *IEEE Trans. Microwave Theory Tech.*, vol. MTT-33, pp. 882–893, October 1985.

6. A. K. Iyer and G. V. Eleftheriades, "Negative refractive index metamaterials supporting 2-D wave propagation," in *IEEE MTT-S International Microwave Symposium Digest*, Seattle, WA, June 2–7 2002, vol. 2, pp. 1067–1070.

7. A. Grbic and G. V. Eleftheriades, "A backward-wave antenna based on negative refractive index L–C networks," in *IEEE AP-S International Symposium*, San Antonio, TX, June 16–21, 2002, vol. 4, pp. 340–343.

8. C. Caloz, H. Okabe, H. Iwai, and T. Itoh, "Transmission line approach of left-handed materials," in *USNC/URSI National Radio Science Meeting*, San Antonio, TX, June 16–21, 2002.

9. A. A. Oliner, "A periodic-structure negative-refractive-index medium without resonant elements," in *USNC/URSI National Radio Science Meeting*, San Antonio, TX, June 16–21, 2002.

10. S. Ramo, J. R. Whinnery, and T. Van Duzer, *Fields and Waves in Communication Electronics*, 3rd ed., John Wiley & Sons, New York, 1994.

11. D. A. Watkins, *Topics in Electromagnetic Theory*, John Wiley & Sons, New York, 1958.

12. D. M. Pozar, *Microwave Engineering*, 2nd ed., John Wiley & Sons, Toronto, 1998.

13. A. Grbic, *Super resolving negative-refractive-index transmission-line lenses*, Ph.D. thesis, The Edward S. Rogers Sr. Department of Electrical and Computer Engineering, University of Toronto, 2005.

14. J. B. Pendry, A. J. Holden, W. J. Steward, and I. Youngs, "Extremely low frequency plasmons in metallic mesostructures," *Phys. Rev. Lett.*, vol. 76, pp. 4773–4776, June 1996.

15. S. Tretyakov, *Analytical Modeling in Applied Electromagnetics*, Artech House, Boston, 2003.

16. J. B. Pendry, A. J. Holden, D. J. Robbins, and W. J. Steward, "Magnetism from conductors and enhanced nonlinear phenomena," *IEEE Trans. Microwave Theory Tech.*, vol. 47, pp. 2075–2084, November 1999.

17. K. C. Gupta, R. Garg, I. Bahl, and P. Bhartia, *Microstrip Lines and Slot Lines*, Artech House, New York, 1996.

18. C. A. Balanis, *Antenna Theory: Analysis and Design*, 2nd ed., John Wiley & Sons, New York, 1997.

19. A. Grbic and G. V. Eleftheriades, "Periodic analysis of a 2-D negative refractive index transmission line structure," *IEEE Trans. Antennas Propag.*, vol. 51, pp. 2604–2611, October 2003.

20. D. Sievenpiper, L. Zhang, R. F. J. Broas, N. G. Alexopolous, and E. Yablonovitch, "High impedance electromagnetic surfaces with a forbidden frequency band," *IEEE Trans. Microwave Theory Tech.*, vol. 47, no. 11, pp. 2059–2074, Nov 1999.

21. K. G. Balmain, A. E. Lüttgen, and P. C. Kremer, "Resonance cone formation, reflection, refraction and focusing in a planar anisotropic metamaterial," *IEEE Antennas Wireless Propag. Lett.*, vol. 1, pp. 146–149, 2002.

22. K. G. Balmain, A. E. Lüttgen, and P. C. Kremer, "Power flow for resonance cone phenomena in planar anisotropic metamaterials," *IEEE Trans. Antennas Propag.*, vol. 51, pp. 2612–2618, October 2003.

23. O. Siddiqui and G. V. Eleftheriades, "Resonance cone focusing in a compensating bilayer of continuous hyperbolic mictrostrip grids," *Appl. Phys. Lett.*, vol. 85, pp. 1292–1294, 2004.

24. A. K. Iyer, P. C. Kremer, and G. V. Eleftheriades, "Experimental and theoretical verification of focusing in a large, periodically loaded transmission line negative refractive index metamaterial," *Opt. Express*, vol. 11, no. 7, pp. 696–708, April 2003.

25. G. V. Eleftheriades, A. K. Iyer, and P. C. Kremer, "Planar negative refractive index media using periodically L–C loaded transmission lines," *IEEE Trans. Microwave Theory Tech.*, vol. 50, no. 12, pp. 2702–2712, December 2002.

26. J. B. Pendry, "Negative refraction makes a perfect lens," *Phys. Rev. Lett.*, vol. 85, pp. 3966–3969, October 2000.

27. D. R. Smith and N. Kroll, "Negative refractive index in left-handed materials," *Phys. Rev. Lett.*, vol. 85, pp. 2933–2936, 2000.

28. R. W. Ziolkowski and E. Heyman, "Wave propagation in media having negative permittivity and permeability," *Phys. Rev. E*, vol. 64, 055625, October 2001.

29. S. A. Ramakrishna, J. B. Pendry, D. Schurig, D. R. Smith, and S. Schultz, "The asymmetric lossy near-perfect lens," *J. Modern Opt.*, vol. 49, pp. 1747–1762, 2002.

30. J. Pacheco, T. M. Grzegorczyk, B.-I. Wu, Y. Zhang, and J. A. Kong, "Power propagation in homogeneous isotropic frequency-dispersive left-handed media," *Phys. Rev. Lett.*, vol. 89, 257401, 2002.

31. J. T. Shen and P. M. Platzman, "Near field imaging with negative dielectric constant lenses," *Appl. Phys. Lett.*, vol. 80, pp. 3286–3288, May 2002.

32. F. Fang and X. Zhang, "Imaging properties of a metamaterial superlens," *Appl. Phys. Lett.*, vol. 82, pp. 161–163, January 2003.

33. S. A. Ramakrishna and J. B. Pendry, "Imaging the near field," *J. Modern Opt.*, vol. 50, pp. 1419–1430, 2003.

34. D. R. Smith, D. Schurig, R. Rosenbluth, S. Schultz, S. A. Ramakrishna, and J. B. Pendry, "Limitations on subdiffraction imaging with a negative refractive index slab," *Appl. Phys. Lett.*, vol. 82, pp. 1506–1508, March 2003.

35. Z. Ye, "Optical transmission and reflection of perfect lenses by left handed materials," *Phys. Rev. B*, vol. 67, 193106, May 2003.

36. A. N. Lagarkov and V. N. Kisel, "Quality of focusing electromagnetic radiation by a plane-parallel slab with a negative index of refraction," *Dokl. Phys.*, vol. 49, pp. 5–10, January 2004.

37. X. S. Rao and C. K. Ong, "Subwavelength imaging by a left-handed superlens," *Phys. Rev. E*, vol. 68, 067601, December 2003.

38. D. Maystre and S. Enoch, "Perfect lenses made with left-handed materials: Alice's mirror," *J. Opt. Soc. Am. A*, vol. 21, pp. 122–131, January 2004.

39. R. Merlin, "Analytical solution of the almost-perfect-lens problem," *Appl. Phys. Lett.*, vol. 84, pp. 1290–1292, February 2004.

40. M. Nieto-Vesperinas, "Problem of image superresolution with a negative-refractive-index slab," *J. Opt. Soc. Am. A*, vol. 21, pp. 491–498, April 2004.

41. R. Marques and J. Baena, "Effect of losses and sispersion on the focusing properties of left-handed media," *Microwave Opt. Technol. Lett.*, vol. 41, pp. 290–294, May 2004.

42. W. C. Chew, "Sommerfeld integrals for left-handed materials," *Microwave Opt. Technol. Lett.*, vol. 42, pp. 369–373, September 2004.

43. K. J. Webb, M. Yang, D. W. Ward, and K. A. Nelson, "Metrics for negative-refractive-index materials," *Phys. Rev. E*, vol. 70, 035602, September 2004.

44. D. R. Smith, D. C. Vier, N. Kroll, and S. Schultz, "Direct calculation of permeability and permittivity for a left-handed metamaterial," *Appl. Phys. Lett.*, vol. 77, pp. 2246–2248, October 2000.

45. D. R. Smith, S. Schultz, P. Markos, and C. M. Soukoulis, "Determination of effective permittivity and permeability of metamaterials from reflection and transmission coefficients," *Phys. Rev. B*, vol. 65, 195104, April 2002.

46. T. Koschny, P. Markos, D. R. Smith, and C. M. Soukoulis, "Resonant and antiresonant frequency dependence of the effective parameters of metamaterials," *Phys. Rev. E*, vol. 68, 065602, December 2003.

47. T. Koschny, M. Kafesaki, E. N. Economou, and C. M. Soukoulis, "Effective medium theory of left-handed materials," *Phys. Rev. Letters*, vol. 93, 107402, September 2004.

48. A. Grbic and G. V. Eleftheriades, "Overcoming the diffraction limit with a planar left-handed transmission-line lens," *Phys. Rev. Lett.*, vol. 92, 117403, March 2004.

49. A. Alù and N. Engheta, "Pairing an epsilon-negative slab with a mu-negative slab: Resonance, tunneling and transparency," *IEEE Trans. Antennas Propag.*, vol. 51, pp. 2558–2571, October 2003.

50. A. Grbic and G. V. Eleftheriades, "Negative refraction, growing evanescent waves, and sub-diffraction imaging in loaded transmission-line metamaterials," *IEEE Trans. Microwave Theory Tech.*, vol. 51, pp. 2297–2305, December 2003.

51. A. Grbic and G. V. Eleftheriades, "Growing evanescent waves in negative refractive index," *Appl. Phys. Lett.*, vol. 82, no. 12, pp. 1815–1817, March 2003.

52. R. Ruppin, "Surface polaritons of a left-handed medium," *Phys. Lett. A*, vol. 277, pp. 61–64, 2000.

53. H. Raether, *Surface Plasmons on Smooth and Rough Surfaces and on Gratings*, Springer-Verlag, New York, 1988.

54. J. B. Pendry and S. A. Ramakrishna, "Refining the perfect lens," *Physica B*, vol. 338, pp. 329–332, 2003.

55. R. Ruppin, "Surface polaritons of a left-handed material slab," *J. Physics: Condensed Matter*, vol. 13, pp. 1811–1819, 2001.

56. F. D. M. Haldane, "Electromagnetic surface modes at interfaces with negative refractive index make a "not-quite-perfect" lens," *cond-mat/0206420*, 2002.

57. B. A. Munk and G. A. Burrell, "Plane-wave expansion for arrays of arbitrarily oriented piecewise linear elements and its applications in determining the impedance of a single linear antenna in a lossy half-space," *IEEE Trans. Antennas Propag.*, vol. AP-27, pp. 331–343, May 1979.

58. H. Y. D. Yang, "Theory of antenna radiation from photonic band-gap materials," *Electromagnetics*, vol. 19, pp. 255–276, May–June 1999.

59. P. C. Clemmow, *The Plane Wave Spectrum Representation of Electromagnetic Fields*, IEEE Press, Piscataway, NJ, 1996.

60. S. A. Cummer, "Simulated causal subwavelength focusing by a negative refractive index slab ," *Appl. Phys. Lett.*, vol. 82, pp. 1503–1505, March 2003.

61. J. B. Pendry and S. A. Ramakrishna, "Near-field lenses in two dimensions," *J. Phys.: Condensed Matter*, vol. 14, pp. 8463–8479, 2002.

62. J. W. Goodman, *Introduction to Fourier Optics*, 2nd ed., McGraw-Hill, New York, 1996.

63. M. W. Feise, P. J. Bevelacqua, and J. B. Schneider, "Effects of surface waves on the behaviour of perfect lenses," *Phys. Rev. B*, vol. 66, 035113, July 2002.

64. I. V. Shadrivov, A. A. Sukhorukov, and Y. S. Kivshar, "Guided Modes in Negative-Refractive-Index Waveguides," *Phys. Rev. E*, vol. 67, 057602, May 2003.

65. B.-I. Wu, T. M. Grzegorczyk, Y. Zhang, and J. A. Kong, "Guided modes with imaginary transverse wave number in a slab waveguide with negative permittivity and permeability," *J. Appl. Phys.*, vol. 93, pp. 9386–9388, June 2003.

66. H. Cory and A. Barger, "Surface-wave propagation along a metamaterial slab," *Microwave Opt. Technol. Lett.*, vol. 38, pp. 392–395, September 2003.

67. A. Alù and N. Engheta, "Guided modes in a waveguide filled with a pair of single-negative (SNG), double-negative (DNG), and/or double-positive (DPS) layers," *IEEE Trans. Microwave Theory Tech.*, vol. 52, pp. 199–210, January 2004.

68. A. Grbic and G. V. Eleftheriades, "Practical limitations of sub-wavelength resolution using negative-refractive-index transmission-line lenses," *IEEE Trans. Antennas Propag.*, 2005.

69. G. Kron, "Equivalent circuits to represent the electromagnetic field equations," *Phys. Rev.*, vol. 64, pp. 126–128, 1943.

70. C. R. Brewitt-Taylor and P. B. Johns, "On the construction and numerical solution of transmission-line and lumped network models of Maxwell's equations," *Int. J. Numerical Methods Eng.*, vol. 15, pp. 13–30, 1980.

71. G. V. Eleftheriades and A. Grbic, "A 3-D negative-refractive-index transmission-line medium," in *IEEE AP-S/URSI International Symposium Digest*, Washington, D.C., July 3–8, 2005.

72. J. S. Nielsen and W. J. R Hoefer, "Generalized dispersion analysis and spurious modes of 2-D and 3-D TLM formulations," *IEEE Trans. Microwave Theory Tech.*, vol. 41, no. 8, pp. 1375–1384, August 1993.

4 Gaussian Beam Interactions with Double-Negative (DNG) Metamaterials

RICHARD W. ZIOLKOWSKI

Department of Electrical and Computer Engineering
University of Arizona,
Tucson, AZ 85721
United States

In this chapter, the interactions of continuous-wave (CW) and pulsed Gaussian beams with double-negative (DNG) metamaterials are considered. The DNG metamaterials are represented as lossy, dispersive Drude model media. Several DNG media including those with $n_{real}(\omega_0) \approx -1$ and $n_{real}(\omega_0) < -1$ at a specified target frequency $f_0 = \omega_0/2\pi$ are considered. Subwavelength focusing of a diverging, normally incident pulsed Gaussian beam with a planar DNG slab is demonstrated. This effect is also used to realize a phase compensator/beam translator system with DPS–DNG pairs. The negative angle of refraction behavior associated with the negative index of refraction exhibited by DNG metamaterials is demonstrated. The transmitted beam resulting from both 3-cycle and CW Gaussian beams that are obliquely incident on a DNG slab are shown to have this property. The scattered fields resulting from Gaussian beams with angles of incidence beyond the critical angle interacting with DNG metamaterial slabs are shown to have a negative Goos–Hänchen (lateral) shift. Focusing of a nearly planar CW Gaussian beam with a concave DNG lens is also discussed. It is shown that the resulting focal region is subwavelength both in the directions transverse and parallel to the beam axis. Several potential applications for these effects in the microwave and optical regimes are highlighted.

4.1 INTRODUCTION

Understanding the interaction of Gaussian beams with media reveals both propagation and scattering properties of those media. Gaussian beams are paraxial solutions of

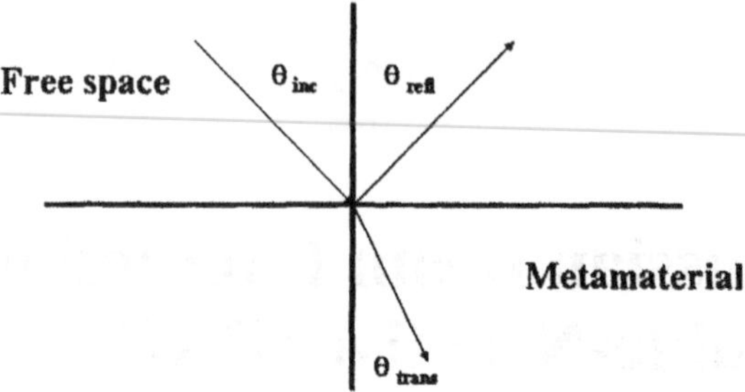

Fig. 4.1 Wave processes involved in scattering from an interface.

the scalar wave equation

$$\left(\nabla^2 + k^2\right) U = 0 \tag{4.1}$$

having the form

$$U(x, y, z) = u(x, y, z)e^{-jkz} \tag{4.2}$$

where the wavenumber $k = \omega\sqrt{\varepsilon}\sqrt{\mu} = 2\pi/\lambda$, the coefficient $u(x, y, z = 0) = \exp[-(x^2 + y^2)/w_0^2]$ has a Gaussian distribution in the plane $z = 0$, and the time convention $\exp(j\omega t)$ is assumed throughout. Thus, the fundamental Gaussian beam is represented as (see, for instance, Ref. 1, pp. 436–439)

$$U(x, y, z) = \frac{w_0}{w(z)} e^{-jkz} e^{-[\rho/w(z)]^2} e^{-j(z/L_R)[\rho/w(z)]^2} e^{-j\tan^{-1}(z/L_R)} \tag{4.3}$$

where the Rayleigh or near-to-far-field distance is

$$L_R = \frac{k\, w_0^2}{2} = \frac{\pi w_0^2}{\lambda} = \frac{A}{\lambda} \tag{4.4}$$

the waist of the beam is

$$w^2(z) = w_0^2 \left[1 + \left(\frac{2z}{kw_0^2}\right)^2\right] = w_0^2 \left[1 + \left(\frac{z}{L_R}\right)^2\right] \tag{4.5}$$

and $\rho = \sqrt{x^2 + y^2}$. This means at $z = L_R$ the waist is $w(L_R) = \sqrt{2}\, w_0$ and the intensity for $\rho = 0$ is $1/2$ of its initial value; that is, the Gaussian beam expands and decreases its amplitude as it propagates. It can be viewed as a superposition of plane waves having wavenumbers primarily in a paraxial cone surrounding the direction of propagation (i.e., $k_z \approx k - k_\rho^2/(2z)$). Consequently, as a Gaussian beam propagates in a medium, one can investigate the rate of the spread of a beam's waist and the rate of the decay of its amplitude to reveal interesting characteristics of that medium.

A material will be denoted throughout as a double positive (DPS) medium if its relative permittivity $\varepsilon_r = \varepsilon/\varepsilon_0$ and permeability $\mu_r = \mu/\mu_0$ are both positive. On the other hand, the relative permittivity and permeability are both negative in a double negative (DNG) medium; that is, $\varepsilon_r < 0$ and $\mu_r < 0$. Consider a semi-infinite slab of metamaterial (an artificially realized material such as a DNG medium) embedded in free space.

Consider a perpendicularly polarized (s-polarized or TE) incident wave (i.e., the electric field is polarized perpendicular to the plane of incidence) that is incident on the slab at an angle of incidence θ_{inc} as shown in Fig. 4.1. The reflected and transmitted waves are known to satisfy the law of reflection and Snell's Law [2]:

$$\theta_{refl} = \theta_{inc} \tag{4.6}$$

$$n_{trans} \sin \theta_{trans} = n_{inc} \sin \theta_{inc} \tag{4.7}$$

where the index of refraction in either medium (i = inc, trans) is given by the expression

$$n_i = \sqrt{\frac{\varepsilon_i}{\varepsilon_0}} \sqrt{\frac{\mu_i}{\mu_0}} = \sqrt{\varepsilon_r}\sqrt{\mu_r} \tag{4.8}$$

where ε_0 and μ_0 are the free space permittivity and permeability. The reflection and transmission coefficients are given by the expressions

$$R = \frac{\eta_{trans} \cos \theta_{inc} - \eta_{inc} \cos \theta_{trans}}{\eta_{trans} \cos \theta_{inc} + \eta_{inc} \cos \theta_{trans}} \tag{4.9}$$

$$T = \frac{2\, \eta_{trans} \cos \theta_{inc}}{\eta_{trans} \cos \theta_{inc} + \eta_{inc} \cos \theta_{trans}} \tag{4.10}$$

where the wave impedance in either medium (i = inc, trans) is

$$\eta_i = \sqrt{\frac{\mu_i}{\varepsilon_i}} \tag{4.11}$$

They thus depend on the angle of incidence and the properties of both media.

We will consider in most cases below a metamaterial that is DNG and is matched to free space. This means that $\varepsilon_r < 0$ and $\mu_r < 0$ so that $n < 0$ and $\eta_{trans} = \eta_{inc}$. Consequently, for normal incidence (i.e., $\theta_{inc} = 0$), one finds $R = 0$ and $T = 1$ for a matched metamaterial. On the other hand, for any oblique angle of incidence onto a DNG interface, Snell's Law indicates that the transmitted angle will be negative. The value of the reflection and transmission coefficients then depends on the index in the DNG region. For any θ_{inc}, one still has $R = 0$ and $T = 1$ when $n_{trans} = -n_{inc}$ since simply $\theta_{trans} = -\theta_{inc}$. However, for example, for $n_{trans} = -6.0\, n_{inc}$ one has $\theta_{trans} = -3.268°$ at $\theta_{inc} = 20°$ so that $R = -0.03$ and $|R|^2 = 9.17 \times 10^{-4}$. The existence and the effects of this negative angle of refraction have been discussed by several groups (e.g., see Refs. 3–11).

Nonetheless, there had been some controversy about this negative angle of refraction [12] despite initial experimental verification [8] with 3-D metamaterial constructs. More recent planar DNG transmission line [13] and related planar refractive cone experiments [14] have more clearly verified this effect. Matching of a DNG metamaterial to free space has been reported [15].

To confirm many of the predicted unusual propagation and scattering properties associated with DNG metamaterials, the interaction of pulsed Gaussian beams with DNG metamaterial slabs has been studied numerically. Because the selected problems were designed relative only to a choice of wavelength, they represent the behavior of DNG metamaterials in the microwave regime as well as in the optical regime. The numerical simulations are obtained with the finite-difference time domain (FDTD) method; the modeling environment is discussed in Section 4.2. The FDTD numerical approach was emphasized because it removes any questions as to the choices associated with signs resulting from the DNG properties. The results for normally incident Gaussian beam interactions with a DNG slab matched to free space are presented in Section 4.3. It will be shown that a planar DNG slab does indeed focus a diverging Gaussian beam. This effect is also used to realize a phase compensator/beam translator system with DPS–DNG pairs. The interactions of an obliquely incident Gaussian beam with a matched DNG slab are considered in Section 4.4. The presence of the negative angle of refraction for the transmitted power flow is clearly demonstrated. Both ultrafast and CW pulsed Gaussian beams are considered. Interesting effects including the generation of strong surface waves and backward waves in the $n_{real}(\omega_0) \approx -1$ case are clearly demonstrated. Demonstration of the Goos–Hänchen effect for Gaussian beams is discussed in Section 4.5. Both DPS and DNG media are considered. The lateral shift that occurs in the DNG case is shown to be the opposite of the one for the DPS case. Many of the simulation results presented in Sections 4.3–4.5 were reported initially in Refs. 16 and 17; the corresponding FDTD movies are available online. Focusing of a nearly planar CW Gaussian beam with a concave DNG lens is considered in Section 4.6. Several suggested practical applications for the negative refraction effects are considered throughout. Conclusions are given in Section 4.7.

4.2 2-D FDTD SIMULATOR

A two-dimensional (2-D) simulation environment with a perpendicularly polarized (s-polarized, TE) field was utilized for these Gaussian beam interaction studies. It was convenient and provided all of the necessary physics. The field components were assumed to be the set: H_x, E_y, and H_z. The same effects have been confirmed with the analogous parallel polarization (p-polarized, TM) field configurations.

As in Refs. 9, 16 and 17, lossy Drude polarization and magnetization models were used to simulate the DNG medium. In the frequency domain, this means the permittivity and permeability were described as

$$\varepsilon(\omega) = \varepsilon_0 \left[1 - \frac{\omega_{pe}^2}{\omega\,(\omega - j\Gamma_e)}\right] \tag{4.12}$$

$$\mu(\omega) = \mu_0 \left[1 - \frac{\omega_{pm}^2}{\omega\,(\omega - j\Gamma_m)}\right] \tag{4.13}$$

The corresponding time domain equations for the polarization, P_y, and the normalized magnetization fields, $M_{nx} = \mu_0 M_x$ and $M_{nz} = \mu_0 M_z$, are

$$\begin{aligned} \partial_t^2 P_y + \Gamma_e P_y &= \varepsilon_0 \omega_{pe}^2 E_y \\ \partial_t^2 M_{nx} + \Gamma_m M_{nx} &= \mu_0 \omega_{pm}^2 M_{nx} \\ \partial_t^2 M_{nz} + \Gamma_m M_{nz} &= \mu_0 \omega_{pm}^2 M_{nz} \end{aligned} \tag{4.14}$$

The normalized magnetization was introduced to make the electric and magnetic field equations completely symmetric. The Drude model is preferred over a Lorentz model for these studies since it provides a much wider bandwidth over which the negative values of the permittivity and permeability can be obtained. This also means that the overall simulation times can be significantly shorter using the Drude models, particularly for low loss media; that is, it will take longer to reach a steady state in the corresponding Lorentz medium case because the resonance region where the permittivity and permeability acquire their negative values will be very narrow. By introducing the electric and magnetic currents

$$\begin{aligned} K_x &= \partial_t M_{nx} \\ K_z &= \partial_t M_{nz} \\ J_y &= \partial_t P_y \end{aligned} \tag{4.15}$$

the field and current equations used to model the beam interaction with a DNG medium become

$$\begin{aligned} \partial_t H_x &= +\frac{1}{\mu_0}\,(\partial_z E_y - K_x) \\ \partial_t K_x + \Gamma_m K_x &= \mu_0 \omega_{pm}^2 H_x \\ \partial_t H_z &= -\frac{1}{\mu_0}\,(\partial_x E_y + K_z) \\ \partial_t K_z + \Gamma_m K_z &= \mu_0 \omega_{pm}^2 H_z \end{aligned} \tag{4.16}$$

$$\begin{aligned} \partial_t E_y &= +\frac{1}{\varepsilon_0}\,[\,(\partial_z H_x - \partial_x H_z) - J_y\,] \\ \partial_t J_y + \Gamma_e J_y &= \varepsilon_0 \omega_{pe}^2 E_y \end{aligned} \tag{4.17}$$

In a DPS medium the corresponding field and current equation set is simply

$$\begin{aligned} \partial_t H_x &= +\frac{1}{\mu_0}\,\partial_z E_y \\ \partial_t H_z &= -\frac{1}{\mu_0}\,\partial_x E_y \\ \partial_t E_y &= +\frac{1}{\varepsilon_0}\,(\partial_z H_x - \partial_x H_z) \end{aligned} \tag{4.18}$$

The equation sets for the DNG medium cases, (4.16) and (4.17), and for the DPS medium cases, (4.18), were solved self-consistently and numerically with the FDTD approach [18,19]; that is, these equations were discretized with a standard leapfrog in time, staggered grid approach. The electric field component was taken at the center of the square cells for integer time steps; the magnetic field components were taken along the cell edges for half-integer time steps. The electric and magnetic currents were located at the cell centers but with their time assignments opposite to the corresponding electric and magnetic field components—that is, the magnetic current components were sampled at integer time steps and the electric current component was sampled at half-integer time steps. This allowed a FDTD stencil that properly simulated the matched medium conditions.

Several matched slab cases were considered. In those cases the parameters for the electric and magnetic Drude models were identical; that is, $\omega_{pe} = \omega_{pm} = \omega_p$ and $\Gamma_e = \Gamma_m = \Gamma$. In all cases, only low loss values were considered by setting $\Gamma = 10^{+8}\ \mathrm{s}^{-1}$. This means that the index of refraction will have the form

$$
\begin{aligned}
n(\omega) &= \left[\frac{\varepsilon(\omega)\mu(\omega)}{\varepsilon_0\,\mu_0}\right]^{1/2} \\
&= 1 - \frac{\omega_p^2}{\omega\,(\omega - j\Gamma)} = 1 - \frac{\omega_p^2}{\omega^2 + \Gamma^2} - j\frac{\Gamma}{\omega}\frac{\omega_p^2}{\omega^2 + \Gamma^2} \\
&\approx 1 - \frac{\omega_p^2}{\omega^2} - j\,\frac{\Gamma\,\omega_p^2}{\omega^3}
\end{aligned}
\tag{4.19}
$$

Consequently

$$
n_{real}(\omega) \approx 1 - \frac{\omega_p^2}{\omega^2} \tag{4.20}
$$

Thus by adjusting the plasma frequency ω_p, one can obtain a desired value of n_{real} at a specified frequency.

In all cases, the center frequency of interest to define the index of refraction was chosen to be $f_0 = 30$ GHz, corresponding to a free-space wavelength $\lambda_0 = 1.0$ cm. This value was selected to connect these results to those presented in Refs. 9, 16, and 17. Note again that all of the results to be presented can be achieved in a similar fashion at any desired set of microwave, millimeter, or optical frequencies with the appropriate frequency values in the Drude models and the corresponding FDTD simulation parameters. For the matched DNG $n_{real}(\omega_0) \approx -1$ cases, these parameters were $\omega_p = 2\pi\sqrt{2}f_0 = 2.66573 \times 10^{11}$ rad/s and, hence, $\Gamma = 3.75 \times 10^{-4}\,\omega_p$. For the matched DNG $n_{real}(\omega_0) \approx -6$ cases, these parameters were $\omega_p = 2\pi\sqrt{7}f_0 = 4.98712 \times 10^{11}$ rad/s and, hence, $\Gamma = 2.01 \times 10^{-4}\,\omega_p$. The real part of the index of refraction for these cases is shown in Fig. 4.2. The input time signals were all multiple cycle m-n-m pulses given by the expressions

$$
f(t) = \begin{cases}
0 & \text{for } t < 0 \\
g_{on}(t)\sin(\omega t) & \text{for } 0 < t < mT_p \\
1 & \text{for } mT_p < t < (m+n)T_p \\
g_{off}(t)\sin(\omega t) & \text{for } (m+n)T_p < t < (m+n+m)T_p \\
0 & \text{for } (m+n+m)T_p < t
\end{cases}
\tag{4.21}
$$

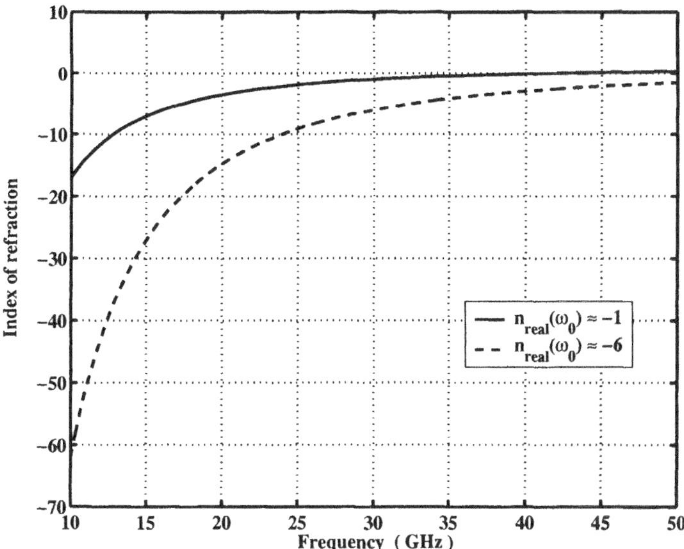

Fig. 4.2 Frequency response of the real part of the index of refraction associated with the lossy Drude models used in the FDTD simulations. After Ref. [16]. Copyright © 2003 Optical Society of America, Inc.

where if we have the terms $x_{on}(t) = t/(mT_p)$ and $x_{off}(t) = [t-(m+n)T_p]/(mT_p)$, then the two time derivative smooth functions are

$$\begin{aligned} g_{on}(t) &= 10x_{on}^3(t) - 15x_{on}^4(t) + 6x_{on}^5(t) \\ g_{off}(t) &= 1 - \left[10x_{off}^3(t) - 15x_{off}^4(t) + 6x_{off}^5(t)\right] \end{aligned} \tag{4.22}$$

These smooth excitation functions generate minimal noise as the waves are introduced into the FDTD simulation region. Each cycle has the period $T_p = 1/f_0$. The function $g_{on}(t)$ goes smoothly from 0 to 1 in m-periods; the function $g_{off}(t)$ goes smoothly from 1 to 0 in m-periods. The function f(t) thus turns on in m-periods, turns off in m-periods, and maintains a constant amplitude for n-periods. All the CW cases below were turned on in 2-cycles and were then held constant for the entire simulation time, T_{total}; that is, $m = 2$ and $(m+n)\,T_p \gg T_{total}$. The ultrafast input signals were $1-1-1$ (3-cycle) pulses. Thus the input time signals had their spectra centered on f_0 with either broad (3-cycle) or narrow (CW) bandwidths.

The simulation space was discretized into squares with a side length $\Delta = \lambda_0/100 = 100\mu$m. The time step was set at 0.95 of the two dimensional Courant value $\Delta/(\sqrt{2}c)$; that is, $\Delta t = 22.39$ ps. The simulation space was truncated with a 10-cell layer Two Time Derivative Lorentz Material (2TDLM) model absorbing boundary condition [20, 21]. The simulation region (z versus x) for the normal incidence beam cases was 830 cells $\times$ 640 cells; for the phase compensator / beam translator case it was 930 cells $\times$ 640 cells, for the oblique incidence cases it was

930 cells × 1040 cells; for the Goos–Hänchen cases it was 520 cells × 1040 cells; and for the concave lens case it was 650 cells × 720 cells. The simulation space was separated into total field and scattered field regions. The Gaussian beams were launched into the total field region using a total field–scattered field (TF–SF) formulation [18, 19]. The Gaussian beam values (4.3) were assigned to the E_y component of the field on the TF–SF boundary. This source approach has been used successfully for a variety of applications (e.g., in Ref. 22).

It is to be noted that the use of this purely numerical simulation approach to study the beam interaction has had several advantages. Very complicated structures, as well as the DNG material itself, could be incorporated into the simulation region. Moreover, narrowband and broadband excitation pulses can be handled in the same simulation environment. However, and most importantly, there are no choices involved in defining derived quantities to explain the wave physics; for example, no wavevector directions nor wave speeds are stipulated *a priori*. The FDTD simulator does not know which way the wave should refract at a DPS–DNG interface or whether it should focus or diverge a beam in a DNG region. It simply calculates what is specified by Maxwell's equations in the various regions. In this manner, it has provided an excellent approach to studying the wave physics associated with DNG metamaterials.

4.3 NORMAL INCIDENCE RESULTS

The normal incidence set of cases considered here deal with the issue of whether a planar DNG slab can focus a diverging Gaussian beam or not. The focal plane of the beam was taken to be the TF–SF boundary. The driving signals were all CW. The spatial distribution of the incident beam is defined by (4.3) as $E_y(x, z_{TF\text{–}SF\ boundary}) = U(x, y = 0, z = 0)$ and thus varies spatially as $\exp\left(-x^2/w_0^2\right)$ on that boundary—that is, its amplitude falls to e^{-1} at its waist w_0. The Gaussian beam that is generated then has many wavevectors associated with it. Wavevectors off the beam axis point away from it for a diverging beam and toward it for a converging beam. Because in all cases the beam was generated in a DPS region, the beam will begin expanding according to (4.3) as soon as the beam leaves the TF–SF boundary.

Because it was expected that a DNG medium would have a negative index of refraction and would focus the beam—that is, it would bend the wavevectors of a diverging beam back toward the beam axis—a strongly divergent beam was sought. A diffraction-limited beam, whose waist was 50 cells ($\lambda_0/2$), was used. This beam thus had a near-to-far field distance $L_R = \pi w_0^2/\lambda_0 = (\pi/4)\,\lambda_0 \approx 79$ cells. Note that the corresponding intensity waist on the TF–SF boundary was $x_{HWHM} = \left[-\frac{1}{2}w_0^2 \ln(1/2)\right]^{1/2} = 0.589\, w_0 \approx 29$ cells. The TF–SF boundary was thus set at $2\lambda_0 = 2L_R = 200$ cells away from the DNG interface. This allowed sufficient distance for the beam to diverge before it hit the interface; that is, its waist at the interface was $w = \sqrt{5}\, w_0 \approx 112$ cells. The DNG slab also had a depth of $2\lambda_0 = 200$ cells; its width was $6\lambda_0 = 600$ cells. Thus if the DNG slab refocuses the

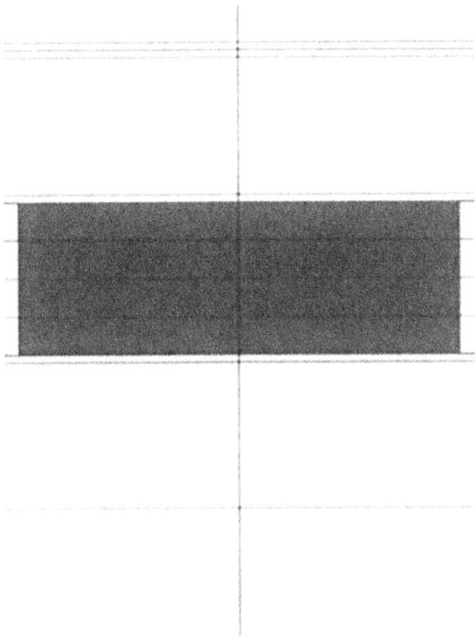

Fig. 4.3 The FDTD simulation geometry for the Gaussian beam normally incident on a DPS or DNG slab. The TF–SF boundary is the second horizontal line and the beam axis is the center vertical line. Electric field sampling points were located at the intersections of the beam axis and the horizontal lines. The location of the slab region in the DNG simulations is shown in gray.

beam, the waist of the beam at the back face of the DNG slab should be approximately (the Drude medium has some small losses) the same as its initial value.

4.3.1 Flat DNG Lenses

To illustrate that the FDTD approach reproduces the expected Gaussian beam propagation properties, consider the results shown in Figs. 4.3 and 4.4. The FDTD simulation space used to investigate beam propagation in free space is shown in Fig. 4.3. The electric field intensity in the simulation region is shown for several instants in time in Fig. 4.4. The grayscale varies from very light to dark to light gray representing small to medium to large intensity values. Twenty-one snapshots in time were captured at equal time intervals for 3000 time steps; the first frame in Fig. 4.3 represents the problem configuration at time zero. This simulation geometry is the one used for both the DPS and DNG cases. The location of the DNG slab is shown in dark gray. The DNG slab's boundary is then superimposed on the frames in Fig. 4.4 to allow for further comparisons to the actual DNG cases. The beam clearly diverges as it propagates in Fig. 4.4. The rate of expansion agrees precisely with the analytical results (4.5). The beam interaction with a matched DNG slab

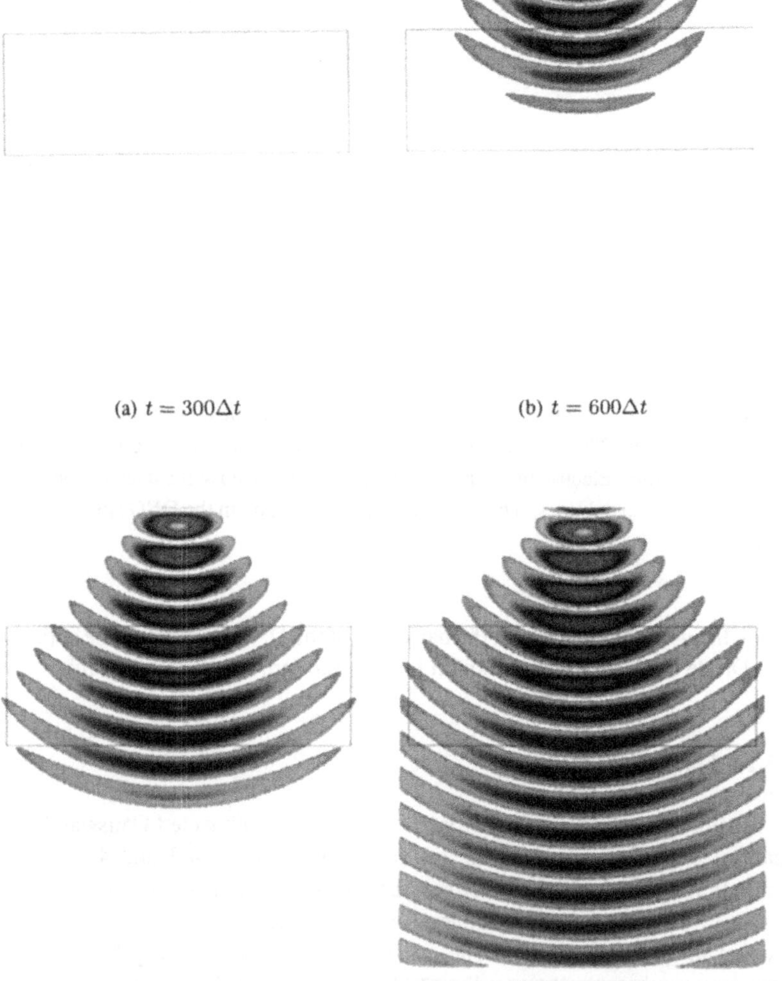

(a) $t = 300\Delta t$ (b) $t = 600\Delta t$

(c) $t = 900\Delta t$ (d) $t = 2700\Delta t$

Fig. 4.4 FDTD-predicted electric field intensity distributions illustrate the expansion of the Gaussian beam as it propagates in a free space region in which $n(\omega) = +1$. After Ref. [16]. Copyright © 2003 Optical Society of America, Inc.

having $n_{real}(\omega_0) \approx -1$ is shown in Fig. 4.5. The DNG slab was $2\lambda_0 = 200$ cells deep and it was $2\lambda_0 = 200$ cells away from the TF–SF plane. The focal plane of the source beam was taken at the TF–SF plane. The simulation was run for 5000 time steps and 21 snapshots were obtained. The frames show clearly that the planar DNG medium turns the diverging wavevectors toward the beam axis and, hence, acts as a lens to focus the beam. The focal plane of the beam in the DNG medium is located at the back face of the DNG slab because, by design, the distance from the TF–SF boundary is equal to the depth of the slab. Since all angles of refraction are the negative of their angles of incidence for the $n_{real}(\omega_0) \approx -1$ slab, the initial beam distribution is essentially recovered at the back face of the slab. The intensity of the electric field along the beam axis and along the front and back faces was sampled at the same time that each field distribution snapshot was obtained. The intensity along the beam axis at $t = 4500\,\Delta t$ is plotted in Fig. 4.6; the intensities along the front and back faces of the DNG slab are given in Fig. 4.7 for $t = 750\,\Delta t$ and $t = 4500\,\Delta t$, respectively. These times were selected because the beam then has its maximum value along the beam axis. The peaking of the beam toward the back face is evident in Fig. 4.6. The beam intensity is seen in Fig. 4.7 to have narrowed from a half width at half maximum of 50 cells $= 0.5\,\lambda_0$ at the front face to 30 cells $= 0.3\,\lambda_0$ at the back face. Thus, as designed, the initial half-width at half-maximum of the intensity is recovered at the back face. Note that the peak intensity is about -0.86 dB or 18% lower than its TF–SF boundary value. This variance stems from the presence of additional wave processes, such as surface wave generation, and from dispersion and loss in the actual Drude model used to define the DNG slab in the FDTD simulations. We note that if, as shown in Ref. 9, a point source is a distance $d/2$ from the front face of an $n_{real}(\omega_0) \approx -1$ slab of depth d, then the first focus of the source is found in the center of the slab a distance $d/2$ from the front face and the second focus is located beyond the slab at a distance $d/2$ from the back face. As the source distance is moved further away from the front face of the slab, the focus in the slab will move closer to the back face of the slab. This would be the anticipated configuration in most practical beam applications where the source is further away from the slab than its depth. However, if one wants to take advantage of the growing evanescent waves in a DNG slab to reconstruct a source at its image focus location beyond the slab and if the slab has even a small amount of loss, the slab will have to be thin to achieve the desired subwavelength focusing there. The source and its image will then have to be very near, respectively, to the front and back faces of the slab. The configuration will then become completely a near-field one; that is, the "perfect lens" situation is thus lost to only near-field configurations when real media with losses are involved. Nonetheless, since the near field occurs for $z \leq L_R$, this distance may be nontrivial for some applications. For instance, with $w_0 \sim \lambda_0$ one has $L_R \sim 3\lambda_0$ so that at ISM frequencies (e.g., at $f_0 = 2.45$ GHz, $L_R \sim 36.7$ cm). This is still an interesting distance, for example, for medical microwave imaging of parts of the body. The corresponding results for the Gaussian beam interacting with the matched DNG slab with $n_{real}(\omega_0) \approx -6$ are shown in Figs. 4.8–4.10. The DNG slab was again $2\lambda_0 = 200$ cells deep and $2\lambda_0 = 200$ cells away from the TF–SF plane. The focal plane of the source beam was again taken at the TF–SF plane. The simulation

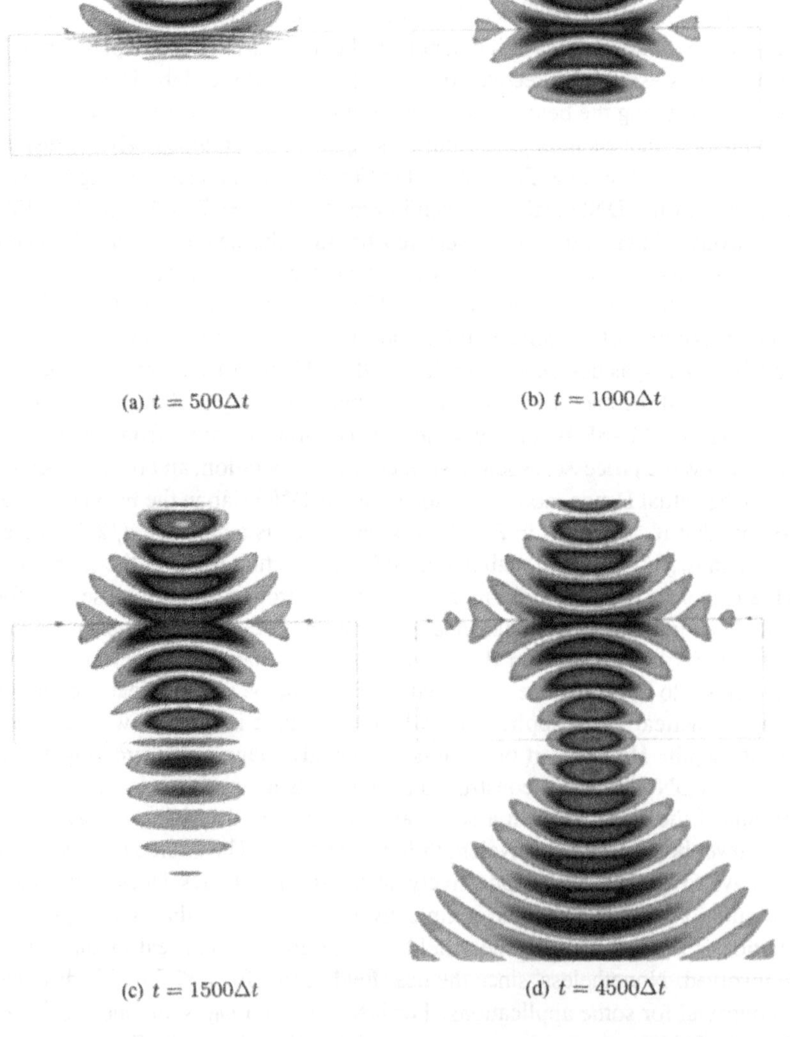

(a) $t = 500\Delta t$

(b) $t = 1000\Delta t$

(c) $t = 1500\Delta t$

(d) $t = 4500\Delta t$

Fig. 4.5 FDTD-predicted electric field intensity distributions illustrate the focusing of the Gaussian beam as it propagates in the $n_{real}(\omega) \approx -1$ DNG slab. Focusing at the back face of the slab is observed. After Ref. [16]. Copyright © 2003 Optical Society of America, Inc.

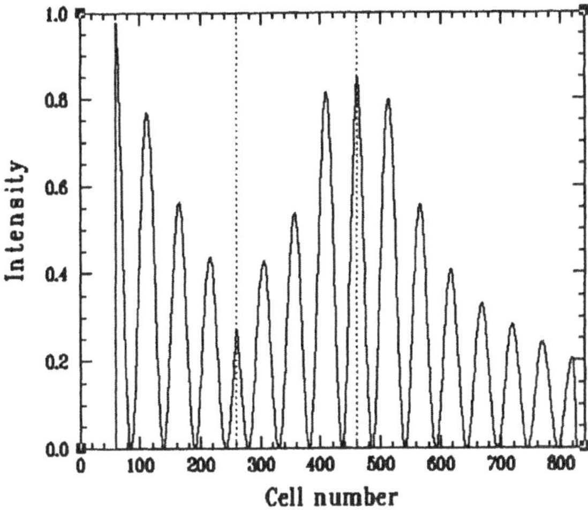

Fig. 4.6 Gaussian beam interaction with a DNG slab having $n_{real}(\omega) \approx -1$. The intensity of the electric field along the beam axis at $t = 4500\Delta t$ is shown. The front and back face locations of the DNG slab are indicated by the dashed vertical lines. The predicted focusing of the electric field intensity at the back face of the DNG slab is apparent. Sharp discontinuities in the derivatives of the field across the DPS–DNG interface are also observed. After Ref. [16]. Copyright © 2003 Optical Society of America, Inc.

was run for 8000 time steps to allow for the (six times) slower speed of the beam in the DNG slab, and 21 snapshots were again obtained. The intensity along the beam axis at $t = 4800\,\Delta t$ is plotted in Fig. 4.9; the intensities along the front and back faces of the DNG slab are given in Fig. 4.10 for $t = 750\,\Delta t$ and $t = 4800\,\Delta t$, respectively. In contrast to the $n_{real}(\omega_0) \approx -1$ case, when the beam interacts with the matched DNG slab with $n_{real}(\omega_0) \approx -6$, there is little focusing observed. The negative angles of refraction dictated by Snell's Law are shallower for this higher magnitude of the refractive index—that is, $\theta_{trans} \approx -\sin^{-1}[\sin\theta_{inc}/6]$. Hence, rather than a strong focusing, the medium channels power from the wings of the beam toward its axis, and maintains its amplitude as it propagates into the DNG medium. The width of the beam at the back face is only slightly narrower yielding only a slightly higher peak value there in comparison to its values at the front face. The strong axial compression of the beam caused by the (factor of 6) decrease in the wavelength in the $n_{real}(\omega_0) \approx -6$ slab is seen in Figs. 4.8–4.10.

Note that in all the DNG cases the beam appears to diverge significantly once it leaves the DNG slab. The properties of the DNG medium hold the beam together as it propagates through the slab. Once it leaves the DNG slab, the beam must begin diverging; that is, if the DNG slab focuses the beam as it enters, the same physics will cause the beam to diverge as it exits. Moreover, there will be no focusing of the power

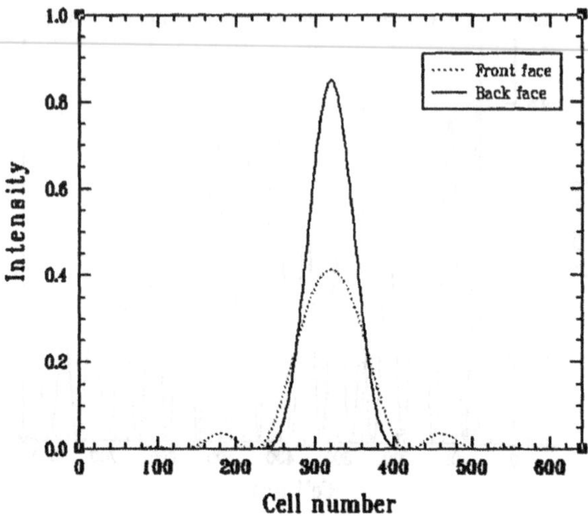

Fig. 4.7 Gaussian beam interaction with a DNG slab having $n_{real}(\omega) \approx -1$. The cross section of the intensity of the electric field along the front face (at $t = 750\Delta t$) and back face (at $t = 4500\Delta t$), orthogonal to the beam axis, are shown. The expected focusing and narrowing of the beam at the back face are observed. After Ref. [16]. Copyright © 2003 Optical Society of America, Inc.

from the wings to maintain the center portion of the beam. The rate of divergence of the exiting beam will be determined by its original value and the properties and size of the DNG medium. Also note that a beam focused into a DNG slab will generate a diverging beam within the slab and a converging beam upon exit from the slab. This behavior has also been confirmed with the FDTD simulator.

One potential application for these results is clearly the use of a matched, flat DNG slab with an index of refraction $n_{real}(\omega_0) \approx -1$ as a lens. There is little reflection loss and the beam is nicely focused. This could have applicability, for instance, in a variety of near-field microwave optical systems. Another potential application is to channel the field into a particular location—for example, to use a large negative index (i.e., $n_{real}(\omega_0) \approx -6$), DNG slab as a superstrate (overlayer) on a detector so that the beam energy would be channeled efficiently onto the detector's face. Most superstrates, being simple dielectrics, defocus the field. Often one includes a curved DPS lens over a detector face to achieve the focusing effect. The flatness of the DNG slab has further advantages in packaging the detectors into an array or a system. Yet another potential application is to combine the negative index properties of the DNG slab with its negative refraction properties to realize a low loss phase compensator/beam translator.

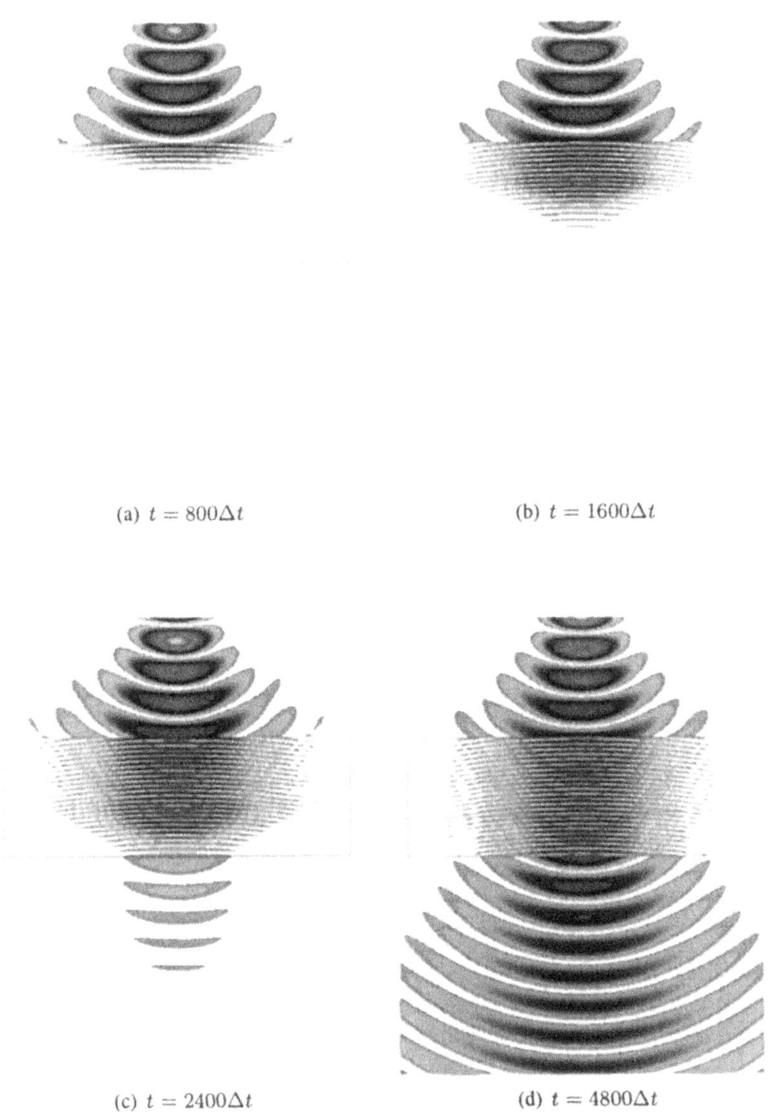

Fig. 4.8 FDTD predicted electric field intensity distribution for the normally incident Gaussian beam interaction with a DNG slab having $n_{real}(\omega) \approx -6$. Channeling of the beam in the DNG slab is observed; the wings of the beam are seen to feed the center of the beam. After Ref. [16]. Copyright © 2003 Optical Society of America, Inc.

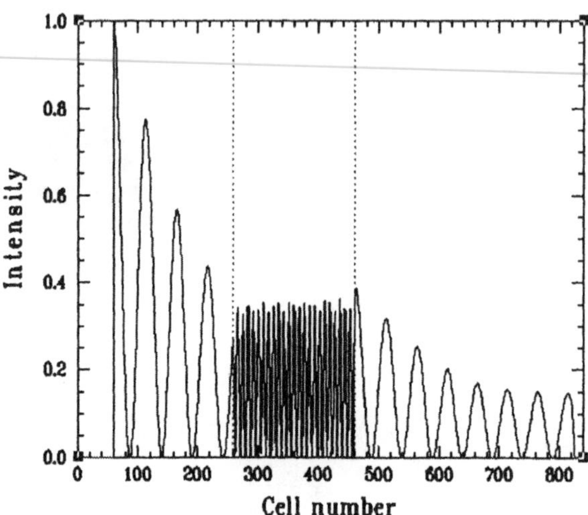

Fig. 4.9 Gaussian beam interaction with a DNG slab having $n_{real}(\omega) \approx -6$. The intensity of the electric field along the beam axis at $t = 4800\Delta t$ is shown. The DNG slab front and back face locations are indicated by the dashed vertical lines. Sharp discontinuities in the derivatives of the field across the DPS–DNG interface and maintenance of the center intensity are observed. After Ref. [16]. Copyright © 2003 Optical Society of America, Inc.

4.3.2 Phase Compensator/Beam Translator

Consider the FDTD geometry shown in Fig. 4.11. The Gaussian beam is again launched from the TF–SF boundary which is $2\lambda_0 = 200$ cells away from the front face of a $2\lambda_0 = 200$ cells-deep DPS slab whose index $n(\omega) = +3$ and that is stacked together with a $2\lambda_0 = 200$ cells-deep DNG slab whose index $n_{real}(\omega_0) \approx -3$. The beam will expand in the DPS slab and will be refocused in the DNG slab. There should be only a small loss in amplitude in this process since the DNG slab is only slightly lossy. Thus the electric field intensity (in principle) could be maintained over the $4\lambda_0 = 400$ cells distance through the slabs. Moreover, the phase of the beam at the output face of the stack will then be the same as its value at the entrance face; that is, the accumulated phase across the DPS–DNG pair of slabs at the excitation frequency is

$$\text{Re}(k_{DPS}d_{DPS}+k_{DNG}d_{DNG})_{\omega=\omega_0} = \frac{\omega_0}{c}[n_{DPS}(\omega_0)+n_{real,\,DNG}(\omega_0)](2\lambda_0) \approx 0 \tag{4.23}$$

The FDTD simulation was run for 8000 time steps; 21 snapshots in time were obtained at equal intervals. The FDTD predicted electric field intensity distributions for this phase compensation/beam translator geometry at selected times are shown in Fig. 4.12. The expansion of the beam in the DPS slab and the refocusing of the

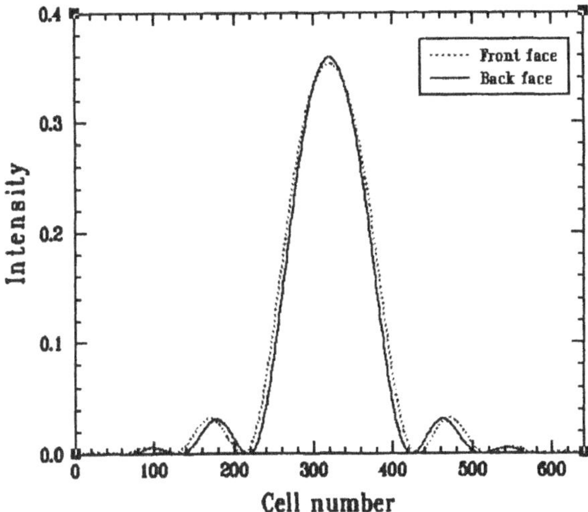

Fig. 4.10 Gaussian beam interaction with a DNG slab having $n_{real}(\omega) \approx -6$. The intensity of the electric field along the front face (at $t = 750\Delta t$) and back face (at $t = 4800\Delta t$), orthogonal to the beam axis, are shown. There is only a slight narrowing in the waist of the beam after its propagation through the entire DNG slab. After Ref. [16]. Copyright © 2003 Optical Society of America, Inc.

beam in the DNG slab are apparent. The values of the electric field intensity along the beam axis are shown in Fig. 4.13 for $t = 8000\ \Delta t$. The values of the electric field intensity transverse to the beam axis at the front face ($t = 3600\ \Delta t$) and at the back face ($t = 8000\ \Delta t$) are shown in Fig. 4.14. The waist of the intensity of the beam is clearly recovered at the back face. There is only a -0.323 dB (7.17%) reduction in the peak value of the intensity of the beam when it reaches the back face. The phase at the entrance and exit faces is the same. The phase compensator thus translates the beam from its front face to its back face with low loss. Using multiple matched DPS–DNG stacks, one could produce a phase-compensated, time-delayed, waveguiding system. Each pair in the stack would act as shown in Figs. 4.12–4.14. Thus the phase compensation/beam translation effects would occur throughout the entire system. Moreover, by changing the index of any of the DPS–DNG pairs, one changes the speed at which the beam traverses that slab pair. Consequently, one can change the time for the beam to propagate from the entrance face to the exit face of the entire DPS–DNG stack. In this manner one could realize a volumetric, low-loss time delay line for a Gaussian beam system.

Fig. 4.11 The FDTD simulation geometry for the phase compensator/beam translator system. The TF–SF boundary (second horizontal line from the top) and the beam axis (center vertical line) are shown. Electric field sampling points were located at the intersections of the beam axis and the horizontal lines. The location of the DPS slab region is shown in dark gray; the location of the DNG slab region is shown in light gray.

4.4 OBLIQUE INCIDENCE RESULTS

The oblique incidence set of cases that have been considered deal with the primary issue of whether a DNG medium will provide a negative angle of refraction or not. Both CW (very large number of cycles) and 3-cycle pulse cases were simulated. The center of the focal plane of the source beam, the focal plane being orthogonal to the beam axis, was the intersection of the beam axis with the TF–SF plane. The beam launched from the TF–SF plane had a waist of 100 cells (λ_0) in that focal plane. The larger beam waist was selected in this study so that the DNG medium did not impact the beam shape as much as the beam propagated through the slab. The TF–SF boundary was set at $3\lambda_0 = 300$ cells away from the DNG interface. This allowed a sufficient distance for the entire beam associated with the 3-cycle pulse to be present in the simulation space before it interacted with the DNG slab. In all cases the angle of incidence of the beam was $\theta_{inc} = 20°$ and the slab depth was $2\lambda_0 = 200$ cells. Recall that for an angle of incidence $\theta_{inc} = 20°$, the amplitude reflection and transmission coefficients are, respectively, $R = 0$ and $T = 1.0$ for

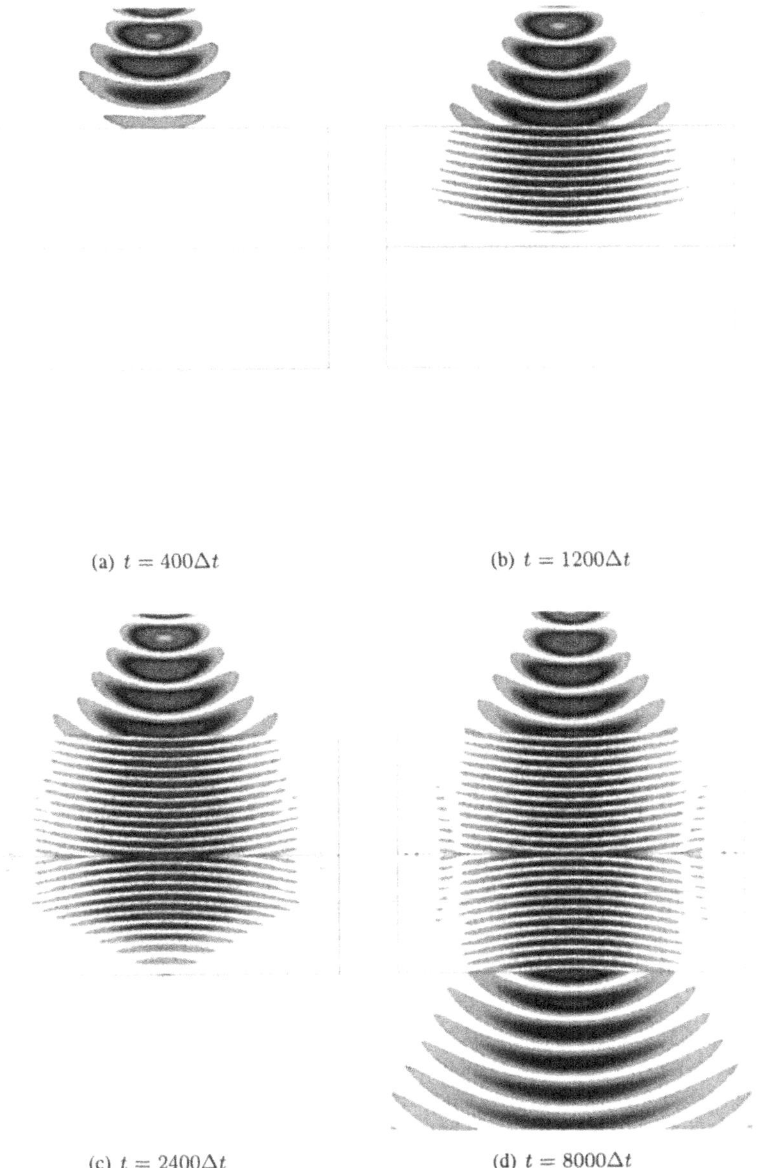

Fig. 4.12 FDTD-predicted electric field intensity distribution for the phase compensator/beam translator system. The Gaussian beam is normally incident on a stack of two slabs, the first being a DPS slab with $n(\omega) = +3$ and the second being a DNG slab with $n_{real}(\omega_0) \approx -3$. The initial beam expansion in the DPS slab is compensated by its refocusing in the DNG slab. The Gaussian beam is translated from the front face of the system to its back face with only a slight attenuation.

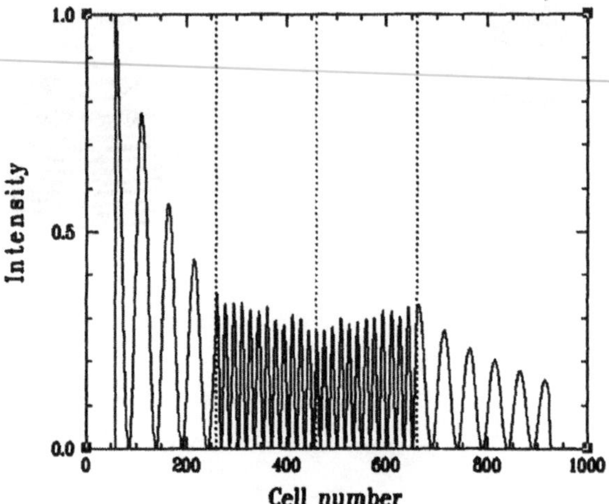

Fig. 4.13 The intensity distribution at $t = 8000\Delta t$ along the beam axis of the phase compensator/beam translator system shows the initial decay of the intensity as the beam expands in the DPS slab (front and back face locations are denoted by the left and middle vertical dashed lines) and its recovery in the DNG slab (middle and right vertical dashed lines). The intensity at the input face is almost completely recovered at the output face.

the $n_{real}(\omega_0) \approx -1$ case and $R = 0.03$ and $T = 0.97$ for the $n_{real}(\omega_0) \approx -6$ case. Hence, the reflected beam intensity will be very small ($|R|^2 = 9.17 \times 10^{-4}$) in comparison to the transmitted beam intensity for the latter case. Consequently, the reflected beam is either absent or is not noticeable in the intensity figures shown below. The configuration of the FDTD simulations is shown in Fig. 4.15. The results for the CW Gaussian beam interacting with the $n_{real}(\omega_0) \approx -1$ DNG slab are shown in Fig. 4.16. The simulation was run for 5000 time steps; 21 snapshots in time were obtained at equal intervals. The negative angle of refraction is clearly seen. The beam was sampled along the front face of the slab and at the plane $2\lambda_0 = 200$ cells from the rear face. It was found that, as predicted by Snell's Law with a negative angle of refraction in the DNG medium, the centroids of the beam at those planes were coincident. Despite the oblique nature of the propagation, the beam did focus the beam in the DNG slab towards its back face. The discontinuities in the derivatives of the fields at the DPS–DNG interfaces are clearly seen (the V-shaped patterns at both interfaces). The results for the 3-cycle pulsed Gaussian beam interaction with the $n_{real}(\omega_0) \approx -1$ DNG slab are shown in Fig. 4.17. The simulation was run for 2500 time steps; twenty-six snapshots in time were obtained at equal intervals. Several interesting effects can be highlighted. Notice that the ultrafast pulse generates a strong surface wave. As predicted in Refs. 23 and 24, surface waves are strongly generated as the beam interacts with the DPS–DNG interface in this $n_{real}(\omega_0) \approx -1$ case. Moreover, as this surface wave propagates away from the interaction region, it

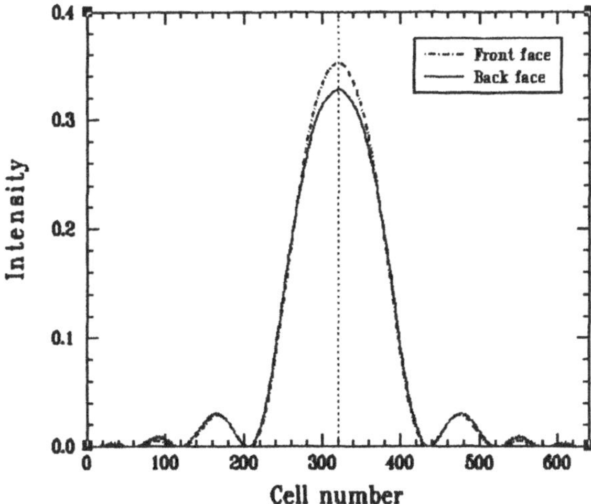

Fig. 4.14 The intensity distribution transverse to the beam axis of the phase compensator/beam translator system at the front face (at $t = 3600\Delta t$) and at the back face (at $t = 8000\Delta t$) shows that the waist of the beam intensity has been recovered as the beam propagates through the entire $4\lambda_0$ long system with only a -0.323 dB attenuation of the peak intensity.

generates a backward wave into the source side of the slab. These backward waves have been observed in, for instance, Ref. 25. Because of the broad bandwidth of the pulse, the dispersive effects of the Drude slab, the large angles involved, and the presence of several distinct wave processes, large distortions in the beam are also present in this example as it propagates through the DNG slab. The results for the CW Gaussian beam interacting with the $n_{real}(\omega_0) \approx -6$ DNG slab are shown in Fig. 4.18. The negative angle of refraction is again clearly seen. However, because of the change in wavelength in the DNG slab, the beam becomes highly compressed along the beam axis. Because the wave speed in the DNG slab correspondingly slows down by a factor of 6, the simulation was run for 8000 time steps. Again, 21 snapshots in time were obtained at equal intervals. The discontinuities in the derivatives of the fields at the DPS–DNG interfaces are again clearly seen. Note that, as also predicted in Ref. 24, surface waves are not strongly generated as the beam interacts with the DPS–DNG interface when $n_{real}(\omega_0) < 0$ and $n_{real}(\omega_0) \neq -1$. In contrast to the $n_{real}(\omega_0) \approx -1$ case, there was little focusing in the DNG slab. Also note that, as anticipated, the reflected beam is not readily apparent in this matched slab case. The results for the 3-cycle pulsed Gaussian beam interaction with the $n_{real}(\omega_0) \approx -6$ DNG slab are shown in Fig. 4.19. The simulation was run for 5000 time steps; 21 snapshots in time were obtained at equal intervals. Several interesting effects can be highlighted. Notice that even with its oblique angle of propagation

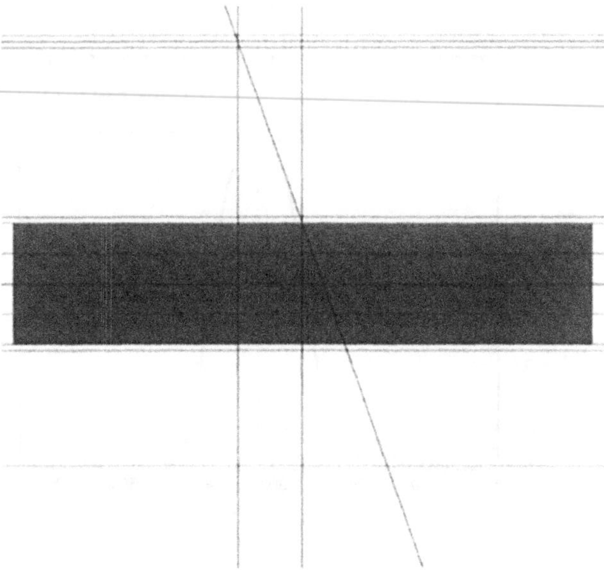

Fig. 4.15 The FDTD geometry for the oblique incidence cases illustrating the negative angle of refraction effects. The TF–SF boundary (second horizontal line from the top) and the beam axis (oblique line) are shown. The centers of the beam at the TF–SF boundary and at the front face of the slab are located at the intersection of the vertical lines with the oblique line. Electric field sampling points were located at the intersections of the right (center) vertical line and the horizontal lines. The location of the DNG slab region is shown in gray.

through the slab and with its high degree of axial compression, the propagation of the pulsed Gaussiam beam is well-behaved. Also note that its takes quite some time for the entire beam to return into the free space medium through the back face of the slab. This dispersive effect causes the ultrafast beam to become quite spread out in time. The beam again quickly expands once the centroid of the beam exits the DNG slab. All of these cases clearly show the presence and effects of the negative angle of refraction realized when an obliquely incident Gaussian beam interacts with a DNG slab. Fine resolution-in-time movies of the behavior of the electric field amplitude (rather than the intensity) in the interaction cases discussed in this section show that the phase propagation is indeed in the direction opposite to the power flow shown in the figures given here. These results thus confirm many of the fundamental properties of Gaussian beam interactions with a DNG medium.

4.5 GOOS–HÄNCHEN EFFECT

It is well known that when a Gaussian beam is obliquely incident beyond the critical angle on an interface from a higher index of refraction DPS medium to a lower one,

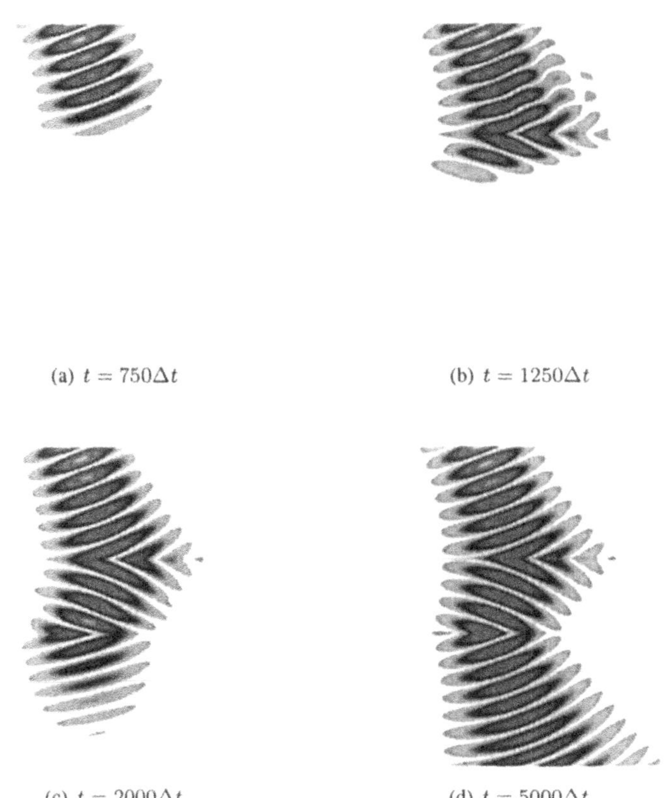

(a) $t = 750\Delta t$

(b) $t = 1250\Delta t$

(c) $t = 2000\Delta t$

(d) $t = 5000\Delta t$

Fig. 4.16 FDTD predicted electric field intensity distribution for the interaction of the CW Gaussian beam that is incident at 20° to a DNG slab having $n_{real}(\omega_0) \approx -1$. A negative angle of refraction equal and opposite to the angle of incidence is observed. After Ref. [16]. Copyright © 2003 Optical Society of America, Inc.

Fig. 4.17 FDTD-predicted electric field intensity distribution for the interaction of the 3-cycle pulsed Gaussian beam that is incident at 20° to a DNG slab having $n_{real}(\omega_0) \approx -1$. A negative angle of refraction of the transmitted pulsed beam is observed. The generation of a backward wave at the front interface is also observed.

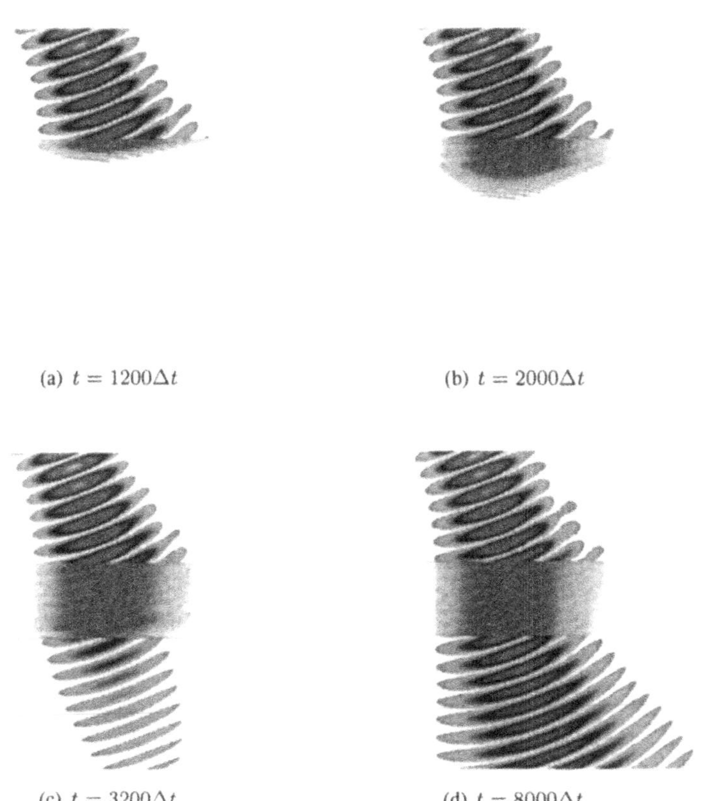

Fig. 4.18 FDTD-predicted electric field intensity distribution for the interaction of the CW Gaussian beam that is incident at 20° to a DNG slab having $n_{real}(\omega_0) \approx -6$. A shallow negative angle of refraction is observed because of the higher magnitude of the index of refraction in the DNG slab. The beam is also compressed axially in the DNG slab because of this higher refractive index. After Ref. [16]. Copyright © 2003 Optical Society of America, Inc.

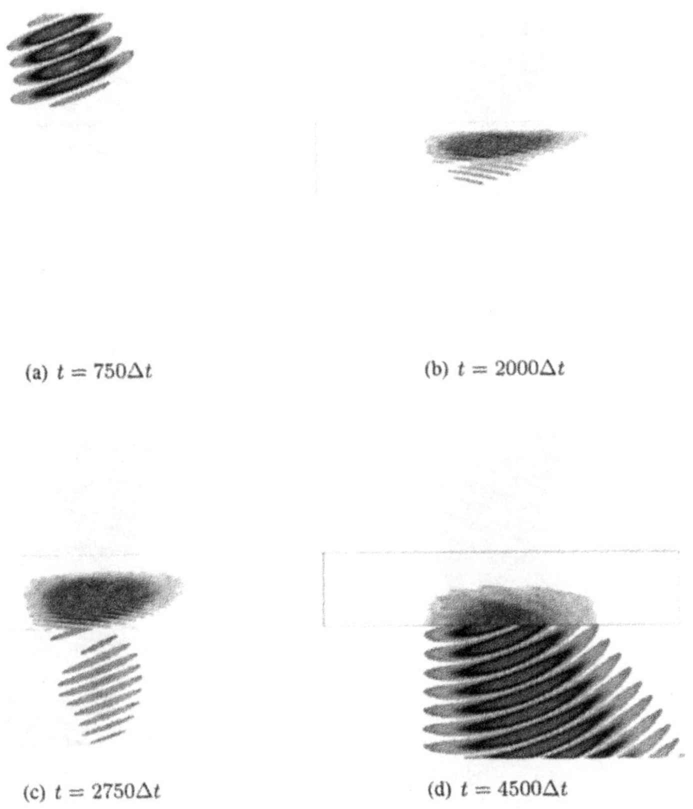

Fig. 4.19 FDTD-predicted electric field intensity distribution for the interaction of the 3-cycle pulsed Gaussian beam that is incident at 20° to a DNG slab having $n_{real}(\omega_0) \approx -6$. A negative angle of refraction of the pulsed beam refracted into the DNG slab is observed. Very large dispersion of the pulsed beam transmitted out of the back face of the DNG slab is also observed.

the centroid of the reflected Gaussian beam will experience a positive lateral shift along the interface from the purely specular reflection point of the centroid of the incident beam [26]. This Goos–Hänchen shift results from the wavevector direction dependence of the reflection coefficient (4.9). Several groups have considered the Goos-Hänchen shift when the second medium is DNG [11, 16, 17, 27, 28]. In particular, if x is the direction along the interface and z is the direction orthogonal to it, then the corresponding wavevector components k_x and k_z are related as $k_z = \sqrt{\omega^2\,\varepsilon\mu - k_x^2}$ in both media. If k_{x0} represents the parallel component of the wavevector corresponding to the center of the incident beam, then an approximate expression for the reflected beam is [26]

$$E_y(x,z) = R(k_{x0})\,e^{-jk_{x0}\Phi'(k_{x0})}\,E_{y,\,inc}(x - \Phi'(k_{x0}), z) \tag{4.24}$$

where the reflection coefficient (4.9) is rewritten in the form

$$R(k_x) = \exp\left[j\,\Phi(k_x)\right] \tag{4.25}$$

Equation (4.25) defines the phase function as $\Phi = -j\,\ln R$; its derivative $\Phi' \equiv \partial\Phi/\partial k_x$ is then readily obtained for use in (4.24).

For an incident DPS medium with $\varepsilon_r(\omega) = +9.0$ and $\mu_r(\omega) = +1.0$ (hence, $n(\omega) = +3$) and for the transmission DPS medium with $\varepsilon_r(\omega) = +3.0$ and $\mu_r(\omega) = +1.0$ (hence, $n(\omega) = +\sqrt{3}$), the critical angle is $\theta_{cr} = \sin^{-1}(n_2/n_1) = 35.26°$. Calculating the lateral shift $\Phi'(k_{x0})$ with a simple Matlab program for an angle of incidence of 40°, one finds that the shift should be $+31.6$ cm for the frequency f_0. For the corresponding FDTD simulations, this would mean that the centroid of the reflected beam would be shifted approximately 32 cells positively along the interface away from the incident beam center. If, on the other hand, the transmission medium is a DNG Drude medium with $\text{Re}[\varepsilon_r(\omega_0)] \approx -3.0$ and $\text{Re}[\mu_r(\omega_0)] \approx -1.0$, hence, $n_{real}(\omega_0) \approx -\sqrt{3}$, then the wavevector components parallel to the interface in either medium must be equal, but the components normal to it are equal in magnitude and opposite in sign. This will cause the phase term $\Phi'(k_{x0})$ to have the opposite sign in the DNG medium case. Thus, one would expect the lateral shift to be $-31.6\ cm$, hence, approximately -32 cells from the incident beam center on the interface. These theoretical results are pictorially represented in Fig. 4.20. The FDTD simulation results for the Goos–Hänchen effect in the DPS and DNG media are shown, respectively, in Figs. 4.21–4.23. The basic FDTD geometry is shown in Fig. 4.21. The CW Gaussian beam was launched with a center frequency f_0 and a $1.0\ \lambda_0$ waist from the TF–SF interface with a 40° angle of incidence in both cases. The incident beam was focused at the center of the front faces of the slabs. The TF–SF interface was $2\lambda_0 = 200$ cells from the front faces of the slabs. The slabs were $2\lambda_0 = 200$ cells deep. The simulations were run for 6000 time steps; 21 snapshots in time were obtained at equal intervals. As shown in Figs. 4.22 and 4.23, the specular-like reflection process for the above-critical-incidence Gaussian beam was realized. The presence of different beam centers is somewhat apparent when both figures are compared. However, to make this comparison quantitative, the electric field intensity distribution measured at $t = 6000\,\Delta t$ along the plane two

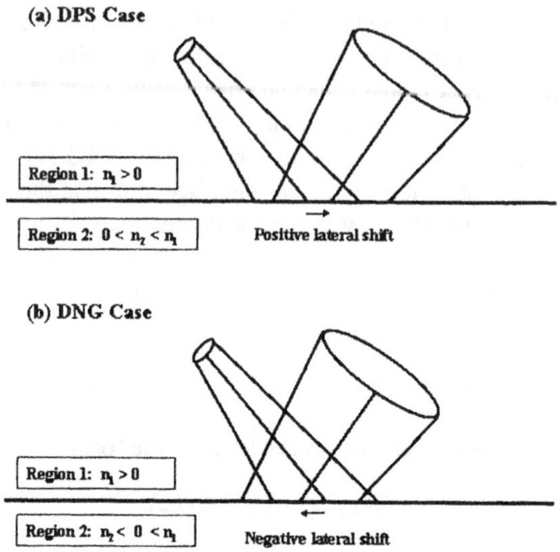

Fig. 4.20 A Gaussian beam obliquely incident from a higher refractive index magnitude medium to a lower one with an angle of incidence beyond the critical angle will generate a reflected beam that experiences (a) a positive Goos–Hänchen lateral shift in a DPS medium and (b) a negative Goos–Hänchen lateral shift in a DNG medium.

cells in front of the TF–SF plane for the DPS and for the DNG cases are shown in Figs. 4.24 and 4.25, respectively. The location in this plane of the initial beam center and the specularly reflected beam center as well as the location of the predicted Goos–Hänchen-shifted beam centers are indicated. Figures 4.24 and 4.25 clearly show the opposite lateral shifts between the DPS and the DNG cases. An analysis of the centroids of the reflected beams yielded a lateral shift of approximately +31 cells in the DPS case and −33 cells in the DNG case, in very reasonable agreement with the predicted values of +32 cells and −32 cells, respectively. The small discrepancy appears to be due to the FDTD sampling location which is slightly in front of the TF–SF plane where the analytical result is obtained. Simulations with finer-resolution FDTD meshes show a decrease in the discrepancy. Note that in the DNG case, there is a time delay for the emergence of the reflected beam from the DNG slab, in agreement with the DNG dispersion behavior characterized in Ref. 29. Also note that there is some wave penetration into both the DPS and DNG slabs as shown in Figs. 4.22 and 4.23, respectively. In the DPS case the penetration occurs with a positive angle of refraction, but in the DNG case it occurs with a negative angle of refraction.

4.6 SUBWAVELENGTH FOCUSING WITH A CONCAVE DNG LENS

We note that the reason that we considered expanding Gaussian beams in all of the cases presented up to this point is the inability of a planar DNG slab to focus a flat

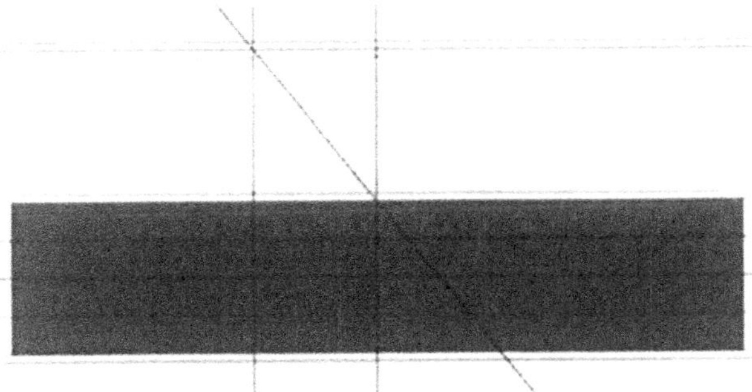

Fig. 4.21 The FDTD geometry for the Goos–Hänchen cases. The TF–SF boundary (second horizontal line from the top) and the beam axis (oblique line) are shown. The centers of the beam at the TF–SF boundary and at the front face of the slab are located at the intersection of the vertical lines with the oblique line. Electric field sampling points were located at the intersections of the center (right) vertical line and the horizontal lines. The location of the DNG slab region is shown in gray.

beam or plane wave. The negative angle of refraction can occur only if there is oblique incidence. To focus a flat Gaussian beam (one with nearly an infinite radius of curvature), one must resort to a curved lens. However, in contrast to focusing (diverging) a plane wave with a convex (concave) lens composed of a DPS medium, one must consider focusing (diverging) a plane wave with a concave (convex) lens composed of a DNG medium.

The FDTD concave DNG lens geometry is shown in Fig. 4.26. A Gaussian beam with a waist of 2 λ_0 = 200 cells was launched from the TF–SF boundary and was normally incident on the concave lens. The TF–SF boundary was 2 λ_0 = 200 cells from the front of the lens. The lens was a DNG medium with $n_{real}(\omega_0) \approx -1$. It was formed by removing a parabolic section from the back side of a slab that was 1.5 λ_0 = 150 cells deep and 6 λ_0 = 600 cells wide. If (x_0, z_0) denotes the location of the focus and f_{DNG} denotes the focal length, the parabolic section was defined by relation

$$z - z_0 = \frac{(x - x_0)^2}{4\, f_{DNG}} - f_{DNG} \tag{4.26}$$

The focal length was set to be $f_{DNG} = \lambda_0 = 100$ cells . The location of the focus was chosen to be at the center of the back face of the slab. Thus, the parabolic section began 0.5 λ_0 = 50 cells into the slab and terminated 1.5 λ_0 = 150 cells from its front face. The full width of the removed parabolic section at the back face was $2 \times 2f_{DNG} = 4\lambda_0 = 400$ cells.

A DPS plano-convex lens of index n_{DPS} with a similar radius of curvature $R = 2f_{DNG} = 2\lambda_0$ (the dark gray region in Fig. 4.26) would have a focus located a distance $f_{DPS} = R/(n_{DPS} - 1) = 2\lambda_0/(n_{DPS} - 1)$ from its back face. Thus, to have the focal point within the very near field , as it is in the DNG case, the index of refraction would have to be very large. In fact, to have it located at the

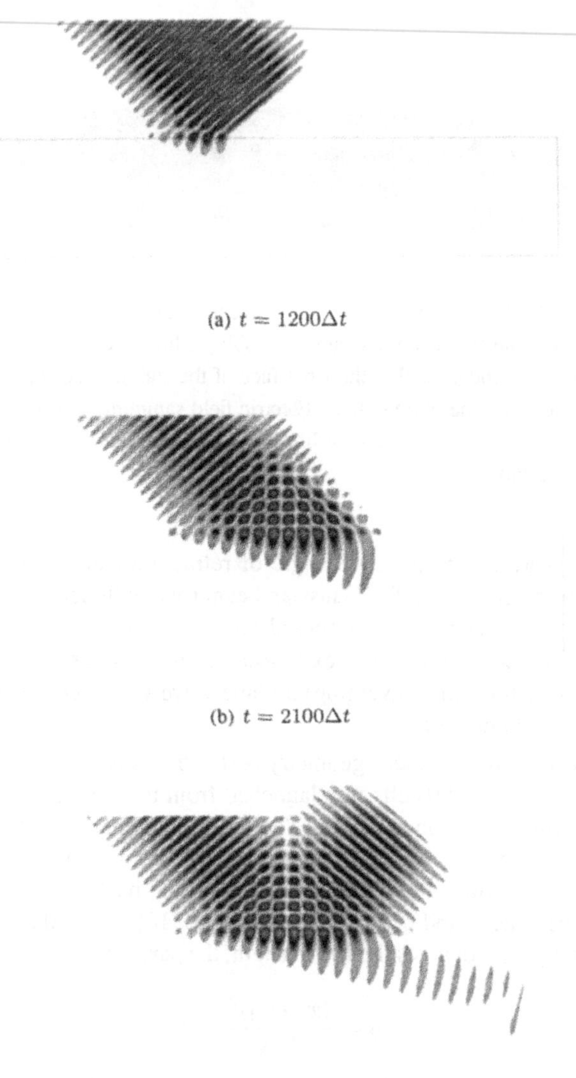

(a) $t = 1200\Delta t$

(b) $t = 2100\Delta t$

(c) $t = 4800\Delta t$

Fig. 4.22 FDTD-predicted electric field intensity distribution for the interaction of the CW Gaussian beam and the DPS slab. The beam is incident at 40° in a DPS medium with $\varepsilon_r(\omega) = +9.0$ and $\mu_r(\omega) = +1$ [hence, $n(\omega) = +3$] onto a DPS slab having $\varepsilon_r(\omega) = +3.0$ and $\mu_r(\omega) = +1$ [hence, $n(\omega) = +\sqrt{3}$]. Some penetration of the beam into the slab occurs with a positive angle of refraction. The reflected beam propagates away from the interface in the total field region through the TF–SF boundary into the scattered field region. After Ref. [16]. Copyright © 2003 Optical Society of America, Inc.

(a) $t = 1200\Delta t$

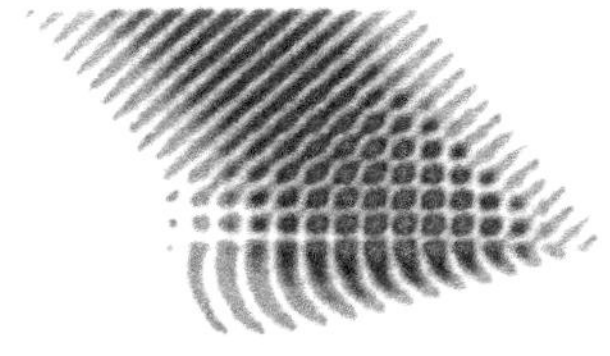

(b) $t = 2100\Delta t$

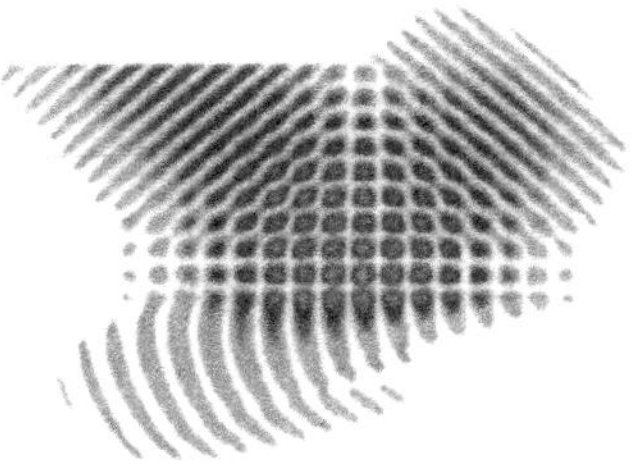

(c) $t = 4800\Delta t$

Fig. 4.23 FDTD-predicted electric field intensity distribution for the interaction of the CW Gaussian beam and the DNG slab. The beam is incident at 40° in a DPS medium with $\varepsilon_r(\omega) = +9.0$ and $\mu_r(\omega) = +1$ [hence, $n(\omega) = +3$] onto a DNG slab having $\mathrm{Re}[\varepsilon_r(\omega_0)] = -3.0$ and $\mathrm{Re}[\mu_r(\omega_0)] = -1$ [hence, $n_{real}(\omega) = -\sqrt{3}$]. Some penetration of the beam into the slab occurs with a negative angle of refraction. The reflected beam propagates away from the interface in the total field region through the TF–SF boundary into the scattered field region.

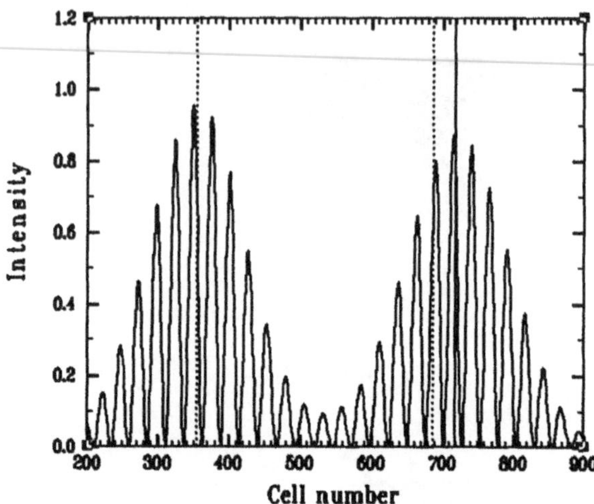

Fig. 4.24 The electric field intensity distribution measured at $t = 6000\Delta t$ at two cells in front of the TF–SF plane for the total internal reflection DPS slab case. The positions of the incident beam center and the specularly reflected beam center are indicated by the dashed vertical black lines. The theoretical positive Goos–Hänchen shift position is indicated by the vertical solid black line. After Ref. [16]. Copyright © 2003 Optical Society of America, Inc.

back face would require $n_{DPS} \to \infty$. This would also mean that very little of the incident beam would be transmitted through such a high-index lens because the magnitude of the reflection coefficient would approach one; that is, the process would become very inefficient. In contrast, the DNG lens achieves a greater bending of the incident waves with only moderate absolute values of the refractive index and can be matched to the incident medium. Moreover, since the incident beam waist occurs at the lens, the expected waist of the focused beam would be $w_{focus} \approx (f_{DPS}/L_R)w_0 = (\lambda_0 f_{DPS})/(\pi w_0) = \lambda_0/[\pi(n_{DPS}-1)$ [30]. Hence, for a normal glass lens $n_{DPS} \approx 1.5$, the waist at the focus would be $w_{focus} \approx \lambda_0/1.57 \approx 64$ cells and the corresponding intensity half-max waist would be $0.589\ w_{focus} \approx 38$ cells. The longitudinal size of the focus is the depth of focus, which for the normal glass lens would be $2(\pi w_{focus}^2/\lambda_0) \approx 257$ cells. Again, to achieve a focus that is significantly subwavelength using a DPS lens, a very large index value would be required and would lead to similar disadvantages in comparison to the DNG lens.

The FDTD-predicted electric field intensity distributions are shown in Fig. 4.27. The simulation was first run for 10000 time steps; 26 snapshots in time were obtained at equal intervals. The selected snapshots illustrate the focusing of the beam by the lens. It was found that the peak intensity in the focal region varies its location periodically. This behavior is illustrated by Figs. 4.27c and 4.27d. A second simulation was also run for 10000 time steps; 21 snapshots in time were obtained at

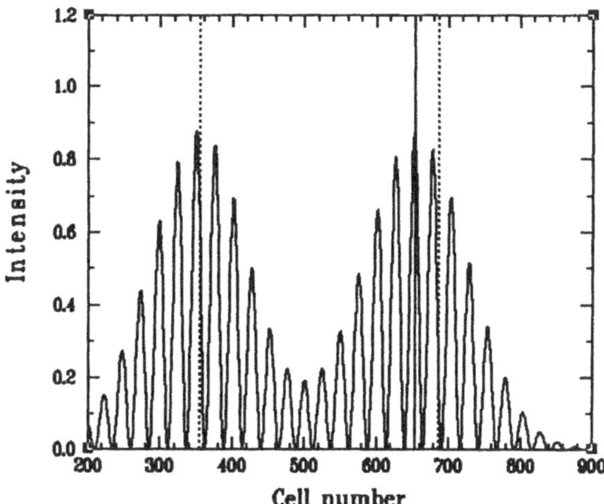

Fig. 4.25 The electric field intensity distribution measured at $t = 6000\Delta t$ at two cells in front of the TF–SF plane for the total internal reflection DNG slab case. The positions of the incident beam center and the specularly reflected beam center are indicated by the dashed vertical black lines. The theoretical negative Goos–Hänchen shift position is indicated by the vertical solid black line.

equal intervals between time steps 9000 and 10000. It was found that the intensity peaked at the focal point, for instance, at $t = 9100\ \Delta t$. The FDTD predicted electric field intensity distribution at that time is shown in Fig. 4.28. In comparison with Figs. 4.27c and 4.27d, the peak occurs in the center of the focal region as expected. The electric field intensity along the beam axis at $t = 9100\ \Delta t$ is shown in Fig. 4.29. The corresponding electric field intensity transverse to the beam axis at the rear face of the lens is shown in Fig. 4.30 and is compared there to the intensity across its front face at $t = 2000\ \Delta t$ (before the waves reflected from the concave section interact with the sample points on the front face). The radius of the focus along the beam axis (half-intensity radius) is measured to be 19 cells $\approx \lambda_0/5$ and along the transverse direction it is 17 cells $\approx \lambda_0/6$. This subwavelength focal region, which is achieved with a matched lens whose index magnitude $|n_{real}(\omega_0)| \approx 1$, is significantly smaller than would be expected from the corresponding, traditional DPS lens.

Even though the focal point is in the extreme near field of the lens, the focal region is nearly symmetrical and has a resolution that is much smaller than a wavelength. Such a subwavelength source has a variety of favorable features that may have applications, for example, in high-resolution imaging with near-field scanning microscopy (NSOM) systems. In particular, the field intensity has been concentrated into a subwavelength region without a guiding structure. It could thus act as a much smaller NSOM aperture source than is available with a typical tapered optical fiber probe and without the associated aperture effects.

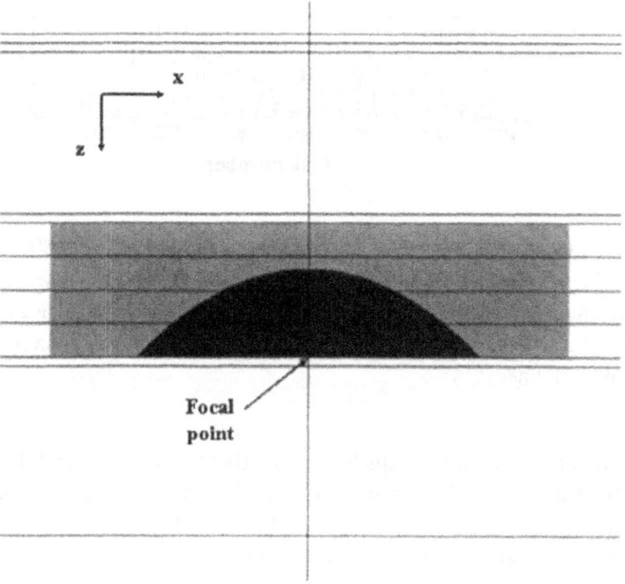

Fig. 4.26 The TF–SF boundary (second horizontal line from the top) and the beam axis (center vertical line) are shown. Electric field sampling points were located at the intersections of the center vertical line and the horizontal lines. The location of the DNG lens region is shown in light gray. The dark gray region is air, as is all of the white region surrounding the lens.

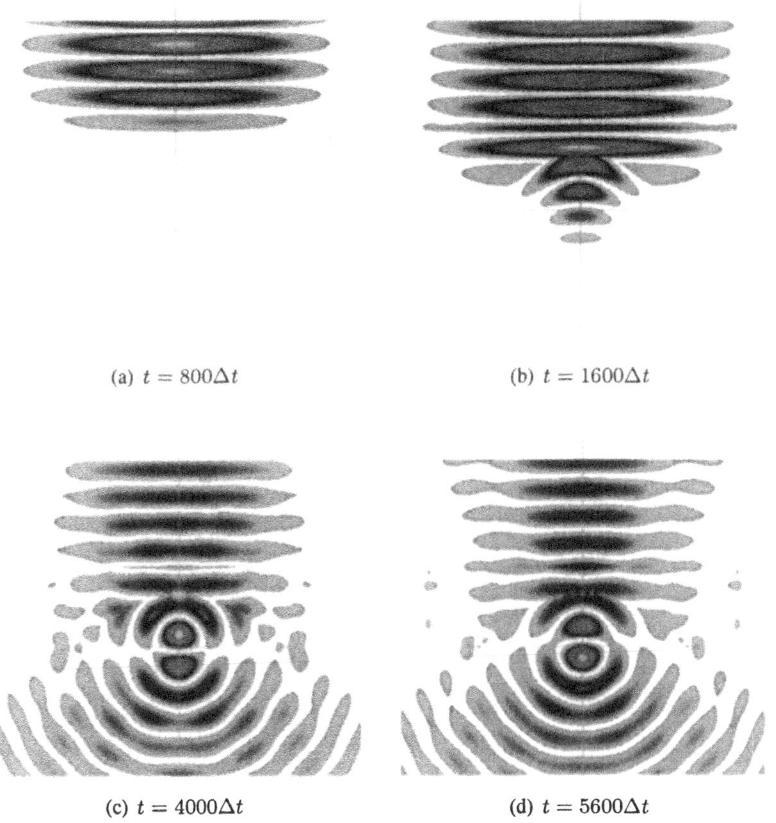

Fig. 4.27 FDTD-predicted electric field intensity distributions for the DNG concave lens. The concave DNG lens focuses the beam as expected. The slab outline and the beam axis are provided for visual references.

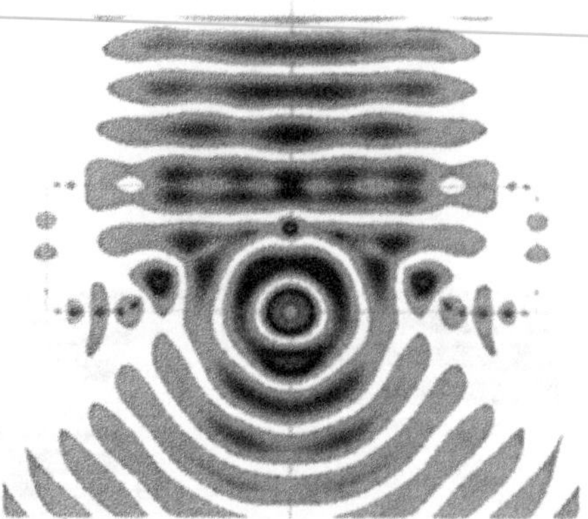

Fig. 4.28 FDTD-predicted electric field intensity distributions for the DNG concave lens at $t = 9100\Delta t$. The peak of the intensity occurs at the predicted focal point.

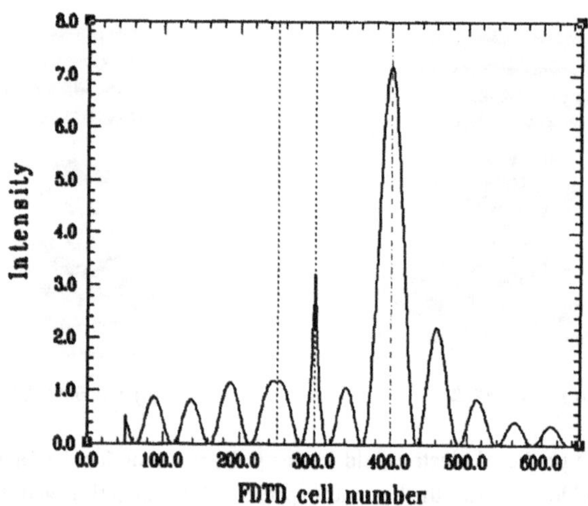

Fig. 4.29 FDTD-predicted electric field intensity distributions for the DNG concave lens at $t = 9100\Delta t$ along the beam axis. The locations of the front and back faces of the lens along the beam axis are defined by the dashed vertical lines. The location of the predicted focus is defined by the dot-dashed vertical line. The peak of the intensity occurs at the predicted focal point.

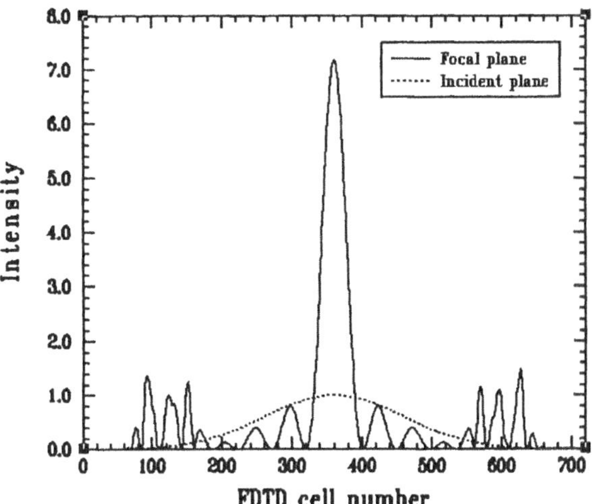

Fig. 4.30 FDTD-predicted electric field intensity distribution for the DNG concave lens at $t = 9100\Delta t$ along the back face and transverse to the beam axis. The corresponding intensity across the front face at $t = 2000\Delta t$ is also shown. Significant focusing of the intensity of the beam at the predicted focal point has been realized with the concave lens.

4.7 CONCLUSIONS

The interactions of Gaussian beams with DPS and DNG slabs were illustrated with numerical FDTD simulations. Both normal incidence and oblique incidence cases were considered. The normal incidence results demonstrated that Gaussian beams can be focused using a planar DNG slab. The oblique incidence results clearly demonstrated the negative refractive angle behavior associated with a DNG-DPS interface. Power flow at the negative angles predicted by Snell's Law was confirmed. The negative refractive effects were realized with both many-cycle (CW) and 3-cycle pulsed Gaussian beams. A negative lateral Goos–Hänchen shift was demonstrated for a beyond-critical-angle Gaussian beam scattering from a DNG slab in comparison to the usual positive lateral Goos–Hänchen shift realized with the corresponding DPS configuration.

A number of interesting applications for DNG media arise from these results. Subwavelength focusing from a planar slab was demonstrated. A phase compensator/beam translator constructed from a stack of DPS–DNG slabs was presented that translated an input beam to essentially the same beam at the output of the system with low loss over several wavelengths. Focusing using a concave DNG lens was illustrated. It was shown that, in contrast to typical DPS lens systems, the focal region can be made to be significantly subwavelength and nearly symmetrical.

A variety of planar and volumetric realizations of DNG metamaterials have been reported in the literature and experiments with them have confirmed various aspects of the DNG effects demonstrated here. Similar physics and engineering aspects of DNG metamaterials have also begun to be considered in the millimeter, terahertz, and optical regimes. However, it should be noted that other approaches to achieve similar effects have also been considered. For instance, the negative refractive angle behavior has already begun to be exploited through the DPS photonic band-gap super-prism effect (e.g., see Refs. 31–33), for applications which include WDM switches and couplers.

Numerical experiments such as those discussed here will be a dominating design environment for any eventual successful exploitation of the DNG metamaterial properties. Their flexibility in handling various types of excitations, material choices, and complex combinations and configurations of those materials provide an advantageous approach to study the physics and engineering aspects of metamaterials. Nonetheless, the eventual usefulness of metamaterials for many potential practical applications will depend greatly on clever fabrication concepts and implementations in those scenarios. The metamaterial area is rich in novel physical effects; their engineering realizations may have profound impact on a number of devices and systems.

Acknowledgments

I would like to thank Professors Keith Balmain and George Eleftheriades for giving me the opportunity to contribute to their book effort. This work was supported in part by DARPA under Contract MDA972-03-100.

REFERENCES

1. J. A. Kong, *Electromagnetic Wave Theory*, John Wiley & Sons, New York, 1986.
2. A. Ishimaru, *Electromagnetic Wave Propagation, Radiation, and Scattering*, Prentice Hall, Englewood Cliffs, NJ, 1991, pp. 36–38.
3. V. G. Veselago, "The electrodynamics of substances with simultaneously negative values of ϵ and μ," *Sov. Phys. Usp.*, vol. 10, pp. 509–514, 1968.
4. D. R. Smith, W. J. Padilla, D. C. Vier, S. C. Nemat-Nasser, and S. Schultz, "Composite medium with simultaneously negative permeability and permittivity," *Phys. Rev. Lett.*, vol 84, pp. 4184–4187, 2000.
5. J. B. Pendry, "Negative refraction makes a perfect lens," *Phys. Rev. Lett.*, vol. 85, pp. 3966–3969, 2000.
6. D. R. Smith and N. Kroll, "Negative refractive index in left-handed materials," *Phys. Rev. Lett.*, vol. 85, pp. 2933–2936, 2000.
7. R. A. Shelby, D. R. Smith, S. C. Nemat-Nasser, and S. Schultz, "Microwave transmission through a two-dimensional, isotropic, left-handed metamaterial," *Appl. Phys. Lett.*, vol. 78, pp. 489–491, 2001.
8. A. Shelby, D. R. Smith, and S. Schultz, "Experimental verification of a negative refractive index of refraction," *Science*, vol. 292, pp. 77–79, 2001.
9. R. W. Ziolkowski and E. Heyman, "Wave propagation in media having negative permittivity and permeability," *Phys. Rev. E* vol. 64, 056625, 2001.
10. C. Caloz, C.-C. Chang and T. Itoh, "Full-wave verification of the fundamental properties of left-handed materials in waveguide configurations," *J. Appl. Phys.*, vol. 90, p. 5483, 2001.
11. J. A. Kong, B.-I. Wu and Y. Zhang, "A unique lateral displacement of a Gaussian beam transmitted through a slab with negative permittivity and permeability," *Microwave Opt. Tech. Lett.*, vol. 33, p. 136, 2002.
12. P. M. Valanju, R. M. Walter, and A. P. Valanju, "Wave refraction in negative-index media: Always positive and very inhomogeneous," *Phys. Rev. Lett.*, vol. 88, 187401, 2002.

13. G. V. Eleftheriades, A. K. Iyer and P. C. Kremer, "Planar negative refractive index media using periodically L–C loaded transmission lines," *IEEE Trans. Microwave Theory Tech.*, vol. 50, pp. 2702–2712, 2002.

14. K. G. Balmain, A. A. Luttgen, and P. C. Kremer, "Resonance cone formation, reflection, refraction and focusing in a planar, anisotropic metamaterial," in *Proceedings of the URSI National Radio Science Meeting*, San Antonio, TX, July 2002, p. 45.

15. R. W. Ziolkowski, "Design, fabrication, and testing of double negative metamaterials," *IEEE Trans. Antennas Propag.* vol. 51, p. 1516, 2003.

16. R. W. Ziolkowski, "Pulsed and CW Gaussian beam interactions with double negative metamaterial slabs," *Opt. Express*, vol. 11, p. 662, 2003.

17. R. W. Ziolkowski , "Pulsed and CW Gaussian beam interactions with double negative metamaterial slabs: errata," *Opt. Express*, vol. 11, p. 1596, 2003.

18. A. Taflove, *Computational Electrodynamics: The Finite-Difference Time-Domain Method*, Artech House, Norwood, MA, 1995.

19. A. Taflove, ed., *Advances in Computational Electrodynamics: The Finite-Difference Time-Domain Method*, Artech House, Norwood, MA, 1998.

20. D. C. Wittwer and R. W. Ziolkowski, "Two time-derivative Lorentz material (2TDLM) formulation of a Maxwellian absorbing layer matched to a lossy media," *IEEE Trans. Antennas Propag.*, vol. 48, p. 192, 2000.

21. D. C. Wittwer and R. W. Ziolkowski, "Maxwellian material based absorbing boundary conditions for lossy media in 3D," *IEEE Trans. Antennas Propag.*, vol. 48, p. 200, 2000.

22. J. B. Judkins, C. W. Haggans, and R. W. Ziolkowski, "2D-FDTD simulation for rewritable optical disk surface structure design," *Appl. Opt.*, vol. 35, p. 477, 1996.

23. M. W. Feise, P. J. Bevelacqua, and J. B. Schneider, "Effects of surface waves on behavior of perfect lenses," *Phys. Rev. B*, vol. 66, 035113, 2002.

24. A. Ishimaru and J. Thomas, "Transmission and focusing of a slab of negative refractive index," in *Proceedings of the URSI National Radio Science Meeting*, San Antonio, TX, July 2002, p. 43.

25. A. Grbic and G. V. Eleftheriades, "Experimental verification of backward-wave radiation from a negative refractive index metamaterial," *J. Appl. Phys.*, vol. 92, p. 5930, 2002.

26. *op. cit.* A. Ishimaru [2], pp. 165–169.

27. J. A. Kong, B.-I. Wu, and Y. Zhang, "Lateral displacement of a Gaussian beam reflected from a grounded slab with negative permittivity and permeability," *Appl. Phys. Lett.*, vol. 80, p. 2084, 2002.

28. I. V. Shadrivov, A. A. Zharov, and Y. S. Kivshar, "Giant Goos–Hänchen effect at the reflection from left-handed metamaterials," *Appl. Phys. Lett.*, vol. 83, p. 2713, 2003.

29. R. W. Ziolkowski and A. Kipple, "Causality and double-negative metamaterials," *Phys. Rev. E*, vol. 68, 026615, 2003.

30. B. E. A. Saleh and M. C. Teich, *Fundamentals of Photonics*, John Wiley & Sons, New York, 1991, pp. 94–95.

31. H. Kosaka, T. Kawashima, A. Tomita, M. Notomi, T. Tamamura, T. Sato, and S. Kawakami, "Superprism phenomena in photonic crystals," *Phys. Rev. B*, vol. 58, no. R10096, 1998.

32. M. Notomi, "Theory of light propagation in strongly modulated photonic crystals: Refractionlike behavior in the vicinity of the photonic band gap," *Phys. Rev. B*, vol. 62, 10696, 2000.

33. C. Luo, S. G. Johnson, J. D. Joannopoulos, and J. B. Pendry, "All-angle negative refraction without negative effective index," *Phys. Rev. B*, vol. 65, no. 201104(R), 2002.

5 Negative Index Lenses

DAVID SCHURIG and DAVID R. SMITH†

Department of Physics
University of California at San Diego
La Jolla, CA 92093
United States

5.1 INTRODUCTION

The index of refraction of a material is a commonly used parameter that describes some of the most fundamental interactions between a material and an electromagnetic wave. A wave incident on the interface between two materials having different refractive indices will have its trajectory bent—or refracted—by an amount determined by the angle of incidence, and the ratio of the refractive indices of the two materials. The refractive index is usually taken with reference to vacuum. Air, for example, has a refractive index close to unity over most of the electromagnetic spectrum, indicating that there is little difference in wave propagation in air versus vacuum.

The phenomenon of refraction enables a material to alter the paths of electromagnetic waves incident on its interface. Because the path of the wave is changed in a manner depending on the angle made with respect to the interface, the interface can be shaped so that specific functions or operations can be performed on the incident wave. One of the most important such operations is focusing, and in this case the shaped material is called a *lens*. Lenses have widespread application and are used at nearly all electromagnetic wavelengths, from radio to optical. Lens design and fabrication (using positive index materials) are mature and sophisticated technologies. Imaging systems having numerous lens elements, each with different surface curvature and material composition, are routinely produced with increasingly greater precision.

However, the ultimate quality that can be realized by a lens system is limited by the materials available. Formerly, all known materials transparent to electromagnetic

†*Present address*: Department of Electrical and Computer Engineering, Duke University, Durham, NC 27708, United States.

radiation have had a positive refractive index. Consequently, all known lenses have had a positive refractive index. The optimization of a lens system that used conventional materials therefore included this inherent constraint. However, this constraint is not fundamental in origin. As first pointed out by Victor Veselago, and as we will discuss, the refractive index of a material can, in principle, be negative. Veselago noted that lens elements produced from negative index materials—if such were ever found—would behave in a very different way from positive index lenses. For example, a convex negative index lens would cause an incident beam to diverge rather than to converge. Likewise, a concave negative index lens would act to focus an incident beam [1].

In addition to changing the nature of convex and concave lenses, Veselago also noted that a *planar* slab with a refractive index equal to minus one could refocus the rays from a nearby source—something not possible with any positive index material. This property of a negative refractive index has proven in recent years to have greater consequences. A Fourier optic analysis of the planar negative index slab reveals that it can produce a focus with greater resolution than suggested by the diffraction limit associated with all previously known passive optical elements. Because of this unexpected property, the planar slab has been called a "perfect lens," although it has little in common with traditional lenses [2].

While the perfect lens does offer a working distance equal to its thickness, it does not possess a focal length and does not focus radiation from distant sources. Since many applications (cameras, telescopes, antennas, etc.) require the ability to focus radiation from distant objects, the detailed behavior of negative index lenses with curved surfaces is of interest. Such lenses can focus far field radiation in the same manner as traditional positive index lenses. Negative refractive index therefore increases the parameter space for lens design and provides several important advantages. Spherical profile lenses composed of negative index media can be more compact, they can be matched to free space, and here we demonstrate that they can also have superior focusing performance.

The monochromatic imaging quality of a lens can be characterized by the five Seidel aberrations: spherical, coma, astigmatism, field curvature, and distortion. These well-known corrections to the simple Gaussian optical formulas are calculated from a fourth-order expansion of the deviation of a wave front from spherical. (A spherical wave front converges to an ideal point focus in ray optics.) The coefficients in this expansion quantify the non-ideal focusing properties of an optical element for a given object and image position. There is an asymmetry of several of the Seidel aberrations with respect to index about zero. Considering that an interface with a relative index of $+1$ is inert and one of relative index -1 is strongly refractive, this asymmetry is not surprising. However, we will conclude that this asymmetry can yield superior focusing properties for negative index lenses. The basis for this assertion is not obvious.

The purpose of this chapter is to explain how to perform a geometric optic analysis of optical systems–particularly lenses–that incorporate isotropic negative index media. Hopefully, this chapter will elucidate all the analytical steps that lead to the conclusions above. Section 5.2 derives Fermat's Principle from Maxwell's equations allowing for the possibility of simultaneously negative electric permittivity and magnetic permeability. This establishes the validity of geometric optics for this case and yields an appropriate generalized definition for refractive index. All the most important results of geometric optics, such as equal angle reflection, Snell's Law, and Gaussian imaging, can be derived from Fermat's Principle. Section 5.3 describes the Gaussian optic results of spherical surfaces and interfaces separating media of opposite refractive index sign. Section 5.4 goes beyond Gaussian optics and describes aberration calculations, with very interesting results for thin lenses.

5.2 GEOMETRIC OPTICS

Maxwell's equations together with scalar response functions, ε and μ, are a valid description of electromagnetic fields in linear isotropic media, regardless of the sign of ε and μ. Negative material response is a well known property of resonant magnetic or electric systems [3], and Maxwell's equations have been accurately describing these systems for quite some time. However, in the past, these systems always had either magnetic *or* electric resonances, and never both in the same frequency range. If only one response function is negative, the wavevector, which obeys

$$\mathbf{k} \cdot \mathbf{k} = \omega^2 \varepsilon \mu$$

must have a significant imaginary part, and the wave solutions

$$e^{i(\mathbf{k} \cdot \mathbf{r} - \omega t)}$$

will not propagate over significant distances. Geometric optics, which is a useful simplification that is accurate only when all material, and field variation length scales are large compared with wavelength, is not then applicable. Thus the question of extending geometric optic concepts like Snell's Law or Fermat's Principle to these systems does not arise. With the recent demonstration of simultaneously negative ε and μ [4] and negative refraction [5], it is important to re-derive Fermat's Principle with attention to the signs of the response functions. Our approach will be to derive Fermat's Principle from the eikonal equation of geometric optics, and is quite different from Veselago's original derivation of Snell's Law from the boundary matching of fields [1].

In this section we will derive a variational principle of the form

$$\delta \int_C f(\varepsilon, \mu)\, ds = 0$$

where f is some, as yet undetermined, function of the material properties and C is the path of a light ray. This equation states that the integral of the function f is stationary with respect to small variations of the path from the physical path of a light ray. We will use only Maxwell's equations, the constitutive relations, and the assumption that a light ray follows the Poynting vector. We will then see the relationship this function f has to the usual definition of refractive index

$$n \equiv \sqrt{\varepsilon\mu} \tag{5.1}$$

Throughout this chapter the square root symbol will denote the positive square root and will only be applied to real positive numbers.

Maxwell's source free equations (in SI units) are

$$\nabla \cdot \mathbf{D} = 0 \tag{5.2a}$$

$$\nabla \cdot \mathbf{B} = 0 \tag{5.2b}$$

$$\nabla \times \mathbf{E} = -\frac{\partial \mathbf{B}}{\partial t} \tag{5.2c}$$

$$\nabla \times \mathbf{H} = \frac{\partial \mathbf{D}}{\partial t} \tag{5.2d}$$

and the constitutive relations are

$$\mathbf{D} = \varepsilon_0 \varepsilon \mathbf{E}$$
$$\mathbf{B} = \mu_0 \mu \mathbf{H}$$

where the unsubscripted ε and μ refer to the relative permittivity and permeability. We will use solutions of the form

$$\mathbf{E} = \mathbf{e}(\mathbf{r})\, e^{ik_0\zeta(\mathbf{r}) - i\omega t} \tag{5.3a}$$

$$\mathbf{H} = \mathbf{h}(\mathbf{r})\, e^{ik_0\zeta(\mathbf{r}) - i\omega t} \tag{5.3b}$$

where $\zeta(\mathbf{r})$ is a real function of position, and $\mathbf{e}(\mathbf{r})$ and $\mathbf{h}(\mathbf{r})$ are potentially complex functions of position. These solutions are completely general since we have arbitrary functions representing both the phase and amplitude of the harmonic field at every point in space. Later, to enter the geometric optic approximation, we will assume that $\mathbf{e}(\mathbf{r})$ and $\mathbf{h}(\mathbf{r})$ are slowly varying amplitude functions, and that the relatively rapidly varying oscillation of these wave solutions is represented by $\zeta(\mathbf{r})$. This latter function, called the eikonal, appears in the phase factor,

$$e^{ik_0\zeta(\mathbf{r})} = e^{i\phi(\mathbf{r})}$$

where we define $\phi(\mathbf{r})$ to be the phase as a function of position. We note that ζ represents the phase in units of the free-space wavelength.

$$\zeta = \frac{\phi}{2\pi}\lambda_0$$

If the phase, ϕ, changes by 2π, then the eikonal, ζ, changes by the free space wavelength, λ_0. We also use the eikonal to define what we mean by a light ray. A light ray is a continuous path that is normal to the surfaces of constant ζ. These surfaces we call the wave fronts or phase fronts.

Substituting (5.3) into (5.2) and using the vector identities,

$$\nabla \cdot \left(\mathbf{A}e^{\psi}\right) = \left(\nabla \cdot \mathbf{A} + \nabla\psi \cdot \mathbf{A}\right) e^{\psi}$$
$$\nabla \times \left(\mathbf{A}e^{\psi}\right) = \left(\nabla \times \mathbf{A} + \nabla\psi \times \mathbf{A}\right) e^{\psi}$$

with some rearrangement we arrive at

$$\nabla\zeta \times \eta_0\mathbf{h} + \varepsilon\mathbf{e} = \frac{1}{ik_0}\nabla\times\eta_0\mathbf{h} \tag{5.4a}$$

$$\nabla\zeta \times \mathbf{e} - \mu\eta_0\mathbf{h} = \frac{1}{ik_0}\nabla \times \mathbf{e} \tag{5.4b}$$

$$\nabla\zeta \cdot \mathbf{e} = \frac{1}{ik_0}\left(\frac{1}{\varepsilon}\nabla\varepsilon \cdot \mathbf{e} + \nabla \cdot \mathbf{e}\right) \tag{5.4c}$$

$$\nabla\zeta \cdot \mathbf{h} = \frac{1}{ik_0}\left(\frac{1}{\mu}\nabla\mu \cdot \mathbf{h} + \nabla \cdot \mathbf{h}\right) \tag{5.4d}$$

where

$$\eta_0 \equiv \sqrt{\frac{\mu}{\varepsilon}}$$

is the impedance of free space. This quantity is just the ratio of the magnitude of the electric to the magnetic field for a free-space plane wave. In the SI system $\eta_0\mathbf{H}$ and $\mathbf{E}$ have the same units.

Now we make the geometric optic approximation. We assume that the spatial variation of the amplitude factors, $\mathbf{e}(\mathbf{r})$ and $\mathbf{h}(\mathbf{r})$, and the material response functions, ε and μ, are slow compared to the phase variation given by the eikonal in the factor $e^{ik_0\zeta(\mathbf{r})}$. All derivatives of the amplitude factors and the material response functions appear on the right-hand side of (5.4), and the derivatives of the eikonal appear on the left-hand side of (5.4). Formally, we arrive at the geometric optic approximation by taking the limit of (5.4) as $\lambda_0 \to 0$—that is, as $k_0 \to \infty$—and obtain

$$\nabla\zeta \times \eta_0\mathbf{h} + \varepsilon\mathbf{e} = 0, \tag{5.5a}$$

$$\nabla\zeta \times \mathbf{e} - \mu\eta_0\mathbf{h} = 0, \tag{5.5b}$$

$$\nabla\zeta \cdot \mathbf{e} = 0, \tag{5.5c}$$

$$\nabla\zeta \cdot \mathbf{h} = 0 \tag{5.5d}$$

Note that this limiting procedure is valid arbitrarily close to a material discontinuity, as long as the interface radius of curvature is large compared to wavelength. If we add the condition that light rays are continuous across discontinuities, as required by the conservation of energy, then geometric optics and Fermat's Principle are still valid. Now we can eliminate the field amplitudes and solve for the eikonal. Solving (5.5b) for $\eta_0\mathbf{h}$ and substituting into (5.5a), we obtain

$$\nabla\zeta \times (\nabla\zeta \times \mathbf{e}) + \varepsilon\mu\mathbf{e} = 0$$

Using the usual vector identity for the double cross product we find

$$\nabla\zeta(\nabla\zeta\cdot\mathbf{e})-\mathbf{e}\,(\nabla\zeta\cdot\nabla\zeta)+\varepsilon\mu\mathbf{e}=0 \tag{5.6}$$

The first term in (5.6) is zero by (5.5c). Since e is not zero everywhere, we arrive at

$$\nabla\zeta\cdot\nabla\zeta=\varepsilon\mu. \tag{5.7}$$

This dimensionless equation is referred to as the eikonal equation of geometric optics. Normally, the right-hand side is replaced by the square of the refractive index, but we will leave it as is until we establish an appropriate definition for this quantity. Up to this point, we have not deviated much from the derivations in Born and Wolf [6] or Kong [7].

The eikonal equation is one of the primary relations we will use in deriving Fermat's principle. The other key ingredient is the time average of the harmonic Poynting vector,

$$\langle\mathbf{S}\rangle=\frac{1}{2}\,\mathrm{Re}\,(\mathbf{E}\times\mathbf{H}^*) \tag{5.8}$$

Substituting (5.3) into (5.8) we find that the phase factor common to both $\mathbf{E}$ and $\mathbf{H}$ drops out, leaving just the (potentially complex) amplitude functions,

$$\langle\mathbf{S}\rangle=\frac{1}{2}\,\mathrm{Re}\,(\mathbf{e}\times\mathbf{h}^*) \tag{5.9}$$

Solving (5.5b) for $\mathbf{h}$ and substituting into (5.9) we obtain

$$\langle\mathbf{S}\rangle=\frac{1}{2\mu\eta_0}\,\mathrm{Re}\,(\mathbf{e}\times(\nabla\zeta\times\mathbf{e}^*))$$

Here we have assumed that μ is real. This is appropriate since we will generally apply geometric optics only to low-loss media, where μ has a negligible imaginary part. Again, using the double cross-product identity, we find

$$\langle\mathbf{S}\rangle=\frac{1}{2\mu\eta_0}\,\mathrm{Re}\,(\nabla\zeta\,(\mathbf{e}\cdot\mathbf{e}^*)-\mathbf{e}^*\,(\mathbf{e}\cdot\nabla\zeta))$$

The first term is real, and the second term is zero by (5.5c), thus

$$\langle\mathbf{S}\rangle=\frac{1}{2\mu\eta_0}\mathbf{e}\cdot\mathbf{e}^*\nabla\zeta$$

The magnitude of this vector is given by

$$|\langle\mathbf{S}\rangle|=\sqrt{\langle\mathbf{S}\rangle\cdot\langle\mathbf{S}\rangle}=\frac{1}{2\,|\mu|\,\eta_0}\mathbf{e}\cdot\mathbf{e}^*\sqrt{\nabla\zeta\cdot\nabla\zeta}.$$

Using the eikonal equation (5.7), this reduces to

$$|\langle\mathbf{S}\rangle|=\frac{1}{2\,|\mu|\,\eta_0}\mathbf{e}\cdot\mathbf{e}^*\sqrt{\varepsilon\mu}$$

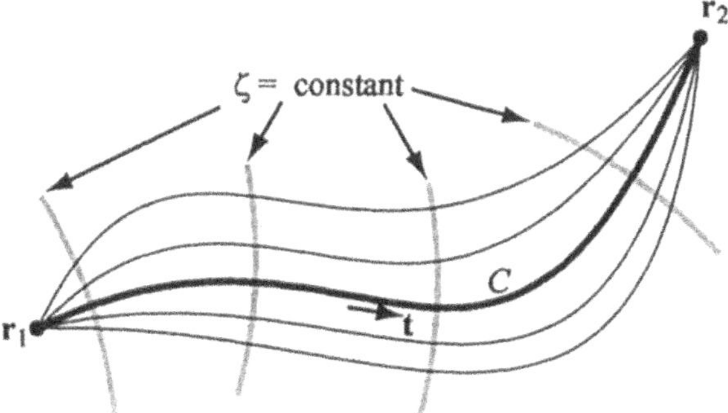

Fig. 5.1 The stationary path, C, between points $\mathbf{r}_1$ and $\mathbf{r}_2$ is normal to the surfaces of constant ζ.

Thus the unit vector is given by

$$\frac{\langle \mathbf{S} \rangle}{|\langle \mathbf{S} \rangle|} = \frac{\operatorname{sign}(\mu)}{\sqrt{\varepsilon \mu}} \nabla \zeta \tag{5.10}$$

Now we see the essence of media with negative material response functions. In a positive material the power flows in the direction of maximum phase advance defined by the gradient of the eikonal, but in a negative material the power flows in the opposite direction.

We will now begin the actual proof of Fermat's Principle. The variational path integral we are interested in gives the phase change between two points, $\mathbf{r}_1$ and $\mathbf{r}_2$, along a light ray represented by a curve, C (Fig. 5.1). From the fundamental theorem of calculus, we have

$$\Delta \zeta \equiv \zeta(\mathbf{r}_2) - \zeta(\mathbf{r}_1) = \int_C d\zeta$$

Using the chain rule to expand the differential we obtain

$$\Delta \zeta = \int_C \nabla \zeta \cdot \frac{d\mathbf{r}}{ds} ds$$

where $\mathbf{r}$ is the position along the curve and s is the path length. The derivative of the position vector with respect to the path length is just a unit vector tangent to the path. If our path is a light ray, then it points in the direction of the Poynting vector. Since a light ray delivers power and must do so in a direction along its path, we obtain

$$\mathbf{t} \equiv \frac{d\mathbf{r}}{ds} = \frac{\langle \mathbf{S} \rangle}{|\langle \mathbf{S} \rangle|}.$$

Now we will examine the first variation of our path integral. This variation represents the change in the value of the integral when we evaluate it along a path infinitesimally nearby the path C and with the same endpoints as C:

$$\delta \int_C \nabla\zeta \cdot \mathbf{t}\, ds = \int_C (\delta\nabla\zeta \cdot \mathbf{t} + \nabla\zeta \cdot \delta\mathbf{t})\, ds \tag{5.11}$$

We note that a nearby curve is displaced transverse to C, this displacement is perpendicular to the gradient of ζ and parallel to the surfaces of constant ζ. Thus we have

$$\delta\zeta = 0$$

and trivially

$$\nabla(\delta\zeta) = 0$$

Switching the order of the variation and the gradient yields

$$\delta\nabla\zeta = 0 \tag{5.12}$$

Thus the first term of the integrand of the right hand side of (5.11) is zero. For evaluating the second term in this integrand, we note that the derivative of the position vector with respect to path length will be a unit vector on any path. On a path nearby to C we will call this vector $\tilde{\mathbf{t}}$. The normalization condition for $\tilde{\mathbf{t}}$ is

$$\tilde{\mathbf{t}} \cdot \tilde{\mathbf{t}} = (\mathbf{t} + \delta\mathbf{t}) \cdot (\mathbf{t} + \delta\mathbf{t}) = 1$$

Expanding we have

$$\mathbf{t} \cdot \mathbf{t} + 2\mathbf{t} \cdot \delta\mathbf{t} + \delta\mathbf{t} \cdot \delta\mathbf{t} = 1$$

If we drop the second-order term and use that fact that, $\mathbf{t} \cdot \mathbf{t} = 1$, we obtain

$$\mathbf{t} \cdot \delta\mathbf{t} = 0$$

From (5.10), $\mathbf{t}$ is parallel (or antiparallel) to $\nabla\zeta$, so we have

$$\nabla\zeta \cdot \delta\mathbf{t} = 0 \tag{5.13}$$

and the second term in the integrand is zero. Thus we have shown that the first variation of the integral (5.11) is zero. The last step in our proof is to rewrite the integrand in terms of ε and μ only, then we will have a useful tool that enables us to determine the paths of light rays between point pairs given only the properties of the intervening materials. Using (5.10) for $\mathbf{t}$, we have

$$\nabla\zeta \cdot \mathbf{t} = \frac{\text{sign}(\mu)}{\sqrt{\varepsilon\mu}} \nabla\zeta \cdot \nabla\zeta$$

If we apply the eikonal equation (5.7), this simplifies to

$$\nabla\zeta \cdot \mathbf{t} = \text{sign}(\mu)\sqrt{\varepsilon\mu}$$

and our variational principle becomes

$$\delta \int_C \text{sign}(\mu)\sqrt{\varepsilon\mu}\, ds = 0$$

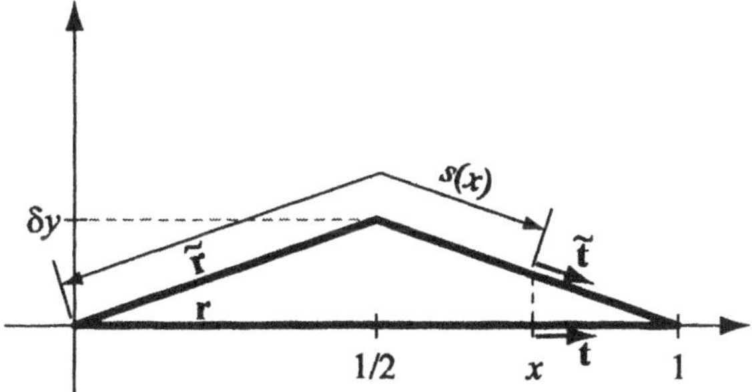

Fig. 5.2 The path $\widetilde{\mathbf{r}}$ deviates from the path $\mathbf{r}$ by an amount δy.

Comparing with the usual form of Fermat's principle

$$\delta \int_C n \, ds = 0 \tag{5.14}$$

we can choose to keep the usual definition of index, (5.1), and write a modified Fermat's principle

$$\delta \int_C \text{sign}\,(\mu)\, n \, ds = 0$$

or redefine the refractive index as

$$n \equiv \text{sign}\,(\mu)\, \sqrt{\varepsilon\mu} \tag{5.15}$$

and maintain the form (5.14). We find the latter more aesthetically pleasing. On the subject of aesthetics, the definition (5.15) has a rather displeasing lack of symmetry with respect to the permittivity and permeability. Noting that geometric optics is only relevant when ε and μ have the same sign we can use an equivalent, symmetric definition for the refractive index.

$$n \equiv \begin{cases} +\sqrt{\varepsilon\mu}, & \varepsilon, \mu \geq 0 \\ -\sqrt{\varepsilon\mu}, & \varepsilon, \mu \leq 0 \end{cases}$$

The integral in (5.14) is often referred to as the *optical path length*, OPL [8], so that Fermat's Principle can also be written rather succinctly as

$$\delta OPL = 0$$

5.2.1 Path Variation Example

Since the arguments that lead up to (5.12) and (5.13) are rather formal, we here give an example, using a specific path and parameterized variation. These are shown in Fig. 5.2. The unvaried path is given by

$$\mathbf{r}\,(x) = x\,\widehat{\mathbf{x}}, \qquad x \in [0, 1] \tag{5.16}$$

and the varied path is given by

$$\widetilde{\mathbf{r}}(x) = x\,\widehat{\mathbf{x}} + f(x)\,\delta y\,\widehat{\mathbf{y}}, \qquad x \in [0,1] \tag{5.17}$$

where the function f,

$$f(x) \equiv \begin{cases} 2x, & 0 \leq x \leq \frac{1}{2} \\ 2 - 2x, & \frac{1}{2} \leq x \leq 1 \end{cases}$$

defines a linear ramp diverging away from the unvaried path along the first half of its length, and converging along the second half. For the unvaried path, the unit tangent vector is given by

$$\mathbf{t} \equiv \frac{d\mathbf{r}}{ds} = \frac{d\mathbf{r}}{dx} = \widehat{\mathbf{x}}$$

since the path length and the x-coordinate are identical. For the varied path we have, more generally,

$$\widetilde{\mathbf{t}} \equiv \frac{d\widetilde{\mathbf{r}}}{ds} = \frac{d\widetilde{\mathbf{r}}}{dx}\frac{ds}{dx}$$

The path length along the varied path is

$$s(x) = \sqrt{x^2 + (2x\delta y)^2} = x\sqrt{1 + 4\delta y^2}$$

and the differential path length is

$$\frac{ds}{dx} = \sqrt{1 + 4\delta y^2} = 1 + 2\delta y^2 + \cdots$$

Thus, to first order we can neglect the difference in path length, and the unit tangent vector is then

$$\widetilde{\mathbf{t}} \approx \frac{d\widetilde{\mathbf{r}}}{dx} = \widehat{\mathbf{x}} + f'(x)\,\delta y\widehat{\mathbf{y}}$$

and the variation in the unit tangent vector is

$$\delta\mathbf{t} \equiv \widetilde{\mathbf{t}} - \mathbf{t} = f'(x)\,\delta y\widehat{\mathbf{y}}$$

which is orthogonal to $\mathbf{t}$,

$$\mathbf{t} \cdot \delta\mathbf{t} = 0$$

just as formally deduced above.

For evaluating $\delta\nabla\zeta(\mathbf{r})$, we perform an expansion of $\nabla\zeta(\widetilde{\mathbf{r}})$

$$\nabla\zeta(\widetilde{\mathbf{r}}) = \nabla\zeta(\mathbf{r}) + \nabla(\nabla\zeta(\mathbf{r})) \cdot \delta\mathbf{r} + \cdots$$

the second term of which is the variation that we seek,

$$\delta\nabla\zeta(\mathbf{r}) = \nabla(\nabla\zeta(\mathbf{r})) \cdot \delta\mathbf{r} \tag{5.18}$$

In this example, the difference in the position vector is given by subtracting (5.16) from (5.17)

$$\delta\mathbf{r} \equiv \widetilde{\mathbf{r}} - \mathbf{r} = f(x)\,\delta y\widehat{\mathbf{y}}$$

Substituting into (5.18) we obtain

$$\delta\nabla\zeta = \frac{\partial}{\partial y}\nabla\zeta\left(\mathbf{r}\right) f\left(x\right)\delta y$$

Interchanging the gradient and the partial y differentiation,

$$\delta\nabla\zeta = \nabla\frac{\partial\zeta\left(\mathbf{r}\right)}{\partial y} f\left(x\right)\delta y \tag{5.19}$$

but on the light ray we have

$$\nabla\zeta\left(\mathbf{r}\right) \equiv \widehat{\mathbf{x}}\frac{\partial\zeta\left(\mathbf{r}\right)}{\partial x} + \widehat{\mathbf{y}}\frac{\partial\zeta\left(\mathbf{r}\right)}{\partial y} + \widehat{\mathbf{z}}\frac{\partial\zeta\left(\mathbf{r}\right)}{\partial z} \propto \mathbf{t} = \widehat{\mathbf{x}}$$

and the gradient of the eikonal has no y-component; the eikonal has no variation in the y-direction since this is parallel to the constant ζ surfacesl; that is, the wave fronts:

$$\frac{\partial\zeta\left(\mathbf{r}\right)}{\partial y} = 0 \tag{5.20}$$

Substituting (5.20) back into (5.19), we see that the variation of the gradient is zero,

$$\delta\nabla\zeta\left(\mathbf{r}\right) = 0$$

5.3 GAUSSIAN OPTICS

5.3.1 Single Surface

As we shall show, the familiar results of *Gaussian optics* apply in the presence of media with negative refractive index. Here we will derive these results using Fermat's Theorem and a notation that we believe will lead to less confusion and sign errors. We feel this is necessary as many descriptions of *Gaussian optics* have complicated and confusing sign conventions even without the possibility of negative index.

Gaussian optics is a first-order approximation of the paths of rays that lie close to an *optic axis*, where the optic axis is a line on which are centered spherical surfaces that define either an interface between media or a reflective surface. The results obtained will be accurate if the angle between a ray and the optic axis is sufficiently small that the length of the ray and its projection on the optic axis are approximately equal, and the ray intersects the spherical surface within a cone angle that is sufficiently small that the cosine of that angle is approximately one. The results obtained are both simple and useful for approximating the behavior of real optical systems that are apertured to relatively small angles around the optic axis [8].

The central result of Gaussian optics is a relationship between pairs of points and a spherical surface. The pairs of points are called *conjugate focal points*. A point source object placed at one member of the pair causes a point image to appear at the other member of the pair. The geometric optic representation of a point source is either a spherically symmetric eikonal,

$$\zeta(\mathbf{r}) = \zeta(|\mathbf{r} - \mathbf{r}_i|)$$

or, equivalently, an infinite set of rays that have a common intersection point. Each one of these rays intersects the spherically symmetric surface and passes through both focal points. Plane waves, which are represented by planar constant eikonal surfaces, or equivalently, by rays that are parallel, are a special case that can be handled by letting the focal point approach infinity.

We will represent the conjugate focal pairs by the Cartesian vectors,

$$\mathbf{r}_1 = x_1\widehat{\mathbf{x}} + y_1\widehat{\mathbf{y}} + z_1\widehat{\mathbf{z}}$$
$$\mathbf{r}_2 = x_2\widehat{\mathbf{x}} + y_2\widehat{\mathbf{y}} + z_2\widehat{\mathbf{z}}$$

where the z-axis will be the optic axis. For the case of a surface that is an interface between media, (i.e., refraction), $\mathbf{r}_1$ will represent the rays to the left of the interface on the optic axis (smaller z) and $\mathbf{r}_2$, the rays to the right (larger z). For the case of a reflecting surface, $\mathbf{r}_1$ and $\mathbf{r}_2$ will represent rays on the same side of the surface. The surface is defined by its center of curvature

$$\mathbf{c} = c\widehat{\mathbf{z}}$$

and a vertex

$$\mathbf{v} = v\widehat{\mathbf{z}}$$

which is an intersection of the sphere with the optic axis. The components of these vectors inherit the usual sign conventions of Cartesian vectors, so there is no arbitrary convention to remember. If there is just one surface, as will be the case for most of this section, we will set the zero of our z-axis to be the vertex of this one surface and simplify our results.

Now we note that a focal point and the set of rays it represents need not lie on the same side of the surface. This is the meaning of real and virtual images. A real focal point is a ray intersection that lies on the same side of the interface as its set of rays. A virtual focal point is a ray intersection that lies on the opposite side of the interface as its set of rays. A virtual focal point is a convergence/divergence that electromagnetic waves do not actually reach, because the waves are interrupted by the interface. If the interface were not present, the waves would converge to and diverge from the focal point. Thus, while a real focal point can be projected onto a screen, a virtual focal point can only be observed through its apparent properties from outside the media in which it appears to occur. The parameter that characterizes the reality/virtuality of a focal point we will call α. This parameter will be plus one for a real focal point and minus one for a virtual focal point. For the case of a single interface with the vertex located at the origin, we have

$$\alpha_1 \equiv -\operatorname{sign}(z_1) \tag{5.21a}$$

$$\alpha_2 \equiv \operatorname{sign}(z_2) \tag{5.21b}$$

so that focal point, $\mathbf{r}_1$, will be real if its z-component is negative, and so on. For the reflector case with rays on the left side of the reflector, we have

$$\alpha_i \equiv -\operatorname{sign}(z_i) \tag{5.22}$$

and for the reflector case with rays on the right side of the reflector we have

$$\alpha_i \equiv \operatorname{sign}(z_i) \tag{5.23}$$

These definitions may seem a bit formal and unnecessary for an intuitive concept, but the result is that one application of Fermat's Principle will handle all of these cases in a systematic way. Also, later, when we tackle the significantly more complex problem of aberrations, this formalism will help make the calculations more conceptually manageable.

Using the above definitions, we can now derive the Gaussian formulas. We wish to find a relationship that will be satisfied by focal point pairs. Fermat's Theorem states that the optical path length between such pairs will be stationary with respect to choice of path. We already know the paths are straight lines inside a homogeneous medium, so the path is completely parameterized by choosing the point on the surface that is the common endpoint of the two straight segments (Fig. 5.3),

$$OPL = \int_C n\, ds = \alpha_1 n_1 l_1 + \alpha_2 n_2 l_2 \tag{5.24}$$

where the integral, which is constant over each of the two segments, has been performed as shown. The coefficients of the geometric path lengths, l_i, are explained as follows. n_i is just the index of the media in which the integration paths lie. α_i must also be present because the integration direction must be reversed for virtual focal points. When both focal points are either real or virtual, the total integration path is from a divergence to a convergence or vice versa, and the integration path can always follow the wave direction. However, if one focal point is real and one virtual, then both focal points are either divergences or convergences. For example, the virtual image of a light source viewed in a flat mirror is also a light source.

We parameterize the surface point by an angle, θ, which is the angle of the surface radius referred to the positive optic axis. According to Fermat, if we take the derivative of this optical path length with respect to θ, we must obtain zero,

$$\frac{dOPL}{d\theta} = \alpha_1 n_1 \frac{dl_1}{d\theta} + \alpha_2 n_2 \frac{dl_2}{d\theta} = 0. \tag{5.25}$$

We use the law of cosines to calculate the (always positive) distance, l.

$$l^2 = R^2 + (z - c)^2 - 2R(z - c)\cos\theta \tag{5.26}$$

where $R \equiv |c|$ is the radius of the surface, the center of which is located at $\mathbf{c} = c\hat{\mathbf{z}}$. One can confirm by checking all the possible cases that this formula is correct

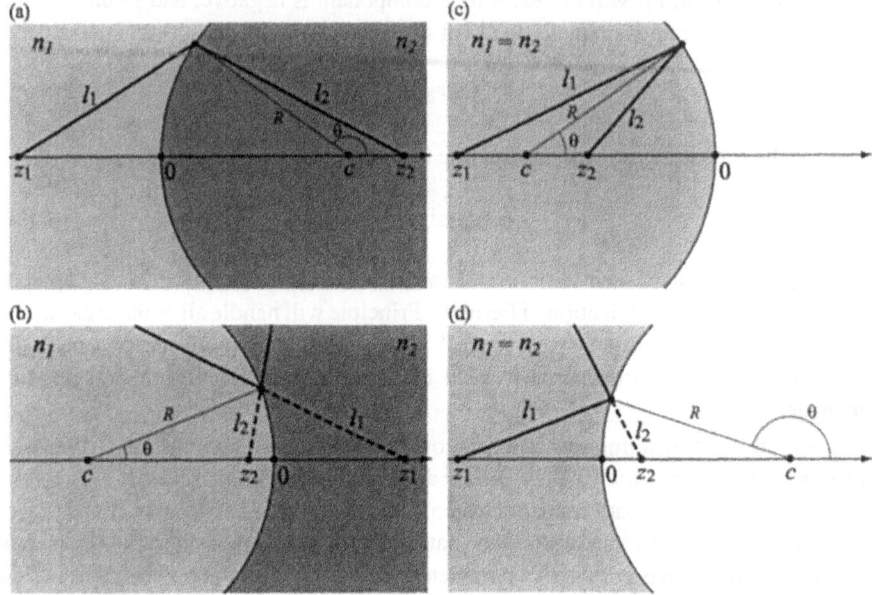

Fig. 5.3 Optical paths for four differnet configurations. (a) A case of refraction. Both focal points 1 and 2 are real and on opposite sides of the surface, which has its center to the right of the vertex. (b) A case of refraction. Both points are virtual. (c) A case of reflection. Both points 1 and 2 are real and on the same side of the surface, which has its center to the left of the vertex. (d) A case of reflection. Focal point 1 is real and 2 is virtual.

regardless of the relative positions of z, c and the vertex. Differentiating (5.26) implicitly with respect to θ, we find

$$2l\frac{dl}{d\theta} = 2R\,(z-c)\sin\theta \tag{5.27}$$

Substituting (5.27) into (5.25), we obtain

$$\frac{\alpha_1 n_1}{l_1}R\,(z_1-c)\sin\theta + \frac{\alpha_2 n_2}{l_2}R\,(z_2-c)\sin\theta = 0$$

which can be rearranged to

$$\alpha_1\frac{z_1}{l_1}n_1\left(1-\frac{c}{z_1}\right) + \alpha_2\frac{z_2}{l_2}n_2\left(1-\frac{c}{z_2}\right) = 0 \tag{5.28}$$

This exact relationship is complicated by the presence of l_1 and l_2 which depend on θ. This equation gives different relationships between focal point pairs for different angles. This is consistent with the fact that spherical surfaces have spherical aberration when not considered in the Gaussian optic limit. In the Gaussian limit, $\cos\theta \approx 1$

and $l \approx |z|$ so that

$$\frac{z}{l} \approx \frac{z}{|z|} = \text{sign}\, z \tag{5.29}$$

Substituting (5.29) into (5.28), we obtain

$$\alpha_1 \,\text{sign}\, z_1 n_1 \left(1 - \frac{c}{z_1}\right) + \alpha_2 \,\text{sign}\, z_2 n_2 \left(1 - \frac{c}{z_2}\right) = 0$$

which is the general result of Gaussian optics applicable to both refraction and reflection. Specializing for refraction by using (5.21) yields

$$-n_1 \left(1 - \frac{c}{z_1}\right) + n_2 \left(1 - \frac{c}{z_2}\right) = 0$$

For reflection we have $n_1 = n_2$, and using either (5.22) or (5.23) gives the same result,

$$\left(1 - \frac{c}{z_1}\right) + \left(1 - \frac{c}{z_2}\right) = 0$$

These results can be written in a more familiar form as

$$\frac{n_2}{z_2} - \frac{n_1}{z_1} = \frac{n_2 - n_1}{c} \tag{5.30}$$

and

$$\frac{1}{z_1} + \frac{1}{z_2} = \frac{2}{c} \tag{5.31}$$

Both (5.30) and (5.31) include the special case of flat surfaces by taking the limit $c \to \infty$. For example, taking this limit of (5.31) yields the familiar result for plane mirrors, $z_2 = -z_1$.

Now we will use these equations to examine some of the unique and interesting properties of refraction across a surface separating media with refractive index of opposite sign. First, we will find the dependence of the surface curvature on relative index, which we define as $n \equiv n_2/n_1$, when both focal points are real, $z_1 < 0$ and $z_2 > 0$. Using (5.30), we obtain

$$c = \frac{n - 1}{n/|z_2| + 1/|z_1|} \tag{5.32}$$

Since n is the index on side 2 relative to side 1, this media will be convex if c is positive and concave if c is negative. We see that when n is positive, the sign of the curvature is completely determined by the factor, $n - 1$. When n is greater than one, the surface must be convex. We all know that a focusing lens must be convex. However, when $0 < n < 1$ the surface must be concave. If this is surprising, it is because materials with index less than one are not commonly used for lenses. (Of course, n is the relative index, so an air lens embedded in glass would be an example of a lens with index less than one.) We also note that this equation correctly predicts a problem for focusing with $n = 1$ media. As we approach the $n = 1$ limit where there is no discontinuity in material properties across the interface, and thus no refraction, the interface radius of curvature must approach zero!

Stranger things occur when the index is less than zero. In this case (5.32) has a pole, and the factor in the denominator may be either positive or negative. This pole, which occurs at $n = -|z_2/z_1|$, indicates that the interface should be flat ($c \rightarrow \infty$) in that case. The possibility of a flat interface, between media of oppositely signed index, creating a real image, was first pointed out by Veselago [1].

The case of refraction at an $n = -1$ interface is directly analogous to reflection. Substituting $n_2/n_1 = -1$ into (5.30), we obtain

$$\frac{1}{z_2} + \frac{1}{z_1} = \frac{2}{c} \tag{5.33}$$

which is identical to (5.31). The difference is that here the two focal points represent rays on opposite sides of the surface whereas for reflection they represent rays on the same side of the surface. A source in front of a flat or convex mirror yields virtual images, but in front of a plane or convex surface on $n = -1$ media, yields real images.

An important special case of Gaussian optics occurs when one focal point is at infinite distance from the surface. One often wants to know how a surface affects parallel light. If we again define $n \equiv n_2/n_1$, using (5.30) we find the following limits.

$$\lim_{z_1 \rightarrow \infty} z_2 = c\frac{n}{n-1}$$

$$\lim_{z_2 \rightarrow \infty} z_1 = -c\frac{1}{n-1}$$

from which the reality parameters are

$$\alpha_2 = \text{sign}\left(c\frac{n}{n-1}\right) \tag{5.34a}$$

$$\alpha_1 = \text{sign}\left(c\frac{1}{n-1}\right) \tag{5.34b}$$

Equation (5.34a) indicates the reality of a focal point when its conjugate represents parallel rays from outside media of index n. Equation (5.34b) applies when the parallel rays originate inside the media.

Using (5.34) we complete Table 5.1. We see from this table that negative index does not exhibit opposite behavior to positive index. With respect to the reality of images, it is actually index between zero and one that behaves oppositely to index greater than one. $n = 1$ is frequently a symmetry point for focusing properties because refractive power goes to zero at this index value.

Up to this point we have only considered on-axis focal points. In Gaussian optics, the positions of focal points along the optic axis (their z-coordinates) can be found independently of their transverse positions (x- and y-coordinates). One may work

Table 5.1 The reality of images created by a single surface acting on parallel rays

Parallel Rays	Surface	$n < 0$	$0 < n < 1$	$1 < n$
outside	convex	real	virtual	real
	concave	virtual	real	virtual
inside	convex	virtual	virtual	real
	concave	real	real	virtual

out the focal points' coordinates along the optic axis first, and then determine their transverse positions afterward. Furthermore, we will only consider optical systems composed of surfaces that are smooth (differentiable) and rotationally symmetric about the optic axis. This rotational symmetry requires that all focal points will lie in the same plane. We will define this plane to be the x–z plane. Since, at this point, we already know the z-components of the focal points, all that remains is to find the x-components. To accomplish this we note the following. *Any* ray linking a pair of focal points will establish the relationship between their transverse positions. We choose the ray that passes through the vertex. This simplifies the problem since a smooth, rotationally symmetric surface is well approximated by a plane for points near the vertex. For the purpose of analyzing small deviations of the path around this particular ray, we will treat the surface as planar.

We begin as before, with the two straight segment optical path. The optical path length is then given by

$$OPL = \alpha_1 n_1 l_1 + \alpha_2 n_2 l_2$$

To apply Fermat's Principle, we will find the path that is stationary with respect to displacements of the x-coordinate of the intersection with the surface around zero, which we will call v_x, the x component of the vertex. (Later this component will be set to zero.) The derivative of the optical path with respect to this parameter must be zero.

$$\frac{dOPL}{dv_x} = \alpha_1 n_1 \frac{dl_1}{dv_x} + \alpha_2 n_2 \frac{dl_2}{dv_x} = 0 \tag{5.35}$$

For our approximately planar interface, the length of the geometric path segments is given by

$$l^2 = z^2 + (v_x - x)^2$$

Differentiating this implicitly, we obtain

$$2l\frac{dl}{dv_x} = 2(v_x - x) \tag{5.36}$$

and setting v_x to zero,

$$\left.\frac{dl}{dv_x}\right|_{v_x=0} = -\frac{x}{l} \tag{5.37}$$

Substituting (5.37) into (5.35) yields

$$\alpha_1 n_1 \frac{x_1}{l_1} + \alpha_2 n_2 \frac{x_2}{l_2} = 0 \tag{5.38}$$

Now we use the Gaussian optic limit, $l \approx |z|$, to obtain

$$\alpha_1 n_1 \frac{x_1}{|z_1|} + \alpha_2 n_2 \frac{x_2}{|z_2|} = 0 \tag{5.39}$$

As above, we can specialize this equation for refraction or reflection. For refraction, we apply (5.21), and (5.39) becomes

$$n_1 \frac{x_1}{z_1} = n_2 \frac{x_2}{z_2} \tag{5.40}$$

For reflection, we use the fact that $n_1 = n_2$, and either (5.22) or (5.23), and (5.39) becomes

$$\frac{x_1}{z_1} + \frac{x_2}{z_2} = 0 \tag{5.41}$$

If z_1 is infinite then x_1 will also be infinite. Equation (5.40) or (5.41) will allow us to calculate the ratio, $\frac{x_1}{z_1}$, from which we can calculate a direction for the rays that are represented by this focal point:

$$\hat{\mathbf{r}}_1 = \frac{\frac{x_1}{z_1}\hat{\mathbf{x}} + \hat{\mathbf{z}}}{\sqrt{\left(\frac{x_1}{z_1}\right)^2 + 1}} \tag{5.42}$$

All rays connected to this focal point are parallel and point in the direction, $\hat{\mathbf{r}}_1$. This can be applied to focal point 2 as well. In fact, as seen from (5.30) or (5.31), both focal points can be at infinite distance, if the surface is flat, $c \to \infty$. For the reflection case, this describes a plane wave reflecting on a plane mirror.

The transverse position of the focal points also has some interesting behavior for negative refractive index. Using, $n \equiv n_2/n_1$, we rewrite (5.40) as

$$\frac{x_2}{x_1} \equiv M_{21} = \frac{1}{n}\frac{z_2}{z_1} \tag{5.43}$$

which also defines the transverse magnification at focal point 2 relative to 1, M_{21}. The sign of the magnification is given by

$$\operatorname{sign}(M_{21}) = -\operatorname{sign}(n)\operatorname{sign}(\alpha_1\alpha_2)$$

From this we can complete Table 5.2, and we note that, with respect to image inversion, negative index has the opposite behavior of positive index.

A very simple result is obtained for the transverse magnification of a flat surface on $n = -1$ media. Letting $c \to \infty$ in (5.33) and substituting into (5.43), we find

$$M_{21} = 1$$

This is part of the story of the perfect lens.

Table 5.2 The relative transverse orientation of images focused by a single surface

Focal Points	$\alpha_1\alpha_2$	$n < 0$	$0 < n$
real/real	1	upright	inverted
real/virtual	-1	inverted	upright
virtual/virtual	1	upright	inverted

5.3.2 Multiple Surfaces

The results of the previous section can be easily extended to describe optical systems with multiple refractive interfaces. The interfaces are described by their centers of curvature

$$\mathbf{c}_i = c_i \widehat{\mathbf{z}}$$

and vertices

$$\mathbf{v}_i = v_i \widehat{\mathbf{z}}$$

The set of material regions separated by these surfaces have refractive indices, n_i, and focal points

$$\mathbf{r}_i = x_i \widehat{\mathbf{x}} + y_i \widehat{\mathbf{y}} + z_i \widehat{\mathbf{z}}$$

Since at most one vertex can be located at the origin, the vertices' coordinates must appear in the refraction formula for the z-coordinate

$$\frac{n_{i+1}}{z_{i+1} - v_i} - \frac{n_i}{z_i - v_i} = \frac{n_{i+1} - n_i}{c_i - v_i} \tag{5.44}$$

and the transverse, x-coordinate,

$$n_i \frac{x_i}{z_i - v_i} = n_{i+1} \frac{x_{i+1}}{z_{i+1} - v_i} \tag{5.45}$$

In these expressions, the ith interface separates the ith and $(i+1)$th material regions. The multiple interface reality parameter does not explicitly appear in the above expressions, but is used below to calculate aberrations.

$$\alpha_{i,j} = \begin{cases} -\operatorname{sign}(z_i - v_j), & i = j \\ \operatorname{sign}(z_i - v_j), & i = j+1 \end{cases} \tag{5.46}$$

The multiple interface reality parameter has two indices. The first index indicates which material region the rays being described lie in. The second index indicates which bounding interface of this material region is being considered. The reality of a focal point depends on the interface considered (e.g., a focal point may be real with respect to its left side interface, but virtual with respect to its right side interface).

Reflective surfaces may be included in multi-interface optical systems by replacing them with interfaces of relative index equal to minus one. In this way the optical axis which is folded back on itself by a mirror may be unwrapped. A suitable

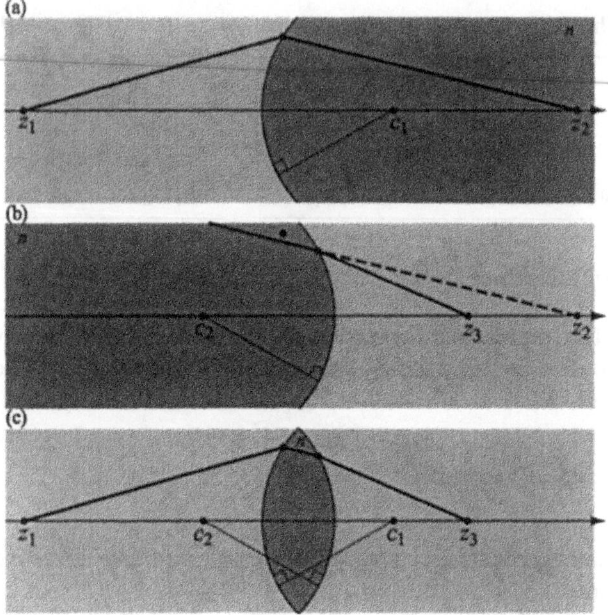

Fig. 5.4 Two surface refraction of a lens. (a) Refraction at surface 1. (b) Refraction at surface 2. (c) Combined refraction. As shown, this is a thick lens, since the two surface vertices are a finite distance apart.

mapping of the indices of refraction, n_i, and the z-coordinates of the focal points will not be described here, but is not difficult to generate. For example, self consistent solutions for multiple reflections between opposing mirrors can yield a Gaussian optical description of a cavity.

5.3.3 Thin Lenses

The simplest multiple surface optical system, and by far the most common, is the thin lens. A thin lens is a combination of two surfaces (Fig. 5.4), where we assume the distance between surface vertices is zero to make the analysis easier. Thin lens formulas are a good first approximation to real lens behavior. A complete description of rays passing through a thin lens would include the locations of three focal points. Usually, we are only interested in the two that are external to the lens. We will refer to the two surfaces as surface 1 and surface 2. Each surface has an associated center of curvature, c_1 and c_2, but since the distance between the vertices is zero, there is only one vertex and we will set it to be at the origin. The three media regions separated by the two surfaces will call regions 1, 2, and 3. Regions 1 and 3, our surrounding media, will have index $n_1 = n_3 = 1$. Region 2 will be our lens media with $n_2 = n$.

Applying (5.44) to each surface we obtain the two equations

$$\frac{n}{z_2} - \frac{1}{z_1} = \frac{n-1}{c_1} \tag{5.47a}$$

$$\frac{1}{z_3} - \frac{n}{z_2} = \frac{1-n}{c_2} \tag{5.47b}$$

Adding (5.47) together, we obtain the familiar lens makers' equation [8]:

$$\frac{1}{z_3} - \frac{1}{z_1} = (n-1)\left(\frac{1}{c_1} - \frac{1}{c_2}\right) \equiv \frac{1}{f} \tag{5.48}$$

From this we can see how to make positive and negative index lenses of the same focal length, f. One way, which retains the same surface curvature magnitude, is as follows. Reflect the value of the index around one, and change the sign of the curvature of both surfaces:

$$n \longrightarrow 2-n$$
$$c_1 \longrightarrow -c_1$$
$$c_2 \longrightarrow -c_2$$

This transformation leads to a thin lens that is unchanged with respect to Gaussian optics, that is, it has the same focal length. The lenses will, however, be different with regard to their optical aberrations and reflectivity or transmissivity. We can also find a relationship for the transverse components. Applying (5.45) to the two surfaces,

$$\frac{x_1}{z_1} = n\frac{x_2}{z_2}$$
$$n\frac{x_2}{z_2} = \frac{x_3}{z_3}$$

Again, we add these together,

$$\frac{x_1}{z_1} = \frac{x_3}{z_3} \tag{5.49}$$

We find that the transverse and longitudinal magnifications are equal.

$$M = \frac{x_3}{x_1} = \frac{z_3}{z_1}$$

The index does not explicitly appear in (5.49). The relationship between the transverse components is only affected by index through the focal length, which determines the relationship between z_1 and z_3.

5.4 ABERRATIONS

Optical systems usually deviate from the approximations that were necessary to derive the simple Gaussian optical formulas. Spherical interfaces do not result in perfectly sharp focal points. It is desirable to quantify the sharpness of a focus so the performance of an optical system can be evaluated or compared with other systems, or optimized by adjusting various geometrical or optical material parameters. This sharpness quantification is carried out by calculating the optical path length of a general ray which passes through the optical system to the desired focal point. This general ray deviates from the Gaussian limit according to parameters that will be defined below.

The optical path length of a reference ray is also calculated. If every general ray had the same optical path length as the reference ray, then the electromagnetic waves represented by these rays would all arrive in phase at the focal point. Thus, at the focal point, the fields would be very large; at other points, where the paths have different optical lengths, the waves would be out of phase and the fields small. In the geometric optical limit, if all rays arrive at a focal point via paths with the same optical length, then that focal point is a perfect point focus. So we quantify the quality of a focus by the difference in path length of a general ray from the reference ray. Specifically, we examine how this path-length difference depends on the parameters that describe the deviation of the general ray from Gaussian optics.

After describing this method of calculating optical aberrations in more detail, we will give the results for one important case, the thin lens. Then we will discuss the unique properties of the aberrations of thin lenses composed of negative index media.

Texts on optical aberrations contain an amazing number of algebraic expressions containing an amazing array of symbols with seemingly arcane meanings to the beginner. This is due in part to the long history of optical analysis. Over the centuries, scientists have produced a great deal of algebraic data. We will take a different approach here by assuming that the reader either (a) only wants to understand conceptually the method of calculating aberrations or (b) wishes to calculate aberration coefficients using a symbolic math computer program. To this end, we will break down the calculation procedure into operations that are either easy to implement or already implemented in such a program and will not reproduce any of the algebraic steps. For all but the simplest optical systems, the algebraic manipulations are quite complex, but can often be handled by a symbolic math program.

The steps to calculating the aberrations of an optical system are as follows. (1) Define the system optic elements. (2) Define aperture stop and find exit pupil. (3) Calculate the positions of the desired focal points. (4) Find the general and reference ray. (5) Calculate the optical path length difference. (6) Expand this difference in a series.

The coefficients in this series give a simple quantitative summary of the quality of the focus generated by the optical system. The lowest order terms in the series (of which there are five) are the well-known Seidel aberrations and are referred to by the names: spherical aberration, coma, astigmatism, field curvature, and distortion. Each of these coefficients also has a geometric interpretation which is discussed in many optics texts, such as Hecht [8] or Mahajan [9], but will not be discussed here.

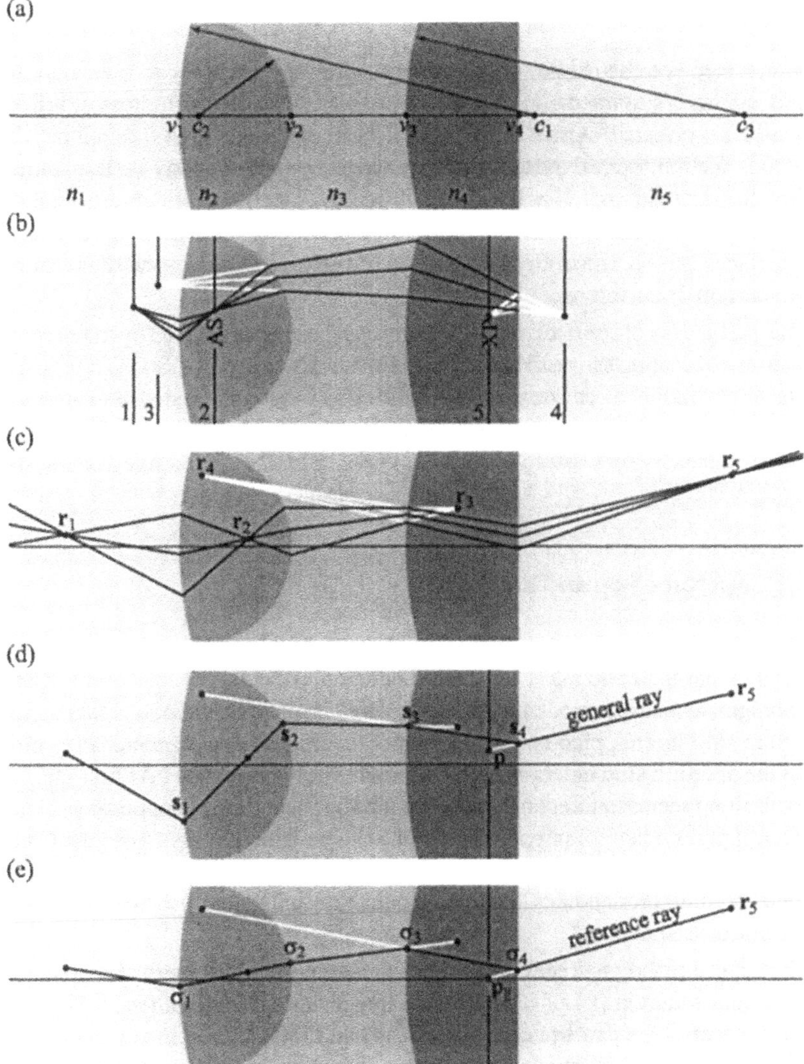

Fig. 5.5 A five-layer optical system. (a) The system is defined by the indices of the layers, n_i, the vertices, v_i, and the centers of curvature, c_i. The fourth surface is flat so that c_4 is at infinity. (b) The aperture stop (aperture 2) and its images are shown. Aperture 1 is real, and apertures 3, 4, and 5 are virtual since they do not appear in their respective layers. The image of the aperture stop in the last layer, 5, is the exit pupil. (c) The focal points, $\mathbf{r}_i$, are shown connected by three rays. $\mathbf{r}_3$ and $\mathbf{r}_4$ are virtual focal points. Virtual rays are shown white. (d) The general ray passes through all the focal points and intersects the exit pupil at point, $\mathbf{p}$. (e) The reference ray passes through all the focal points and intersects the exit pupil on the optic axis at point $\mathbf{p}_{||}$.

5.4.1 System Optic Elements

The first step is to define the optical system. For our purposes an optical system is some number, N, of media layers, which possess homogenous indices of refraction, n_i, and are oriented normal to a single linear optic axis. (As mentioned in the previous section, optical systems that are folded by mirrors may be transformed to simple linear systems.) The layers are joined at interfaces which are rotationally symmetric about the optic axis and are often spherical surfaces. In the latter case the surfaces can be defined by the position of there vertices, v_i, and the positions of there centers of curvature, c_i.

If a surface is not spherical, it will, in any case, be approximated by a sphere for the purposes of locating the positions of the Gaussian focal points. So even for aspheric surfaces the radius of curvature at the optic axis is always needed, as is the vertex. For an aspheric, other parameters may also be required to define the surface, such as the eccentricity for a conic section. A system definition for a five-layer system is shown in Fig. 5.5a.

5.4.2 Aperture Stop and Exit Pupil

The aperture stop is the object that limits the extent to which light rays deviate laterally from the optic axis. This could be the outer radius of lens or an adjustable diaphragm, as in a camera. In any case, we will assume that all points that define the aperture stop lie in a plane normal to the optic axis. It is not immediately obvious why the aperture stop enters into the calculation of aberrations. At first glance, the aperture stop seems to affect only the amount of light entering the optical system, but, in fact, it plays a key role in parameterizing the deviation of rays from the Gaussian limit. In addition to this role, the position of the aperture stop can enable trade-offs of one aberration for another. This can be quite advantageous in some cases, but will not be discussed here.

The aperture stop is a real object. If we imagine that the aperture stop is illuminated, light scattering off of it will form images associated with each layer in our optical system. We can use equations (5.44) and (5.45) to find these sets of image points, which may be either real or virtual. The images of the aperture stop in the first and last layer are called the *entrance pupil* and *exit pupil*, respectively [8]. By convention, it is the exit pupil that is used for aberration calculations. It is worth noting that, in the Gaussian limit, a ray that passes through the center of the aperture stop also passes through the center of the exit pupil, and a ray which grazes the rim of the aperture stop (a peripheral ray) also grazes the rim of the exit pupil. Figure 5.5b shows our optical system with the aperture stop (AS) and its images, including the exit pupil (XP), which in this case is a virtual object whose apparent position is within the second to last layer.

Since the calculations of Gaussian optics and its corrections (aberrations) apply to rays in the local neighborhood of the optic axis, which is always free of obstruction, it is clear that specifics of the aperture stop such as radius or shape are irrelevant.

What is relevant is the position of the plane of the aperture stop, which gives us the position of the plane of the exit pupil along the optic axis. The point of intersection, $\mathbf{p}$, of a ray with the exit pupil plane is the parameter we seek. It is one of the two parameters that we will use to expand the deviation of rays from the Gaussian optic limit. When both these parameters lie close to the optic axis, the perfect focusing of Gaussian optics is nearly achieved. When these parameters deviate from the optic axis, we find aberrations from the point focus.

A significant simplification occurs for an optical system composed of a single thin lens. In this case, the aperture stop will frequently be located at the plane of the lens—that is, the common vertex of both the surfaces of the lens. By (5.48) we see that the aperture stop and both of its Gaussian images will all lie at this same common position. Now the parameter, $\mathbf{p}$, is just the intersection of a ray with the lens plane, which is both simple and intuitive.

5.4.3 Focal Points

The procedure for calculating the Gaussian focal points, $\mathbf{r}_i$, for a multiple surface optical system is described above. In aberration calculations, the last focal point, $\mathbf{r}_N$, plays a special role. $\mathbf{r}_N$ is the other parameter used to expand the deviation of rays from the Gaussian optic limit. We note here that the equations that give the relationships between the focal points (5.44) and (5.45) can be solved by a symbolic math program. Probably, the program will even handle the values of infinity used to represent flat surfaces or parallel rays from a distant source. As mentioned above, if any of the surfaces are aspherical, they are approximated by spherical surface near the optic axis. The positions of the Gaussian focal points are based on these spherical approximations.

5.4.4 General and Reference Ray

Next we find the general and reference rays. These rays are not the rays one would compute in ray tracing; they do not obey Snell's Law or Fermat's Principle, exactly. These are Gaussian rays, and thus they always pass through the Gaussian focal point. A Gaussian ray can be completely defined by the set of focal points plus one additional point. Conceptually, we begin finding the straight line drawn through the last focal point, $\mathbf{r}_N$, and a general point on the exit pupil, $\mathbf{p}$, Fig. 5.5d. This line intersects the last surface S_{N-1} at some point $\mathbf{s}_{N-1}$. The resulting line segment between $\mathbf{r}_N$ and $\mathbf{s}_{N-1}$ is one segment of the general ray. One can write a general function for a symbolic math program that performs this function and represent it as follows:

$$\mathbf{s} = f(\mathbf{p}, \mathbf{q}; S) \tag{5.50}$$

where f is a function that returns $\mathbf{s}$, the intersection of a surface S and the line through two arbitrary points $\mathbf{p}$ and $\mathbf{q}$. For a spherical surface, S will be represented by the vertex v and the center of curvature c. For computing the general ray we begin with

$$\mathbf{s}_{N-1} = f(\mathbf{p}, \mathbf{r}_N; S_{N-1}) \tag{5.51}$$

as described above. The general ray then refracts at this surface and continues along a line that passes through the next focal point, $\mathbf{r}_{N-1}$. $\mathbf{s}_{N-1}$ and $\mathbf{r}_{N-1}$ now define a line which will intersect S_{N-2} at $\mathbf{s}_{N-2}$:

$$\mathbf{s}_{N-2} = f\left(\mathbf{s}_{N-1}, \mathbf{r}_{N-1}; S_{N-2}\right) \tag{5.52}$$

We then continue, in reverse, along our optical system using

$$\mathbf{s}_{i-1} = f\left(\mathbf{s}_i, \mathbf{r}_i; S_{i-1}\right) \tag{5.53}$$

until we arrive at $\mathbf{s}_0$. Only the first operation is unique in that it used the general point on the exit pupil, $\mathbf{p}$, as the first surface point. The reference ray is calculated in exactly the same way, except that the reference ray passes through the center of the exit pupil (an also the center of the aperture stop). We will use the notation $\mathbf{p} = \mathbf{p}_\perp + \mathbf{p}_\parallel$, where $\mathbf{p}_\parallel$ is the component of $\mathbf{p}$ along the optic axis and thus points to the center of the exit pupil, and $\mathbf{p}_\perp$ is the lateral position of the point $\mathbf{p}$ within the exit pupil and referred to the optic axis. Then the surface intersection points of the reference ray, σ_i, are given by

$$\sigma_{N-1} = f\left(\mathbf{p}_\parallel, \mathbf{r}_N; S_{N-1}\right)$$

and

$$\sigma_{i-1} = f\left(\sigma_i, \mathbf{r}_i; S_{i-1}\right)$$

The deviation of the general ray from the reference ray is set by the deviation of the general point on the exit pupil, $\mathbf{p}$, from the center of the exit pupil, $\mathbf{p}_\parallel$. This deviation is given by $\mathbf{p}_\perp$, and thus the components of $\mathbf{p}_\perp$ together with x_N become the basis of the series expansion parameters.

For the case when z_i is infinite, we may need a slightly different function

$$\mathbf{s} = \tilde{f}\left(\mathbf{p}, \hat{\mathbf{r}}; S\right)$$

that finds the intersection of the surface S and a line that passes through $\mathbf{p}$ and points in direction $\hat{\mathbf{r}}$. The successive use of the functions f or $\tilde{f}$ is the step that generates the most algebraic complexity. For an optical system with many interfaces, the complexity may overwhelm the symbolic math program. Use of any symmetries in calculating the $\mathbf{s}_i$ may help. Allowing some parameters of the system to be numeric may help. Often it is desirable to see the analytical dependence of the aberration coefficients on an optical system parameter, but if not, let it be numeric. Numerics can always be combined with each other and simplified more easily. All parameters in this calculation can be numeric except the components of $\mathbf{p}_\perp$, and x_N since we wish to perform an analytical series expansion with these parameters.

5.4.5 Optical Path-Length Difference

We now come to the penultimate step in our procedure, calculating the optical path-length difference of the general and reference ray. We generalize the optical

Table 5.3 Simplification of the optical path length[a]

α_{01}	α_{11}	Order
-1	$+1$	$\mathbf{r}_1$ $\mathbf{s}_0$ $\mathbf{s}_1$
$+1$	$+1$	$\mathbf{s}_0$ $\mathbf{r}_1$ $\mathbf{s}_1$
$+1$	-1	$\mathbf{s}_0$ $\mathbf{s}_1$ $\mathbf{r}_1$

[a]The reality parameters are calculated for each possible ordering of the surface intersections, $\mathbf{s}_1$, $\mathbf{s}_2$, and the focal point, $\mathbf{r}_1$.

path-length formula for single surfaces to multiple surfaces as follows. For the general ray we have

$$OPL = \sum_{i=0}^{N-1} \alpha_{ii} n_i \left|\mathbf{r}_i - \mathbf{s}_i\right| + \alpha_{i+1,i} n_{i+1} \left|\mathbf{s}_i - \mathbf{r}_{i+1}\right| \tag{5.54}$$

which is just a generalization of (5.24) using the multiple surface reality parameter (5.46).

Examining the first two terms in the sum (5.54), we will explain an important simplification:

$$OPL = \alpha_{00} n_0 \left|\mathbf{r}_0 - \mathbf{s}_0\right| + \alpha_{10} n_1 \left|\mathbf{s}_0 - \mathbf{r}_1\right| + \alpha_{11} n_1 \left|\mathbf{r}_1 - \mathbf{s}_1\right| + \alpha_{21} n_2 \left|\mathbf{s}_1 - \mathbf{r}_2\right| + \cdots \tag{5.55}$$

We can collapse the two terms in n_1 as follows. First, we note that $\mathbf{s}_0$, $\mathbf{r}_1$, and $\mathbf{s}_1$ are collinear since $\mathbf{s}_0$ was the intersection of S_0 and the line through and $\mathbf{r}_1$ and $\mathbf{s}_1$. Using Table 5.4.5, we see each possibility for the relative order of these points along the positive z-direction, and the corresponding values for the reality parameter. In each case the sum of the two terms in n_1 simplify, resulting in

$$OPL = \alpha_{00} n_0 \left|\mathbf{r}_0 - \mathbf{s}_0\right| + n_1 \left|\mathbf{s}_0 - \mathbf{s}_1\right| + \alpha_{21} n_2 \left|\mathbf{s}_1 - \mathbf{r}_2\right| + \cdots$$

Applying this simplification to (5.54), we obtain

$$OPL = \alpha_{00} n_0 \left|\mathbf{r}_0 - \mathbf{s}_0\right| + \sum_{i=0}^{N-2} n_i \left|\mathbf{s}_i - \mathbf{s}_{i+1}\right| + \alpha_{N,N-1} n_N \left|\mathbf{s}_{N-1} - \mathbf{r}_N\right|$$

We can similarly obtain the optical path length for the reference ray

$$OPL_{ref} = \alpha_{00} n_0 \left|\mathbf{r}_0 - \sigma_0\right| + \sum_{i=0}^{N-2} n_i \left|\sigma_i - \sigma_{i+1}\right| + \alpha_{N,N-1} n_N \left|\sigma_{N-1} - \mathbf{r}_N\right|$$

and then the optical path-length difference is

$$\begin{aligned}\Delta OPL &\equiv OPL - OPL_{ref} \\ &= \alpha_{00} n_0 \left(\left|\mathbf{r}_0 - \mathbf{s}_0\right| - \left|\mathbf{r}_0 - \sigma_0\right|\right) + \sum_{i=0}^{N-2} n_i \left(\left|\mathbf{s}_i - \mathbf{s}_{i+1}\right| - \left|\sigma_i - \sigma_{i+1}\right|\right) \\ &\quad + \alpha_{N,N-1} n_N \left(\left|\mathbf{s}_{N-1} - \mathbf{r}_N\right| - \left|\sigma_{N-1} - \mathbf{r}_N\right|\right)\end{aligned}$$

It may be desirable to break up this sum to do the series expansion. It is most likely beneficial to compute coefficients for corresponding pairs of segments of the general and reference rays. Only these pairs will be small for small values of the expansion parameters.

The first and last terms in the sum correspond to ray segments that connect to the first focal point, $\mathbf{r}_0$, and the last focal point, $\mathbf{r}_N$. In these terms the focal points appear explicitly. This is a problem if these focal points lie at infinity. In this case we have calculated only a direction for the focal points, $\widehat{\mathbf{r}}_i$. Using this direction, we can calculate the path-length difference for this case. For example, if the first focal point lies at infinity we have

$$\Delta OPL_0 = \alpha_{00} n_0 \left(|\mathbf{r}_0 - \mathbf{s}_0| - |\mathbf{r}_0 - \sigma_0| \right) = n_0 \left(\sigma_0 - \mathbf{s}_0 \right) \cdot \widehat{\mathbf{r}}_0$$

5.4.6 Expand the Difference

Now we come to the last step, computing the expansion. We have the expression, ΔOPL, that we wish to expand, so let us discuss the expansion parameters. Due to the rotational symmetry of our optical system, the path length difference can only depend on the expansion parameters in z-axis rotationally invariant combinations. (Recall that our choice of using the x-axis for lateral deviations was arbitrary.) These rotationally invariant combinations are the three inner products of $\mathbf{p}_\perp$ and $\mathbf{r}_{N\perp}$, which are the general point on the exit pupil relative to the pupil's center, and the lateral deviation of the last focal point from the optic axis. We define parameters to represent these invariants:

$$\begin{aligned} \beta &\equiv \mathbf{p}_\perp \cdot \mathbf{p}_\perp = p_x^2 + p_y^2 \\ \gamma &\equiv \mathbf{p}_\perp \cdot \mathbf{r}_{N\perp} = p_x x_N \\ \delta &\equiv \mathbf{r}_{N\perp} \cdot \mathbf{r}_{N\perp} = x_N^2 \end{aligned}$$

We need to invert these expressions to substitute this new expansion parameter set for the old one:

$$\begin{aligned} p_x &= \gamma \delta^{-1/2} \\ p_y &= \left(\beta - \gamma^2 \delta^{-1} \right)^{1/2} \\ x_N &= \delta^{1/2} \end{aligned}$$

Now we substitute into the expression for ΔOPL and do the expansion:

$$\begin{aligned} \Delta OPL\left(p_x, p_y, x_N\right) &= \Delta OPL\left(\gamma \delta^{-1/2}, \left(\beta - \gamma^2 \delta^{-1} \right)^{1/2}, \delta^{1/2} \right) \\ &= \sum_{l,m,n=0}^{\infty} C_{lmn} \beta^l \gamma^m \delta^n \end{aligned}$$

By the rotational symmetry of our system, the first nonzero terms are second order in β, γ, δ and thus fourth order in $p_\perp$ and $r_{N\perp}$. These aberrations are shown, with

Table 5.4 The lowest-order aberrations are second order in the rotational invariants, α, β, γ, and fourth order in $p_\perp$ and $r_{N\perp}$ [a]

β	γ	δ	Aberration	$p_\perp$	$r_{N\perp}$	$\cos(\theta)$
2	0	0	spherical	4	0	0
1	1	0	coma	3	1	1
0	2	0	astigmatism	2	2	2
1	0	1	field curvature	2	2	0
0	1	1	distortion	1	3	1

[a] The aberrations, with their traditional names, are shown in traditional ordering, from highest to lowest in $p_\perp$.

their traditional names, in Table 5.4. The ordering is from highest to lowest in $p_\perp$, which is the ordering of greatest to least impact for a system that uses significant aperture (compared with surface radius of curvature) to focus objects that are close to the optic axis. In a system where field of view is the key figure of merit (for example, in a survey telescope), the reverse ordering may be more appropriate.

5.4.7 Example: Thin Lenses

Now we examine the lowest-order aberrations for the important case of thin spherical lenses, with the aperture stop in the plane of the lens. These coefficients can be found by the above procedure.

$$C_{200} = -\frac{1}{32f^3 n\,(n-1)^2}\left[n^3 + (n-1)^2\,(3n+2)\,p^2 + 4\,(n+1)\,pq + (n+2)\,q^2\right] \tag{5.56a}$$

$$C_{110} = -\frac{1-p}{8f^3 n\,(n-1)}\left[(2n+1)\,(n-1)\,p + (n+1)\,q\right] \tag{5.56b}$$

$$C_{020} = -\frac{(1-p)^2}{8f^3} \tag{5.56c}$$

$$C_{101} = -\frac{(1-p)^2}{16f^3 n}\,(n+1) \tag{5.56d}$$

$$C_{011} = 0 \tag{5.56e}$$

These coefficients are called the Seidel aberrations. Also appearing in these expressions are p, the position factor, and q, the shape factor, where we follow the definitions of Mahajan [9]. The position factor is given by

$$p \equiv 1 - \frac{2f}{z_3}$$

where f is the focal length and z_3 is the image position. Through the thin spherical lens imaging equation,

$$\frac{1}{z_3} - \frac{1}{z_1} = \frac{1}{f} = (n-1)\left(\frac{1}{c_1} - \frac{1}{c_2}\right)$$

where z_1 is the object position and c_1 and c_2 are the centers of curvature, the position factor is directly related to the magnification,

$$M = \frac{z_3}{z_1} = \frac{p+1}{p-1}$$

The shape factor is given by

$$q \equiv \frac{c_2 + c_1}{c_2 - c_1}$$

A lens with a shape factor of 0 is symmetric, and ± 1 is a plano-curved lens. Using the shape and position factor, all thin spherical lens configurations are described.

We will first examine the very important case of a source object at infinite distance. This is a position factor of -1. We are left with two parameters that can be used to reduce aberrations, n and q. We will set the value of q to eliminate one of the aberrations and compare the remaining aberrations as a function of index. We will restrict our attention to moderate values of index. At large absolute values of index, the aberrations approach the same value independent of sign, but dielectric lenses with high index have significant reflection coefficients due to the impedance mismatch to free space.

The usual ordering of the aberrations is from highest to lowest in the order of r, the aperture coordinate. This is the ordering of most image degradation to least if one is forming images with significant lens aperture, but small to moderate image size, which is a common occurrence in applications. Thus, spherical aberration is an obvious target for elimination. However, there are no roots of C_{200} for values of index greater than one, which is why this aberration is referred to as spherical aberration, since it appears to be inherent to spherical lenses. The usual practice is to eliminate coma (the next in line), and it so happens that the resulting lens has a value for the spherical aberration that is very near the minimum obtainable. Adjusting the shape factor, q, is often called lens bending. If we bend the lens for zero coma—that is find the roots of C_{110} with respect to q—we obtain

$$q_c = \frac{(2n+1)(n-1)}{n+1} \tag{5.57}$$

We plug this value for q and $p = -1$ into (5.56) and plot the remaining three nonzero aberration coefficients as well as q_c in Fig. 5.6. We note that there are two values of index where $q = 1$, which represent a plano-concave/convex lens. Setting (5.57) equal to one, we obtain

$$n^2 - n - 1 = 0. \tag{5.58}$$

the roots of which are the ubiquitous golden ratios, $n = \phi \simeq 1.62$ and $n = 1 - \phi \simeq -0.62$. We also note that there is a window of index values near $n = -0.7$ where

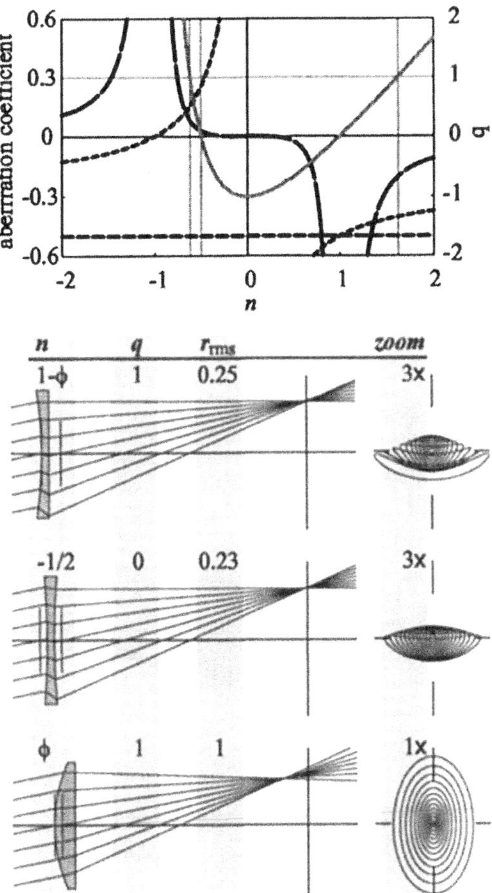

Fig. 5.6 Top plot shows spherical aberration (long dash), astigmatism (short dash), field curvature (very short dash), and shape factor (light gray) as a function of index for a lens focusing an object at infinity and bent for zero coma. Thin gray vertical lines indicate properties for lenses shown in ray tracing diagrams (bottom), meridional profile (left), and image spot (right). Incident angle is 0.2 radians and lenses are $f/2$. Index, shape factor, relative rms spot size, and spot diagram zoom are shown tabularly. In the meridional profile, optic axis and Gaussian image plane are shown as well as lens principle planes (short vertical lines). In the spot diagram, Gaussian focus is at the center of cross hairs. Reprinted figure with permission from Ref. [10]. Copyright © 2004 by the American Physical Society.

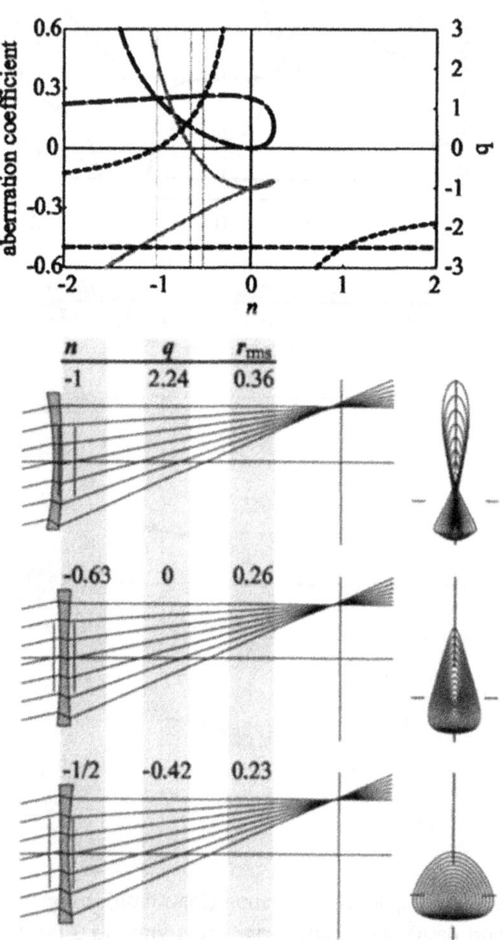

Fig. 5.7 All as in Fig. 5.6, except the following. The lens is bent for zero spherical aberration. Coma is shown with medium-length dashing. Spot size, r_{rms}, is relative to bottom lens spot in Fig. 5.6. All spot diagrams are at the same scale. Reprinted figure with permission from Ref. [10]. Copyright © 2004 by the American Physical Society.

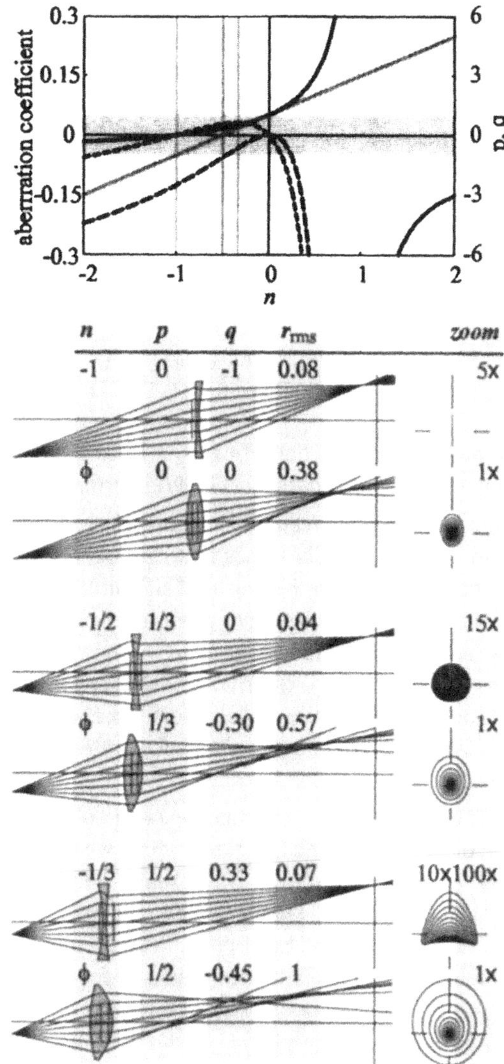

Fig. 5.8 All as in Fig. 5.6, except the following. Lens configuration with object and image at finite positions and bent for zero spherial aberration and coma. Position factor is shown dark gray. Real image object pairs only occur when position factor is in shaded region, $|p| < 1$. Lens pairs are $f/1.23$, $f/1.08$, and $f/0.90$ and have magnifications -1, -2, and -3. In the second-to-last spot diagram, horizontal ($10\times$) and vertical ($100\times$) zooms are not equal. Reprinted figure with permission from Ref. [10]. Copyright © 2004 by the American Physical Society.

both the spherical aberration and field curvature are small. There is no equivalent window in positive index.

Several ray tracing diagrams with both meridional rays and ray spot diagrams are shown for specific values of index in Fig. 5.6. The reference lens has index ϕ, which is close to typical values used in visible optical lenses and near enough to $n = 1$ for reasonably low reflection. The lenses of negative index shown are in fact closer to $n = -1$, which is the other index which permits perfect transmission, so this is a fair comparison. The negative index lenses all show significantly tighter foci than the positive index lens.

If we attempt to bend a lens with $p = -1$ to obtain zero spherical aberration, we obtain the two solutions

$$q_s = \frac{2\left(n^2 - 1\right) \pm n\sqrt{1 - 4n}}{n + 2}$$

These expressions have real values only for $n \leq 1/4$, so an implementation of such a lens (embedded in free space) is not possible with normal materials.

It is a surprising and significant result that negative index permits an entire family of spherical aberration free spherical lenses that can focus a distant object to a real focus (Fig. 5.7). The solution with the negative sign in the expression for q_s (solid curves) has less coma for moderate negative values of index, so ray tracing diagrams are shown for that solution. We note that at $n = -1$, the field curvature is also zero; thus this lens has only two of the five Seidel aberrations, namely, coma and astigmatism. For a positive index reference we use the zero coma, $n = \phi$ lens from above. Here again, negative index lenses achieve a tighter focus than a comparable positive index lens.

Now we examine the case of $|p| < 1$—that is, a real object and real image both at finite position. Since p and q are both free parameters, we can conceivably eliminate two aberrations. If we eliminate spherical aberration and coma, the resulting lens is called *aplanatic* [6]. It is a well-known, though incorrect, result that a spherical lens can only have *virtual* aplanatic focal pairs. The correct statement is that only negative index spherical lenses can have *real* aplanatic focal pairs.

If we set C_{200} and C_{110} to zero and solve for p and q, we obtain four solutions, the two nontrivial ones are given by

$$p_{sc} = \mp\frac{n + 1}{n - 1} \tag{5.59a}$$

$$q_{sc} = \pm\left(2n + 1\right) \tag{5.59b}$$

We will focus on the solution with a minus sign for p and the plus sign for q. This solution has smaller aberrations for lens configurations that magnify an image. The other solution is better for image reduction. Inserting the expressions (5.59) into (5.56) we have plotted the two remaining nonzero coefficient as well as the values of p_{sc} and q_{sc} (Fig. 5.8). Ray diagrams are shown for lenses with magnifications of -1, -2, and -3. Also shown is a reference positive index lens for each. The reference lenses (which cannot be aplanatic) are of moderate index, ϕ, with the same magnification and $f/\#$ as the lenses they are compared to. They are bent for zero coma but also have spherical aberration near the minimum possible for the configuration. Again the negative index lenses produce superior foci.

The lens of index -1 and magnification -1 is particularly interesting. At this index value the field curvature is also zero. This remarkable lens configuration has only one of the five Seidel aberrations, astigmatism. This is confirmed by ray tracing, which shows a one-dimensional "spot" at the image plane. This is perfect focusing in the sagittal plane. Perfect focusing also occurs in the meridional plane, in front of sagittal focus.

One may ask why this asymmetric lens, $q = -1$, performs so well in a symmetric configuration, $p = 0$. This lens can be equivalently viewed as a biconcave doublet with one component having index -1 and the other having index 1—that is, free space. Driven by this observation, we found that all biconcave doublets with arbitrary indices of $\pm n$ have identical focusing properties. The only observable difference is in the internal rays, which are always symmetric about the planar interface but make more extreme angles at higher index magnitude.

Using the current optical system design paradigm, aberrations are minimized by combining elements with coefficients of opposite sign. However, more elements mean greater complexity and cost. Taking advantage of an expanded parameter space that includes negative index can reduce the number of required elements—possibly even to one.

REFERENCES

1. V. G. Veselago, "The electrodynamics of substances with simultaneously negative values of ϵ and μ," *Sov. Phys. Usp.*, vol. 10, pp. 509–514, 1968.

2. J. B. Pendry, "Negative refraction makes a perfect lens," *Phys. Rev. Lett.*, vol. 85, pp. 3966–3969, 2000.

3. C. Kittel, *Introduction to Solid State Physics*, 7th ed., John Wiley & Sons, New York, 1996.

4. D. R. Smith, W. Padilla, D. C. Vier, S. C. Nemat-Nasser, and S. Schultz, "Composite medium with simultaneously negative permeability and permittivity," *Phys. Rev. Lett.*, vol. 84, pp. 4184–4187, 2000.

5. R. A. Shelby, D. R. Smith, and S. Schultz, "Experimental verification of a negative index of refraction," *Science*, vol. 292, pp. 77–79, 2001.

6. Max Born and Emil Wolf, *Principles of Optics*, 6th ed., Pergamon Press, Oxford, 1993.

7. J. A. Kong, *Electromagnetic Wave Theory*, 2nd ed., Wiley-Interscience, New York, 1990.

8. Eugene Hecht, *Optics*, 3rd ed., Addison-Wesley, MA, 1998.

9. Virendra N. Mahajan, *Optical Imaging and Aberrations*, vol. I, 1st ed., SPIE, Bellingham, Washington, 1998.

10. D. Schurig and D. R. Smith, "Negative index lens aberrations," *Phys. Rev. E*, vol. 70, 065601, 2004.

6 Planar Anisotropic Resonance-Cone Metamaterials

KEITH G. BALMAIN and ANDREA A. E. LÜTTGEN

The Edward S. Rogers Sr. Department of Electrical and Computer Engineering
University of Toronto
Toronto, Ontario, M5S 3G4
Canada

6.1 INTRODUCTION

A "metamaterial" in the present context is taken to mean an interconnected electrical network of various components or basic elements such as capacitors, inductors, resistors, metal wires (or strips), transmission-line segments, waveguide segments, diodes, and transistors. Although the metamaterial is made of discrete elements, these elements are usually densely packed and periodic such that the interelement spacing is decidedly smaller than a free-space wavelength for the frequency range in use. Although the basic elements are usually small, in general they cannot always be viewed as infinitesimal (as they would be in the context of circuit theory) because their dimensions can be an appreciable fraction of a wavelength and their interactions can be electromagnetic. The consequence of this is that the metamaterial generally tends to behave as if it were a continuous material, albeit one with electromagnetic properties quite unlike those of more familiar materials such as dielectrics or semiconductors. The evolution of isotropic metamaterial research has followed the now well-known progression beginning with the 1968 continuous-medium theoretical postulate of Veselago [1], Pendry's work on lens-like focusing in 2000 [2], experimental verification [3], and theoretical/experimental development of a two-dimensional (2-D) L–C loaded-transmission-line model [4].

In principle, the metamaterial could be one-, two-, or three-dimensional, but the choices become more limited as one gets closer to applications. The three-dimensional (3-D) case is sufficiently complicated to make its construction and testing definitely challenging, especially at microwave frequencies. The one-dimensional (1-D) case is fundamentally a transmission line and therefore by definition it cannot exhibit the anisotropy featured in this chapter, so the realistic choice for the present chapter narrows rapidly to the 2-D case. This is not a severe constraint because the 2-D case enables relatively easy use of well-known, standard, printed-circuit-board techniques for construction, thus ensuring both ready access to the circuit board for experimental electrical testing and straightforward computational modeling.

Regarding anisotropy, many know that the most familiar medium that is highly electrically anisotropic is a magnetized plasma (often termed a "magnetoplasma") in which electrons move under the influence of steady magnetic fields and oscillating electric fields. This is the medium that enables radio wave propagation in the ionosphere and reflection from it, as well as electromagnetic wave acceleration of charged particles in a nuclear fusion reactor. Early theoretical work on antenna fields influenced by ionospheric anisotropy was carried out by Bunkin [5], Kogelnik [6], and Kuehl [7] and it led to sharply focused near and far fields measured in the laboratory and called "resonance cones" by Fisher and Gould [8], thus launching an area of study that continues actively to the present [9]. A relationship between these plasma media and metamaterials emerges from the plasma permittivity matrix whose elements can be either positive or negative, with the positive elements suggesting metamaterials made of capacitors and the negative elements suggesting metamaterials made of inductors [10]. To represent strong anisotropy, this leads directly to both computational and physical metamaterial models consisting of orthogonal interconnections of capacitors and inductors.

This parallel between the plasma and the metamaterial does have a weakness because, in the plasma, the rotation (that is, gyration) of the charged particles around the ambient steady magnetic field gives rise to off-diagonal elements in the permittivity matrix which are not present in a network of orthogonal capacitors and inductors. However, in plasma representations of the near fields of small sources such as short antennas [11], the off-diagonal terms are often neglected, thus producing a permittivity matrix with only diagonal elements, two of which are identical, which amounts to the "uniaxial" approximation often referred to in crystal optics (see Chapter 7 in [12]). It is this approximation that carries over readily into the field of metamaterial electromagnetics because of its usefulness in representing large, distributed networks of capacitors and inductors, as has been shown in earlier papers [13, 14], and as will be seen in the sections to follow.

6.2 HOMOGENEOUS ANISOTROPIC-MEDIUM ANALYSIS

Consider a uniaxial anisotropic medium that is spatially homogeneous yet electrically anisotropic in two dimensions y and z and is further constrained such that all field quantities are independent of the x-coordinate under the condition that $\partial/\partial x \equiv 0$. This same constraint will also be imposed on the sources of the fields, sources which are infinitely long, filamentary, electric and magnetic current densities designated $\mathbf{J}$ and $\mathbf{M}$, of constant magnitude and phase over their entire lengths. Thus, in rectangular coordinates, Maxwell's equations take the form

$$\begin{bmatrix} 0 & -\partial/\partial z & \partial/\partial y \\ \partial/\partial z & 0 & 0 \\ -\partial/\partial y & 0 & 0 \end{bmatrix} \begin{bmatrix} H_x \\ H_y \\ H_z \end{bmatrix} = j\omega\varepsilon_0 \begin{bmatrix} \varepsilon_{xx} & 0 & 0 \\ 0 & \varepsilon_{yy} & 0 \\ 0 & 0 & \varepsilon_{zz} \end{bmatrix} \begin{bmatrix} E_x \\ E_y \\ E_z \end{bmatrix} + \begin{bmatrix} J_x \\ J_y \\ J_z \end{bmatrix} \tag{6.1}$$

$$\begin{bmatrix} 0 & -\partial/\partial z & \partial/\partial y \\ \partial/\partial z & 0 & 0 \\ -\partial/\partial y & 0 & 0 \end{bmatrix} \begin{bmatrix} E_x \\ E_y \\ E_z \end{bmatrix} = -j\omega\mu_0 \begin{bmatrix} H_x \\ H_y \\ H_z \end{bmatrix} - \begin{bmatrix} M_x \\ M_y \\ M_z \end{bmatrix} \tag{6.2}$$

These equations can be written in two entirely separate groups which are therefore independent groups, as follows, the first group being

$$-\partial H_y/\partial z + \partial H_z/\partial y = j\omega\varepsilon_0\varepsilon_{xx}E_x - J_x \tag{6.3}$$

$$\partial E_x/\partial z = -j\omega\mu_0 H_y - M_y \tag{6.4}$$

$$-\partial E_x/\partial y = -j\omega\mu_0 H_z - M_z \tag{6.5}$$

and the second group being

$$\partial H_x/\partial z = j\omega\varepsilon_0\varepsilon_{yy}E_y + J_y \tag{6.6}$$

$$-\partial H_x/\partial y = j\omega\varepsilon_0\varepsilon_{zz}E_z + J_z \tag{6.7}$$

$$-\partial E_y/\partial z + \partial E_z/\partial y = -j\omega\mu_0 H_x - M_x \tag{6.8}$$

In (6.3)–(6.5), the fields can propagate in the y- and z-directions only, and the only electric field component is E_x. Thus these fields may be termed *transverse electric* or *TE* referred to the direction of propagation. In (6.6)–(6.8), the fields are similarly constrained to propagate in the y- and z-directions only, and the only magnetic field component is H_x. Thus these fields may be termed *transverse magnetic* or *TM* referred to the direction of propagation. Because the present chapter concerns anisotropic media, our consideration becomes limited to the TM fields given by (6.6)–(6.8) in which anisotropy enters whenever ε_{yy} and ε_{zz} are different.

In (6.6)–(6.8), for simplicity, the source will now be limited to the magnetic current density M_x, which will be further constrained to be filamentary with $M_x = K\delta(y)\delta(z)$, where K is the total magnetic current. With the electric current densities now set to zero, equations (6.6) and (6.7)—when combined with (6.8)—readily permit elimination of E_y and E_z to yield

$$(\varepsilon_{zz})^{-1}\partial^2 H_x/\partial y^2 + (\varepsilon_{yy})^{-1}\partial^2 H_x/\partial z^2 + \omega^2\mu_0\varepsilon_0 H_x = j\omega\varepsilon_0 M_x \tag{6.9}$$

This equation has the general form of a Helmholtz equation with a source term on the right-hand side. If we are concerned with near fields only, we can employ a low-frequency/long-wavelength (quasi-static) approximation as in [11] and set the third term on the left-hand side equal to zero, giving

$$(\varepsilon_{zz})^{-1}\partial^2 H_x/\partial y^2 + (\varepsilon_{yy})^{-1}\partial^2 H_x/\partial z^2 = j\omega\varepsilon_0 K\delta(y)\delta(z) \tag{6.10}$$

If the permittivities ε_{yy} and ε_{zz} are made different to produce anisotropy but are both positive, the above partial differential equation is elliptic and the effects of anisotropy are relatively weak. However, if the two permittivities are of opposite sign, the partial differential equation is hyperbolic and the influence of anisotropy becomes very strong. In fact, equation (6.10) then takes on the same general form as a wave equation but one that involves only space coordinates rather than the familiar wave equation in space–time coordinates.

Solution of equation (6.10) can proceed through the use of the Fourier transform pair

$$\tilde{f}(\mathbf{k}) = \iint_{-\infty}^{\infty} f(\mathbf{r}) e^{-j\mathbf{k}\cdot\mathbf{r}} dy dz \tag{6.11}$$

$$f(\mathbf{r}) = \frac{1}{(2\pi)^2} \iint_{-\infty}^{\infty} \tilde{f}(\mathbf{k}) e^{j\mathbf{k}\cdot\mathbf{r}} dk_y dk_z \tag{6.12}$$

Application of (6.11) to (6.10) yields the k-space equation

$$\left[(k_y^2/\varepsilon_{zz}) + (k_z^2/\varepsilon_{yy})\right] \tilde{H}_x(\mathbf{k}) = -\, j\omega\varepsilon_0 K \tag{6.13}$$

By solving algebraically and applying the inverse transform (6.12), one gets

$$H_x(\mathbf{r}) = \frac{-j\omega\varepsilon_0 K}{(2\pi)^2} \iint_{-\infty}^{\infty} \frac{e^{j\mathbf{k}\cdot\mathbf{r}}}{(k_y^2/\varepsilon_{zz}) + (k_z^2/\varepsilon_{yy})} dk_y dk_z \tag{6.14}$$

The anisotropy factor a now can be introduced such that

$$a^2 = \varepsilon_{yy}/\varepsilon_{zz} \tag{6.15}$$

Without loss of generality, a can be taken to have a positive real part and z can be assumed to be positive.

The denominator in the integrand of (6.14) can be factored to give

$$H_x(\mathbf{r}) = \frac{-j\omega\varepsilon_0 K \varepsilon_{zz} a^2}{(2\pi)^2} \iint_{-\infty}^{\infty} \frac{e^{jk_z z} dk_z}{(k_z + jak_y)(k_z - jak_y)} e^{jk_y y} dk_y \tag{6.16}$$

The calculus of residues can be employed to give

$$H_x(\mathbf{r}) = \frac{-j\omega\varepsilon_0 K \varepsilon_{zz} a}{4\pi} \int_{-\infty}^{\infty} \frac{e^{-ak_y z}}{k_y} e^{jk_y y} dk_y \tag{6.17}$$

and symmetry invoked to give

$$H_x(\mathbf{r}) = \frac{-j\omega\varepsilon_0 K \varepsilon_{zz} a}{2\pi} \int_{0}^{\infty} \frac{e^{-ak_y z}}{k_y} \sin(k_y y)\, dk_y \tag{6.18}$$

With the aid of the Fourier sine transform numbered (6.2) in [15], one gets

$$H_x(\mathbf{r}) = \frac{\omega\varepsilon_0\varepsilon_{zz}Ka}{2\pi}\arctan\left(\frac{y}{az}\right) \tag{6.19}$$

From equations (6.6)–(6.8),

$$E_y = j\frac{K}{2\pi}\frac{y}{a^2z^2+y^2} \tag{6.20}$$

$$E_z = j\frac{K}{2\pi}\frac{a^2z}{a^2z^2+y^2} \tag{6.21}$$

With $\hat{\mathbf{y}}$ and $\hat{\mathbf{z}}$ as unit vectors, the displacement current density $\mathbf{J}^d$ can be derived as

$$\begin{aligned}\mathbf{J}^d &= j\omega\mathbf{D}\\ &= j\omega\left(\hat{\mathbf{y}}D_y + \hat{\mathbf{z}}D_z\right)\\ &= j\omega\varepsilon_0\left(\hat{\mathbf{y}}\varepsilon_{yy}E_y + \hat{\mathbf{z}}\varepsilon_{zz}E_z\right)\end{aligned} \tag{6.22}$$

The essential features of equations (6.19) to (6.22) can be displayed by considering the highly anisotropic special case with

$$\varepsilon_{yy} = +1 \tag{6.23}$$

$$\varepsilon_{zz} = -1 \tag{6.24}$$

For this case, (6.19) exhibits a logarithmic singularity along $y = z$ which is the characteristic surface of the hyperbolic partial differential equation deduced from (6.10). In three dimensions, if the fields originated from a point source, this characteristic surface would be a cone widely known in the context of anisotropic plasma physics as the "resonance cone," a term which was apparently first used by Fisher and Gould [8] and which is in current use—for example, by James [9]. It is convenient to continue to use this terminology in the present work in which the fields are essentially 2-D.

It is of particular interest to calculate the displacement current density vector $\mathbf{J}^d$ in the direction parallel to the surface of the resonance cone. This can be done by defining new unit vectors $\hat{\mathbf{u}}$ parallel to the cone surface and $\hat{\mathbf{v}}$ perpendicular to it, setting

$$\hat{\mathbf{u}} = \left(\hat{\mathbf{y}} + \hat{\mathbf{z}}\right)/\sqrt{2} \tag{6.25}$$

$$\hat{\mathbf{v}} = \left(\hat{\mathbf{y}} - \hat{\mathbf{z}}\right)/\sqrt{2} \tag{6.26}$$

together with the corresponding inverse relations

$$\hat{\mathbf{y}} = \left(\hat{\mathbf{u}} + \hat{\mathbf{v}}\right)/\sqrt{2} \tag{6.27}$$

$$\hat{\mathbf{z}} = \left(\hat{\mathbf{u}} - \hat{\mathbf{v}}\right)/\sqrt{2} \tag{6.28}$$

From (6.22), this leads to

$$\mathbf{J}^d = \frac{\omega\varepsilon_0 K}{2\sqrt{2}\pi}\left[\hat{\mathbf{u}}\left(\frac{1}{z-y}\right) + \hat{\mathbf{v}}\left(\frac{1}{z+y}\right)\right] \tag{6.29}$$

which shows that the parallel ($\hat{\mathbf{u}}$) component of the displacement current density abruptly changes sign (reverses phase) as it passes through its maximum magnitude on the resonance cone axis at $y = z$. To put it another way, under hyperbolic conditions, the near field of the infinitely long magnetic current filament is characterized by contiguous, counterflowing streams of displacement current running along either side of the resonance cone axis and parallel to it. In this way, the highly anisotropic medium generates within itself a near-field energy-transport mechanism that is analogous to a two-conductor transmission line. Further, because this result stems from a basic Green's function in the theory of highly anisotropic media, similar phenomena should be observable in a wide range of situations involving anisotropic electromagnetic metamaterials, an assertion that will be tested in the following sections of this chapter.

6.3 FREE-STANDING ANISOTROPIC-GRID METAMATERIAL

The grid to be simulated is shown in Fig. 6.1a, excited at one corner by a V-shaped dipole center-fed with a 1-V source. The computer program used in the simulation is a thin-wire code employing piecewise-sinusoidal expansion and testing functions [16].

The L and C parameters are chosen to produce a high-magnetic-field resonance cone extending corner-to-corner from the feed point along the diagonal line, as shown in Fig. 6.1b. Figure 6.2 shows in simulation that the phase variation of the current flowing along the diagonal corresponds within 2% to propagation at the velocity of light in vacuum. Figure 6.3 shows that the currents to either side of the diagonal propagate at very nearly the same phase velocity, but more significant is their phase difference which is within 10 degrees of being exactly out-of-phase. What we see here is an example of the counterflowing currents already noted in Section 6.2 of this chapter for a homogeneous anisotropic medium, a medium quite unlike the present thin anisotropic grid which is certainly not homogeneous. To wrap up the present section, we note in Fig. 6.4 that the three currents on and adjacent to the diagonal are similar in magnitude, displaying only small undulations indicative that the degree of edge mismatch is correspondingly small.

6.4 ANISOTROPIC GRID OVER INFINITE GROUND

As shown in Fig. 6.5a, for the purpose of simulation the anisotropic L–C grid is positioned over an infinite ground plane. Each grid wire junction is connected to ground by a vertical wire that contains either a voltage generator or a resistor, depending on the requirement. Excitation of the grid is provided by a 1-V source connected between the grid corner and the ground plane. Around the periphery of the grid, power absorption is provided by load resistors inserted in the vertical wires. Elsewhere under the grid, high-value resistors are inserted in the vertical

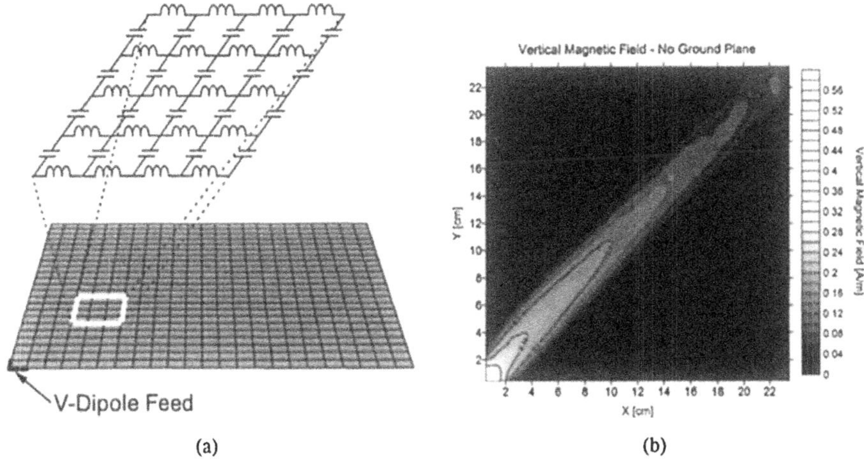

Fig. 6.1 (a) A free-standing or "floating," anisotropic, L–C loaded grid; that is, a grid with no ground plane present. A typical interior subregion is shown in the inset. The grid is excited by a V-shaped, center-fed 1.75-GHz dipole formed around the lower-left half of the square cell in the lower-left corner of the grid. The grid is 24 cells square, each cell being 1 cm square. The wire radius is 0.1 mm. The lumped loads per 1-cm wire segment are L = 25.32 nH, C = 0.25 pF, damping resistance 3.2 Ω. (b) Contour plot of vertical (z-directed) magnetic field strength magnitude at the center of each cell for the configuration in Fig. 6.1a, displaying a resonance cone at a 45-degree angle along the line $x = y$ determined by the equality of the capacitive and inductive reactance magnitudes in each unit cell.

wires, their only purpose being to allow approximate calculation of the open-circuit grid-to-ground voltage. This voltage is derived from the vertical currents calculated from running the moment-method computer program. Otherwise, these resistors have negligible effect on the results of the simulation. A typical example of the results is shown in Fig. 6.5b as a contour plot of the grid-to-ground voltage in the vicinity of a 45-degree corner-to-corner resonance cone. The phase of the current flowing along the diagonal is shown in Fig. 6.6, in which the phase progression is that of a forward wave propagating at a velocity 4% slower than the velocity of light in vacuum. Figure 6.7 shows the phase of the currents parallel and immediately adjacent to the diagonal, currents that differ in phase by 209 degrees close to the source, increasing somewhat with distance. In other words, these currents are approximately out-of-phase. Figure 6.8 shows all three currents to be similar in magnitude.

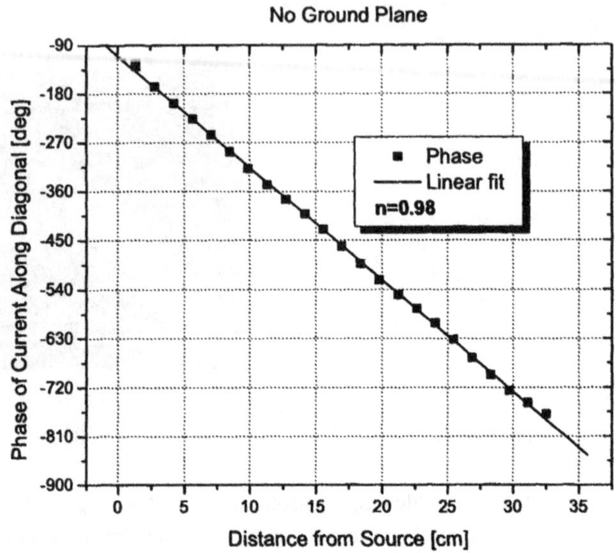

Fig. 6.2 Phase of current flowing along the grid diagonal which is the center line of the resonance cone. It indicates a forward wave with refractive index of 0.98 propagating away from the source (increasingly negative phase).

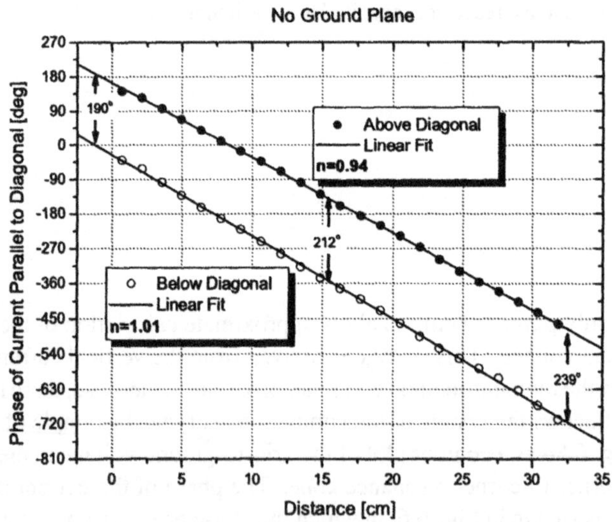

Fig. 6.3 Phase of currents flowing parallel to the diagonal, along the two lines of available data points closest to the diagonal, designated on the plot as being "above" and "below" the diagonal. Note that these currents are approximately out-of-phase.

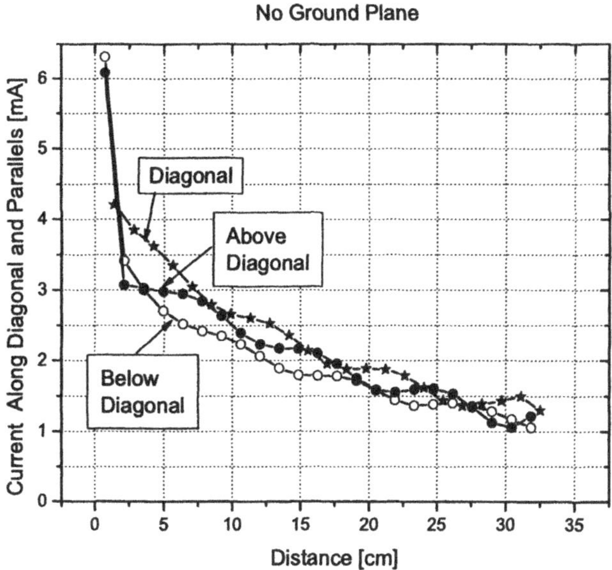

Fig. 6.4 Magnitude of currents flowing along the grid diagonal and adjacent to it.

6.5 ANISOTROPIC GRID WITH VERTICAL INDUCTORS, OVER INFINITE GROUND

The configuration with vertical inductors is shown in Fig. 6.9a, the purpose being to enable significant vertical current flow hence vertically polarized radiation in directions along the ground plane or elevated above it. Previously, the use of inductors to ground had been used successfully to generate radiation from a one-dimensional transmission-line antenna design as shown by Grbic and Eleftheriades (see [17] and [18]). For the present simulation, Fig. 6.9b shows the grid-to-ground voltage with a resonance cone along the diagonal line clearly in evidence. Figure 6.10 is a plot of the current phase for the current flowing along the diagonal, with the refractive index indicating a fast forward wave. Figure 6.11 shows the phase of the currents flowing parallel to and immediately adjacent to the diagonal, currents that differ in phase by 205 degrees close to the source (thus being approximately out-of-phase), with phase difference increasing somewhat with distance and showing considerable similarity to Fig. 6.7 except for the fast-wave property similar to that of Fig. 6.10. Figure 6.12 shows all three currents to be similar in magnitude as in the cases considered in Sections 6.3 and 6.4. However, the present case includes vertically-oriented inductors to ground so their currents deserve special attention on account of their contributions to radiation. The vertical current phase beneath the diagonal (center of the resonance cone) is presented in Fig. 6.13, in which the refractive index of 0.29 indicates a fast wave. Comparable refractive indices are calculated beneath the lines of data points to either side of the diagonal (Fig. 6.14). Comparable vertical

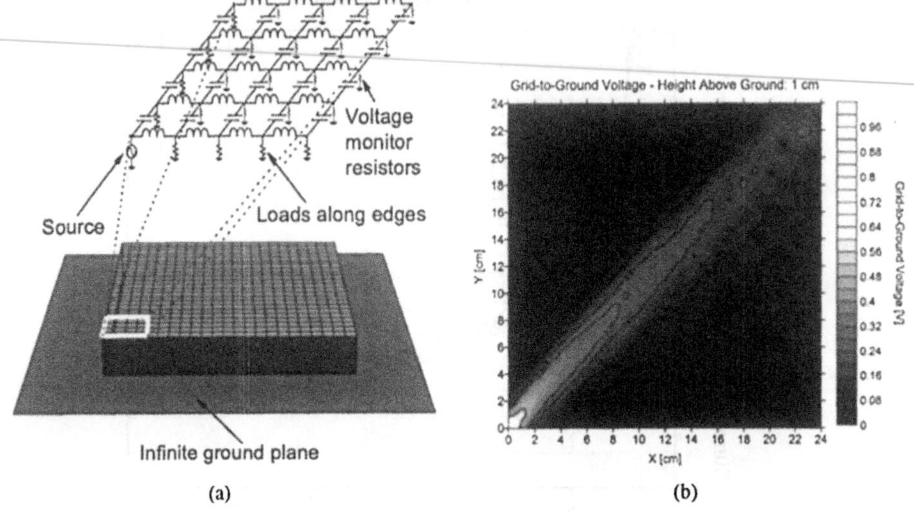

Fig. 6.5 (a) Anisotropic grid positioned over an infinite ground plane. The source is a 1-V 1.80-GHz voltage generator inserted in the vertical wire segment between grid and ground, located at the lower-left corner of the grid. The grid is 1 cm above ground. Resistive loads of 150 Ω per segment end are inserted in the vertical wire segments between grid and ground around the grid periphery, to minimize reflections from the grid edge. High-resistance resistors (50 MΩ) are inserted in the vertical wire segments between all interior grid-wire intersections and ground, to enable deduction of grid-to-ground voltages from segment currents. (b) Contour plot of grid-to-ground voltage showing a resonance cone concentrated along the grid diagonal, for the configuration of Fig. 6.5a.

current magnitudes are noted beneath the resonance cone (Fig. 6.15) with high values limited to the feed region. In Fig. 6.16, the horizontal-plane radiation pattern exhibits a major lobe in a direction approximately broadside to the resonance cone, a result that is consistent with the forward phase progression across the resonance cone that is apparent in Figs. 6.13 and 6.14.

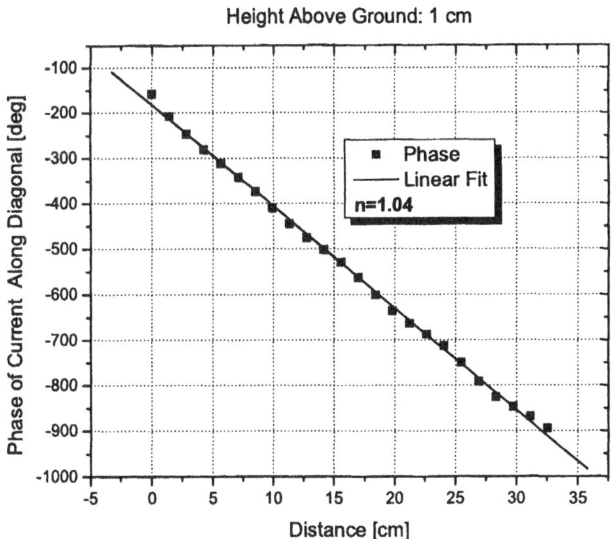

Fig. 6.6 Phase of current flowing along the grid diagonal as a function of distance from the source. It indicates a forward wave with refractive index of 1.04 propagating away from the source.

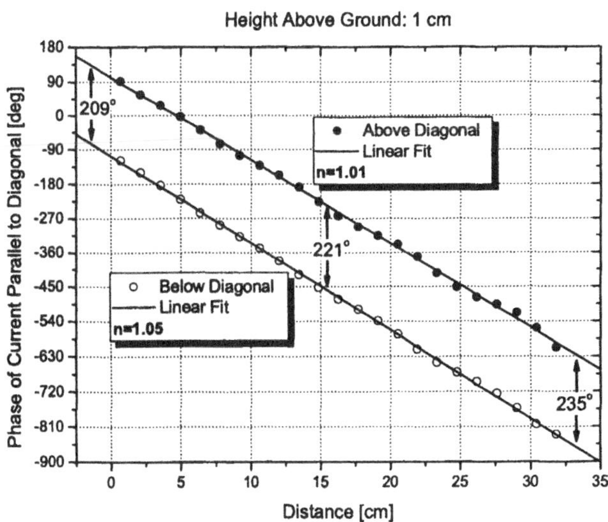

Fig. 6.7 Phase of currents flowing parallel to the diagonal as in Fig. 6.3. Note that these currents, when compared with those in Fig. 6.3, remain approximately out-of-phase.

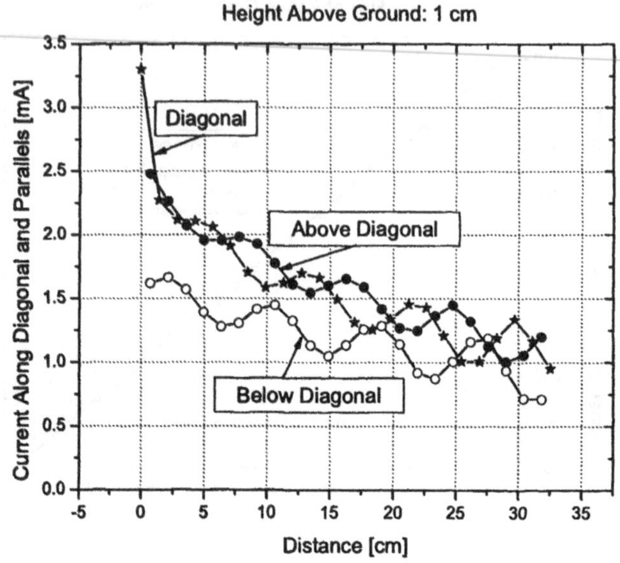

Fig. 6.8 Magnitude of currents flowing along the grid diagonal and adjacent to it.

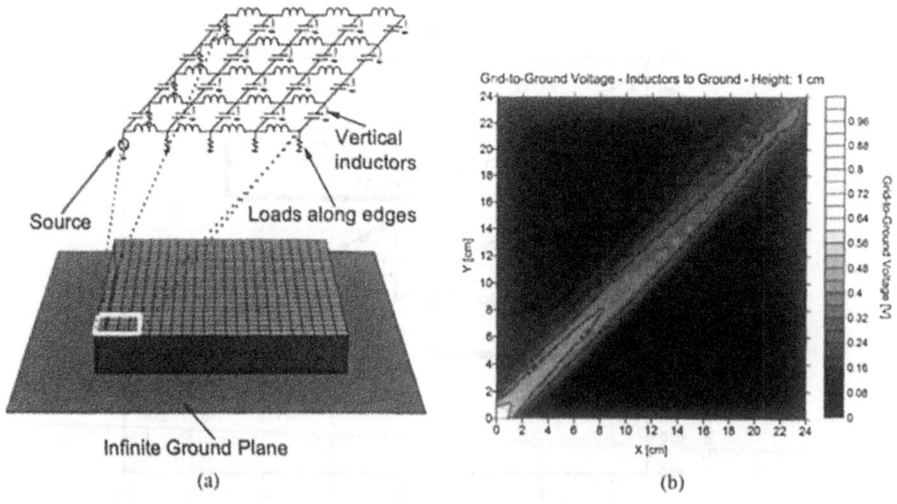

Fig. 6.9 (a) Anisotropic grid over an infinite ground plane as in Fig. 6.5 but now excited at 1.65 GHz and having inductors (33 nH per segment end) from all interior grid-wire intersections to ground. (b) Contour plot of grid-to-ground voltage, for the configuration of Fig. 6.9a.

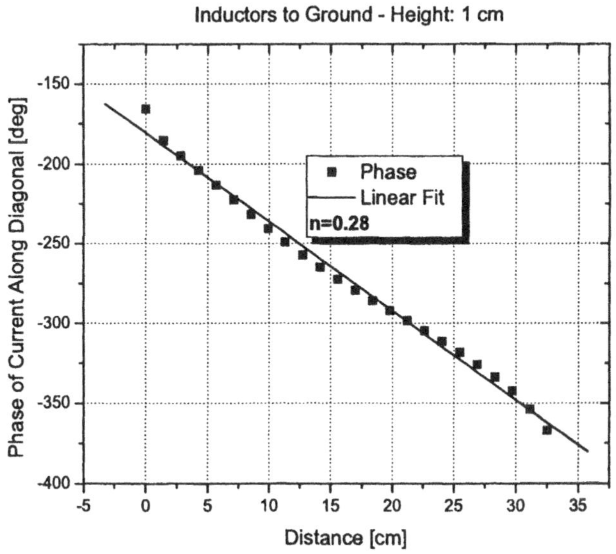

Fig. 6.10 Phase of current flowing along the grid diagonal for the configuration of Fig. 6.9.

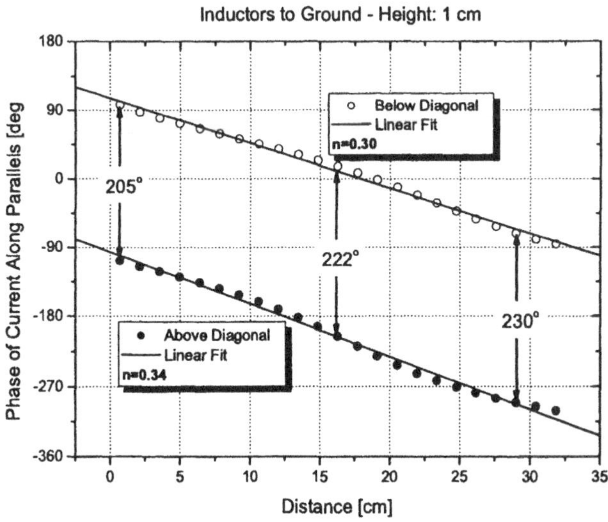

Fig. 6.11 Phase of currents flowing parallel to the diagonal.

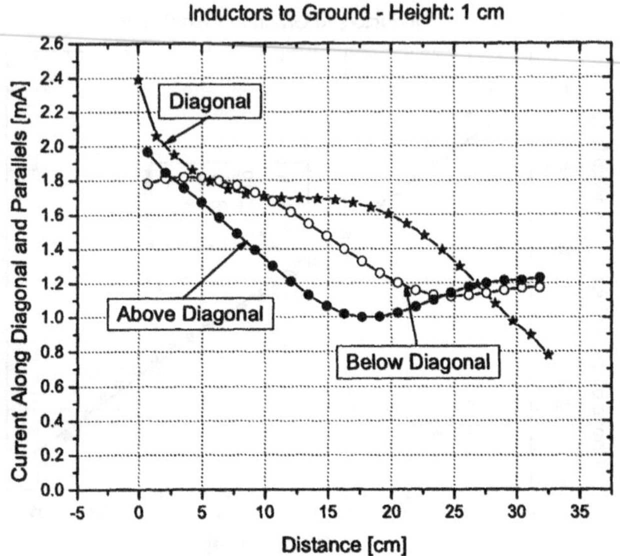

Fig. 6.12 Magnitude of currents flowing along the grid diagonal and parallel to it.

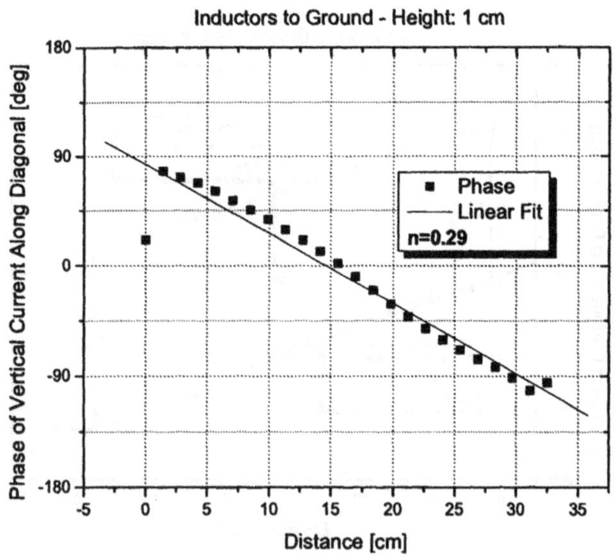

Fig. 6.13 Phase of vertical currents flowing beneath the anisotropic grid's diagonal.

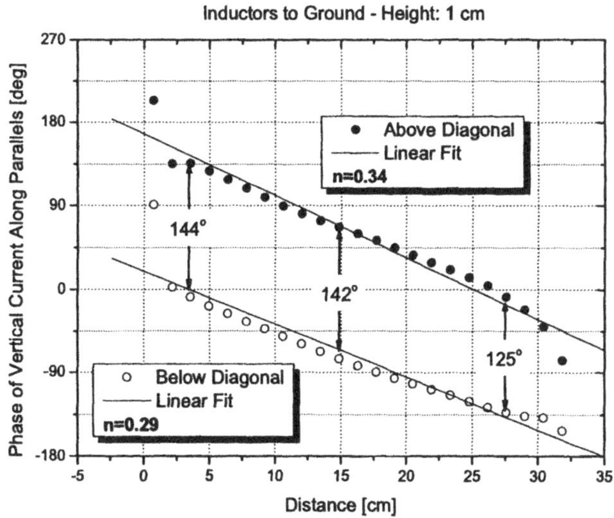

Fig. 6.14 Phase of vertical currents flowing beneath lines parallel to the diagonal of the anisotropic grid.

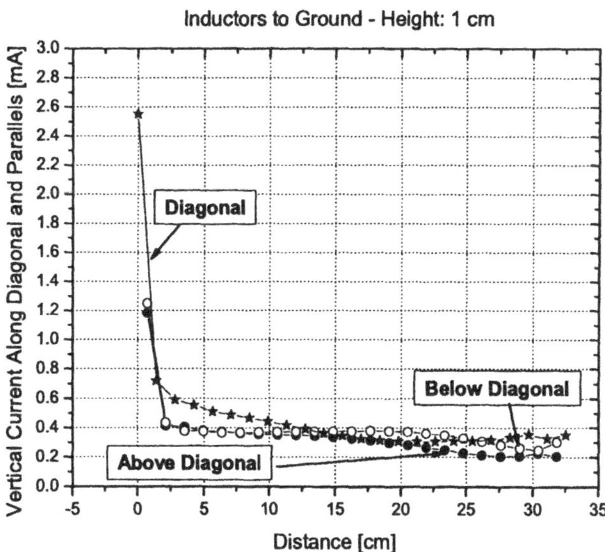

Fig. 6.15 Magnitude of vertical currents flowing beneath both the diagonal of the anisotropic grid and the adjacent parallel lines.

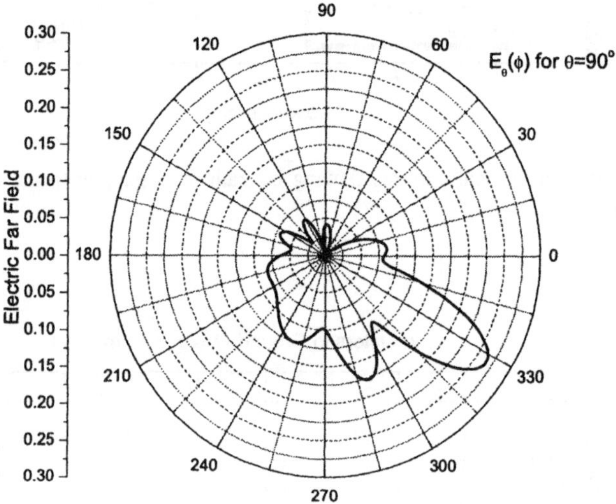

Fig. 6.16 Vertically polarized, horizontal-plane radiation pattern with main lobe in a direction approximately broadside to the resonance cone.

6.6 CONCLUSIONS

For an infinitely long, filamentary magnetic current line source, the quasi-static near fields have been calculated, showing clearly the high-field resonance cone extending outward from the source at an angle determined by the ratio of the permittivity matrix elements of opposite sign. Of special interest is the displacement current density flowing parallel to the resonance cone, which abruptly reverses sign as it passes through infinite magnitude exactly on the cone, bringing us to the conclusion that the resonance cone near field consists of a pair of counterflowing, contiguous displacement currents. While this has been found for a particular case, that case is a fundamental Green's function in the theory, suggesting that this conclusion may be quite generally applicable. The significance lies in the inescapable analogy with a two-conductor transmission line, which explains the ease with which even a small source in such a medium can inject power into the near field and from there into the far field. This makes plausible the finding of a number of authors that even a very small antenna in such a medium exhibits an unexpectedly large radiation resistance, some being so startled by this finding that they referred to it as an "infinity catastrophe" (see page 753 in the book by Felsen and Marcuvitz [12]).

Having seen this basic example of counterflowing currents in a homogeneous, uniaxially anisotropic medium, we have tested its applicability to anisotropic metamaterials by considering in simulation an evolutionary sequence. The sequence consisted of a free-standing anisotropic grid, such a grid over and parallel to a ground plane, and this same grid over ground but with each grid wire intersection connected to ground by a vertical inductor-loaded wire to carry radiating currents. In each case, under excitation by a small source, the anisotropic grid exhibited a resonance cone and carried currents parallel to the cone that were approximately out-of-phase, so that they could reasonably be described as counterflowing. It therefore appears that counterflowing currents and related effective coupling to a radiation field are likely to be found in a wide variety of resonance-cone metamaterials.

Acknowledgments

This work was supported in part by Defence Research and Development Canada (Ottawa), and in part by the Natural Sciences and Engineering Research Council of Canada.

REFERENCES

1. V. G. Veselago, "The electrodynamics of substances with simultaneously negative values of ϵ and μ," *Soviet Phys. Usp.*, vol. 10, no. 4, pp. 509–514, January–February 1968.

2. J. B. Pendry, "Negative refraction makes a perfect lens," *Phys. Rev. Lett.*, vol. 85, no. 18, pp. 3966–3969, 30 October 2000.

3. R. A. Shelby, D. R. Smith, and S. Schultz, "Experimental verification of a negative index of refraction," *Science*, vol. 292, no. 5514, pp. 77–79, April 6, 2001.

4. G. V. Eleftheriades, A. K. Iyer, and P. C. Kremer, "Planar negative refractive index media using periodically L-C loaded transmission lines," *IEEE Trans. Microwave Theory Tech.*, vol. 50, no. 12, pp. 2702–2712, December 2002.

5. F. V. Bunkin, "On radiation in anisotropic media," *Sov. Phys. JETP*, vol. 5, no. 2, pp. 277–283, September 1957.

6. H. Kogelnik, "On electromagnetic radiation in magneto-ionic media," *J. Res. Nat'l. Bureau of Standards-D. Radio Propag.*, vol. 64D, no. 5, pp. 515–523, September–October 1960.

7. H. H. Kuehl, "Electromagnetic radiation from an electric dipole in a cold anisotropic plasma," *Phys. Fluids*, vol. 5, no. 9, pp. 1095–1103, September 1962.

8. R. K. Fisher and R. W. Gould, "Resonance cones in the field pattern of a short antenna in an anisotropic plasma," *Phys. Rev. Lett.*, vol. 22, no. 21, pp. 1093–1095, May 26, 1969.

9. H. G. James, "Electrostatic resonance-cone waves emitted by a dipole in the ionosphere," *IEEE Trans. Antennas Propag.*, vol. AP-48, no. 9, pp. 1340–1348, September 2000.

10. K. G. Balmain and G. A. Oksiutik, "RF probe admittance in the ionosphere: Theory and experiment," in *Plasma Waves in Space and in the Laboratory*, Vol. 1, Edinburgh University Press, Edinburgh, 1969, pp. 247–261.

11. K. G. Balmain, "The impedance of a short dipole antenna in a magnetoplasma," *IEEE Trans. Antennas Propag.*, vol. AP-12, no. 5, pp. 605–617, September 1964.

12. L. B. Felsen and N. Marcuvitz, *Radiation and Scattering of Waves*, IEEE Press, New York, 1994 (originally published by Prentice-Hall in 1973).

13. K. G. Balmain, A. A. E. Lüttgen, and P. C. Kremer, "Resonance cone formation, reflection, refraction and focusing in a planar anisotropic metamaterial," *IEEE Antennas Wireless Prop. Lett.*, vol. 1, no. 7, pp. 146–149, 2002.

14. K. G. Balmain, A. A. E. Lüttgen, and P. C. Kremer, "Power flow for resonance cone phenomena in planar anisotropic metamaterials," *IEEE Trans. Antennas Propag.*, vol. 51, no. 10, pp. 2612–2618, October 2003.

15. A. Erdélyi, W. Magnus, F. Oberhettinger, and F. G. Tricomi, *Tables of Integral Transforms*, Vol. 1, McGraw-Hill, New York, 1954.

16. M. A. Tilston and K. G. Balmain, “A multiradius, reciprocal implementation of the thin-wire moment method,” *IEEE Trans. Antennas Propag.*, vol. AP-38, no. 10, pp. 1636–1644, October 1990.

17. A. Grbic and G. V. Eleftheriades, “A backward-wave antenna based on negative refractive index L-C networks,” in *2002 IEEE Antennas and Propagation Society International Symposium Digest*, vol. 4, pp. 340–343.

18. A. Grbic and G. V. Eleftheriades, “Experimental verification of backward-wave radiation from a negative refractive index material,” *J. Appl. Phys.*, vol. 92, no. 10, pp. 5930–5935, November 15, 2002.

7 Negative Refraction and Subwavelength Imaging in Photonic Crystals

CHIYAN LUO and JOHN D. JOANNOPOULOS

Department of Physics and Center for Materials Science and Engineering
Massachusetts Institute of Technology
Cambridge, MA 02139
United States

7.1 INTRODUCTION

The propagation of light waves in complex optical media has remained a topic of considerable interest ever since the introduction of diffraction gratings. Its simplest description has been presented in the form of the so-called *effective medium theory*, in which a composite metamaterial is approximated to be a *uniform* medium with "effective" values of optical constants, that is, permittivity ϵ and permeability μ. Recent work by Pendry, Smith, and co-workers [1–5] opened a new window in the effective medium theory. A new class of composite metallic structures, which can be described by an effective uniform medium with $\epsilon < 0$ and $\mu < 0$, was predicted and experimentally realized in the microwave regime. These composite media have come to be known as "left-handed metamaterials," which was predicted by Veselago to lead to a variety of unusual electromagnetic phenomena a long time ago [6]. One of the most notable properties of these materials is the possibility of negative refraction, in which the refractive index n of the material must be taken to be the *negative* square root of the product of ϵ and μ, $n = -\sqrt{\epsilon\mu}$, if the conventional Snell's Law of refraction is applied. Another intriguing proposal associated with left-handed metamaterials, discovered recently by Pendry [4], is the possibility of a perfect lens, which in principle allows light to be focused down to a spot size much smaller than the wavelength and beyond the classical diffraction limit. Such novel findings have generated much excitement in the old discipline of electromagnetism and become of great interest to the general scientific community. Several experiments from different laboratories have been reported confirming the existence of negative refraction in these artificial structures and supporting the viewpoint of an effective uniform material with $\epsilon < 0$ and $\mu < 0$ [7–10].

The aim of the present work is to discuss the two unconventional light phenomena enabled by left-handed materials, namely, negative refraction and subwavelength focusing, in general photonic crystals. *Photonic crystals* (PCs) refer to a fairly broad class of composite optical materials, and a brief introduction to them is provided in Section 7.1.1. In contrast to the effective medium theory of left-handed metamaterials, in which the composite structure is regarded as effectively *uniform* and the unusual properties of light are identified as the consequences of ϵ and $\mu < 0$, our approach shows that similar anomalies can occur as a direct result of the photonic crystal effects by taking the effects of the microstructures into full account. The underlying physical principles discussed in this work are based on complex Bragg scattering phenomena and are very different from those in a left-handed metamaterial. The refractive properties of photonic crystals based on the present viewpoint are also known as *photonic crystalline optics*, and they have been discussed by several authors in recent literature in parallel to the work on left-handed materials [11–13]. Here we address the optics for both propagating and evanescent waves in photonic crystals, following our recent papers on photonic crystalline superlenses [14–17]. We give certain general criteria under which these anomalous phenomena are possible in general photonic crystals and present several specific numerical examples. Our findings reveal that negative refraction and subwavelength imaging are very general phenomena beyond the concept of left-handed materials. For example, they can be expected with dimension-scaled dielectric photonic crystals at optical frequencies, whereas the optical regime is inaccessible by simple scaling for the current metallic left-handed structures due to the large loss in metals . Another important implication is that the designs of structures supporting these unusual effects can be much simpler and flexible, owing to the universality of photonic crystals. In particular, effects inside a two-dimensional (2-D) can be entirely captured by a 2-D arrangement, and three-dimensional (3-D) crystals enables truly 3-D phenomena. Very recently, a number of experimental work have been reported verifying the basic predictions of our work [18, 19].

7.1.1 Introduction to Photonic Crystals

In this section we give a very brief introduction to the subject of photonic crystals. Interested readers are referred to the original papers [20, 21] and reviews [22] for more detailed discussions of the topics presented here.

Photonic crystals were first discussed by Yablonovitch in 1987 with a famous analogy of photons to electrons: a lattice of electromagnetic scatterers can tailor the properties of light in much the same way as crystalline solids do to electrons. In particular, when the lattice constants are on the order of the wavelength of light and the scattering strength of each scatterer is strong, the propagation of light waves inside such a lattice will be strongly modified by the photonic lattice structure. The basic problem is then to determine what the new photonic modes are inside such a lattice, usually specified by a position-dependent, periodic permittivity $\epsilon(\mathbf{r})$. Maxwell's equations can be cast in a stationary state form as

$$\Theta \mathbf{F} = \frac{\omega^2}{c^2}\mathbf{F} \tag{7.1}$$

where Θ is a position-dependent Hermitian operator containing $\epsilon(\mathbf{r})$, $\mathbf{F}$ is the electromagnetic (EM) field, and ω is the frequency of the stationary state (for example, when $\mathbf{F}$ is the magnetic field $\mathbf{H}$, $\Theta = \nabla \times \epsilon^{-1}(\mathbf{r})\nabla \times$). The form in equation (7.1) provides an explicit analogy to Schrödinger's equation for electrons in solids: Θ corresponds to the periodic atomic potentials and the square frequency ω^2/c^2 corresponds to the energy eigenvalue. The solution to equation (7.1) in a photonic crystal can be classified using the fundamental concepts in the band theory of electrons. According to *Bloch's Theorem*, the eigenmode $\mathbf{F}$ can be made *Bloch-periodic* with a *Bloch wavevector* $\mathbf{k}$ that lies within a *Brillouin zone*. The eigenfrequency ω then emerges as a function of $\mathbf{k}$ and band index n: $\omega = \omega_n(\mathbf{k})$, which maps out the *photonic band structure* as $\mathbf{k}$ is varied throughout the Brillouin zone. A *photonic band gap* can result for a frequency range in which no eigenmodes are allowed. Actual photonic band structures are usually calculated using numerical techniques similar to those developed in the studies of the electronic band structure, such as the plane-wave expansion [23]. On the other hand, there exist a number of physical differences between EM eigenmodes in photonic crystals and electron Bloch waves in crystals. Of particular importance is the extremely wide generality of the principles of photonic crystals. This is because there are no fundamental lengthscales in Maxwell's equations, and the band structure may be scaled to arbitrary frequencies provided that the photonic lattice is correspondingly scaled. A system designed at the microwave regime can therefore be "scaled" to designs at other wavelengths (e.g., the optical regimes) using identical parameters in fractions of wavelength. This is vastly different from the electronic case in which the Bohr radius sets a natural lengthscale for the quantum system. Moreover, as the EM waves are vector fields, the photonic band structure contains information about light polarization and is richer than its scalar-wave counterpart in the electronic problems. Finally, it is worth noting that numerical results of photonic band structure are essentially *exact* within the linear response approximation and can be more reliable than those of electronic band structures, which are almost always complicated by the effects of electron–electron interaction and Fermi statistics.

Photonic crystals have been proposed to revolutionize the generation and propagation of light waves. Indeed, the existence of a photonic band gap in a range of frequencies prohibits light propagation in all possible directions. This has led not only to novel photon–atom bound states that alter the fundamental physics of light emission, but also to the concept of basic building blocks for optical materials that can be used to construct devices for practical applications. Deliberate defect structures inside the band gap adds a new design dimension for versatile light control. Ultrahigh-Q cavities may be introduced as point defects in a perfect crystal to realize resonance channel add–drop filters and ultra-low threshold lasers. Channels for efficient light transportation are formed by line defects which can guide light through extremely sharp corners. Furthermore, by combining the index-guiding mechanism of slab waveguides with two-dimensional photonic band-gap effects, control and manipulation of light in full three dimensions can be realized with present planar lithographic techniques. These developments echo with the past band-gap engineering in semiconductor electronics and suggest that photonic crystals may be used to tailor the properties of light in much the same way.

The rest of this work is divided into two parts. In this first part we address the problem of how negative refraction of propagating waves occurs in photonic crystals. We then include evanescent waves and study their influence on wave transmission in detail in the second part, where it is demonstrated that photonic crystals can give rise to the spectacular phenomena of Veselago–Pendry superlensing. In the concluding section, the main ideas of this work are briefly summarized.

7.2 NEGATIVE REFRACTION IN PHOTONIC CRYSTALS

7.2.1 Analysis of Refraction in Uniform Materials

Let us first revisit the problem of light refraction on the interface between free space and a uniform material with permittivity ϵ and permeability μ, under a coordinate system whose z-axis is perpendicular to the interface. The incident wave of a definite incident direction can be represented as a plane wave of a constant frequency (CW) ω. In this simple case, all the propagating optical modes inside the material are plane waves, and the system is invariant under an arbitrary translation in the Oxy plane. The refraction process at the interface will thus conserve the wavevector component parallel to the interface. We can then analyze the problem in terms of the dispersion contours—that is, the contours formed by the wavevectors of all the allowed propagating waves inside a medium corresponding to the same frequency. These contours are just circles $\mathbf{k}^2 = \epsilon\mu\omega^2/c^2$, with a radius $\left|\sqrt{\epsilon\mu}\right|\omega/c$ in the material. The refracted waves can be found on the dispersion contours (or equifrequency dispersion contours) by conserving the surface-parallel component of the wavevector from the incident wave. Geometrically, the refracted waves are obtained from the points on the free-space contour representing the incident wave by drawing a line perpendicular to the interface and finding its intersections with the material dispersion contour. In this way, a set of possible refracted wavevectors can be obtained, whose directions represent the possible directions of the phase advancement in the material. Finding the direction of refraction requires a little bit of care as this direction is usually associated with that of the energy flow. For a CW wave, this direction is given by the *group velocity* $\mathbf{u} = \partial\omega/\partial\mathbf{k}$ of the refraction wave—that is, the gradient direction on the dispersion contour. This gradient direction in this example is the same as that of the wavevector in a normal medium with $\epsilon > 0$ and $\mu > 0$, but is antiparallel to that of the wavevector for left-handed materials because of their

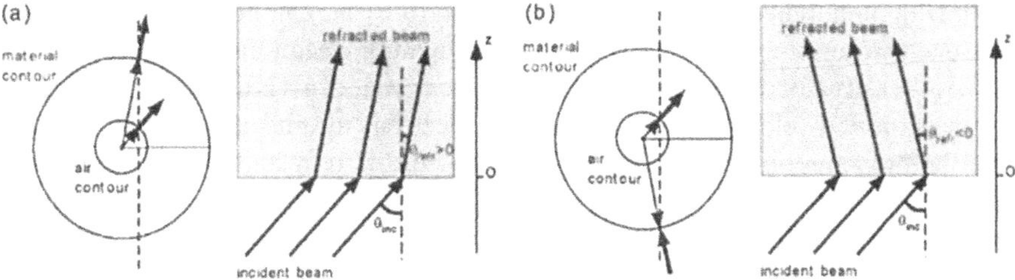

Fig. 7.1 Illustration of refraction in uniform media. (a) Positive refraction in a medium with $\epsilon > 0$ and $\mu > 0$. (b) Negative refraction in a medium with $\epsilon, 0$ and $\mu < 0$. Thin arrows stand for wavevector directions, and thick arrows indicate the direction of energy flow.

*negative group velocities.** Moreover, a refracted wave must propagate away from the interface, and thus we must have $\mathbf{u} \cdot \hat{\mathbf{z}} > 0$. Combining these analyses, we arrive at the situation depicted in Fig. 7.1. In particular, for a left-handed material, both the parallel component of the refracted group velocity and the z-component of the refracted wavevector are reversed compared to a normal medium. This is precisely captured by a *negative index of refraction* $n = -\sqrt{\epsilon\mu}$, a natural consequence of left-handed materials.

7.2.2 Analysis of Refraction in Photonic Crystals

The analysis in Section 7.2.1 is general and can in fact be easily extended to the case of a photonic crystal, with little changes in essence. A photonic crystal system is invariant under translations of an integral number of lattice constants, and it supports CW propagating waves that can be specified by a Bloch wavevector $\mathbf{k}$ and a band index n. Thus the dispersion contours can be formed by all the Bloch wavevectors within the first Brillouin zone at the same frequency. For a semi-infinite crystal, the symmetry reduces to the translations of an integral number of surface periods parallel to the interface, and correspondingly the conserved quantity in a refraction process is the component of the Bloch wavevector parallel to the interface within the *surface Brillouin zone.* An equivalent and perhaps easier alternative is provided by the extended-zone scheme, in which the bulk-crystal dispersion contours consist of all the wavevectors, including all that differ from the Bloch wavevectors in the first Brillouin zone by a reciprocal lattice vector. The refraction process may then be analyzed by matching the wavevectors inside the bulk crystal to that of the incident wave and also reducing the solution wavevectors to the first Brillouin zone to find the refracted Bloch modes. To obtain the refraction directions, the Poynting vectors needs to be calculated for each of these modes. This procedure is facilitated by the

*It can be easily verified from Maxwell's equations that the Poynting's vector $\mathbf{E} \times \mathbf{H}$ is antiparallel to the wavevector $\mathbf{k}$ if $\epsilon < 0$ and $\mu < 0$.

following important result: The group velocity $\mathbf{u} = \partial\omega/\partial\mathbf{k}$ continues to represent the energy transport vector (i.e., the average Poynting vector divided by the average energy density) for every Bloch mode inside the crystal. This is the photonic analogue of the expected velocity equation for Bloch electrons in solid state physics [24] and can be proven along similar lines [25]. Hence, the refracted beams travel along the gradient directions on the dispersion contour. The extension of the geometric refraction analysis from uniform materials to photonic crystals is thus complete.

In general, the refraction in photonic crystals can be more complicated than in uniform media. A single incident wave can produce multiple refracted Bloch-wave orders, if there are multiple branches of dispersion contours at the same frequency or if the interface is not along major crystal symmetry directions. This results in a scenario similar to simple Bragg scattering in the elementary case of diffraction gratings. Moreover, the presence of strongly scattering crystal structures can modify the dispersion relation of light so much that the shapes of the dispersion contours are far from being circular. The gradient directions on these contours and thus the refraction directions can be very different from those expected in a uniform material. Furthermore, the strength of each refraction wave can be quite different from either refraction in uniform media or simple diffractions. In the extreme case, the occurrence of a photonic band gaps can eliminate entirely waves on certain directions. These considerations illustrate that photonic crystal structures can give rise to nontrivial effects in refraction, and we will refer to them as *complex Bragg scattering*. In the following, we examine in detail how the complex Bragg scattering in photonic crystals gives rise to negative refraction.

As a specific example, we consider a 2-D square lattice of air holes in a dielectric background $\epsilon = 12.0$, with a lattice constant a and a hole radius $r = 0.35a$. Light waves propagating in the plane of 2-D periodicity can be classified into the usual TE and TM polarizations, and for simplicity let us look at waves having a magnetic field parallel to the cylinders (TE modes). The photonic band structure of this crystal can be calculated numerically using plane wave expansion, and it is shown as the dispersion contours in Fig. 7.2. Here we observe that due to the negative-definite effective mass $\partial^2\omega/\partial k_i \partial k_j$ at the M point, the frequency contours are *convex* in the vicinity of M and have inward-pointing group velocities. According to our approach, for frequencies that correspond to all-convex contours near M, negative refraction occurs as illustrated in Fig. 7.3. The distinct refracted propagating modes are determined by the conservation of the frequency and the wavevector component parallel to the air/photonic-crystal interface. If the interface normal is along ΓM [(11) direction] and the contour is everywhere convex, then an incoming plane wave from air will couple to a single Bloch mode that propagates into this crystal on the *negative* side of the surface normal. Negative refraction is thus realized in the first band of the photonic crystal. To verify this, we proceed with a computational experiment using finite-difference time-domain (FDTD) simulations. Here, a continuous-wave (CW) Gaussian beam of frequency $\omega_0 = 0.195(2\pi c/a)$ and a half-width $\sigma = 1.7(2\pi c/\omega_0)$ is launched at 45° incidence toward the (11) crystal surface of our photonic crystal and subsequently refracts into it. A snapshot of this refraction process is shown in Fig. 7.4,

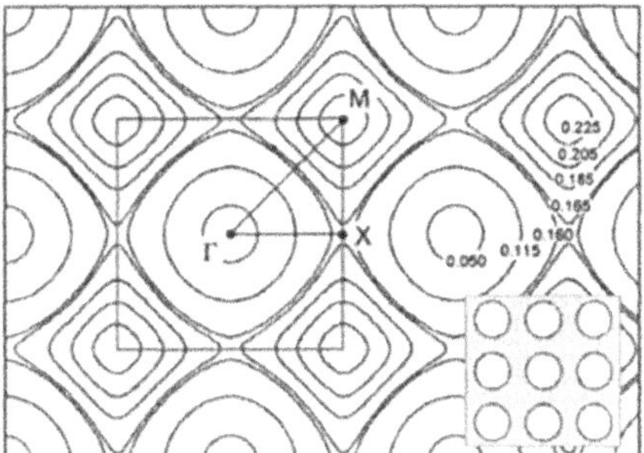

Fig. 7.2 Several constant-frequency contours of the first band of a model photonic crystal, drawn in the repeated zone scheme. Frequency values are in units of $2\pi c/a$. Reprinted figure with permission from Ref. [14]. Copyright © 2002 by the American Physical Society.

which clearly demonstrates that the overall electromagnetic energy indeed travels on the reversed side of the surface normal. If we look closely at the refracted field patterns in the crystal, we can see that there are constant-phase regions lying on parallel straight lines and forming "phase fronts" in the photonic crystal. However, there are multiple ways of constructing parallel lines connecting these discrete regions, corresponding to multiple choices of phase-front definitions. This reflects the fact that, in a photonic crystal, $\mathbf{k}$ is only defined up to a reciprocal lattice vector $\mathbf{G}$. For the present situation, we choose the phase fronts for the refracted beam to be the set of constant-phase lines with the largest wavelength in the crystal, which corresponds to the smallest $|\mathbf{k}|$ and hence the unique $\mathbf{k}$ representing the Bloch phase in the first Brillouin zone. The normals of the phase-fronts defined in this way now points toward the *positive* side of the surface normal; that is, the phase velocity exhibits *positive* refraction in our example.

It is clear from this example that the notion of a backward wave or a left-handed material is in fact not a prerequisite for negative refraction. The physics of our example differ from those of left-handed materials in that the lowest band now has $\mathbf{k} \cdot \partial\omega/\partial\mathbf{k} \geq 0$ everywhere within the first Brillouin zone, meaning that the group velocity is never opposite to the Bloch wavevector $\mathbf{k}$. In this sense, we are operating in a regime of forward waves and positive effective index. Such a scenario can also happen in a right-handed medium with hyberbolic dispersion relations, such as those induced by anisotropy. For example, the TE modes in a uniform nonmagnetic medium with dielectric tensor

$$\tilde{\epsilon} = \begin{pmatrix} \epsilon_1 & 0 \\ 0 & \epsilon_2 \end{pmatrix} \tag{7.2}$$

(with $\epsilon_1 > 0$ and $\epsilon_2 < 0$) have a dispersion relation $k_2^2/\epsilon_1 - k_1^2/|\epsilon_2| = \omega^2/c^2$. Similar negative refraction will then happen on the (01) surface. Again, the phase velocity always makes an acute angle with the group velocity.

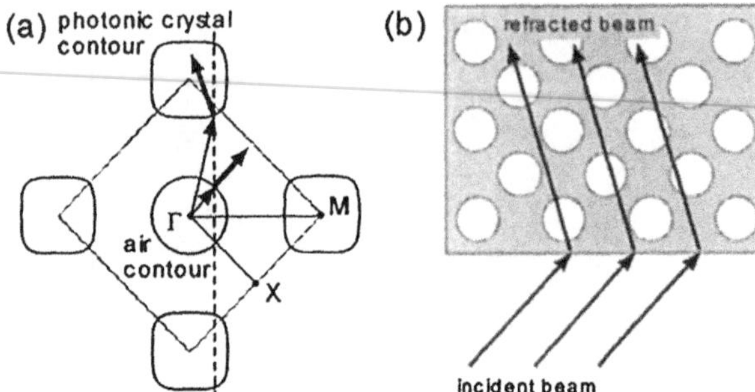

Fig. 7.3 (a) Negative-refracted beams constructed from dispersion contours and conservation of interface-parallel wavevector. Thick arrows indicate group-velocity directions, and thin arrows stand for phase-velocity directions. (b) Schematic diagram of refraction rays in the crystal. Reprinted figure with permission from Ref. [14]. Copyright © 2002 by the American Physical Society.

7.2.3 Dispersion Contours of 2-D Photonic Crystals

So far, our discussion has been centered on a particular polarization (TE) in a particular band of a particular photonic crystal. It is straightforward to extend similar arguments to other systems: the key quantity of interest is the dispersion contours of a photonic crystal. For reference purposes, we give in Figs. 7.5–7.8 the dispersion contours of the first few bands in common 2-D dielectric photonic crystals. The normal-incident transmission coefficients through a finite slab of such photonic crystals for interface termination along major symmetry directions are also given adjacent to the band structures in these figures. These transmission curves exhibit oscillatory features when the frequency is inside a band, which vanishes once the frequency is inside a band gap. The oscillatory features are similar to the Fabry–Perot effect and become more rapid for thicker slabs. The coupling strength of external plane waves to Bloch photon modes can be qualitatively inferred from the transmission spectrum by the average transmission value over these oscillations. The dispersion contours given in these figures are the photonic analogy of the Fermi surface of metals in electron band theory [13]. In the first few bands of most crystals, the shapes of dispersion contours are relatively simple with one or two branches, but these shapes can become exceedingly complex for higher bands. In this work, we will be mostly concerned with simple shapes and circular-like dispersion surfaces; the use of the high-curvature, strongly anisotropic part of the dispersion surfaces are demonstrated in the so-called superprism effect [11]. For square lattices, negative refraction is most easily realized near the top of the first or the second band where the photonic "effective mass" is negative. Near the top of the first band, the dispersion contours becomes all-convex around the corner of the first Brillouin zone M-point, and negative refraction can be realized on the (11) interface exactly as what we presented above. Near the top of the second band, the dispersion contour becomes all-convex around a corner of

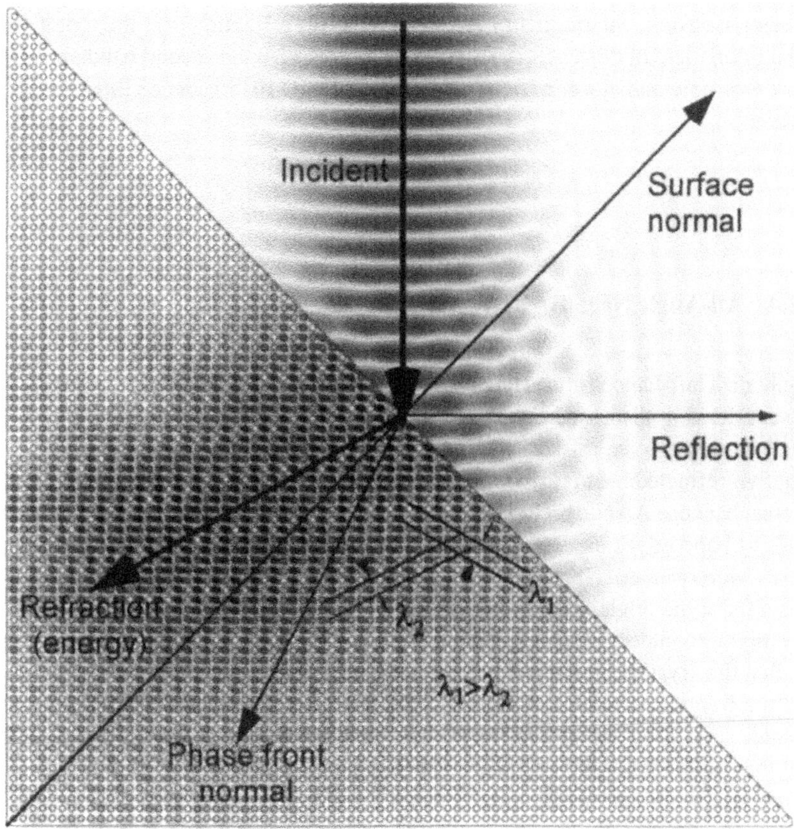

Fig. 7.4 FDTD simulation of negative refraction in a photonic crystal. Shown is a snapshot of the magnetic field perpendicular to the plane (darker shades for larger magnitude). The dielectric boundaries are outlined. The arrows and texts illustrate the various important. Also shown in straight lines are two possible ways of construction phase fronts from the field pattern. The set of phase fronts with the maximum wavelength (λ_1) is chosen to be that of the refracted beam.

the second band, which is Γ in the reduced-zone scheme. Although the shape of the contours appears circular and may suggest an isotropic behavior resembling that of a left-handed material, there is usually a marked difference between transmission values on (10) and (11) interfaces at the frequencies corresponding to the top of the second band, as shown for example in Fig. 7.6. The (10) surface is usually favored in the second band. For triangular lattices, because the first Brillouin zone is closer to being isotropic, the frequency range for negative refraction near the top of the first band is smaller. In this case, the dispersion contours in the second band appear to be more isotropic and closer to being circular, and the (10) incidence interface is again favored.

7.2.4 All-Angle Negative Refraction

To obtain a left-handed-material behavior and to set the foundation for the phenomena of superlensing discussed later in this work, we will look further for the condition of all-angle negative refraction; that is, for all incident angles one obtains a single negative-refracted beam inside the photonic crystal. This is also the case of practical interest because AANR precludes modes with very small group velocities, which are close to band edges and are generally difficult to couple to from an external plane wave. Moreover, AANR also eliminates the effect of total *external* reflection, which exists for some angles if the absolute value of the effective index is less than unity and might be undesirable in some applications. To realize AANR, sufficient criteria are that the crystal contours be both *all-convex* and *larger* than the vaccum contours, and the frequency be below $0.5(2\pi c/a_s)$ where a_s is the surface periodicity of the crystal. It can be seen from the results displayed here that in 2-D the structures that may fulfill the AANR requirements are square-lattice crystals in their first band or triangular-lattice crystals in their second band (TM polarization only). This can actually be understood from the following intuitive argument. In the periodic-zone scheme, the dispersion contour for the first few bands of photonic crystals may be constructed by (a) joining all the spherical contours of an effective *uniform* medium which are centered on the reciprocal lattice sites and (b) rounding the sharp parts of the joint surface across Brillouin zone boundaries. For a given Brillouin zone corner C, we expect that the more neighboring reciprocal-lattice sites C has, the stronger the resulting rounding effect, and the easier it is for the dispersion contours to become all-convex around C. Thus, a rough rule to choose the optimum geometric lattice for negative refraction is just to maximize the number N of C's nearest-neighbor reciprocal lattice sites. If negative refraction is to be realized in the first band, then C is a corner of the first Brillouin zone, and a 2-D square lattice ($N = 4$) performs better than a 2-D triangular lattice ($N = 3$). If negative refraction is to be realized in the bands after folding once, then C is a corner of the second Brillouin zone (Γ in this case), and the triangular lattice with $N = 6$ is thus preferred.

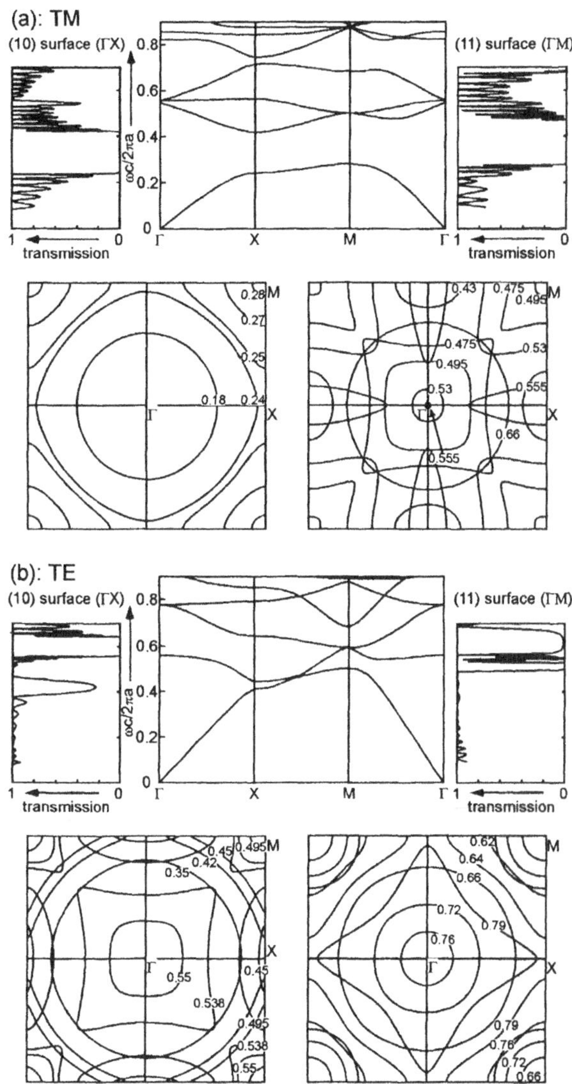

Fig. 7.5 Band structure, transmission coefficients, and dispersion contours for the first few bands of a 2-D square lattice of rods in air. The rods have a dielectric constant $\epsilon = 12$, the lattice constant is a, and the rod radius is $0.2a$.

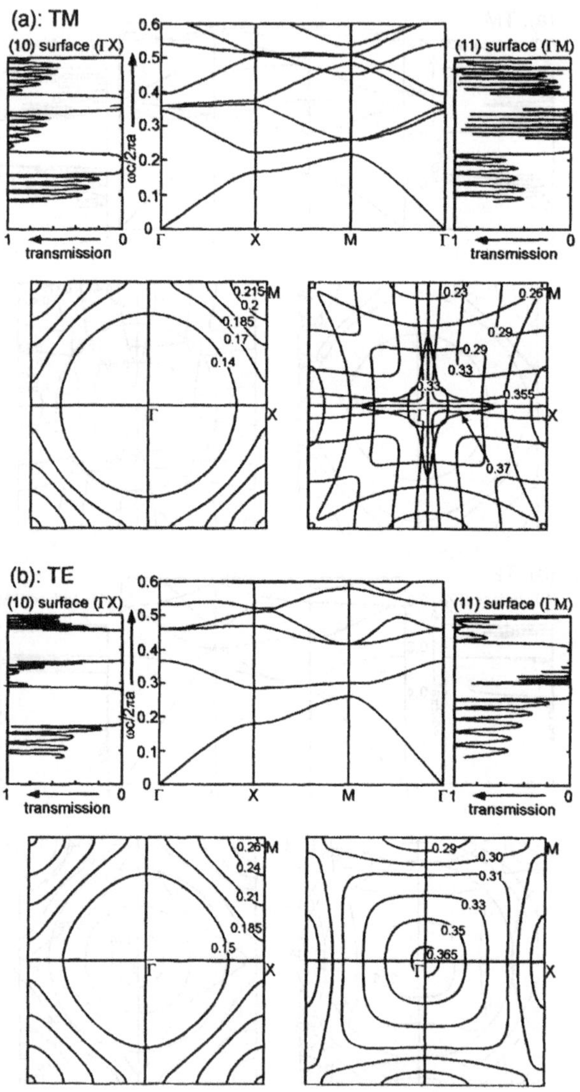

Fig. 7.6 Band structure, transmission coefficients, and dispersion contours for the first few bands of a 2-D square lattice of holes in a high-index background with dielectric constant $\epsilon = 12$. The lattice constant is a and the hole radius is $0.4a$.

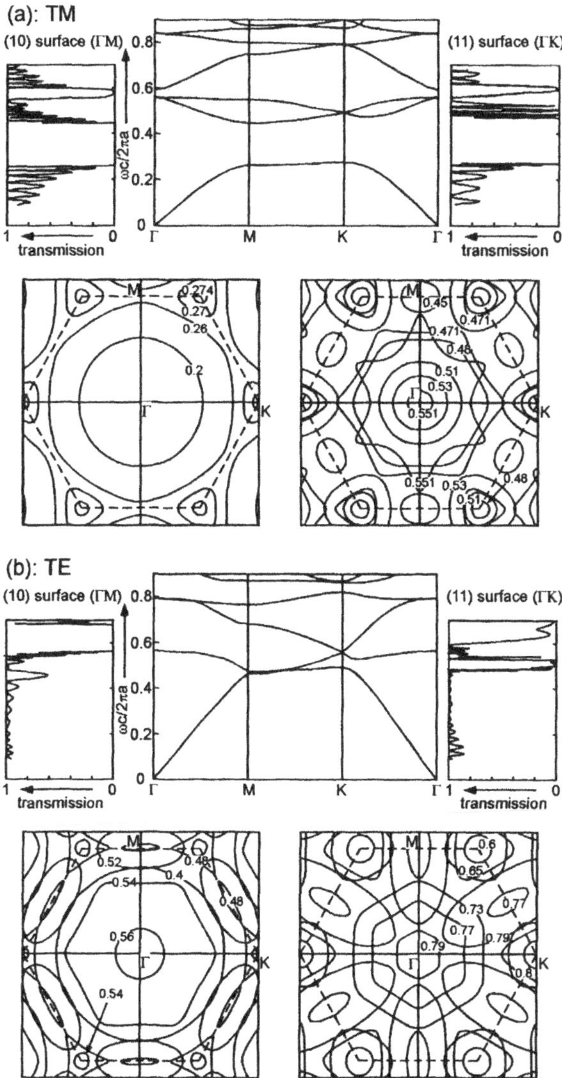

Fig. 7.7 Band structure, transmission coefficients, and dispersion contours for the first few bands of a 2-D triangular lattice of rods in air. The rods have a dielectric constant $\epsilon = 12$, the lattice constant is a, and the rod radius is $0.2a$.

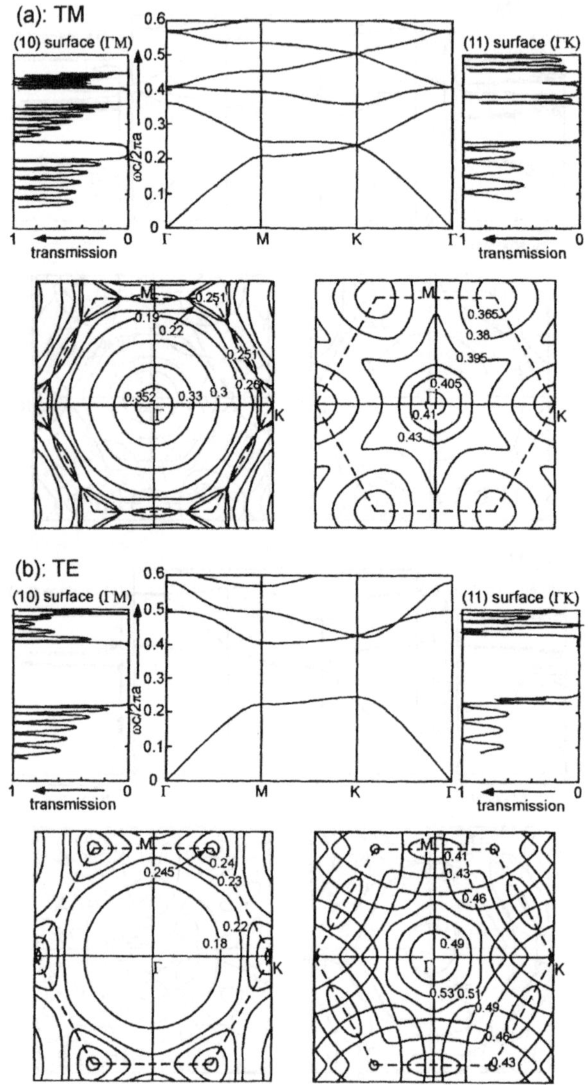

Fig. 7.8 Band structure, transmission coefficients, and dispersion contours for the first few bands of a 2-D triangular lattice of holes in a high-index background with dielectric constant $\epsilon = 12$. The lattice constant is a and the hole radius is $0.4a$.

7.2.5 Negative Refraction in Three-Dimensionally Periodic Systems

Is it possible to generalize the intuition in Section 7.2.4 to three-dimensional (3-D) situations? If so, for negative refraction to be realized in the fundamental (i.e. the first two) bands, a simple cubic (sc) reciprocal lattice with $N = 8$ should be used, resulting in a sc crystal with (111) surface termination. If negative refraction in the bands after folding once are considered, the face-centered cubic (fcc) *reciprocal* lattice which has $N = 12$ should be chosen, giving a body-centered cubic (bcc) structure in real space. It is, however, important to keep in mind that the intuition and arguments here are mostly based on scalar-wave assumptions, and the vectorial nature of light waves will usually bring significant complications in a 3-D situation. For example, the dispersion surface in even the first two bands becomes the two branches of one complex connected surface, a situation similar to that in optics of anisotropic crystals. The simple picture of refraction considered above is thus usually accompanied by the occurrence of the peculiar phenomena familiar in the optics of crystals, such as bifringence and conical refraction. Nevertheless, in the geometric lattice of bcc crystals having $N = 12$, we have found a particular configuration enabling the simple picture of single-beam AANR in full 3-D. This consists of a bcc lattice of large low-index voids in a high-index background. A configuration that can give good performance is cubic air voids of size $0.75a$ (sides parallel to those of the conventional bcc cell whose size is a) in germanium at infrared wavelengths with $\epsilon = 18$ (qualitatively similar band structure can result in slightly different designs—for example, spherical air voids in Si with $\epsilon = 12$, which may present a less challenging task in terms of fabrication). The band structure of this photonic crystal is shown in Fig. 7.9. There is a large frequency range in the third band within which the dispersion surface forms a single all-convex, near-spherical surface and is larger than that of free space, as shown in Fig. 7.9b. This situation is similar to the second band of 2-D triangular lattices, and it is important to find out the particular surface orientation on which the crystal modes can couple to external plane waves with a significant efficiency. The interesting point here is that there is only one band in this frequency range in full 3-D, and thus the refraction phenomena will be strongly polarization-dependent. For example, along the (001) direction, the two degenerate polarizations of normal-incidence radiation in free space and the single-degenerate photonic-crystal mode belong to different irreducible representations of the surface symmetry group. As a result, they do not couple with each other, and thus the (001) surface should *not* be used as the interface for negative refraction. Instead, the (101) interface, whose symmetry group has different irreducible representations for the two polarizations, should be used. Along the normal direction on this interface, the crystal mode favors coupling with waves polarized along the $(10\bar{1})$ and does not couple with those polarized along (010). This dependence of coupling efficiency on polarization direction illustrates another important difference between photonic crystals in their higher bands and an *isotropic*, *uniform* left-handed material with a negative refractive index.

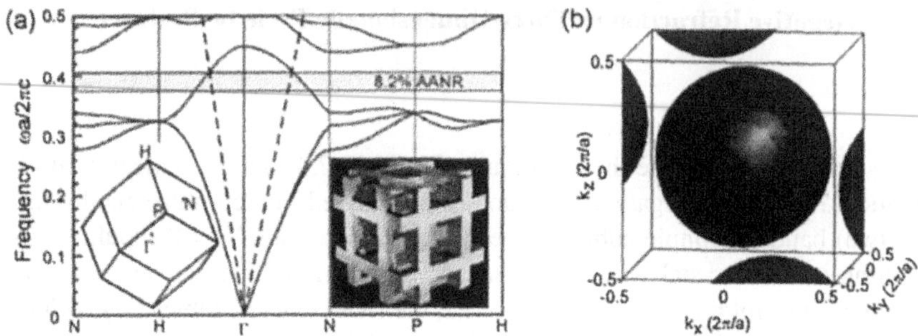

Fig. 7.9 (a) Band structure (solid line) of a 3-D photonic crystal for negative refraction. The shaded region is the AANR frequency range for this crystal. The dashed lines are light lines along ΓH and ΓN. Insets are the shape and special symmetry vertices of the first Brillouin zone and a computer rendering of the crystal in real space. (b) The dispersion contour of the crystal in (a) at $\omega = 0.407(2\pi c/a)$ in a periodic zone scheme. Reprinted with permission from Ref. [15]. Copyright © 2002 American Institute of Physics.

7.2.6 Case of Metallic Photonic Crystals

It should become clear from the discussion in previous sections that negative refraction is an unambiguous physical effect that can take place in 2-D and 3-D dielectric photonic crystal systems. The principles presented here are general and can be extended to other systems. We now discuss metallodielectric photonic crystals as a particular example of alternative possible systems. These crystals are made by inserting metallic elements into a dielectric background or an all-dielectric photonic crystal. They provide another common implementation of the photonic-crystal concept through the strong scattering off metallic surfaces and are specially useful at microwave frequencies where most metals can be treated as lossless. Such systems were studied in detail in the past from the viewpoint of forbidden band gaps, and it is the properties of propagating waves in these systems that are of interest here.

It is noteworthy that the left-handed metamaterial studied by Smith et al. was also made with periodic arrangement of metallic elements. In this regard, there are important differences between the physics of a left-handed material, which is based on the simultaneous negativity of ϵ and μ, and those of the metallic photonic crystals here, which takes place by means of complex Bragg scattering, as mentioned in the general introduction in Section 7.1. Moreover, while the left-handed metamaterials currently have 2-D functionalities but require an intrinsically 3-D analysis, the metallic photonic crystals present a much simpler concept in design: A 2-D analysis suffices for all 2-D effects, and a 3-D crystal can realize truly 3-D phenomena. Compared to its all-dielectric counterpart, a metallic photonic crystal is also interesting because electromagnetic fields are strongly expelled from the inside of the metals. For crystals containing ideal metals, which have a penetration depth of zero and are adopted here for simplicity, they can represent the voids-in-high-index all-dielectric structures in the theoretical limit of *infinite* dielectric contrast. The required background refractive indices can thus be lowered and make AANR accessible to a broader range of materials. For example, for TM-polarized waves traveling in a square lattice of metallic rods embedded within a background dielectric $\epsilon_b = 9$, the band structure shown in Fig. 7.10 already exhibits an appreciable phase-space region where the dispersion surface becomes all-convex around the M point. This situation can be reproduced using rods of radii ranging from $0.1a$ to $0.3a$. In contrast, for the corresponding all-dielectric system (the metallic rods replaced by air voids) with the same modest filling ratio, there is no complete photonic band gap, the rounding of the dispersion surface does not occur until fairly close to the gap edge, and the anomalous photon propagation effects are weak until $r > 0.3a$. Moreover, the metallic photonic crystal does not require a background material. If no high-index materials are used and the background is free space, the photon frequencies will need to be scaled by a factor of $\sqrt{\epsilon_b} = 3$. Negative refraction will exist in a large frequency range, although only for a limited range of incident angles.* For metallodielectric systems in 3-D, we have found that a bcc lattice of nonoverlapping metallic spheres in a background dielectric can produce AANR in full 3-D for a background permittivity of $\epsilon = 3$ and sphere diameter of $0.85a$, a being the side length of the conventional bcc cell. It is worth noting that this 3-D crystal has very important advantages over an all-dielectric structure in achieving AANR: The index requirement for the background matrix is quite low and can be satisfied by many materials, and straightforward fabrication procedures at microwave length scales are available at present using layer-by-layer stacking of solid matrix holding metallic spheres.

7.2.7 Summary

In this part we have studied negative refraction of light in various 2-D and 3-D photonic crystals and demonstrated that negative refraction can arise in general photonic crystals without requiring a backward-wave to be formed. We focused on the method of dispersion surfaces and explained how it can be applied to understand and predict some of the anomalous properties concerning light propagation (e.g. AANR). In practice, as can be seen from Fig. 7.4 and Figs. 7.5–7.8, the refraction processes also occur with an appreciable coupling efficiency, usually greater than or on the order of 50%, except close to a band edge. Thus, as long as we work in the AANR frequency range and use the appropriate crystal interface, the coupling efficiency will not pose a serious difficulty in negative refraction. It is noteworthy that this transmission value is almost three orders of magnitude larger than the transmission realized in the left-handed materials. We will discuss next in more detail the unusual transmission property for metamaterials and photonic crystals.

*The increased frequencies will make the crystal phase space much smaller than that of free space and destroy AANR.

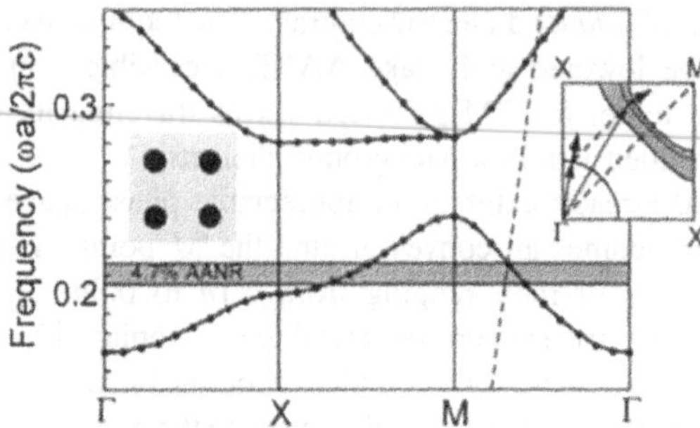

Fig. 7.10 First few TM bands of a 2-D square lattice of metallic cylinders in dielectric computed by FDTD. The photonic dispersion relations are indicated by the circles connected by solid lines. The broken line is the light line centered on the M point. The shaded region is the AANR frequency range. The left inset is a schematic illustration of the photonic crystal (filled circle stands for metals in a background dielectric). The right inset is a portrayal of the Brillouin zone and the refraction in wavevector space. Air modes and photonic-crystal modes are marked out on their respective dispersion surfaces. The long and thin arrows indicate the phase velocity $\mathbf{k}$, and the short and thick arrows indicate the group (energy) velocity $\partial\omega/\partial\mathbf{k}$. The shaded areas represent the phase space corresponding to the AANR frequency range.

7.3 SUBWAVELENGTH IMAGING WITH PHOTONIC CRYSTALS

Early in the 1960s, when Veselago first studied the electrodynamics of materials having negative values of ϵ and μ, he found that for a material having $\epsilon = -1$ and $\mu = -1$, the refraction of light is particularly interesting. He observed that the refraction angle must be exactly negative of the incident angle, and the reflection coefficient is exactly zero for all incident angles. Veselago thus proposed that all the light rays emitted by a point optical source in front of a flat slab of left-handed material must be exactly refocused, first inside the slab and then behind the slab. In 2000, Pendry considered this problem further and included evanescent waves in his analysis. He discovered a rather striking transmission property of the slab and made the intriguing prediction that negative refractive indices can lead to a *perfect lens*. For completeness, here we briefly review the Veselago–Pendry analysis.

7.3.1 Veselago–Pendry Left-Handed Lens

We use a coordinate system as shown in Fig. 7.11 where the z-axis is perpendicular to the slab interface. Instead of a point source, we first consider incident waves as plane waves incident on a semi-infinite left-handed material, with a transverse wavevector k along the x-axis and polarized perpendicular to the x–z plane, as shown in Fig. 7.11a. As discussed along with Fig. 7.1b, the refraction direction can be determined from Snell's Law, in which the negative values of ϵ and μ leads to a negative refractive index $n = -\sqrt{\epsilon\mu}$. For $\epsilon = -1$ and $\mu = -1$, $n = -1$, and the refraction angle

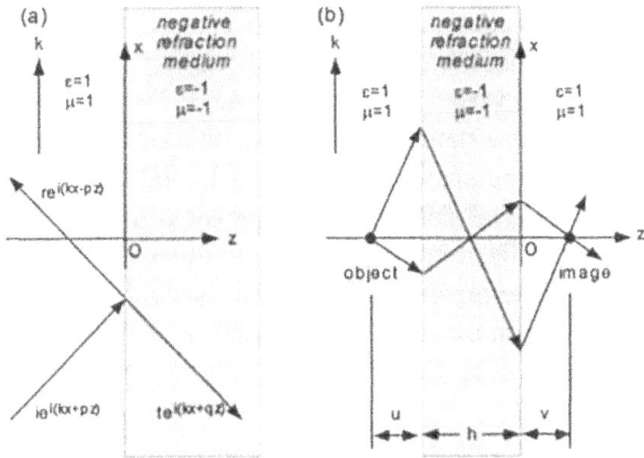

Fig. 7.11 Schematic illustration of the perfect imaging effect of a flat slab of left-handed material with $\epsilon = -1$ and $\mu = -1$. (a) Transmission of a single plane wave into a semi-infinite left-handed material. (b) A point image placed in front of the slab is focused to an image behind the slab.

becomes the exact negative of the incident angle. In terms of the (x, z) components of the wavevectors, the incident wave is specified as $(k, p = \sqrt{\omega^2/c^2 - k^2})$, and the refraction wave is (k, q). Let the amplitude of the incident, reflected, and refracted waves be i, r, and t, and matching the electric and magnetic fields across the interface gives

$$E_i + E_r = E_t \tag{7.3}$$

$$\frac{pc}{\omega}(E_i - E_r) = \frac{qc}{\mu\omega} E_t \tag{7.4}$$

With $\epsilon = -1$ and $\mu = -1$, $q = -\sqrt{\epsilon\mu\omega^2/c^2 - k^2} = -p$ for propagating waves $(k < \omega/c)$, and the solution to the above system is

$$E_r = 0 \tag{7.5}$$

$$E_t = E_i \tag{7.6}$$

that is, there is no reflection. In this way, Veselago showed that the imaging effect of a slab of left-handed materials as shown in Fig. 7.11 satisfies $u + v = h$, and is aberration-free in the sense of geometric optics. However, if $k^2 > \epsilon\mu\omega^2/c^2$, p has a positive imaginary part representing evanescent waves, and $q = -p$ would imply that such waves become growing in left-handed materials, a possibility first noted by Pendry. Such a growing field is of course prohibited in a semi-infinite left-handed material by the boundary condition at infinity, indicating that the extension of the Veselago analysis to evanescent waves is nontrivial. Thus in equations (7.3)–(7.4), we must reverse the sign of q, so that for evanescent waves $q = +i\sqrt{k^2 - \omega^2/c^2} = p$ to maintain a decaying wave as $z \to \infty$. Equation (7.4) thus reduces to

$$E_i - E_r = -E_t \tag{7.7}$$

which is compatible with equation (7.3) only if $E_i = 0$. Therefore, an electromagnetic state whose amplitudes are exponentially decaying away from the interface can exist without any incident waves for for all $k > \omega/c$, if the left-handed material has $\epsilon = -1$ and $\mu = -1$. Such a state is known as an interface photon state, and any incident evanescent light will pump energy into this state indefinitely. Pendry considered further a left-handed material slab of a finite thickness h and calculated the interaction between the two localized interface states. In this case, the left-handed material region is limited to a finite region, and the growing field can exist. A transmission calculation of an evanescent incident wave of transverse wavevector $k > \omega/c$ for the slab in Fig. 7.11b using similar field-matching on the boundary gives

$$E_y = \begin{cases} E_i e^{-\sqrt{k^2-\omega^2/c^2}(z+h)} e^{ikx} & (z \leq -h) \\ E_i e^{\sqrt{k^2-\omega^2/c^2}(z+h)} e^{ikx} & (0 \geq z \geq -h) \\ E_i e^{\sqrt{k^2-\omega^2/c^2}(h-z)} e^{ikx} & (z \geq 0) \end{cases} \tag{7.8}$$

Thus the left-handed slab can provide an amplification factor of $e^{\sqrt{k^2-\omega^2/c^2}h}$, effectively restoring the field amplitude over distance h travelled in vacuum. As long as $\epsilon = -1$ and $\mu = -1$ holds, this analysis can be applied in an identical manner to waves of all polarizations and all incident wavevectors. For a system imaging a point source as shown in Fig. 7.11b, the condition $u + v = h$ guarantees that on the image plane all the evanescent waves are restored to exactly the same amplitude as they are on the source plane. Pendry thus concluded that such a left-handed material slab forms a *perfect* lens. Because evanescent waves correspond to the fine details on the source plane that are normally lost during propagation through any classical imaging system, Pendry's analysis seems to offer a novel avenue of high-resolution imaging beating the classical diffraction limit.

While the focusing effect appears to be a straightforward consequence of the ray optics, the feasibility of an aberration-free, perfect image is a theoretically controversial issue. In particular, it was pointed out that the perfect lens effect is very fragile and can be easily destroyed by a small amount of absorption loss in a realistic system. Moreover, even in the ideal lossless limit considered by Pendry, a spatial region $(-v \leq vf \leq v)$ with highly divergent field amplitudes exists, where the amount of time required to build up the electromagnetic energy density tends to infinity. Moreover, when evanescent waves are important and the subwavelength features are of central interest, the usual effective medium model of the metamaterial places severe constraints on the lattice constant a: It must be smaller than the subwavelength details one is seeking to resolve. The question of whether and to what extent the Pendry effect occurs in the more general case of photonic crystals is the subject of the following investigation.

The first issue is whether a slab of a negative refractive photonic crystal can focus in the sense of Veselago. We use the FDTD method to perform numerical experiments on the AANR crystals studied in the previous sections. For the 2-D case, we use a (11)-orientated section of dielectric photonic crystals as shown in Fig. 7.12. Depicted here are snapshots of the electromagnetic field for a CW point source

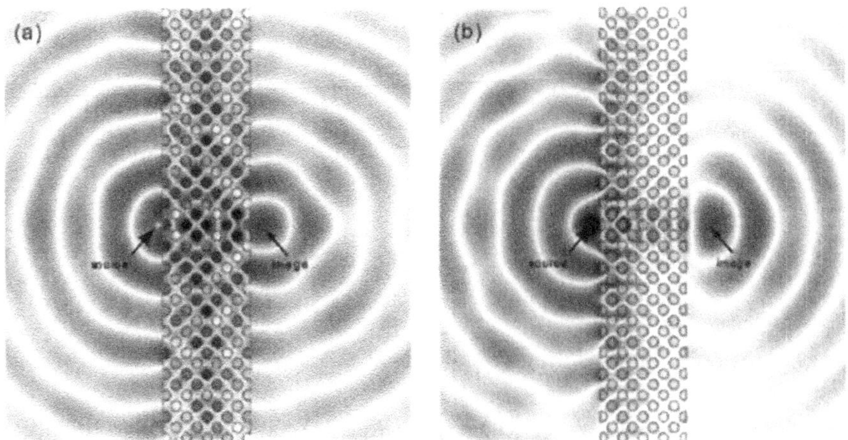

Fig. 7.12 FDTD simulation of the focusing effect of 2-D AANR photonic crystals (lattice constant a). (a) H_z field for the TE modes (air holes of diameter $0.7a$ in dielectric $\epsilon = 12$). (b) E_z field for the TM modes (dielectric rods ($\epsilon = 14$) of diameter $0.6a$ in air). Darker shades correspond to larger field magnitudes. Reprinted figure with permission from Ref. [14]. Copyright © 2002 by the American Physical Society.

placed in front of the slab. When the frequency is chosen to lie within the lowest AANR frequency range, images corresponding to the focus of the point source behind the slab are clearly discernible in both TE and TM cases. In Fig. 7.13 we also show the simulation results for a point dipole transmitted through the 3-D AANR crystal studied in the previous section. A significant portion (roughly 27% out of a possible 50%) of the total dipole radiation goes through the slab and becomes refocused into a wavelength-sized image below the slab, demonstrating that the Veselago focusing effect can occur in full 3-D. If the dipole is pointing along the wrong polarization (010), then most of the radiation will be reflected and the focusing action is rather weak.

Furthermore, the focused images can exhibit subwavelength resolutions. As an example, in the 2-D TE case the time-averaged intensity distribution around the focus shows a transverse size of only 0.67λ. We would like to understand this last possibility in detail by studying the transmission of evanescent waves through a slab of such photonic crystals to see how Pendry's analysis is modified in this case. It is important to note that the transmission of evanescent waves considered here differs fundamentally from its conventional implication of energy transport, since evanescent waves need not carry energy in their decaying directions. Evanescent transmission amplitudes is thus allowed to greatly exceed unity, just as in Pendry's perfect lens effect. We discuss two mechanisms linking amplification of evanescent waves to the existence of bound slab photon states. These bound states are decoupled from the continuum of propagating waves; thus our findings are distinct from the effect

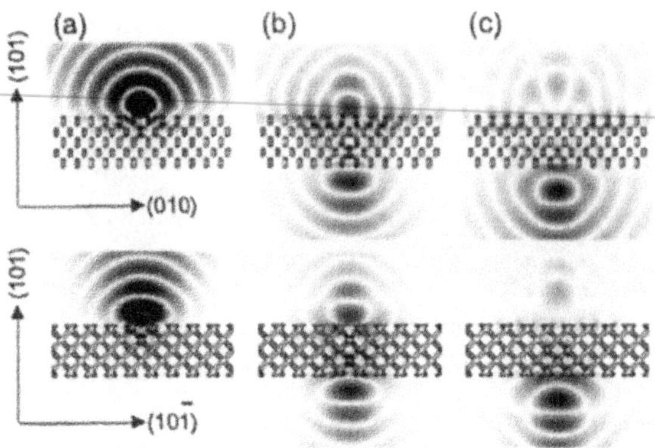

Fig. 7.13 Six 2-D snapshots of electric field along $(10\bar{1})$ during an FDTD simulation of the focusing effect for the 3-D AANR photonic crystal in Fig. 7.9. A point dipole pointing along $(10\bar{1})$ is used. (a)–(c) Slices through the two perpendicular planes containing (101) direction, taken from the instant when the dipole reaches its peak to the instant when the image reaches its peak. The dielectric structure is outlined.

of Fano resonances in electromagnetism, recently studied in the context of patterned periodic structures and surface-plasmon assisted energy transmission. We will show that, as for the problem of negative refraction, the effect of subwavelength imaging exists for photonic crystals and does not in general require a negative refractive index. Moreover, a cutoff for the transverse wavevector of evanescent waves that can be amplified exists naturally for photonic crystals, and no divergence will occur at large transverse wavevectors for photonic crystals and physical metamaterials. Furthermore, we study the detailed image pattern for a 2-D crystal and demonstrate a subtle and very important interplay between propagating and evanescent waves.

7.3.2 Origin of Near-Field Amplification

As a first step, we consider the transmission of a light wave through a lossless dielectric structure as shown in Fig. 7.14. The structure is periodic in the transverse direction, and the incident light has a definite frequency ω (with a free-space wavelength $\lambda = 2\pi c/\omega$) and has a transverse wavevector $\mathbf{k}$. A finite slab is assumed to have a thickness of h and a mirror symmetry with respect to $z = -h/2$ as shown in Fig. 7.14a. The transmission through the slab can be conceptually constructed by first considering the transmission through a single air/photonic-crystal interface (Fig. 7.14b), and then summing up all the contributions as light bounces back and forth inside the slab. The incident field $\mathbf{F}_{in}$, the reflected field $\mathbf{F}_{refl}$, and the transmission field $\mathbf{F}_{trans}$ are formally related to each other by the transmission matrix t and the reflection matrix r through $\mathbf{F}_{trans} = t\mathbf{F}_{in}$ and $\mathbf{F}_{refl} = r\mathbf{F}_{in}$, where all fields are column vectors expressed in the basis of eigenmodes of the corresponding medium with transverse wavevector/Bloch-wavevector $\mathbf{k}$. The overall transmission through

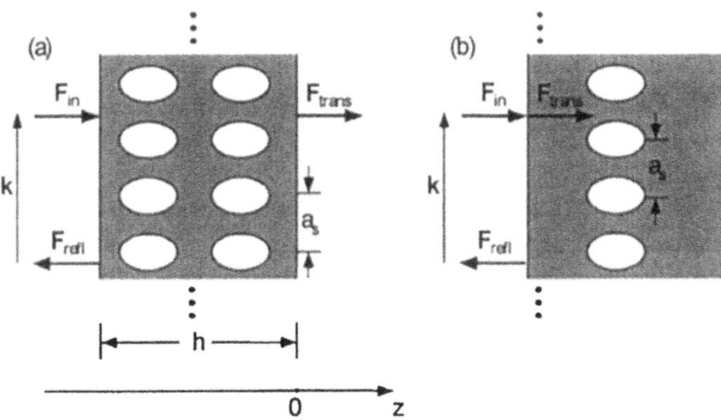

Fig. 7.14 Illustration of lightwave transmission through a photonic crystal. (a) Transmission through a slab of photonic crystal. $\mathbf{F}_{in}$ and $\mathbf{F}_{refl}$, are measured at the left interface of the slab ($z = -h$) and $\mathbf{F}_{trans}$ is measured at the right interface ($z = 0$). (b) Transmission through a single air/photonic-crystal interface. $\mathbf{F}_{in}$, $\mathbf{F}_{refl}$, and $\mathbf{F}_{trans}$ are all measured at the interface. $\mathbf{k}$ is the transverse wavevector and a_s is the surface periodicity. Reprinted figure with permission from Ref. [17]. Copyright © 2003 by the American Physical Society.

the slab can be written as

$$t = t_{p-a}\left(1 - T_{\mathbf{k},p} r_{p-a} T_{\mathbf{k},p} r_{p-a}\right)^{-1} T_{\mathbf{k},p} t_{a-p} \tag{7.9}$$

In equation (7.9), t_{a_p} and t_{p-a} are the transmission matrices through the individual interfaces from air to the photonic crystal and from the photonic crystal to air, respectively, r_{p-a} is the reflection matrix on the crystal/air interface, and $T_{\mathbf{k},p}$ is the translation matrix that takes the fields from $z = -h$ to $z = 0$ inside the crystal. When h is an integral multiple of the crystal z-period, $T_{\mathbf{k},p}$ is diagonal with elements $e^{ik_z h}$, where k_z is the z-component of the Bloch-wavevector $\mathbf{k} + k_z\hat{\mathbf{z}}$ of the crystal eigenmode with $\Im k_z \geq 0$. We now discuss the possibility of amplification in t—for example, in the zeroth-order diagonal element t_{00} describing waves with no change in the transverse wavevector.

In general, equation (7.9) describes a transmitted wave that is exponentially small for large enough $|\mathbf{k}|$. This may be seen from the special case of a slab of a uniform material with permittivity ϵ and permeability μ, where all the matrices in equation (7.9) are built from the basis of a single plane wave and thus reduce to a number. In particular, $T_{\mathbf{k},p} = \exp(ik_z h) = \exp(ih\sqrt{\epsilon\mu\omega^2/c^2 - |\mathbf{k}|^2}) = \exp(-h\sqrt{|\mathbf{k}|^2 - \epsilon\mu\omega^2/c^2})$, which becomes exponentially small as $|\mathbf{k}|$ goes above $\sqrt{\epsilon\mu}\omega/c$. Equation (7.9) thus becomes a familiar elementary expression:

$$t_{00} = \frac{t_{p-a} t_{a-p} e^{ik_z h}}{1 - r_{p-a}^2 e^{2ik_z h}} \tag{7.10}$$

Equation (7.10) has an exponentially decaying numerator, while for fixed r_{p-a} the denominator approaches 1. Thus, waves with large enough $|\mathbf{k}|$ usually decay during transmission, in accordance with their evanescent nature.

There exist, however, two mechanisms by which the evanescent waves can be greatly amplified through transmission, a rather unconventional phenomenon. The first mechanism, employed by Pendry as in Section 7.3.1, is based on the fact that the reflection and transmission coefficients across individual interfaces can become *divergent*, which we refer to as *single-interface resonance*. For example, under the conditions of single-interface resonance, $t_{a-p}, t_{p-a}, r_{p-a} \rightarrow \infty$, and in the denominator of equation (7.10), the term $r_{p-a}^2 \exp(2ik_z h)$ dominates over 1. In this limit,* equation (7.10) becomes

$$t_{00} = \frac{t_{p-a} t_{a-p}}{-r_{p-a}^2} e^{-ik_z h} \tag{7.11}$$

The divergences in the numerator and denominator of equation (7.11) cancel each other, and the net result is that for large $|\mathbf{k}|$,

$$t_{00} = \exp(-ik_z h) = \exp(-h\sqrt{|\mathbf{k}|^2 - \epsilon\mu\omega^2/c^2})$$

leading to amplification of exactly the right degree to focus an image. The same arguments can be applied to the general case of equation (7.9), as long as k_z is regarded as the z-component of the eigenwavevector with the *smallest* imaginary part, which produces the dominant term in $e^{ik_z h}$. As elements of r_{p-a} grow sufficiently large, the matrix product $T_{\mathbf{k},p} r_{p-a} T_{\mathbf{k},p} r_{p-a}$ dominates over the identity matrix. Since in this case, the matrix under inversion in equation (7.9) scales as $\exp(-2\Im k_z h)$ and the rest scales as $\exp(-\Im k_z h)$, the amplification behavior is still present in a general element of t. The transmission t_{00} can be represented by equation (7.11) with the coefficients to the exponential replaced by smooth functions of ω and $\mathbf{k}$.

The second mechanism for enhancement of evanescent waves relies on a direct divergence in the overall transmission—that is, an *overall resonance*. This is clear from equation (7.9) for the uniform medium, whose denominator becomes zero when $1 - r_{p-a}^2 \exp(2ik_z h) = 0$, which is the condition for transverse guiding via total internal reflection. An evanescent incident wave can satisfy this condition exactly. It thus holds that a direct divergence can exist in the overall transmission of evanescent waves without any accompanying single-interface resonance, and therefore finite and strong amplification of evanescent waves results when the incident wave does not exactly satisfy but is sufficiently close to the resonance condition. In this case, there is no upper limit on the amplification and the transmission can even exceed that prescribed by equation (7.11); hence there is the potential to form an image provided that the correct degree of amplification is induced. These arguments are also valid in

*Although equations (7.9) and (7.10) are derived with the summation of an infinite series that breaks down in this limit, it can be shown by analytical continuation arguments that these equations remain valid in the sense that a vanishing small amount of loss exists in the system.

equation (7.9), at or near the singular points of $1 - T_{\mathbf{k},p} r_{p-a} T_{\mathbf{k},p} r_{p-a}$ whose inverse occurs in the transmission. If we write the relation between the light frequency ω and wavevector $\mathbf{k}$ at these singular points as $\omega = \omega_0(\mathbf{k})$, then close to such a resonance the transmission t_{00} is described by

$$t_{00} = \frac{C_0(\omega, \mathbf{k})}{\omega - \omega_0(\mathbf{k})} \tag{7.12}$$

where $C_0(\omega, \mathbf{k})$ is a smooth function of ω and $\mathbf{k}$. For a given ω, the issue is then to design photonic crystals with the appropriate dispersion relation $\omega_0(\mathbf{k})$ so that (7.12) approximates the required amount of amplification.

Both mechanisms of evanescent wave amplification here involve some divergences in the transmission process. Similar to that discussed in Section 7.3.1, such a divergence physically means that energy is being pumped indefinitely by the incident wave into the transmitted and reflected fields, whose amplitudes increase in time without limit. Equivalently, a finite field inside the structure can be produced by zero incident field; that is, it is a *bound* (guided) electromagnetic mode. A bound photon mode on the air/photonic-crystal interface leads to a single-interface resonance, and a bound photon state inside the slab leads to an overall resonance. In the overall resonance case, the dispersion relation of the bound photon mode is just $\omega = \omega_0(\mathbf{k})$ in equation (7.12). A similar equation can also be used to represent the zeroth-order term in t_{p-a} close to a single-interface resonance:

$$t_{p-a,00} = \frac{C_{p-a}(\omega, \mathbf{k})}{\omega - \omega_{p-a}(\mathbf{k})} \tag{7.13}$$

with the single-interface bound photon dispersion relation $\omega_{p-a}(\mathbf{k})$ and a smooth function $C_{p-a}(\omega, \mathbf{k})$. C_{p-a} and C_0 here represent the coupling strength between the incidence wave and the respective bound photon state. It is instructive to compare these two amplification mechanisms by their applicable ranges. In an ideal material slab with $\epsilon(\omega_{sp}) = -1$ and $\mu(\omega_{sp}) = -1$, every evanescent wave is amplified by a single-interface resonance. For a slab, r_{p-a} diverges for any incident $\mathbf{k}$ and no overall resonance happens. However, both ϵ and μ are necessarily dispersive, and the condition $\epsilon = -1$ and $\mu = -1$ can occur only at a single surface-plasmon frequency ω_{sp}. Detuning from ω_{sp} we can satisfy the guiding condition

$$1 - r_{p-1}^2(\omega_\pm) e^{2ik_z h} = 0 \tag{7.14}$$

at two separate frequencies $\omega_\pm$, above and below ω_{sp}, reflecting the fact that the surface photon states on the two interfaces of the slab interact with each other forming symmetric and antisymmetric combinations. In the general case, equation (7.14) can be satisfied and bound photon states inside the slab form even without the prior existence of interface states—that is, without r_{p-a} diverging. Thus both mechanisms may be available to amplify evanescent waves. To have a single-interface resonance in equation (7.9), it is required that the term associated with single-interface reflections dominate over 1. This can be expressed as

$$|\omega - \omega_{p-a}(\mathbf{k})| \ll |C_{p-a}| e^{-\Im k_z h} \tag{7.15}$$

For an overall resonance, the condition becomes

$$|\omega - \omega_0(\mathbf{k})| \approx |C_0| e^{-\Im k_z h} \tag{7.16}$$

in order to produce an amplification magnitude similar to that in equation (7.11). We note that C_{p-a} and C_0 are roughly on the same order of magnitude if the bound photon modes inside the slab are constructed from combinations of the surface photon states. It is thus clear that in the general case, amplification of evanescent waves requires operating much closer to an exact resonance in the single-interface-resonance mechanism than in the overall-resonance mechanism. In addition, the overall resonance can in principle happen near a bulk-guided mode that is not evanescent inside the photonic-crystal slab. Thus, in general, amplification is more easily achieved using an overall resonance than using a single-interface resonance. In the following, therefore, we make primary use of the second resonance mechanism and realize amplification of evanescent waves in the manner discussed here.

Thus amplification can arise from the coupling between the incident evanescent field and bound photon states with an infinite lifetime, which usually exist below the light line.* With Bragg scattering, in a periodic structure the range of wavevector region below the light line is limited by the first surface Brillouin zone. What happens to transmission of evanescent waves whose wavevectors lie beyond the first surface Brillouin zone and become folded back into the light cone? In this case, the associated slab photon resonance mode changes from a *bound* state to a *leaky* state, and its frequency $\omega_0(\mathbf{k})$ becomes *complex*—that is, $\omega_0(\mathbf{k}) \to \omega_0(\mathbf{k}) - i\gamma(\mathbf{k})$, with ω and γ real. This situation is described by the element t_{nn} of t with $n \neq 0$ and n indexing the surface diffraction orders. When the incidence $\mathbf{k}$ is sufficiently close to that of a leaky photon mode the transmission becomes

$$t_{nn} = \frac{C_0(\omega, \mathbf{k})}{\omega - \omega_0(\mathbf{k}) + i\gamma(\mathbf{k})} \tag{7.17}$$

which always has a finite magnitude. In principle, $|t_{nn}|$ can also reach values larger than unity provided that γ is small enough. However, as n goes away from 0 the spatial variation in the incident wave becomes more rapid. The leaky photon state, on the other hand, always maintains a constant field profile with variations on a fixed spatial scale, roughly that of each component in a cell of the crystal. Hence, for n sufficiently far from 0, C_0 in (7.17), determined by the overlap between the incident wave and the slab photon modes, must always approach zero, and so must $|t_{nn}|$. The numerical results presented later in this work indicate that, for the structure considered here, the transmission for evanescent waves coupling to leaky modes with $n \neq 0$ is *always* small and the possible amplification effect can be ignored.

We have thus shown that an amplified transmission of evanescent waves at a given ω and $\mathbf{k}$ is not restricted to materials with $\epsilon < 0$ or $\mu < 0$ only, and can be achieved

*In certain occasions, due to reasons of modal symmetry, they can also appear in discrete locations above the light line.

by coupling to bound photon states in general. Another feature of this approach is that, with the single-interface resonance-amplification mechanism, the reflection coefficient r can vanish, but in the overall-resonance mechanism here an amplified transmission process also implies an amplified reflected evanescent field in general. Since the latter mechanism is used in our numerical calculations below, most of the effects that arise due to the transmitted evanescent waves should also be expected in the reflected waves as well. These might lead to nontrivial consequences, for example a feedback on the emitting source. In this work, we assume the source field is generated by some independent processes and ignore the potential influences of these effects.

7.3.3 Photonic-Crystal Superlenses

We now consider the problem of Pendry superlensing at a given frequency ω in photonic crystals. An ideal point source emits a coherent superposition of fields $\mathbf{F}_{source}(\mathbf{k})$ of different transverse wavevectors $\mathbf{k}$, with $|\mathbf{k}| < \omega/c$ being propagating waves and $|\mathbf{k}| > \omega/c$ being evanescent waves. When such a point source is placed on the z-axis, the optical axis, at $z = -h - u$, the image intensity distribution in $z \geq 0$ becomes

$$I_{image}(\mathbf{r}) = \left| \int d\mathbf{k} B^T \cdot T_{\mathbf{k}_n,a}(z) t(\mathbf{k}) T_{\mathbf{k}_n,a}(u) \mathbf{F}_{source}(\mathbf{k}) \right|^2 \tag{7.18}$$

In equation (7.18), $T_{\mathbf{k}_n,a}(z)$ is the translation matrix in air that takes the fields through a distance z, $\tau_{\mathbf{k}}$ is a polarization vector, $\mathbf{r}_t$ stands for transverse coordinates, $\mathbf{k}_n = \mathbf{k} + \mathbf{G}_n$, with $\mathbf{G}_n$ indexing the surface reciprocal vectors, and $\mathbf{k}_n$ is the transverse wavevectors in the air basis. $B^T = (\ldots, \tau_{\mathbf{k}} \exp(i\mathbf{k}_n \cdot \mathbf{r}_t), \ldots)$ is a row vector representing polarizations and phases of the air plane-wave basis, and its dot product with the column vector of the transmitted field amplitude produces the complex field amplitude. The integral is carried out over the first surface Brillouin zone. In the case of a uniform material, it is over the entire transverse $\mathbf{k}$-plane.

Conventional lenses only image the portion of the propagating incident field with $|\mathbf{k}| < k_M$ for $k_M < \omega/c$, limited by the numerical aperture. Pendry's *perfect* lens, on the other hand, not only focuses all propagating waves with negative refraction but also amplifies all evanescent waves, so that all Fourier components of the source field reappear perfectly in the image plane. Here, we use the term *superlens* to denote a negative-refractive slab that not only focuses all propagating waves by negative refraction without the limitation of a finite aperture, but also amplifies at least *some* evanescent waves in a continuous range beyond that of the propagating waves. In this context, *superlensing* refers to the unconventional imaging effects due to the presence of the additional near-field light. The phenomena considered here thus contain the main features of Pendry's perfect lens. In general, however, the magnitude of transmission will not reproduce exactly that required for perfect image recovery, and the resulting image will be *imperfect* and possess quantitative aberrations. We focus our attention on 2-D situations (the x–z plane) where most of the quantities can be treated as scalars. $F_{source}(k) = 1$ is used for all k's, appropriate for a point source in 2-D.

Our starting point here is the focusing by negative refraction of all propagating waves with $k < \omega/c$ (AANR). All propagating waves with $k < \omega/c$ can thus go through the photonic-crystal slab with transmission of order unity and thereby focus into a real image (i.e. an intensity maximum) behind the slab. Superlensing requires amplified transmission for an additional range of transverse wavevectors $\omega/c < k < k_M$. In Pendry's perfect lens, $k_M = \infty$. The important difference for superlensing with a photonic crystal is that k_M is in general finite. This is clear from the discussion in Section 7.3.2, where a finite high cutoff to the transmission spectrum results from the Bragg scattering of light to leaky photon modes. This finite k_M makes the image reconstruction process through a photonic-crystal superlens no longer divergent even in the lossless case. Physically, the amplification of evanescent waves requires near-resonance coupling and the resulting growth of an approximate bound photon state during transmission. Amplification of larger k components thus requires exponentially higher energy density in the bound photon mode and an exponentially longer time interval to reach a steady state. Our numerical results are mainly calculated in the frequency domain and therefore represent the steady-state behavior after the transients have died away in *finite* time.

To actually realize amplification of evanescent waves, one must design the photonic-crystal structure carefully so that equation (7.16) holds for *all* evanescent waves with $\omega/c < k < k_M$. For large k, $k_z = +\sqrt{\omega^2/c^2 - k^2}$ has a large positive imaginary part, and therefore the operation frequency ω should be very close to the resonance frequency $\omega_0(k)$ of a bound photon state for the amplification to cancel the decay outside the crystal. This means that for large k, $\omega_0(k)$ must approach a "flat" line near ω within the AANR range, as shown in Fig. 7.15. In general, there are two classes of bound photon modes within a photonic-crystal slab. One consists of those guided by the slab as a whole, similar to the guided modes in a uniform dielectric slab. The other class includes those guided by the air/slab interfaces; that is, they are linear combinations of interface states, which decay exponentially both in air and in the crystal away from the slab interfaces. Although both classes of bound modes can be employed to achieve amplification for a given k, most of the wavevector region $\omega/c < k < \pi/a_s$ within the AANR frequency range involves a partial photonic band gap and can only accommodate the interface photon states. Furthermore, the interface states are known to depend on the fine details of the interface structure (e.g., the interface termination position) and can be tuned to be any frequency across the band gap at least for a single k-point. Thus, the slab interface photon states are attractive candidates for achieving flat bound photon bands within the AANR range for superlensing. We will give one 2-D design that meets this goal by simply adjusting the crystal-interface termination position.

What is the ultimate limit to the imaging resolution of a photonic-crystal superlens? The image of a point source can be interpreted as an intensity peak on the constant-z plane of AANR focusing, and the resolution of such a peak can be measured by the distance between the nearest minima around this intensity peak. Taking into account

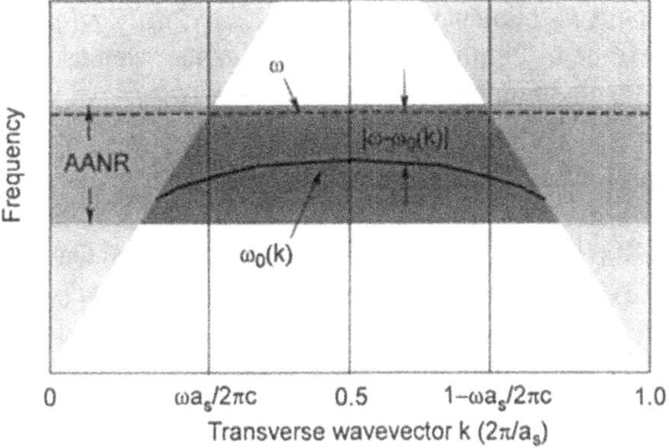

Fig. 7.15 Schematic illustration of superlensing in photonic crystals. The light-shaded region is the light cone, and the dark shaded region is the AANR frequency range. The curve marked $\omega_0(k)$ outside the light cone is a bound photon band inside a photonic-crystal slab. The operating frequency ω is marked by the broken line. Amplification of evanescent waves requires that $|\omega - \omega_0(k)|$ be small for $k > \omega/c$, which in turn requires that the $\omega_0(k)$ curve be *flat*. In this repeated zone scheme, a bound photon state may only exist in the range $\omega/c < k < 2\pi/a_s - \omega/c$, which imposes an upper cutoff for superlensing using photonic crystals. Reprinted figure with permission from Ref. [17]. Copyright © 2003 by the American Physical Society.

the decay of evanescent light as it travels in air, the image field can be estimated using a simplified model. In this model, we assume unit total transmission from the source plane to the image plane for $|k| < k_M$, and zero transmission for $|k| > k_M$. The intensity profile on the image plane then reads

$$I_{image}(x) = \left| \int_{-k_M}^{k_M} e^{ikx} dk \right|^2 = \frac{4\sin^2(k_M x)}{x^2} \tag{7.19}$$

which has a peak at $x = 0$ with a transverse size of $\Delta = 2\pi/k_M$, measured by the distance between the first zeroes around the peak. This image size is zero in a material with $\epsilon = -1$ and $\mu = -1$, since $k_M = \infty$, leading to the interpretation of a *perfect* image. In a photonic crystal, quantitative estimates of the minimum possible Δ (i.e., maximum possible k_M) may be obtained by looking at Fig. 7.15. In the best situation, all the interface modes with $k < 0.5(2\pi/a_s)$ can be used directly for amplification if they satisfy (7.16),* which gives $k_M \geq 0.5(2\pi/a_s)$. Since this estimate ignores the strong Bragg-scattered wave components in the interface states near the surface Brillouin zone edge, it is a conservative estimate. From Fig. 7.15, we also deduce the maximum wavevector of a "flat" interface band below the light line

*The nonzero slope of (7.16) at the surface Brillouin zone edge is usually negligible.

that can be coupled to at frequency ω to be $(1 - \omega a_s/2\pi c)(2\pi/a_s) = 2\pi/a_s - \omega/c$, which is an overestimate. Putting these results together, we thus obtain the ultimate resolution limit of a photonic-crystal superlens to be

$$\frac{a_s \lambda}{\lambda - a_s} < \Delta < 2a_s \tag{7.20}$$

According to the Rayleigh criteria, the minimum feature size that can be resolved by such a device is $\Delta/2$. Thus, the resolution of a photonic-crystal superlens at a single frequency is only limited by its surface periodicity a_s instead of the wavelength of operation λ. Our considerations also give a guide line for designing high-resolution superlenses: For a given wavelength, the smaller the a_s (i.e., the lower in the band structure one operates with AANR), the better the resolutions will be. In principle, by using sufficiently large dielectric constants and contrast in its constituents, a photonic-crystal superlens can be designed to operate at a wavelength arbitrarily larger than a_s. A similar superlensing trend is also suggested with localized plasmon polariton resonances in metallic photonic crystals. If, furthermore, a sufficiently flat interface band is achieved in the AANR frequency range of such a photonic-crystal slab by manipulating its interface structures, imaging arbitrarily exceeding the diffraction limit is possible. Therefore, there is no theoretical limit to superlensing in general photonic crystals. In practice, of course, available materials, material losses, and unavoidable imperfections in interface structures will limit the performance of such superlenses.

It must be noted that the image of a superlens considered here is substantially different from conventional real images of geometric optics. Conventional real images always correspond to an intensity maximum—that is, a peak in the intensity distribution both in x- and z-directions. When only the propagating waves are transmitted through the superlens, they similarly produce an intensity maximum at the image of AANR in $z > 0$. The position of this image may be estimated by paraxial geometric optics around the z-axis. However, when evanescent waves are included, they bring distortions to the image and the new intensity maximum is no longer at the position of the AANR focusing. A simple illustration of this phenomenon is provided by our simplified cutoff model. The full expression of the image in this model with a high cutoff $k_M > \omega/c$ can be written as

$$\begin{aligned} I_{image}(x,z) &= \left| \int_{-\frac{\omega}{c}}^{\frac{\omega}{c}} e^{ikx + i\sqrt{\frac{\omega^2}{c^2} - k^2}(z-v)} dk \right. \\ &\quad \left. + \left(\int_{-k_M}^{-\frac{\omega}{c}} + \int_{\frac{\omega}{c}}^{k_M} \right) e^{ikx - \sqrt{k^2 - \frac{\omega^2}{c^2}}(z-v)} dk \right|^2 \end{aligned} \tag{7.21}$$

where $z = v$ is assumed to be the focusing plane of AANR. Inside the absolute value sign of equation (7.21), the first term has a constructive interference at $z = v$ and represents an intensity maximum there, but the second term always always displays asymmetric amplitude distributions in z around $z = v$. Thus, for $k_M > \omega/c$ the overall intensity distribution no longer has a maximum at the AANR focusing plane.

The detailed image pattern has a sensitive dependence on the interplay between the propagating and evanescent waves.

For k_M slightly above ω/c, the strength of evanescent waves is comparable to that of propagating waves. An intensity maximum still exists in the region $z > 0$, but is shifted away from $z = v$ toward the slab. This intensity maximum thus appears as a real image similar to conventional optics. In general it will have a transverse size smaller than $2\pi/k_M$, the transverse size of the peak in the plane $z = v$. This situation may be called the *moderate subwavelength limit*, and the image resolution will always be a fraction of the wavelength.

When k_M exceeds a certain threshold, evanescent waves begin to dominate the image pattern. In this case, an intensity maximum completely disappears in the region $z > 0$ where the intensity along z-axis becomes monotonically decreasing with z. The k_M threshold for this behavior can be estimated to be about $k_{M,th} = 1.35\omega/c$, with a transverse size of the intensity maximum about half a wavelength at this threshold, using the simplified model equation (7.21). This crude qualitative estimate can be compared to the following numerical results. The case of superlensing with $k_M > k_{M,th}$ can thus be called the *extreme subwavelength limit*. From another viewpoint, at $\omega/c << k_M$, $\lambda >> a_s$, and if we also assume that the slab thickness h is small compared to λ, the system may be regarded to be in the near-static limit. The absence of an intensity maximum in $z > 0$ may then be understood simply by the elementary fact that in electrostatics/magnetostatics potentials can never reach local extrema in a sourceless spatial region. Thus, in the extreme subwavelength limit the imaging effect of a superlens is in the transverse direction only. Compared to a conventional lens, whose image intensity generally follows a power-decaying law, the superlens has a characteristic region between the slab and the image where an exponentially growing intensity distribution exists.

A related effect is that the intensity generally becomes higher for $k_M > \omega/c$ than for $k_M \leq \omega/c$. This occurs in the simplified model equation (7.21) due to the addition of evanescent wave components. Note also that in this model no exact resonant divergence is present in the transmission, and that the intensity enhancement effect is prominent in the region between the slab and the plane $z = v$. A more general situation occurs when the operating frequency ω falls inside the narrow frequency range of the bound photon bands, so that a distinct number of bound mode with near-zero group velocities can be excited on resonance. The contribution from one such exact resonance pole at $\omega = \omega_0(k_0)$ to the transmission in the limit of a vanishingly small amount of loss may be estimated as

$$\int dk \frac{C_0(\omega, k)}{\omega_0(k_0) - \omega_0(k) + i0^+} \approx \frac{1}{(\partial\omega/\partial k)\,|_{k=k_0}} \int dk \frac{C_0(\omega, k)}{k_0 - k + i0^+} \tag{7.22}$$

The integral over k, though not suitable for analytical evaluation in general, can usually be regarded as having a finite value and depending on the detailed behavior in $C_0(\omega, k)$. The influence of each pole on the transmitted image is thus inversely proportional to the group velocity of the bound photon state at the resonance, and is strongest for modes with the smallest group velocities. When coupled to flat interface bands with very small group velocities, the image field has a wide spatial distribution in $z > 0$ dominated by the pattern of excited bound photon modes, sometimes extending beyond the plane $z = v$. Thus, we can call this regime of superlensing *enhanced surface resonance*. On the one hand, such an effect could be useful in applications where a large field amplitude is desired. On the other hand, since a surface resonance is a *delocalized* distribution, there is very little information contained within the image about the transverse location of the source. This is a subtle point to be avoided in imaging applications.

In many experimental situations, light intensity is the quantity that is responsible for most physical effects and can be measured directly. Both a subwavelength transverse resolution and a spatial region with high intensity can thus serve as direct experimental evidence of superlensing. From the viewpoint of applications, imaging in the transverse direction below the diffraction limit is sufficient and desirable for many situations, such as sensing/detecting or strong focusing for active phenomena. Our considerations indicate the possibility of a variety of image patterns impossible in conventional geometric optics in the image of a superlens, based on the interplay between near-field and far-field light. With a photonic crystal, a flexible superlens may be constructed in which all of these physical effects are readily observable.

7.3.4 Numerical Results

In this section, we study superlensing in photonic crystals numerically. We focus on a square lattice of "+"-shaped air voids oriented along the (11) direction in a lossless dielectric $\epsilon = 12$, with the various sizes specified in Fig. 7.16a. The lattice constant of the square lattice is a, and the surface periodicity is $a_s = \sqrt{2}a$. This structure possesses a similar band structure to the one studied in Section 7.2 and also allows for efficient numerical calculations of transmission through a finite slab in the frequency domain. The TE polarization is assumed, and similar results can be expected for TM modes as well.

7.3.4.1 Surface Band Structure The bound photon bands on the photonic-crystal slab are calculated by planewave expansion using the supercell approach. The results are presented in Fig. 7.16a. Below the light cone and inside the region of the projected infinite-crystal band structure, the modes are bound photon states guided by the slab as a whole. The modes inside the partial photonic band gap are the interface states guided around the air/slab interfaces. The field profiles of the symmetric and antisymmetric combinations of the surface modes on the two interfaces are also shown in Fig. 7.16b. Deep in the gap where the confinement is strong, the splitting between these two bands becomes small and the two bands merge into one curve on Fig. 7.16a. The crystal thickness h and the associated surface termination position are chosen so that the interface modes are two flat, nearly degenerate bound photon bands near the frequency $\omega = 0.192(2\pi c/a)$ within the AANR frequency range. This situation thus approximately realizes is in Fig. 7.15 and is well-suited for achieving superlensing.

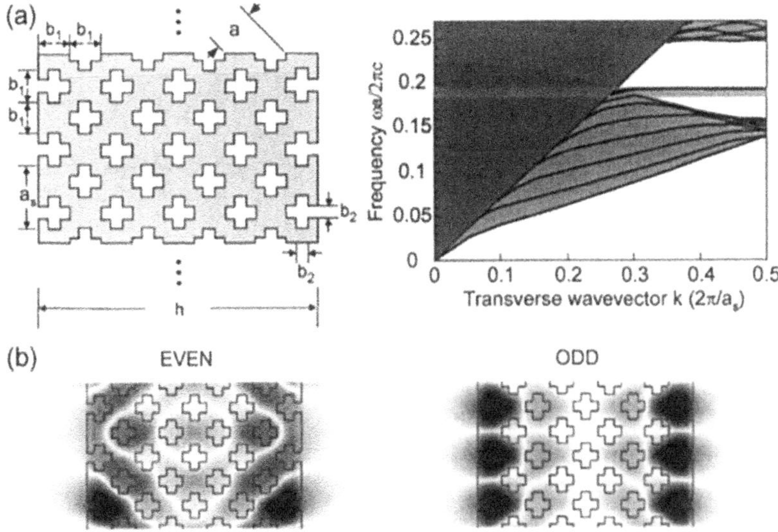

Fig. 7.16 Bound TE photon modes in a 2-D photonic crystal slab. (a) Left panel: The photonic-crystal slab used in calculation. The parameters are $a_s = \sqrt{2}a$, $b_1 = 0.5a_s$, $b_2 = 0.2a_s$ and $h = 4.516a_s$. Right panel: The calculated band structure of bound photon modes, plotted on top of interface-projected band structure (the lightly filled region bounded by curves). The dark-filled region indicates the light cone. The shaded rectangular region is the AANR frequency range of this photonic crystal. (b) Distribution of the magnetic field perpendicular to the plane for the surface photonic modes at $k = 0.45(2\pi/a_s)$. Left and right panels represent odd and even mirror symmetries. Darker shades indicate larger magnitudes of the magnetic field. Reprinted figure with permission from Ref. [17]. Copyright © 2003 by the American Physical Society.

7.3.4.2 Transmission Spectrum The transmission calculations can be performed in the frequency domain with the scattering-matrix method. To compare the results with those obtained from eigenmode computation by planewave expansion, we fix the incident wavevector and calculate the frequency spectrum of the transmission. The calculated transmission data are presented on a logarithmic scale in Fig. 7.17a. The pronounced peaks in the spectrum represent resonant excitation of the bound photon states by the incident evanescent radiation and approach infinity in the limit of continuous numerical sampling points in frequency. From the comparison between the transmission peaks and the interface band structure in Fig. 7.17b, we obtain excellent agreement between the two numerical methods. Near each resonance, the transmission of evanescent waves reaches large amplification values well exceeding unity, providing the basis of superlensing.

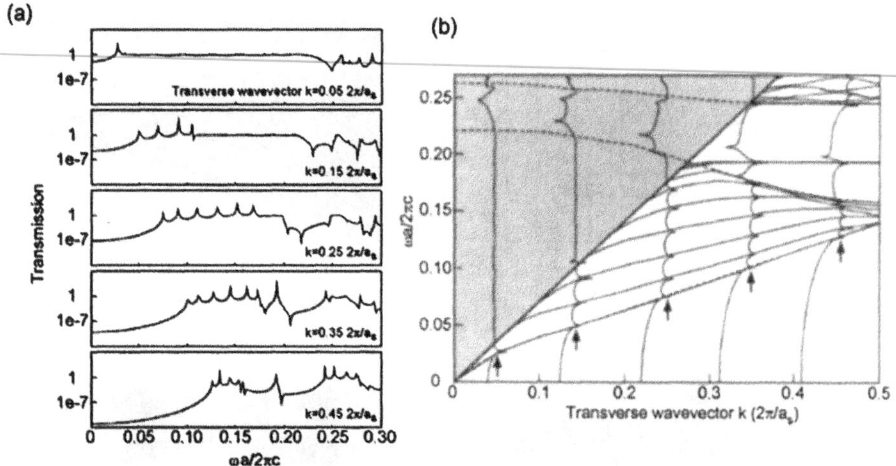

Fig. 7.17 Frequency spectrum of transmission and links to bound slab photon modes. (a) Zeroth-order transmission ($|t_{00}|^2$) versus frequency through the photonic-crystal slab in Fig. 7.16 for various transverse wavevectors, plotted on a logarithmic scale. (b) The transmission curves in (a) plotted on the bound photon band structure of the slab. The arrows indicate the transverse wavevector for each transmission curve. The shaded region is the light cone. Reprinted figure with permission from Ref. [17]. Copyright © 2003 by the American Physical Society.

7.3.5 Image Patterns of a Superlens

We calculate the transmission as a function of incident wavevector k at a fixed frequency ω close to that of the interface modes, using the method of Section 7.3.4.2. The complex transmission data for these planewaves are then summed up at each z according to equation (7.18) to obtain the image generated by a point-dipole source placed at $(x, z) = (0, -h - u)$. In all calculations, $u = 0.1a_s$ is used, and the k sampling points range from $k = -5(2\pi/a_s)$ to $k = 5(2\pi/a_s)$ in steps of $0.001(2\pi/a_s)$, to model the continuous range of $-\infty < k < \infty$. This finite resolution corresponds to a finite transverse overall dimension of the structure of 1000 periods and is sufficient for illustrating the prominent features in the image pattern.* Our results are summarized in Figs. 7.18, 7.19, and 7.20. The frequency is shifted by only $0.001(2\pi c/a)$ from one figure to the next. In all cases the transmission for propagating waves is nearly the same. However, large differences in the field patterns for $z > 0$ can be observed, suggesting that a fine control over the transmission of evanescent waves is possible.

*The transmission near exact resonances in an infinite structure in principle requires a much higher computational resolution.

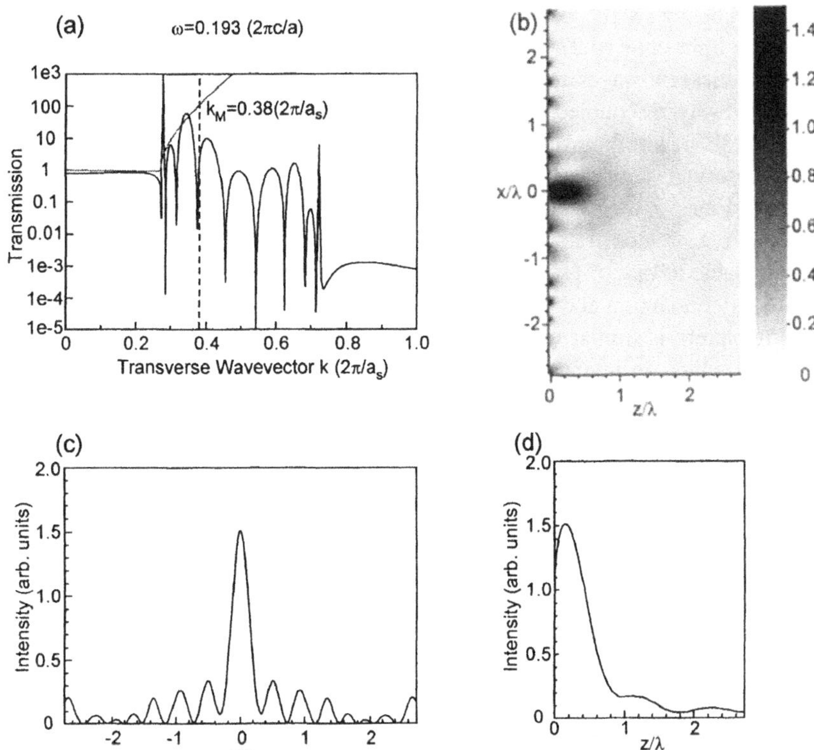

Fig. 7.18 Transmission and intensity distribution in the image space for $\omega = 0.193\ (2\pi c/a)$, for a photonic crystal slab in Fig. 7.16 illuminated by a point source. (a) $|t_{00}|^2$ for $k < 0.5(2\pi/a_s)$) plotted on a logarithmic scale versus the incident transverse wavevector. The gray curve indicate the transmission required for perfect image reconstruction at the AANR focusing plane. The effective high cutoff of the transverse wavevector k_M is marked out in broken lines. (b) The intensity distribution in real space. The right interface of the slab is at $z = 0$. (c) The cross section of (b) in the plane of $z = 0.6a_s = 0.16\lambda$. (d) The cross section of (b) with $x = 0$. Reprinted figure with permission from Ref. [17]. Copyright © 2003 by the American Physical Society.

For $\omega = 0.193(2\pi c/a)$ (Fig. 7.18), the operating frequency is outside the frequency range of the interface bands. The calculated transmission shown in Fig. 7.18a exhibit smooth behavior throughout the whole range of wavevectors. Note that the magnitude of transmission oscillates around order unity for the evanescent waves for $0.27(2\pi/a_s) < k < 0.73(2\pi/a_s)$, but for $k > 0.73(2\pi/a_s)$ it drops precipitously to a low level below 1×10^{-3}. This confirms our expectation that the bound photon band below the light cone ($0.27 < ka_s/2\pi < 0.73$) should lead to amplified transmission of evanescent waves and that the amplification effect should disappear once the evanescent wave is coupled back into the continuum. In the calculated image shown in Fig. 7.18b, a clear intensity maximum at $(x, z) = (0, 0.6a_s = 0.16\lambda)$ in free space can be observed. Cross sections of this maximum in both the x- and the z-axes are shown in Figs. 7.18c and 7.18d, respectively. The transverse x-size of this peak is $0.66\lambda < \lambda$, demonstrating that the contribution of evanescent waves to imaging is comparable to that of propagating waves. This situation, however, still produces an intensity maximum and is therefore in the moderate subwavelength regime. The imaging pattern is similar to what we obtained previously using the FDTD method before, in which an intensity maximum was identified to be a real image. Meanwhile, in the present case, the geometric image plane of AANR focusing calculated from constant-frequency contours of this photonic crystal is at $z = 1.4a_s = 0.38\lambda$. In that plane, the intensity distribution in x is similar to that in Fig. 7.18c, with a transverse size $\Delta = 0.71\lambda$. This value corresponds to an effective high cutoff $k_M \approx 1.4\omega/c = 0.38(2\pi/a_s)$, which is close to the approximate threshold k_M value in the simplified model obtained in Section 7.3.3. We also plot the amplification required to restore the source *perfectly* at the image plane in red lines in Fig. 7.18a, which can be compared to the actual transmission data and the effective cutoff k_M. Although the focusing is not perfect, the range $\omega/c < k < k_M$ roughly indicates the interval in which the actual transmission follows the behavior in the ideal case.

If ω is decreased slightly to $\omega = 0.192(2\pi c/a)$ (Fig. 7.19), the frequency falls within the narrow range of the interface-band frequencies. The transmission increases dramatically, and pairs of peaks in the transmission spectrum occur, representing excitation of interface-state combinations of even and odd mirror symmetry. These interface modes have large amplitudes, as evidenced by the compressed shade table in Fig. 7.19b and the exponential decay of intensity along z-axis in Fig. 7.19d, and they now completely dominate the image. The focusing effect of propagating waves becomes insignificant compared to this strong background. If the field distribution in a plane of constant z is measured, an example shown in Fig. 7.19c, many closely spaced, near-periodic strong peaks appear, in striking contrast to the familiar appearance of a focused optical image. Here, this pattern of intensity distribution persists for increasing z in the near field and even appears on the focusing plane of AANR $z = 1.1a_s = 0.31\lambda$. Due to the exponential decay of intensity along the z-axis and the *delocalized* field distribution in the transverse direction, neither the z-coordinate nor the transverse location of the source can be easily retrieved from this image pattern. which is hence undesirable for imaging purposes. We infer the effective cutoff wavevector by the width of the central peak $\Delta = 1.8a_s = 0.49\lambda$ on the plane of AANR image and obtain $k_M = 0.56(2\pi/a_s)$, as marked out on Fig. 7.19a where

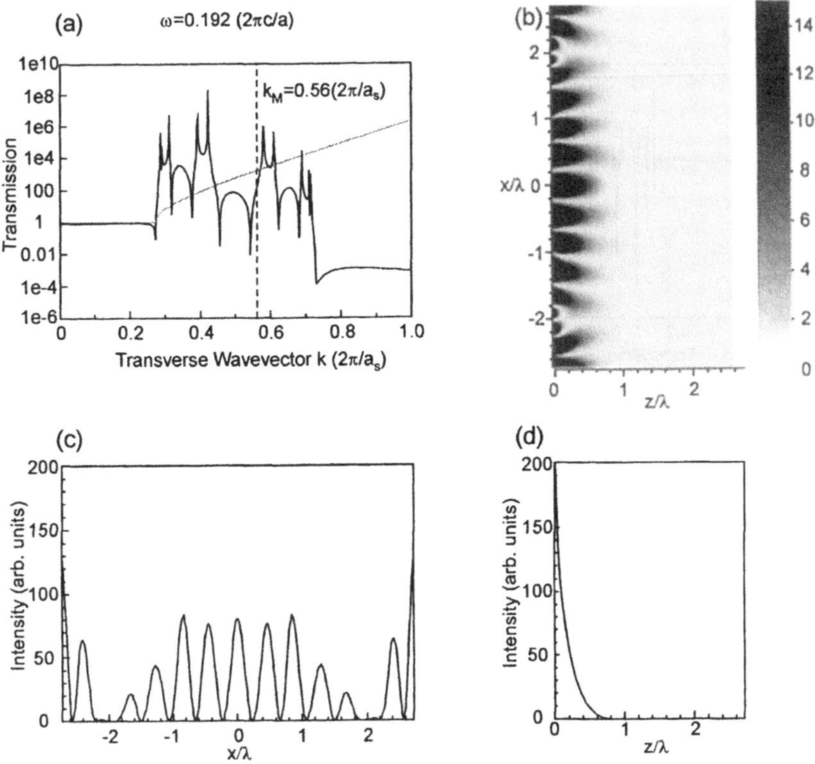

Fig. 7.19 (a)–(d) Numerical results similar to those shown in Fig. 7.18 for $\omega = 0.192\ (2\pi c/a)$. Part c is a plot in the plane of $z = 0.6a_s = 0.16\lambda$. Reprinted figure with permission from Ref. [17]. Copyright © 2003 by the American Physical Society.

the transmission curve for *perfect* image reconstruction is also plotted. It is evident that the actual transmission deviates significantly from the ideal case.

An image pattern with intermediate behavior between these two scenarios can occur, for example, if we take ω to be $\omega = 0.191(2\pi c/a)$ (Fig. 7.20). This frequency is outside the interface-band frequency range, and consequently the transmission becomes smooth again. Amplified evanescent waves are still present in the image space, which create an exponentially decaying intensity profile along the z-axis as shown in Fig. 7.20d. In contrast to the case in Fig. 7.19, a distinct intensity peak can now appear within the plane of $z = a_s = 0.27\lambda$ shown in Fig. 7.20c, with a size significantly smaller than wavelength. Here we have actually achieved $\Delta = 0.45\lambda$ at approximately the same location as calculated from AANR. This image size is consistent with the general prediction of equation (7.20), in which $a_s\lambda/(\lambda - a_s) = 0.37\lambda$ and $2a_s = 0.54\lambda$ for the present photonic crystal. We infer the high cutoff wavelength k_M in this case to be $k_M = 2.2\omega/c = 0.6(2\pi/a_s)$, corresponding to the extreme subwavelength limit. In this limit, there is no intensity

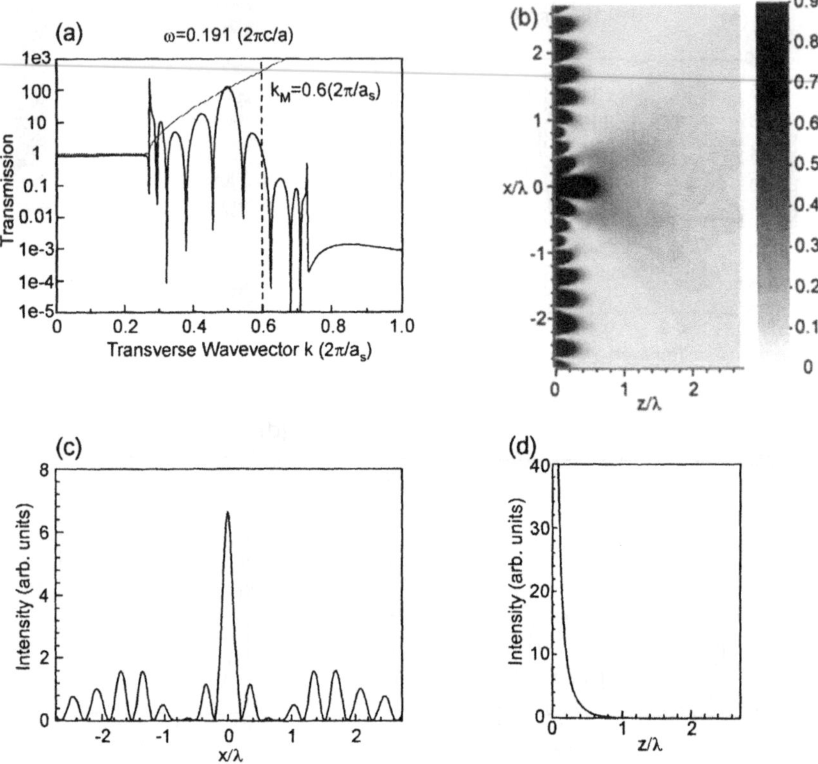

Fig. 7.20 (a)–(d) Numerical results similar to those shown in Fig. 7.18 for $\omega = 0.191\ (2\pi c/a)$. Part (c) is a plot in the plane of $z = a_s = 0.27\lambda$. Reprinted figure with permission from Ref. [17]. Copyright © 2003 by the American Physical Society.

maximum in $z > 0$, and the calculated transmission for $k < k_M$ also displays similar trend in k-dependence as the ideal transmission for perfect image recovery. We have therefore found a superlensing situation quite similar to that considered in Pendry's perfect lens, in the present case with an upper cutoff and without requiring negative-index materials.

These computational results establish that superlensing is possible with carefully designed photonic crystals and demonstrate large modifications to the image field distribution due to the presence of evanescent light. In the example structure, each of these image patterns can appear in a narrow frequency range inside that of AANR, as summarized in the detailed interface band structure in Fig. 7.21.

7.3.6 Discussion

For completeness, we show in Fig. 7.22 the calculated near-field intensity distributions in $z > 0$ for a point source of various frequencies throughout the first photonic

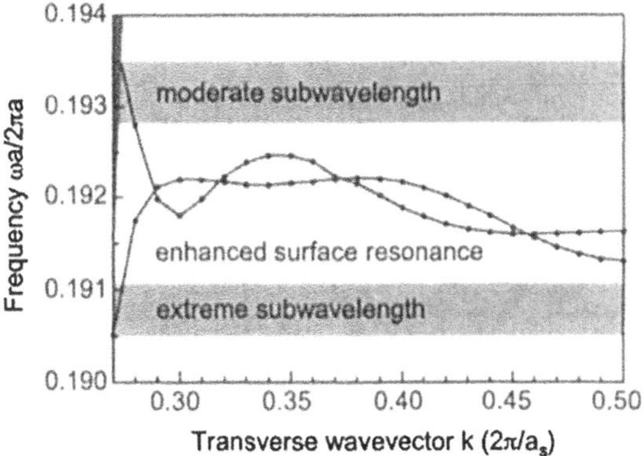

Fig. 7.21 Detailed surface band structure and its influence on subwavelength imaging. The filled circles are the divergence peaks in the calculated transmission through the structure in Fig. 7.17. The dark-shaded area in the upper left corner is the light cone. The frequency range for imaging with none or moderate subwavelength contribution is from 0.1928 to 0.1935. The frequency range for extreme subwavelength superlensing is from 0.1905 to 0.1911. The region between them is the region of the flat surface bands for enhanced surface resonance. The particular sequence of these three frequency ranges here are due to the shape of the surface bands and can be different in other systems. Reprinted figure with permission from Ref. [17]. Copyright © 2003 by the American Physical Society.

band, with all other parameters the same as those in Section 7.3.5. It is clear, that for frequencies lower than the AANR range ($\omega = 0.050, 0.100$, and $0.145(2\pi c/a)$), since most of the propagating waves do not experience negative refraction and are not focused, a broad background peak is always present in the transverse direction. An interesting feature to observe is that $\omega = 0.145(2\pi c/a)$ is close to the band edge where there are many flat bands of guided modes that can be resonantly excited. Consequently, significant subwavelength features appear on the broad background behind the slab. However, the overall resolution is now determined by the background, which is spatially broad and does not correspond to a subwavelength imaging effect. For frequencies above the AANR range ($\omega = 0.194$, and $0.210(2\pi c/a)$, since some of incident propagating radiation from air will experience total external reflection, the transverse resolution is always limited to be larger than or equal to the operating wavelength. All these can be compared to $\omega = 0.193(2\pi c/a)$, where the extraordinary superlensing enhancement in both the imaging resolution and intensity is shown. From this analysis, we conclude that *the only frequencies at which one can observe superlensing are inside the AANR range and close to a flat surface band.*

The above discussion has focused on ideal situations with no material absorption of light or structural imperfections. In practice, material losses are always present, which means that no transmission considered here will be truly infinite. In general, appreciable material losses will impose severe limitations to the amplification of evanescent waves, in a manner similar to that of the intrinsic energy leakage rate of

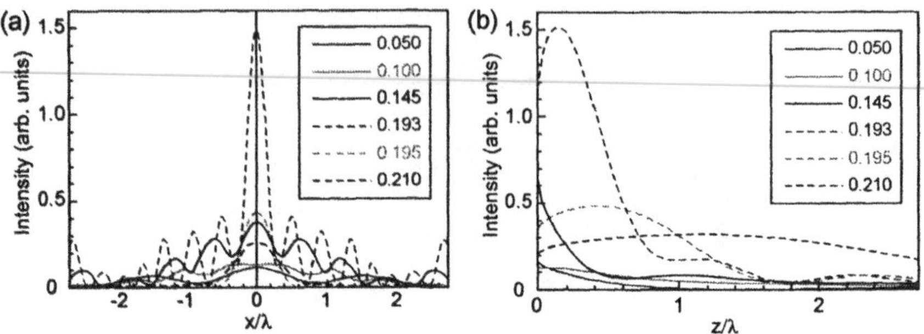

Fig. 7.22 Numerical results of the imaging for various frequencies throughout the first photonic band for the structure in Fig. 7.16. (a) Intensity distribution along the transverse direction, commonly measured at $z = 0.5a_s$ for several frequencies shown as insets. The transverse intensity distribution at larger z values has a similar background but weaker near-field modulations. (b) Intensity distribution along the z-axis for the shown frequencies. The inset numbers are the frequencies corresponding to each curve, in units of $(2\pi c/a)$. Reprinted figure with permission from Ref. [17]. Copyright © 2003 by the American Physical Society.

a crystal mode above the light line. It is also of course expected that, in the limit of extremely small material loss, our findings about the image of a superlens will remain valid. As an example, we show the calculated focusing effect in slightly lossy photonic crystals in Fig. 7.23. The losses are modeled as a positive imaginary part on the permittivity ϵ of the dielectric host, and results are calculated at the extreme subwavelength frequency $\omega = 0.191(2\pi c/a)$ for ϵ starting from $\epsilon = 12 + 0.01i$ up to $12 + 0.05i$. As the losses increase, the strength of the transmitted near-fields is attenuated, and the subwavelength features in the central image peak gradually disappear. It is clear that a resolution at or below $\Delta = 0.5\lambda$ for a localized intensity peak in x is still achievable if $\epsilon \leq 12 + 0.01i$. The effects of surface imperfections on subwavelength imaging can also be qualitatively analyzed. We consider these defects to occur only on a length scale that is smaller than a lattice constant, and thus much smaller than the operating wavelength, with correspondingly little influence on propagating waves. Since the transmission of evanescent waves depends critically on the bound interface photon states, which in turn depend sensitively on the surface structure, imperfections are expected to be most influential on the crystal surface. Their effects may thus be minimized by improving the surface quality. Another kind of structural imperfection is a finite lateral size of the crystal. We have applied the FDTD method to such finite systems and have found that, for a 20-period-wide slab, a focusing resolution around $\Delta = 0.6\lambda$ can be still obtained. These considerations suggest that the effects described in this work should be observable in realistic situations.

Our discussion on the image of a 2-D superlens can be put to experimental verification in various ways. In the microwave regime, the TM modes in 2-D photonic crystals can be realized by a 2-D photonic-crystal slab sandwiched between

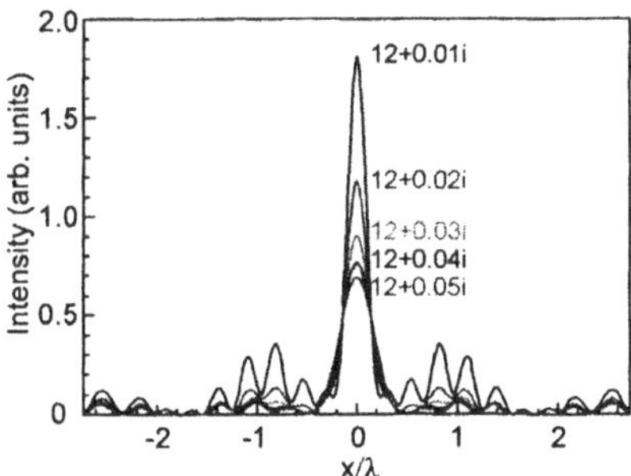

Fig. 7.23 Calculated transverse intensity distribution for imaging with lossy photonic crystals. Each curve correspond to the intensity calculated for a different host material permittivity as shown in the insets. The crystal and the point source are otherwise identical to those in Fig. 7.20. The intensity is plotted in the plane $z = a_s$ at the frequency $\omega = 0.191(2\pi c/a)$. Reprinted figure with permission from Ref. [17]. Copyright © 2003 by the American Physical Society.

two metal plates. At optical wavelengths, 2-D crystals may be obtained by replacing the metallic components with multilayer films with a large gap, or simply by index guiding. A more interesting extension of these phenomena would be to a full 3-D system. For example, the resolution of focusing in 3-D with infinite aperture and without evanescent waves is still limited by the wavelength λ, while the surface periodicity discussed in equation (7.20) should be replaced by the reciprocal of the minimum radius of the surface Brillouin zone. We show here in Fig. 7.24 the results of the computed bound photon modes of a slab of the 3-D photonic crystal studied in Section 7.2.5. As discussed there, this photonic crystal enables AANR in full 3-D and is most effective for waves polarized along $(10\bar{1})$. The interface band structure along ΓK and ΓM computed here, complicated as it may seem at first sight, bears a striking similarity to the TE and TM slab polariton bands of a dispersive negative-index materials when the polarization is taken into account. For the particular interface termination shown in Fig. 7.24, it is possible to obtain surface states within the AANR range of this photonic crystal. Since there is still a vast amount of freedom in tuning the fine details of the crystal surface structure without breaking the periodicity, it can be further expected that particular designs exist which lead to flat surface bands and can enable superlensing in full 3-D. This tunability and flexibility in our approach should make photonic crystals a powerful and beautiful candidate in manipulating and focusing light on subwavelength scales.

7.4 CONCLUSIONS

In this work we have explained in detail the principles of negative refraction and subwavelength imaging and have presented specific examples of these anomalous

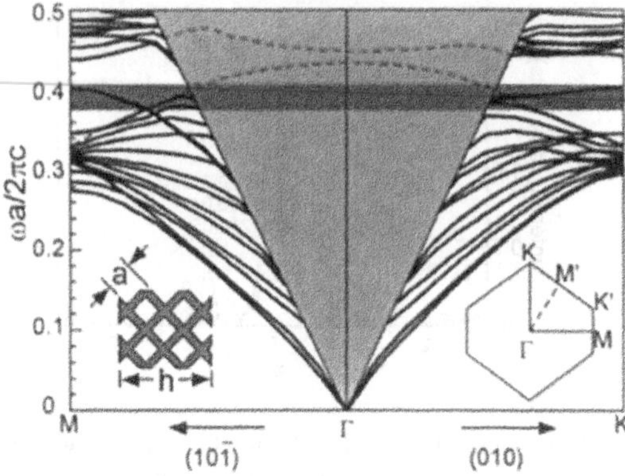

Fig. 7.24 Bound photon modes and projected band structure for a 3-D photonic crystal capable of AANR. The solid lines are bound photon modes, and the broken lines are the outlines for projected photonic band structure on the surface Brillouin zone. The light shaded region is the light cone. The dark-shaded range is the AANR frequency range. A cross section of the crystal and the surface Brillouin zone are shown as insets. The thickness of the photonic-crystal slab is $h = 3.47a$. Reprinted figure with permission from Ref. [17]. Copyright © 2003 by the American Physical Society.

optical phenomena in photonic crystal systems. Our discussions are general and cover 2-D/3-D dielectric/metallic systems. For propagating waves, photonic crystals demonstrate an example of a class of metamaterial in which negative refraction occurs without an effective negative index. For evanescent waves, photonic crystals also permit existence of special bound electromagnetic states which can be used to amplify evanescent waves and focus light to subwavelength resolutions. We have discussed in detail the peculiar features that arise from the interplay of propagating and evanescent light in the images of Veselago–Pendry superlensing. We hope that this work serves as a useful guide for the readers to understand the fundamentals of photonic crystals and to discover novel applications of these new materials on their own.

Acknowledgments

We wish to thank our collaborators Steven G. Johnson and J. B. Pendry for their invaluable contributions to all aspects of the work reviewed here.

REFERENCES

1. J. B. Pendry, A. J. Holden, W. J. Stewart and I. Youngs, "Extremely low frequency plasmons in metallic mesostructures," *Phys. Rev. Lett.*, vol. 76, pp. 4773–4776, 1996.

2. J. B. Pendry, A. J. Holden, D. J. Robbins and W. J. Stewart, "Magnetism from conductors and enhanced nonlinear phenomena," *IEEE Trans. Microwave Theory Tech.*, vol. 47, pp. 2075–2084, 1999.

3. D. R. Smith, W. J. Padilla, D. C. View, S. C. Nemat-Nasser and S. Schultz, "Composite medium with simultaneously negative permeability and permittivity," *Phys. Rev. Lett.*, vol. 84, pp. 4184–4187, 2000.

4. J. B. Pendry, "Negative refraction makes a perfect lens," *Phys. Rev. Lett.*, vol. 85, pp. 3966–3969, 2000.

5. R. A. Shelby, D. R. Smith and S. Schultz, "Experimental verification of a negative index of refraction," *Science*, vol. 292, pp. 77–79, 2001.

6. V. G. Veselago, "The electrodynamics of substances with simultaneously negative values of ϵ and μ," *Sov. Phys. Usp.*, vol. 10, pp. 509–514, 1968.

7. R. A. Shelby, D. R. Smith, S. C. Nemat-Nasser and S. Schultz, "Microwave transmission through a two-dimensional, isotropic left-handed metamaterial," *Appl. Phys. Lett.*, vol. 78, pp. 489–491, 2001.

8. R. Marques, J. Martel, F. Mesa and F. Medina, "Left-handed-media simulation and transmission of em waves in subwavelength split-ring-resonator-loaded metallic waveguides," *Phys. Rev. Lett.*, vol. 89, p. 183901, 2002.

9. C. G. Parazzoli, R. B. Greegor, K. Li, B. E. C. Koltenbah and M. Tanielian, "Experimental verification and simulation of negative index of refraction using Snell's Law," *Phys. Rev. Lett.*, vol. 90, p. 107401, 2003.

10. A. A. Houck, J. B. Brock and I. L. Chuang, "Experimental observations of a left-handed material that obeys Snell's Law," *Phys. Rev. Lett.*, vol. 90, p. 137401, 2003.

11. H. Kosaka, T. Kawashima, A. Tomita, M. Notomi, T. Tamamura, T. Sato and S. Kawakami, "Superprism phenomena in photonic crystals," *Phys. Rev. B*, vol. 58, pp. R10096-R10099, 1998.

12. B. Gralak, S. Enoch and G. Tayeb, "Anomalous refractive properties of photonic crystals," *J. Opt. Soc. Am. A*, vol. 17, pp. 1012–1020, 2000.

13. M. Notomi, "Theory of light propagation in strongly modulated photonic crystals: Refractionlike behavior in the vicinity of the photonic band gap," *Phys. Rev. B*, vol. 62, pp. 10696–10705, 2000.

14. C. Luo, S. G. Johnson, J. D. Joannopoulos and J. B. Pendry, "All-angle negative refraction without negative index," *Phys. Rev. B*, vol. 65, p. 201104(R), 2002.

15. C. Luo, S. G. Johnson and J. D. Joannopoulos, "All-angle negative refraction in a three-dimensionally periodic photonic crystal," *Appl. Phys. Lett.*, vol. 81, pp. 2352–2354, 2002.

16. C. Luo, S. G. Johnson, J. D. Joannopoulos and J. B. Pendry, "Negative refraction without negative index in metallic photonic crystals," *Opt. Express*, vol. 11, pp. 746–754, 2003.

17. C. Luo, S. G. Johnson, J. D. Joannopoulos and J. B. Pendry, "Subwavelength imaging in photonic crystals," *Phys. Rev. B*, vol. 68, p. 045115, 2003.

18. E. Cubukcu, K. Aydin, E. Ozbay, S. Foteinopolou and C. M. Soukoulis, "Subwavelength resolution in a two-dimensional photonic-crystal-based superlens," *Phys. Rev. Lett.*, vol. 91, p. 207401, 2003.

19. E. Cubukcu, K. Aydin, E. Ozbay, S. Foteinopolou and C. M. Soukoulis, "Electromagnetic waves: Negative refraction by photonic crystals," *Nature (London)*, vol. 423, pp. 604–605, 2003.

20. E. Yablonovitch, "Inhibited spontaneous emission in solid-state physics and electronics," *Phys. Rev. Lett.*, vol. 58, pp. 2059–2062, 1987.

21. S. John, "Strong localization of photons in certain disordered dielectric superlattices," *Phys. Rev. Lett.*, vol 58, pp. 2486–2489, 1987.

22. J. D. Joannopoulos, R. D. Meade and J. N. Winn, *Photonic Crystals: Molding the Flow of Light*, Princeton University Press, Princeton, 1995.

23. K. M. Ho, C. T. Chan and C. M. Soukoulis, "Existence of a photonic gap in periodic dielectric structures," *Phys. Rev. Lett.*, vol. 65, pp. 3152–3155, 1990.

24. N. W. Ashcroft and N. D. Mermin, *Solid State Physics*, Harcourt Brace College Publishers, New York, 1976.

25. A. Yariv and P. Yeh, *Optical Waves in Crystals: Propagation and Control of Laser Radiation*, John Wiley & Sons, Toronto, 1984.

8 Plasmonic Nanowire Metamaterials

ANDREY K. SARYCHEV and VLADIMIR M. SHALAEV

School of Electrical and Computer Engineering
Purdue University
West Lafayette, IN 47907
United States

8.1 INTRODUCTION

The optical properties of nanostructured materials have been intensively studied during the last decade. Among particularly important problems in this field are the focusing and guiding light on nanometer scales beyond the diffraction limit of the conventional far-zone optics. In object imaging, the near-field part of the radiation contains all information about the scatterer. As the distance from the object increases, the evanescent portion of the scattered field exponentially decays, resulting in information loss on the "fine" (subwavelength) features of the scatterer. The usual way to solve this problem suggests either using shorter wavelengths or measuring in the near zone; both these methods have their limitations. A new way to solve this imaging problem has been proposed by Pendry, who further developed earlier studies on negative refraction [1,2]. According to Pendry, when the scattered light passes through a material with a negative refractive index (specifically, it should be equal to -1), the evanescent components of the scattered field grow exponentially, allowing the restoration of the scatterer image with subwavelength resolution. Despite the obvious importance of such a superlens, it is worth noting here that possible applications for materials with negative refraction can go far beyond this idea. This is because the refractive index enters most of optical "laws" so that the possibility of its sign reversion can result in their serious revision and new exciting applications resulting from this.

Smith, Padilla, Vier, and Schultz [3] have demonstrated such negative-index materials (also referred to as double-negative or left-handed materials, LHMs, because the electric, magnetic vectors and the wavevector form a left-handed system, in this case) in the microwave range. (For recent references see the special issue of *Optics*

Express [4] and Refs. 5 and 6). In our earlier papers [4, 7], we proposed a first LHM (based on a nanowire composites [8]) that can have negative refraction in the near-IR and visible spectral ranges. A similar nanowire system was later considered by Panina et al. [9].

In this chapter we discuss the electrodynamics of nanowires materials and study the behavior of nanowire plasmon modes. We also describe how nanowire composites can be used for developing LHMs *in the near-IR and visible* parts of the spectrum.

The rest of the chapter is organized as follows. In the next section we discuss the interaction of a single metal nanowire with an electromagnetic wave (we refer to such a wire as a "conducting stick"). Section 8.3 describes the effective properties of composites comprising conducting sticks. In Section 8.4 we present computer simulations for the local electromagnetic field in stick composites. Section 8.5 discusses the magnetic response of two parallel conducting sticks and effective properties of composites comprising pairs of such sticks. In Section 8.6 we show that the forward and backward scattering by planar nanowire system can be characterized by their effective dipole and magnetic moments. Section 8.7 summarizes our results.

8.2 ELECTRODYNAMICS OF A SINGLE METAL NANOWIRE

Composite materials containing conducting sticks dispersed in a dielectric matrix have new and unusual properties at high frequencies. When frequency ω increases, the wavelength $\lambda = 2\pi c/\omega$ of an external electromagnetic field can became comparable in size with the stick length $2a$. In this case, one might think that the sticks act as an array of independent micro-antennas and an external wave should be scattered in all directions. Yet, we show that composite materials have well-defined dielectric and magnetic properties at high frequencies. Such "effective-medium" description is possible because a very thin conducting stick interacts with an external field like a dipole. Therefore, we can still use the effective dielectric constant ε_e or effective conductivity $\sigma_e = -i\omega\varepsilon_e/4\pi$ to describe the interaction of stick composites with an external electromagnetic wave. However, we note the formation of large stick clusters near the percolation threshold may result in scattering.

Since conducting stick composites are supposed to have effective parameters for all concentrations p outside the percolation threshold, we can use the percolation theory to calculate the effective conductivity σ_e. However, the theory has to be generalized to take into account the nonquasi-static effects. The problem of effective parameters of composites beyond the quasi-static limit has been considered in Refs. [8] and [10–15]. It was shown there that the mean-field approach can be extended to find the effective dielectric constant and magnetic permeability at high frequencies. Results of these considerations can be briefly summarize as follows. One first finds the polarizability for a particle in the composite illuminated by an electromagnetic wave (the particle is supposed to be embedded in the "effective medium" with dielectric constant ε_e). Then, the effective dielectric permittivity ε_e is determined by the self-consistent condition requiring that the averaged polarizability of all particles should vanish. Thus, for the nonquasi-static case, the problem is reduced to calculation of the polarizability of an elongated conducting inclusion. That is, we consider the retardation effects resulting from the interaction of a conducting stick with the electromagnetic wave scattered by the stick.

The diffraction of electromagnetic waves on a conducting stick is a classical problem of the electrodynamics. A rather tedious theory for this process is presented in several textbooks [16, 17]. We show below that the problem can be solved analytically in the case of very elongated sticks when the aspect ratio a/b is so large that $\ln(a/b) \gg 1$.

We consider a conducting stick of length $2a$ and radius b illuminated by an electromagnetic wave. We suppose that the electric field in the wave is directed along the stick and the stick is embedded in a medium with $\varepsilon = 1$. The external electric field excites in the stick and electric current $I(z)$, where z is the coordinate along the stick, measured from its midpoint. The dependence $I(z)$ is nontrivial when the wavelength λ is of the order of or smaller than the stick length. There is also a nontrivial charge distribution $q(z)$ along the stick in this case. The charge distribution $q(z)$ determines the polarizability of the stick. To find $I(z)$ and $q(z)$, we introduce the potential $U(z)$ of the charges $q(z)$ distributed over the stick surface. From the equation for the electric charge conservation we obtain the following formula:

$$\frac{dI(z)}{dz} = i\omega q(z) \tag{8.1}$$

which relates the charge per unite length $q(z)$ and the current $I(z)$. Note that the electric charges q are induced by the external field E_0 and they can be expressed as the divergence of the polarization vector, $q = -4\pi \operatorname{div} \mathbf{P}$. Then, the polarization $\mathbf{P}$ can be included in definition of the electric displacement $\mathbf{D}$ so that $\operatorname{div} \mathbf{D} = 0$. However, in calculating the high frequency field in a conducting stick it is convenient to explicitly consider charges generated by the external field.

To find an equation for the current $I(z)$ we treat a conducting stick as a prolate conducting spheroid with semiaxes a and b. The direction of the major axis is supposed to coincide with direction of the electric field $\mathbf{E}_0 \exp(-i\omega t)$ in the incident wave. The electric potential of the charge $q(z)$ is given by the following solution to Maxwell's equations (see, e.g., Ref. [17], p. 377):

$$U(z) = \oint \frac{[q(z')/2\pi\rho(z')] \exp(ik\,|\mathbf{r} - \mathbf{r}'|)}{|\mathbf{r} - \mathbf{r}'|} ds' \cong \int_{-a}^{a} \frac{q(z') \exp(ik\,|z - z'|)}{\sqrt{(z - z')^2 + \rho(z)^2}} dz' \tag{8.2}$$

where the integration in the first integral is performed over the surface of the stick, $\mathbf{r}$ and $\mathbf{r}'$ are two points on the surface of the stick with the coordinates z and z', respectively, $\rho(z) = b\sqrt{1 - z^2/a^2}$ is the radius of the cross section at the coordinate z, and $k = \omega/c$ is the wavevector of the external field. In transition to the second expression in equation (8.2), we neglect terms of the order of $\rho(z)/a < b/a \ll 1$. We divide the last integral in equation (8.2) into two parts, setting $q(z') \exp(ik\,|z - z'|) =$

$q(z) + [q(z') \exp(ik|z - z'|) - q(z)]$, that is,

$$U(z) \cong q(z) \int_{-a}^{a} \frac{dz'}{\sqrt{(z - z')^2 + \rho(z)^2}} + \int_{-a}^{a} \frac{q(z') \exp(ik|z - z'|) - q(z)}{|z - z'|} dz' \tag{8.3}$$

The first integral in equation (8.3) is given by

$$\int_{-a}^{a} \frac{dz'}{\sqrt{(z - z')^2 + \rho(z)^2}} = 2 \ln\left(2\frac{a}{b}\right) \tag{8.4}$$

The second integral in equation (8.3) has no singularity at $z = z'$, and therefore its value is $\sim q(z)$, which is an odd function of the coordinate z. We assume for simplicity that $q(z)$ is proportional to z and in this approximation

$$\int_{-a}^{a} \frac{q(z') \exp(ik|z - z'|) - q(z)}{|z - z'|} dz' = 2\, q(z) \left[-e^{i\,a\,k} + \mathrm{Ei}(a\,k)\right] \tag{8.5}$$

where the $\mathrm{Ei}(x)$ function is defined as

$$\mathrm{Ei}(x) = \int_0^x \left[\exp(it) - 1\right]/t\, dt \tag{8.6}$$

By substituting equations (8.4) and (8.5) in equation (8.3), we obtain

$$U(z) = q(z)/C \tag{8.7}$$

where the capacitance C is given by

$$C = \frac{1}{2\left[\ln(2a/b) - e^{i\,a\,k} + \mathrm{Ei}(a\,k)\right]} \tag{8.8}$$

The capacitance C takes the usual value $C = 1/\left[2\ln(2a/b) - 2\right]$, in the quasi-static limit $ka \to 0$. The retardation effects result in additional terms in equation (8.8) that have small magnitudes in comparison with the leading logarithmic term. This result is obtained within the logarithmic accuracy: its relative error is on the order of $1/\ln(a/b)$, and the ratio a/b is assumed so large that its logarithm is also large.

By substituting equation (8.7) into equation (8.1), we obtain the following equation:

$$\frac{dI(z)}{dz} = i\omega C U(z) \tag{8.9}$$

which relates the current $I(z)$ and the surface potential $U(z)$. The electric current $I(z)$ and electric field $E(z)$ on the stick surface are related by the usual Ohm's Law

$$E(z) = RI(z) \tag{8.10}$$

where R is the impedance per unit length. Since the stick is excited by the external field $E_0 \exp(-i\omega t)$ which is parallel to its axis, the electric field $E(z)$ is equal to

$$E(z) = E_0 - \frac{dU(z)}{dz} + i\frac{\omega}{c} A_z(z) \tag{8.11}$$

We consider now the vector potential $A_z(z)$ induced by the current $I(z)$ flowing in the stick and obtain $A_z(z)$ by the same procedure as was used to estimate the potential U. Thus, with the same logarithmic accuracy we find

$$\begin{aligned} A_z(z) &= \frac{1}{c}\int_{-a}^{a} \frac{I(z')\exp\left(ik\,|z-z'|\right)}{\sqrt{(z-z')^2+\rho(z)^2}}\,dz' \\ &\simeq \frac{2}{c}I(z)\ln\left(\frac{2a}{b}\right)+\frac{1}{c}\int_{-a}^{a}\frac{I(z')\exp(ik\,|z-z'|)-I(z)}{|z-z'|}\,dz' \end{aligned} \tag{8.12}$$

where c is the speed of light. To estimate the second integral in equations (8.12), we approximate the current $I(z)$, which is an even function of z, as $I(z) = I(0)\left[1-(z/a)^2\right]$; thus we obtain for $z \ll a$ that

$$\frac{1}{c}\int_{-a}^{a}\frac{I(z')\exp\left(ik\,|z-z'|\right)-I(z)}{|z-z'|} \simeq \frac{1}{c}I(z)\left[-1+l\left(ka\right)\right] \tag{8.13}$$

where the function $l(x)$ is given by

$$l(x) = \left[2+2e^{ix}\left(ix-1\right)+x^2\right]x^{-2}+2\,\mathrm{Ei}(x) \tag{8.14}$$

By substituting equation (8.13) in equation (8.12), we obtain the following relation between the vector potential and current:

$$A_z(z) = \frac{L}{c}I(z) \tag{8.15}$$

where L is the inductance per unit length,

$$L \simeq 2\ln\left(\frac{2a}{b}\right)-1+l(ka) \tag{8.16}$$

The last term in equation (8.16) is much smaller than the first one when $2\ln(2a/b) \gg 1$. Nevertheless, we keep this term since, as we show below, it plays an important role in the electromagnetic response. Equation (8.16) is invalid near the ends of the stick; however, in calculating the polarizability, this region is unimportant.

By substituting equations (8.11) and (8.15) in equation (8.10), we obtain the following form of Ohm's Law:

$$-\frac{dU(z)}{dz} = \left(R-i\frac{\omega L}{c^2}\right)I(z)-E_0 \tag{8.17}$$

To obtain a closed equation for the current $I(z)$, we differentiate equation (8.9) with respect to z and substitute the result into equation (8.17) for $dU(z)/dz$. Thus we obtain

$$\frac{d^2I(z)}{dz^2}+i\omega C\left[\left(R-i\frac{\omega L}{c^2}\right)I(z)-E_0\right]=0 \tag{8.18}$$

with the boundary conditions requiring the vanishing current at the ends of the stick,

$$I(-a) = 0, \qquad I(a) = 0 \tag{8.19}$$

A solution for equation (8.18) gives the current distribution $I(z)$ in a conducting stick irradiated by an electromagnetic wave. Then we can calculate the charge distribution and the polarizability of the stick.

As mentioned, we consider the conducting stick as a prolate spheroid with semi-axes such that $a \gg b$. To determine the impedance R in equation (8.18) we recall that the cross-section area of a spheroid at coordinate z is equal to $\pi b^2[1-(z/a)^2]$; thus we have the following expression for the impedance:

$$R = \frac{1}{\pi b^2[1-(z/a)^2]\sigma_m^*} \tag{8.20}$$

where σ_m^* is the renormalized stick conductivity taking into account the skin effect. We assume that the conductivity σ_m changes due to the skin effect in the same way as the conductivity of a long wire of radius b (see, e.g., Ref. [18], Section 61),

$$\sigma_m^* = \sigma_m f(\Delta), \qquad f(\Delta) = \frac{1-i}{\Delta}\frac{J_1[(1+i)\Delta]}{J_0[(1+i)\Delta]} \tag{8.21}$$

where J_0 and J_1 are the Bessel functions of the zeroth and first order, respectively, and the parameter Δ is equal to the ratio of the stick radius b and the skin depth,

$$\Delta = b\sqrt{2\pi\sigma_m\omega}/c \tag{8.22}$$

When the skin effect is weak (i.e., $\Delta \ll 1$) the function $f(\Delta) = 1$ and the renormalized conductivity σ_m^* is equal to the stick conductivity $\sigma_m^* = \sigma_m$. In the opposite case of a strong skin effect ($\Delta \gg 1$), the current I flows within a thin skin layer at the surface of the stick. Then equation (8.22) gives $\sigma_m^* = (1-i)\sigma_m/\Delta \ll \sigma_m$.

For further consideration, it is convenient to rewrite equations (8.18) and (8.19) in terms of the dimensionless coordinate $z_1 = z/a$ and dimensionless current $I_1 = I/\left(\sigma_m^*\pi b^2 E_0\right)$. We introduce the dimensionless relaxation parameter

$$\gamma = 2i\frac{b^2\,\pi\,\sigma_m^*}{a^2\,C\,\omega} = \varepsilon_m^*\left[g + \frac{b^2}{a^2}\left(1 - e^{i\,a\,k} + \mathrm{Ei}(a\,k)\right)\right] \tag{8.23}$$

where $\varepsilon_m^* = i4\pi\sigma_m^*/\omega$ is the renormalized dielectric constant for metal, and

$$g = (b/a)^2\left[\ln(2a/b) - 1\right] \tag{8.24}$$

is the depolarization factor for a very prolate ellipsoid (see, e.g., Ref. [18], Section 4). We also introduce the dimensionless frequency

$$\Omega\,(ak) = ak\sqrt{LC} \simeq ak\left[1 + \frac{1 + e^{i\,a\,k}\left(-1 + i\,a\,k + a^2\,k^2\right)}{2\,(ak)^2\log(2\,a/b)}\right] \tag{8.25}$$

In transition to the second equality in equation (8.25) we assume that $\log(2\,a/b) \gg 1$. By substituting parameters γ and Ω in equations (8.18) and (8.19), we obtain the following equation for the dimensionless current:

$$\frac{d^2 I_1(z_1)}{dz_1^2} + \left[-\frac{2}{\gamma\left(1 - z_1^2\right)} + \Omega^2\right] I(z_1) + \frac{2}{\gamma} = 0 \qquad (8.26)$$

$$I_1(-1) = I_1(1) = 0$$

To understand the physical meaning of equation (8.26), let us consider first the quasi-static case when the skin effect is negligible and $ka \ll 1$. Then, it follows from equations (8.23) and (8.26) that $\Omega^2\gamma \sim \Delta \ll 1$. Therefore, we can neglect the second term in the square brackets in equation (8.26) and find the current,

$$I_1(z_1) = \frac{1 - z_1^2}{1 + \gamma} \qquad (8.27)$$

and electric field E_m inside a conducting stick as [see equations (8.10) and (8.20)]

$$E_m = \frac{1}{1 + \varepsilon_m g} E_0 \qquad (8.28)$$

As anticipated, the electric field E_m is uniform and coincides with the quasi-static internal field in a prolate conducting spheroid. Note that we assume that $\varepsilon_m \gg 1$.

In the opposite case of a strong skin effect, the product $\Omega^2\gamma \sim \Delta$ is much larger than unity: $\Omega^2\gamma \gg 1$. Therefore, we can neglect the first term in the square brackets in equation (8.26) and find that

$$I_1(z_1) = \frac{2}{\Omega^2\gamma}\left[\frac{\cos(\Omega z_1)}{\cos(\Omega)} - 1\right] \qquad (8.29)$$

As follows from this equation the current has maxima when $\cos(\Omega) \approx 0$, which corresponds to the well-known antenna resonance [16, 17] at wavelengths $\lambda = \lambda_n = 2a/(2n+1)$, with $n = 0, 1, 2, \ldots$.

In the general case of arbitrary Ω and γ, the solution for equation (8.26) cannot be expressed as a finite set of known special functions [19]. Still, this equation can be readily solved numerically. The numerical integration of equation (8.26) shows that for large enough wavelengths ($\lambda > \lambda_1$), the solution can be approximated by the simple equation

$$I_1(z_1) = \frac{1 - z_1^2}{1 + \gamma\cos(\Omega)} \qquad (8.30)$$

which is an interpolation for equations (8.27) and (8.29).

When the current I is known, we can calculate the specific polarizability P_m of a conducting stick: $\alpha_m = (VE_0)^{-1}(\varepsilon_m - 1)\int E_m\, dV$, where $V = 4\pi ab^2/3$ is the volume of the stick. Assuming $\varepsilon_m \gg 1$, we obtain

$$\alpha_m = \frac{\varepsilon_m^*}{1 + \gamma\cos(\Omega)} \qquad (8.31)$$

Above, we assumed that the stick is aligned with the electric field of the incident electromagnetic wave. Stick composites can be formed by randomly oriented rods. In this case, we have to modify equation (8.31) for the stick polarizability. We consider a conducting stick directed along the unit vector $\mathbf{n}$ and suppose that the stick is irradiated by an electromagnetic wave with the electric field

$$\mathbf{E} = \mathbf{E}_0 \exp\left[i\left(\mathbf{k} \cdot \mathbf{r}\right)\right] \tag{8.32}$$

where $\mathbf{k}$ is the wavevector inside the composite. The current I in a strongly elongated stick is excited by the component of the electric field, which is parallel to the stick

$$\mathbf{E}_{\parallel}(z) = \mathbf{n}\left(\mathbf{n} \cdot \mathbf{E}_0\right) \exp\left[i\left(\mathbf{k} \cdot \mathbf{n}\right) z\right] \tag{8.33}$$

where z is the coordinate along the stick.

The field $\mathbf{E}_{\parallel}$ averaged over the stick orientations is aligned with the external field $\mathbf{E}_0$ and has the following magnitude:

$$E_0^*(z) = \frac{E_0}{(kz)^2}\left[\frac{\sin(kz)}{kz} - \cos(kz)\right] \tag{8.34}$$

The current in the stick is a linear function of the filed $\mathbf{E}_{\parallel}$. Since the average field $\mathbf{E}_{\parallel}$ is aligned with $\mathbf{E}_0$, the current averaged over the stick orientations is also parallel to the external field $\mathbf{E}_0$.

To obtain the current $\langle\langle I(z)\rangle\rangle$ averaged over the stick orientations and the average stick polarizability $\langle\langle P_m\rangle\rangle$, we substitute the field $E_0^*(z)$ given by equation (8.34) into equation (8.18) for the field E_0. Hereafter, the sign $\langle\langle\ldots\rangle\rangle$ stands for the average over stick orientations. Then, the current $\langle\langle I\rangle\rangle$, the polarizability $\langle\langle P_m\rangle\rangle$, and the effective dielectric permittivity depend on frequency ω and, in addition, on the wavevector k. This means that a conducting stick composite is a medium with spatial dispersion. This result is easy to understand, if we recall that a characteristic scale of inhomogeneity is the stick length $2a$, which can be of the order of or larger than the wavelength. Therefore, it is not surprising that the interaction of an electromagnetic wave with such composite has a nonlocal character and, therefore, the spatial dispersion is important. One can expect that additional waves can be be excited in the composite in the presence of strong spatial dispersion.

Below we consider wavelengths such that $\lambda > \lambda_2$; therefore, we can expand $E_0^*(z)$ in a series as

$$E_0^*(z) = \frac{E_0}{3}\left(1 - \frac{(kz)^2}{10}\right) \tag{8.35}$$

and restrict ourselves to the first term, when considering the dielectric properties. Since the average field is given by $E_0^*(z) = E_0/3$, then the average current is equal to $\langle\langle I\rangle\rangle = I/3$, where the current I is defined by equation (8.30). As a result, the stick polarizability averaged over the orientations is equal to $\langle\langle P_m\rangle\rangle = P_m/3$, where P_m is given by equation (8.31).

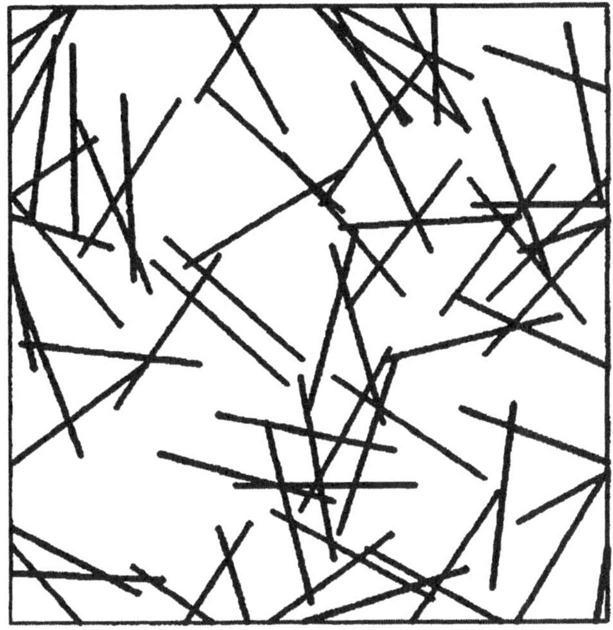

Fig. 8.1 Conducting stick composite.

8.3 CONDUCTING STICK COMPOSITES: EFFECTIVE MEDIUM APPROACH

Here we consider composites that contain very elongated conducting inclusions, "sticks," embedded in a dielectric host with dielectric constant ε_d as shown in Fig. 8.1. The sticks are randomly distributed in the host. The problem here is to calculate the macroscopic dielectric response of such a composite. Metal–dielectric composites, where conducting inclusions are very elongated, can have various important applications (see, e.g., Refs. [8, 13, 15, 20–22] and references therein). Here we show that conducting stick composites can be employed as metamaterials with tunable effective dielectric and magnetics properties.

To calculate the effective properties of a composite, we use a self-consistent approach known as the effective medium theory (EMT) [23–25]. The EMT has the virtue of mathematical and conceptual simplicity, and it is a method that provides a quick insight into the effective properties of metal–dielectric composites. Usually, the EMT is based on the assumption that metal and dielectric grains are embedded in the same homogeneous effective medium whose properties should be determined self-consistently. This assumption should be modified to take into account intrinsic structures of conducting stick composites.

Let us consider a small domain of the composite with the size $l \sim b \ll a$. The probability that the domain contains a conducting stick is estimated as $p(l) \sim$

$l^3N(p) \sim b^3N(p)$, where $N(p)$ is the stick concentration. The probability $p(l)$ is small even for the concentrations p corresponding to the percolation threshold $p_c \sim b/a$ [8], where it is estimated as $p(l) \sim b^3N(p_c) \sim (b/a)^2 \ll 1$. Therefore, the dielectric constant of such a domain is equal to ε_d, with the probability close to unity. On the other hand, we can prescribe the bulk effective dielectric constant ε_e to the domain with the size l much larger than the stick length $2a$. Thus we obtain that the local dielectric constant $\varepsilon(l)$ depends on the scale under consideration: for $l < a$, the dielectric constant $\varepsilon(l)$ is equal to $\varepsilon(0) = \varepsilon_d$ and, for $l > a$, $\varepsilon(l) = \varepsilon_e$. We use a simple assumption that a conducting stick is surrounded by a medium with the dielectric constant given by

$$\begin{aligned} \varepsilon(l) &= \varepsilon_d + (\varepsilon_e - \varepsilon_d)l/a, \quad l < a \\ \varepsilon(l) &= \varepsilon_e, \quad l > a \end{aligned} \tag{8.36}$$

We can summarize the main assumptions for our effective-medium theory suggested first in Ref. [8] as follows:

1. Each conducting stick is embedded in the effective medium with the dielectric constant $\varepsilon(l)$ that depends on the scale l as described by equation (8.36). The value of ε_e is to be determined self-consistently.

2. The dielectric regions are assumed to be spherical and they are embedded in the effective medium with the dielectric constant ε_e.

3. The effective permittivity ε_e is determined by the condition that the polarizability averaged over all inclusions should vanish [10–12].

Since sticks are randomly oriented, the dielectric regions of the composite are supposed to have spherical shapes, as assumed above. The specific polarizability of a dielectric region is given then by the known quasi-static equation (see, e.g., Ref. [23])

$$\alpha_d = \frac{3(\varepsilon_d - \varepsilon_e)}{2\varepsilon_e + \varepsilon_d} \tag{8.37}$$

The polarizability of a conducting stick embedded in the effective medium (8.36) is obtained from equation (8.31), by replacing in the numerator ε_m^* with $\varepsilon_m^*/\varepsilon_e$. The scale dependence of the local dielectric constant in equation (8.36) results in a modification of the the parameter γ [see equation (8.23)] to

$$\tilde{\gamma} = \frac{\varepsilon_m^*}{\varepsilon_d}\left[\tilde{g} + \frac{b^2}{a^2}\left(1 - e^{ix} + \mathrm{Ei}(x)\right)\right]$$

where $\tilde{g} = (b/a)^2\left[\ln(1 + 2a\varepsilon_d/b\varepsilon_e) - 1\right]$ and $x = \sqrt{\varepsilon_d}ka$ [8]. Then the condition that the average polarizability should vanish gives the following equation:

$$\langle\langle 4\pi\alpha_m\rangle\rangle + (1-p)\,4\pi\alpha_d = \frac{1}{3}p\frac{\varepsilon_m^*}{\varepsilon_e}\frac{1}{1+\tilde{\gamma}\cos\Omega} + 3\frac{\varepsilon_d - \varepsilon_e}{2\varepsilon_e + \varepsilon_d} = 0 \tag{8.38}$$

where p is the volume concentration of the conducting sticks and the sign $\langle\langle\cdots\rangle\rangle$ denotes the average over the orientations.

To understand the composite properties at high frequencies, we consider a solution of equation (8.38) for the stick concentration p below the percolation threshold $(b/a)^2 \ll p \ll b/a$. Assuming that $\varepsilon_d \ll |\varepsilon_e| \ll a/b$, we obtain the explicit equation for the effective dielectric permittivity

$$\varepsilon_e \simeq \varepsilon_d \frac{2}{9} p \frac{a^2}{b^2} \frac{1}{[\ln(2a/b) - e^{ix} + \mathrm{Ei}(ka)] \cos[\Omega(ka)] + \varepsilon_d/\varepsilon_m^*} \tag{8.39}$$

where functions $\mathrm{Ei}(x)$ and $\Omega(x)$ are defined in equations (8.6) and(8.25), respectively.

We consider now the effective dielectric permittivity ε_e for the case of a perfect metal ($|\varepsilon_m| \to \infty$). Then the electromagnetic field does not penetrate in a metal and, as follows from equation (8.21), and the renormalized conductivity ε_m^* also tends to infinity. We neglect the second term in the denominator of equation (8.39) and obtain that the effective permittivity ε_e has maxima when $\mathrm{Re}\,\Omega = \pi/2 + n\pi, n = 0, 1{,}2\ldots$, which approximately corresponds to the wavelengths $\lambda_n = \left(4a/\sqrt{\varepsilon_d}\right)/(1+2n)$.

Now we consider the behavior of the effective dielectric constant near the lowest resonance frequency $\omega_0 = \pi c/\left(2a\sqrt{\varepsilon_d}\right)$. By expanding the denominator of equation (8.39) in a power series of $\omega - \omega_0$ and taking into account that $\varepsilon_m^* \to \infty$ and $\ln(a/b) \gg 1$, we obtain

$$\varepsilon_e \simeq \varepsilon_d p \frac{4}{9\pi \log(2a/b)} \frac{a^2}{b^2} \frac{1}{(\omega_0^* - \omega)/\omega_0 - i\gamma} \tag{8.40}$$

where $\omega_0^* = \omega_0 \left[1 + 2(\pi - 2)/\left(\pi^2 \log(2a/b)\right)\right] \approx \omega_0$ and the loss factor $\gamma = \left(\pi^2 - 4\right)/\left[\pi^2 \log(2a/b)\right] \ll 1$.

It is interesting to point out that the effective dielectric constant is independent of the metal conductivity ε_m, as it should be for the limiting case when the electromagnetic field does not penetrate to the metal. At the resonance frequency $\omega = \omega_0^*$ the real part of ε_e changes its sign and it becomes negative when $\omega > \omega_0$. The imaginary part of ε_e has a maximum at the resonance and its magnitude

$$\varepsilon_e''(\omega_0) \simeq \varepsilon_d \frac{4\pi}{9(\pi^2 - 4)} p \frac{a^2}{b^2} \tag{8.41}$$

does not depend on the conductivity of the sticks. We obtain that the imaginary part of the effective dielectric constant does not vanish for composites with perfectly conducting sticks. The presence of the effective losses, in this case, is due to the excitation of the internal modes in the composite. When ε_m and the dielectric host have no losses, the amplitudes of these modes continuously increase with time. In real composites, there are always some losses in the conducting sticks as well as in the dielectric host. Therefore, the internal field should stabilize at some large values. Thus, one can anticipate the existence of giant local fields in conducting stick composites.

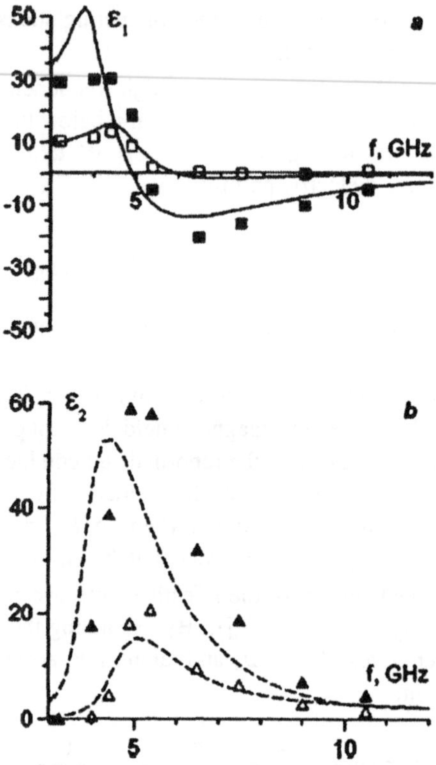

Fig. 8.2 Real (a) and imaginary (b) parts of the permittivity for a composite filled with aluminum-coated fibers 20 mm long (thickness $\sim 1\ \mu$m). The fiber volume concentration is 0.01% and 0.03%. Points indicate experimental data and line describe theoretical results.

Microwave metamaterials with negative dielectric permittivity were first obtained earlier [13,20]. In Fig. 8.2 we present experimental and theoretical results obtained in Ref. [13] for the microwave dielectric function of composites containing very thin aluminum microwires. In such metamaterials the real part of ε_e becomes negative for the frequency above the resonance as seen in Fig. 8.2.

8.4 CONDUCTING STICK COMPOSITES: GIANT ENHANCEMENT OF LOCAL FIELDS

We consider now the field distribution in thin (~ 10 nm) but relatively long ($\sim 1\ \mu$m) metal sticks. A problem of field distribution around such metal particles can hardly be solved analytically. We describe a numerical model based on the discrete dipole approximation (DDA) following our papers [4,7]. This approach was first introduced by Purcell and Pennypacker [26].

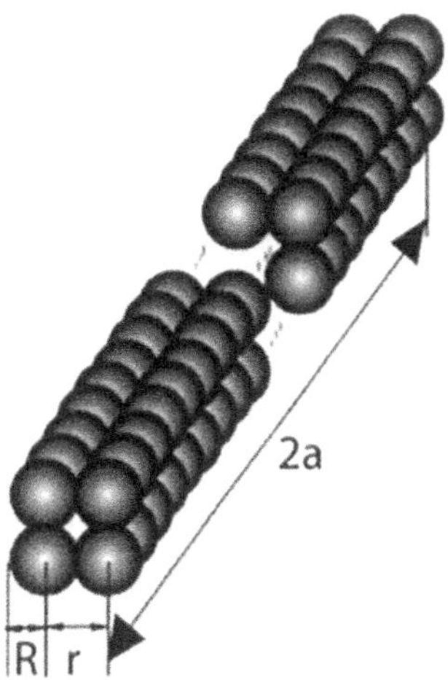

Fig. 8.3 Long stick modeled by chains of spheres. After Ref. [4]. Copyright © 2003 Optical Society of America, Inc.

In the DDA approximation, a conducting stick is represented by a large amount of small spherical particles of some radius R as shown in Fig. 8.3. Each particle is placed in a node of a cubic lattice with period a. The position of individual particles is denoted by $\mathbf{r}_i$. It is supposed that the particle radius R is much smaller than the wavelength λ so that interactions of particles are well described by their dipole moments $\mathbf{d}_i$. Each particle is subjected to an incident field $\mathbf{E}_0$ and to the field scattered by all other particles. Therefore, the dipole moments of particles are coupled to the incident field and to each other and can be found solving the following coupled-dipole equations (CDEs):

$$\mathbf{d}_i = \alpha \left[\mathbf{E}_0 \left(\mathbf{r}_i \right) + \sum_{j \neq i} \hat{G}(\mathbf{r}_i - \mathbf{r}_j)\mathbf{d}_j \right] \tag{8.42}$$

where α is the polarizability of a particle, $\mathbf{E}_0(\mathbf{r}_i)$ is the incident field at point $\mathbf{r}_i$, and $\hat{G}(\mathbf{r}_i - \mathbf{r}_j)\mathbf{d}_j$ gives the field produced by dipole $\mathbf{d}_j$ at the point $\mathbf{r}_i$ and $\hat{G}(\mathbf{r}_i - \mathbf{r}_j)$ is the free-space dyadic Green's function:

$$\begin{aligned} G_{\alpha\beta} &= k^3[A(kr)\delta_{\alpha\beta} + B(kr)r_\alpha r_\beta] \\ A(x) &= [x^{-1} + ix^{-2} - x^{-3}]\exp(ix) \\ B(x) &= [-x^{-1} - 3ix^{-2} + 3x^{-3}]\exp(ix) \end{aligned} \tag{8.43}$$

with $\hat{G}\mathbf{d} \equiv G_{\alpha\beta}d_\beta$. The Greek indices represent Cartesian components and the summation over the repeated indices is implied. The polarizability α of an individual dipole is given by Lorentz–Lorenz formula with the radiative correction introduced

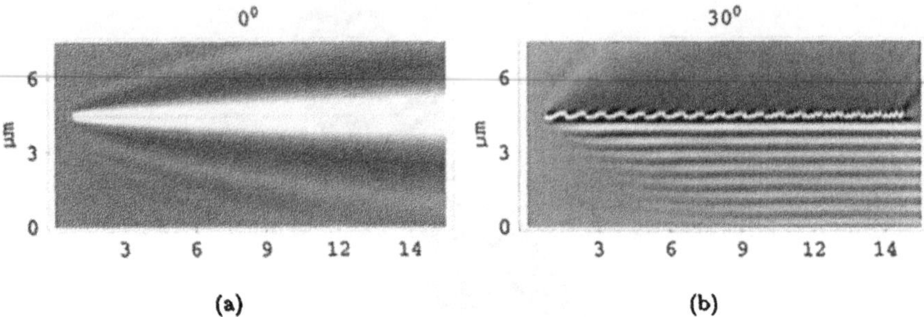

Fig. 8.4 EM field distribution for a long needle. The wavelength of incident light is 540 nm. The angle between the wavevector of incident light and the needle is (a) 0° and (b) 30°. After Ref. [4]. Copyright © 2003 Optical Society of America, Inc.

by Draine [27]:

$$\alpha_{LL} = R^3 \frac{\varepsilon_m - 1}{\varepsilon_m + 1}, \qquad \alpha = \frac{\alpha_{LL}}{1 - i\,(2/3)\,(kR)^3\,\alpha_{LL}} \tag{8.44}$$

Results of calculations for (8.42) depend on the intersection ratio r/R—that is, the ratio of the distance between neighboring particles and its radius (see Fig. 8.3). We choose the ratio as $r/R \approx 1.66$ to reproduce the quasi-static polarizability of an elongated metal ellipsoid.

In our numerical simulations [4, 7], a single nanostick was represented by four parallel chains of spherical particles to take into account the skin effect (see Fig. 8.3). Specifically, we consider the field distribution in the vicinity of a conducting stick with roughly $2b = 30$ nm thickness and $2a \approx 15\ \mu$m long, illuminated by a plane wave with the wavelength of 540 nm. Our results, shown in Fig. 8.4, clearly identify the interference pattern between irradiation and the plasmon polariton wave excited on the metal surface. Similar interference patterns were observed in experiments [28, 29]. Note that the electromagnetic field is concentrated around the wire surface, which suggests the possibility to use nanowires as nano waveguides.

Simulations for shorter sticks ($2a = 480$ nm) presented in Fig. 8.5 also show the existence of sharp plasmon resonances [4, 7] when the wavelength of the light is a multiple of surface plasmon (half) wavelengths. The enhancement of the local field intensity in the resonance can reach the magnitude of 10^3. The spatial area where the field concentrates is highly localized around the nanowire, and it can be as small as 100 nm. This plasmon resonance is narrowband, with the spectral width in a single nanowire about 50 nm, which corresponds to the discussed above equations (8.39) and (8.40).

In a composite with metal sticks randomly distributed in a dielectric substrate, the metal–dielectric transition occurs at a significantly smaller metal concentration than in the case of percolation films formed by spherical particles. In the 2-D case of a

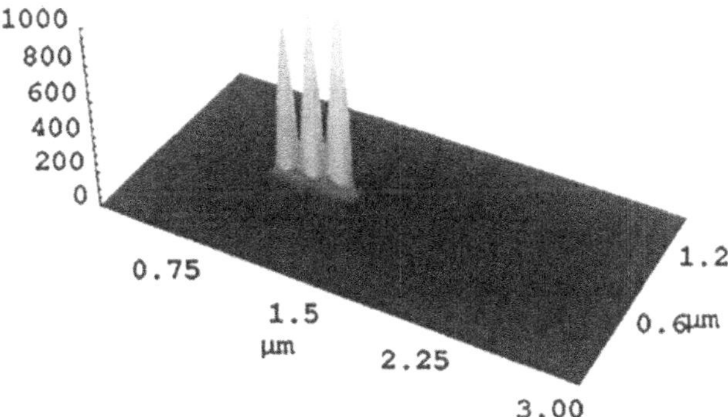

Fig. 8.5 The intensity distribution of the electric field at surface plasmon polariton resonance in a silver nanowire excited by a plane electromagnetic wave. The angle between the nanowire and the wavevector of the incident light is 30 degrees. The wavevector and **E** vector of the incident irradiation are in the plane of the figure; the needle length is 480 nm. After Ref. [4]. Copyright © 2003 Optical Society of America, Inc.

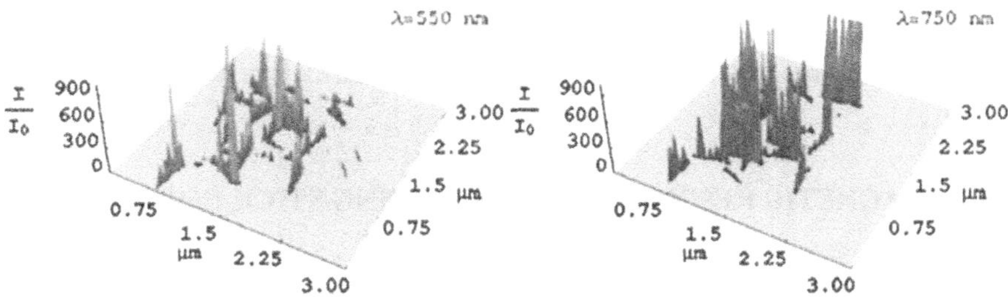

Fig. 8.6 Field distribution in nanowire percolation Ag composite for the incident wavelength of 550 nm (left) and 750 nm (right). In both figures the case of normal incidence with $(E||x)$ is considered. After Ref. [4]. Copyright © 2003 Optical Society of America, Inc.

wire composite, the percolation threshold is close to the inverse of the stick aspect ratio [8], and hence it can be arbitrary small for sufficiently long sticks.

We simulate composites by a random distribution of identical metal nanowires over a dielectric surface. In these simulations, the length of individual nanowires is given by $2a = 480$ nm, while their diameter is 30 nm. Figure 8.6 shows the intensity $|E|^2$ of the local electric field at wavelengths $\lambda = 540$ and 750 nm. Our simulations exhibit the existence of localized plasmon modes in such composites. Similar to localization of quasi-static plasmon modes [30], the localization of plasmon-polaritons bounded in the metal nanowires leads to large enhancement of local optical fields. Our simulations suggest that the local intensity enhancement factor reaches 10^3.

Our simulations also show that plasmon modes cover a broad spectral range. The incident field at a given wavelength excites small resonant parts of the percolation

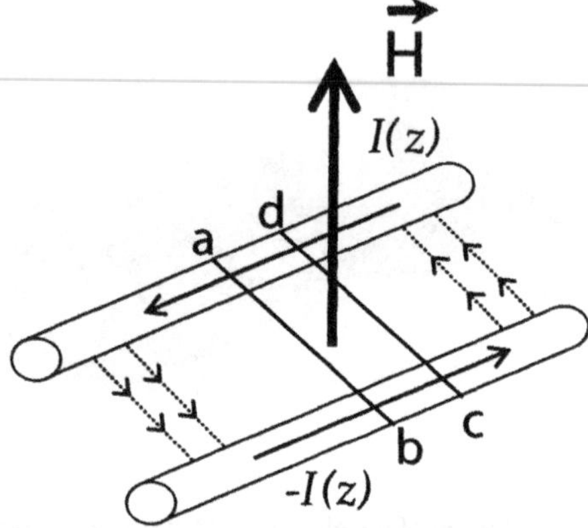

Fig. 8.7 Current in the two-stick circuit excited by external magnetic field **H**. The displacement currents, "closing" the circuit, are shown by dashed lines.

system, resulting in a large enhancement of the local fields in these elements. In our case, such resonating elements can be single nanowires or groups of nanowires. Different clusters of wires resonate at different frequencies and all together cover a broad spectral range where the stick composite has plasmon modes.

8.5 MAGNETIC RESPONSE OF CONDUCTING STICK COMPOSITES

We consider now a metal–dielectric composite consisting of pairs of parallel metal sticks embedded in a dielectric host. We assume that the volume concentration p of the sticks is less than the percolation threshold $p < p_c \sim b/a \ll 1$. We also suppose that neither the sticks nor the dielectric have magnetic properties. One might think that the composite has no magnetic properties under such conditions. Indeed, the magnetic response of a single conducting stick is small even at high frequencies (Ref. [18], Section 59). Since we have concentration $p \ll 1$, one could anticipate that the response of the entire composite is also small.

In reality, as we show below, the composite may have a giant magnetic response at some frequencies. The reason for such a behavior is the resonant response of the stick pairs to a high-frequency magnetic field. The external magnetic field excites electric currents in these stick pairs. The magnetic moments for the currents flowing in the stick pairs result in the magnetic response of the composite. Consider a pair of the sticks and suppose that an external magnetic field $H = H_0 \exp(-i\omega t)$ is applied perpendicular to their plane. This field excites a circular current I in the system of two parallel sticks. The circular current I flows in one stick in one direction and in the opposite direction in another stick as shown in Fig. 8.7. The displacement currents flowing between the two sticks close the circuit. The considered two-stick circuit acts as the well-known two-wire transmission line excited by an external magnetic field. The current I in the two-wire line can be calculated from the Telegrapher's equation (see, e.g., Refs. [17] and [31]). The electrodynamics processes in the line of two wires, separated by a distance d are determined by the impedance per unit length,

$$Z = \frac{2}{\sigma_m^* \pi b^2} - i\frac{\omega}{c^2} L_2 \tag{8.45}$$

where σ_m^* and b are the renormalized stick conductivity [see equation (8.21)] and radius, respectively. The parameter L_2 is the self-inductance per unit length for a system of two parallel straight wires having a cross section of radius b. We define the inductance L_2, following the procedure used to define the inductance of a single stick [see derivation of equation (8.16)]. Thus we obtain

$$L_2 = 4\ln(d/b) - (d/a)^2 + \frac{1}{6}(dk)^2 \left[3 + 4iak + 6\log(2a/d)\right] \tag{8.46}$$

where d is the distance between the axes of the wires. Another important parameter is the mutual capacity per unit length C_2 of two wires. The approach that has been used to define a capacitance of a single stick [see equation (8.8)] results in the following equation for C_2

$$C_2 = \frac{\varepsilon_d}{4\log(d/b) - 3(d/a)^2 + (dk)^2 \left[2\log(2a/d) - 1\right]/2} \tag{8.47}$$

where ε_d is the permittivity of the dielectric host. The capacitance C_2 determines the value of the displacement currents flowing between the two wires. Following the procedure described in Section 8.2, we introduce the current I as the current in a single stick. This current depends on the coordinate z along the stick. We also introduce the potential difference $U(z)$ between the two sticks. Using Faraday's Law

$$\oint_{(a,b,c,d)} \mathbf{E}\, d\mathbf{l} = i\frac{\omega}{c}\iint_S \mathbf{H}\, ds \tag{8.48}$$

where $S = d \times dz$ is the area restricted by the contour (a, b, c, d) as shown in Fig. 8.7, we find

$$-\frac{dU(z)}{dz} = ZI(z) + idkH_0 \tag{8.49}$$

The current $I(z)$ depends on the coordinate z since it can go out from one stick and come into another stick. The second equation for $I(z)$ and $U(z)$ is obtained from the charge conservation law. Considering the currents in the sticks and the displacement current between them we find

$$\frac{dI(z)}{dz} = i\omega C_2 U(z) \tag{8.50}$$

The combination of equations (8.49) and equation (8.50) gives the second-order differential equation for the current,

$$\frac{d^2 I(z)}{dz^2} = -g^2 I(z) + \frac{C_2 d\omega^2}{c} H_0 \tag{8.51}$$
$$-a < z < a, \quad I(-a) = I(a) = 0$$

where the parameter g equals

$$g^2 = k^2 \left[1 + \frac{1}{\log(d/b)} \left(\frac{d^2}{2\,a^2} + \frac{(dk)^2}{4} + \frac{i}{6}\,ak(dk)^2 \right) - \frac{8C_2}{(kb)^2 \varepsilon_m^*} \right] \tag{8.52}$$

and $\varepsilon_m^* = i4\pi\sigma_m^*/\omega$ is the renormalized metal permittivity [see equation (8.21)]. Note that we still assume that $d/a \ll 1$ and $dk \ll 1$. For the perfect metal ($|\varepsilon_m^*| \to \infty$), the parameter g does not depend on the metal properties. We solve equation (8.51) for the current $I(z)$ and calculate the magnetic moment $\mathbf{m}$ for the circuit in the two sticks,

$$\mathbf{m} = \frac{1}{2c} \int [\mathbf{r} \times \mathbf{j}(\mathbf{r})]\, d\mathbf{r} \tag{8.53}$$

where $\mathbf{j}(\mathbf{r})$ is the density of the current in the two conducting sticks and the density of the displacement currents. Integration in equation (8.53) goes over the two conducting sticks as well as over the space between them where the displacement currents are flowing. From equations (8.51)–(8.53) we obtain the magnetic moment for the system of two sticks:

$$m = 2H_0 a^3 C_2 (kd)^2 \frac{\tan(ga) - ga}{(ga)^3} \tag{8.54}$$

Let us now estimate quantitatively the effective magnetic permeability μ_e of the conducting stick composite. We suppose that the stick pairs are oriented in one direction. Taking into account the definition of the effective magnetic permeability $\mu_e \mathbf{H}_0 = \mathbf{H} + 4\pi\mathbf{M}$, where $\mathbf{M}$ is the magnetic moment per unite volume, we obtain from equation (8.54) the following equation for the component of μ_e perpendicular to the pairs:

$$\mu_e \approx 1 + 4\pi n \frac{m}{H_0} \approx 1 + 4\pi p \frac{a}{b} C_2 a d k^2 \frac{\tan(ga) - ga}{(ga)^3} \tag{8.55}$$

where n is the density of the stick pairs, $p = bdan$ is the volume concentration of the pairs, and parameters C_2 and g are given by equations (8.47) and (8.52), respectively. The effective magnetic permeability μ_e of the conducting stick composite is shown in Fig. 8.8 for the concentration $p = 0.2$. The permeability μ_e reaches its maximum at the resonance and becomes negative for the wavelength below the resonance. The length of a pair $2a = 400$ nm is much smaller than the resonance wavelength $\sim 2\,\mu$m. Therefore, the spatial dispersion effects, discussed at the end of Section 8.2, can be neglected and the composite has a well-defined magnetic permeability. Thus, composite materials formed by pairs of metal nanowires can act as left-handed material with negative refraction in the optical range.

8.6 PLANAR NANOWIRE COMPOSITES

In the sections above, we considered the response of conducting stick composites to the electric and magnetic fields. In this section, we consider a planar composite comprising regular array of pairs of parallel nanowires (see Fig. 8.9), which is

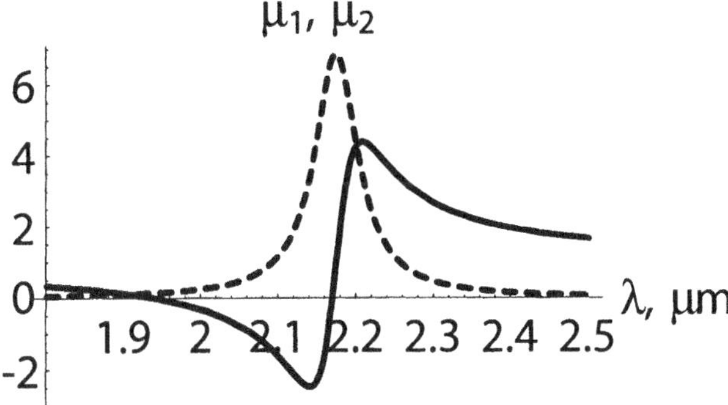

Fig. 8.8 Optical magnetic permeability $\mu = \mu_1 + \mu_2$ (μ_1, continuous line; μ_2, dashed line) of the composite containing pairs of silver sticks; $a = 200$ nm, $d = 50$ nm, $b = 10$ nm.

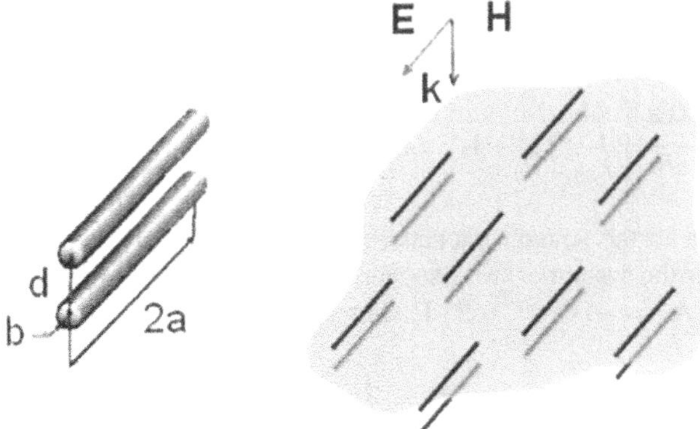

Fig. 8.9 A layer of pairs of parallel nanowires. After Ref. [4]. Copyright © 2003 Optical Society of America, Inc.

illuminated by a plane electromagnetic wave impinging perpendicular to the plane of the composite. First, we show that in the far zone the field scattered by pairs of nanowires can be approximated by the effective dipole and magnetic moments even when the size of the pair is comparable with the wavelength λ of the incident light. Then we consider the optical properties of a layer of such nanowire pairs.

Electric and magnetic fields at the distance R away from the nanowire pair with dimensions $2a \times d \times 2b$ (see Fig. 8.9) are derived from the vector potential $\mathbf{A}$ that for large distances, $R \gg \lambda, b_1, b_2, d$, takes the following standard form:

$$\mathbf{A} = \left(e^{ikr}/cR\right) \int e^{-ik(\mathbf{n}\cdot\mathbf{r})} \mathbf{j}(\mathbf{r})\, d\mathbf{r}$$

where $\mathbf{j}(\mathbf{r})$ is the current density inside the nanowires and vector $\mathbf{n}$ is the unit vector in the observation direction. We introduce the vector $\mathbf{d}$ directed from one nanowire to another, and assume that the coordinate system has its origin in the center of the system so that the centers of the wires have the coordinates $\mathbf{d}/2$ and $-\mathbf{d}/2$, respectively. The electromagnetic wave is incident in the plane of the wires perpendicular to them (see Fig. 8.9), that is, the wavevector $\mathbf{k} \parallel \mathbf{d}$. Then, the vector potential $\mathbf{A}$ can be written as

$$\mathbf{A} = \frac{e^{ikR}}{cR}\left[e^{-\frac{ik}{2}(\mathbf{n}\cdot\mathbf{d})}\int_{-b_1}^{b_1} e^{-\frac{ik}{2}(\mathbf{n}\cdot\mathbf{z})}\mathbf{j}_1(\rho)\, d\rho + e^{\frac{ik}{2}(\mathbf{n}\cdot\mathbf{d})}\int_{-b_1}^{b_1} e^{-ik\mathbf{n}\cdot\mathbf{z}}\mathbf{j}_2(\rho)\, d\rho\right] \tag{8.56}$$

where $\mathbf{j}_1$ and $\mathbf{j}_2$ are the currents in the wires, and $\mathbf{z}$ is the coordinate along the wires ($\mathbf{z} \perp \mathbf{d}$). As known, the dipole component is dominated in scattering by a thin antenna even for an antenna size comparable to the wavelength (see, e.g., Ref. [34]). Therefore we can approximate the term $e^{-ik\mathbf{n}\cdot\mathbf{z}}$ in equation (8.56) by unity. Note that for the forward and backward scattering, which are responsible for the effective properties of a medium, this term is exactly equal to one.

We consider the system where the distance d between the wires is much smaller than the wavelength and expand equation (8.56) in a series over d. This results in

$$\mathbf{A} = \frac{e^{ikR}}{cR}\left[\int_{-b_1}^{b_1} (\mathbf{j}_1 + \mathbf{j}_2)\, d\rho - \frac{ik}{2}(\mathbf{n}\cdot\mathbf{d})\int_{-b_1}^{b_1} (\mathbf{j}_1 - \mathbf{j}_2)\, d\rho\right] \tag{8.57}$$

The first term in the square brackets in equation (8.57) gives the effective dipole moment $\mathbf{P}$ for the system of two nanowires and its contribution to the scattering can be written as $\mathbf{A}_d = -ik\left(e^{ikR}/R\right)\mathbf{P}$, where

$$\mathbf{P} = \int \mathbf{p}(\mathbf{r})\, d\mathbf{r} \tag{8.58}$$

and $\mathbf{p}$ is the local polarizations. The integration in equation (8.58) is over the volume of both wires. The second term in equation (8.57) gives the magnetic dipole and quadrupole contributions to the vector potential:

$$\mathbf{A}_{mq} = \frac{ike^{ikR}}{R}\left[[\mathbf{n}\times\mathbf{M}] - \frac{\mathbf{d}}{2c}\int_{-b_1}^{b_1} (\mathbf{n}\cdot(\mathbf{j}_1 - \mathbf{j}_2))\, d\rho\right] \tag{8.59}$$

where $\mathbf{M}$ is the magnetic moment of two wires,

$$\mathbf{M} = \frac{1}{2c}\int [\mathbf{r}\times\mathbf{j}(\mathbf{r})]\, d\mathbf{r} \tag{8.60}$$

and the integration is over the volume of the wires as in equation (8.58).

We show now results of our numerical simulations for the optical properties of gold nanowires (Fig. 8.10). According to our simulations, both the dielectric and magnetic moments excited in the nanowire system are opposite to the excited field when the wavelength of the incident field is below resonance. Thus, in this frequency range, a composite material based on parallel nanowire pairs have the dielectric permittivity

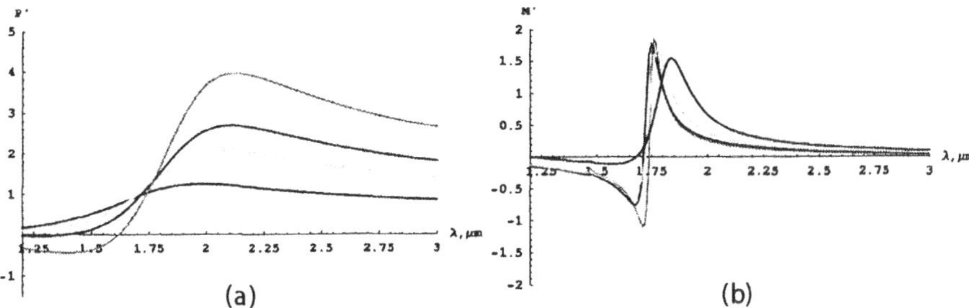

Fig. 8.10 Dielectric (a) and magnetic (b) moments in nanowire pairs as functions of wavelengths. The distance between the nanowires in the pairs is varied: $d = 0.15\,\mu$m (1), $d = 0.23\,\mu$m (2), $d = 0.3\,\mu$m (3), and $d = 0.45\,\mu$m (4); for all plots, $a = 0.35\,\mu$m and $b = 0.05\,\mu$m. The moments are normalized to the unit volume. After Ref. [4]. Copyright © 2003 Optical Society of America, Inc.

and magnetic permeability both negative and thus the composite acts as a left-handed material. These results are in good qualitative agreement with equations (8.31) and (8.54) which were derived for the case of needles with a high aspect ratio.

We consider now the transmittance and reflectance of a planar nanowire composite when an electromagnetic wave impinges normal to its plane. We take into account the dipole $\mathbf{P}$ and magnetic $\mathbf{M}$ moments given by equations (8.58) and (8.60), respectively, since they are responsible for the main contribution to forward and backward scattering. The second term in equation (8.59) describes a quadrupole contribution, which vanishes for the forward direction. (see discussion in Ref. [4]).

Maxwell's equations for the composite can be written in the following form

$$\operatorname{curl}\mathbf{E} = ik\mathbf{H}, \qquad \operatorname{curl}\mathbf{H} = \frac{4\pi}{c}\mathbf{j} - ik\mathbf{E} \tag{8.61}$$

where $\mathbf{j}$ is the current in the nanowires. We split the current $\mathbf{j}$ in two parts $\mathbf{j} = \mathbf{j}_P + \mathbf{j}_M$. Here $\mathbf{j}_M$ is the circular current in the nanowire pair. This current can be presented as $\mathbf{j}_M = c\operatorname{curl}\mathbf{m}$ with the vector $\mathbf{m}$ vanishing outside the composite. Then equations (8.61) can be rewritten as

$$\operatorname{curl}\mathbf{E} = ik\left(\mathbf{H}' + 4\pi\mathbf{m}\right), \qquad \operatorname{curl}\mathbf{H}' = \frac{4\pi}{c}\mathbf{j}_P - ik\mathbf{E} \tag{8.62}$$

where $\mathbf{H}' = \mathbf{H} - 4\pi\mathbf{m}$. We suppose that $z = 0$ is the principal plane of the composite and the electromagnetic wave is incident along the z-axis. We average equations (8.62) over the $\{x, y\}$ plane and integrate them over the space between the two reference planes placed in front ($z = -a$) and behind ($z = a$) the composite. The distance a is chosen so that $d \ll a \ll 1/k$. After the integration, equations (8.62) take the following form

$$\mathbf{E}_2 - \mathbf{E}_1 \cong ik4\pi dp\mathbf{M}_1, \quad \mathbf{H}_2 - \mathbf{H}_1 = -ik4\pi dp\mathbf{P}_1 \tag{8.63}$$

where $\mathbf{E}_1 = \mathbf{E}(-a)$, $\mathbf{E}_2 = \mathbf{E}(a)$, $\mathbf{H}_1 = \mathbf{H}(-a)$, $\mathbf{H}_2 = \mathbf{H}(a)$; $\mathbf{P}_1 = \mathbf{P}/(b_1 b_2 d)$ and $\mathbf{M}_1 = \mathbf{M}/(b_1 b_2 d)$ are the dipole and magnetic moments of the nanowire pairs. These moments are given by equations (8.58) and (8.60) which are normalized to the volume of the pairs, p is the filling factor, that is, the ratio of the area covered by the nanowires and the total area of the film.

The moments $\mathbf{P}_1$ and $\mathbf{M}_1$ are proportional to the effective electric and magnetic fields, respectively. For the dilute case ($p \ll 1$) considered here we can write $\mathbf{P}_1$ and $\mathbf{M}_1$ as $4\pi p \mathbf{P}_1 = \varepsilon(\mathbf{E}_2 + \mathbf{E}_1)/2$ and $4\pi p \mathbf{M}_1 = \mu(\mathbf{H}_2 + \mathbf{H}_1)/2$, where the coefficients ε and μ are the effective dielectric constant and magnetic permeability of the nanowire composite. Then equations (8.63) take the following form:

$$\mathbf{E}_2 - \mathbf{E}_1 \cong ikd\mu(\mathbf{H}_1 + \mathbf{H}_2), \quad \mathbf{H}_2 - \mathbf{H}_1 = -ikd\varepsilon(\mathbf{E}_1 + \mathbf{E}_2) \tag{8.64}$$

We match equation (8.64) at $z = -a$ with the plane wave solution

$$E = E_0[\exp(ikz) + r\exp(-ikz)]$$

that holds in front of the film ($z < -a$) and match equations (8.64) at $z = a$ with the solution $E = E_0 t\exp(ikz)$ that holds behind the film. E_0 is the amplitude of the impinging wave, r and t are reflection and transmission coefficients, respectively. This matching results in two equations for r and t. Solutions to these equations allow us to find the reflection R and transmittance TT coefficients of the nanowire composite in the following form

$$R = \left|\frac{2dk(\varepsilon - \mu)}{(-2 + idk\varepsilon)(-2 + idk\mu)}\right|^2, \quad T = \left|\frac{4 + d^2k^2\varepsilon\mu}{(-2 + idk\varepsilon)(-2 + idk\mu)}\right|^2 \tag{8.65}$$

When $\varepsilon = \mu$, the reflectance vanishes while the transmittance is given by $T = |(2 + idk\varepsilon)/(2 - idk\varepsilon)|^2$. If $\varepsilon = \mu$ and it is a real number, the reflectance $T = 1$. Still, the interaction of the electromagnetic wave with the composite results in the phase shift $2\arctan(dk\varepsilon/2)$ for the transmitted wave. The phase shift is positive if $\varepsilon = \mu > 0$ and the shift is negative when $\varepsilon = \mu < 0$. The last case corresponds to a left-handed material. Thus, a negative phase of the transmitted electromagnetic wave indicates the left-handedness of the composite.

8.7 CONCLUSIONS

We presented a detailed study of the electrodynamic properties of metal–dielectric composites consisting of elongated conducting inclusions—conducting sticks—embedded in a dielectric host. Conducting stick composites have new and unusual properties at high frequencies when surface plasmon-polaritons are excited in the sticks. The effective dielectric permittivity has strong resonances at some frequencies. The real part vanishes at the resonance and acquires negative values for frequencies above resonance. The dispersion behavior does not depend on the stick conductivity and takes the universal form when the stick conductivity tends to infinity. The surface plasmon-polaritons can be localized in such composites.

We show that composites consisting of nonmagnetic inclusions may have a large magnetic response in the optical spectral range. The effective magnetic response is strong in a composite comprising pairs of parallel nanowires and results from the collective interactions of the nanowires in pairs with an external magnetic field. A giant paramagnetic response can occur in this case, in some frequency ranges. The composite materials based on plasmonic nanowires can have a negative refractive index and thus act as left-handed materials in the optical range of the spectrum.

Acknowledgments

The authors acknowledge useful contributions from and discussions with Drs. Podolskiy and Drachev. This work was supported in part by NSF grants ECS-0210445 and DMR-0121814.

REFERENCES

1. J. B. Pendry, "Negative refraction makes a perfect lens," *Phys. Rev. Lett.*, vol. 85, no. 18, pp. 3966–3969, 2000.

2. V. G. Veselago, "The electrodynamics of substances with simultaneously negative values of ϵ and μ," *Sov. Phys. Usp.*, vol. 10, no. 4, pp. 509–514, 1968.

3. D. R. Smith, W. J. Padilla, D. C. Vier, S. C. Nemat-Nasser, and S. Shultz, "Composite medium with simultaneously negative permeability and permittivity," *Phys. Rev. Lett.*, vol. 84, no. 18, pp. 4184–4187, 2000.

4. V. A. Podolskiy, A. K. Sarychev, and V. M. Shalaev, "Plasmon modes and negative refraction in metal nanowire composites," *Opt. Express*, vol. 11, no. 7, pp. 735–745, 2003.

5. A. A. Houck, J. B. Brock, and I. L. Chuang, "Experimental observations of a left-handed material that obeys Snell's Law," *Phys. Rev. Lett.*, vol. 90, 137401, April 3, 2003.

6. C. G. Parazzoli, R. B. Greegor, K. Li, B. E. C. Koltenbah, and M. Tanielian, "Experimental verification and simulation of negative index of refraction using Snell's Law," *Phys. Rev. Lett.*, vol. 90, 107401, March 11, 2003.

7. V. A. Podolskiy, A. K. Sarychev, and V. M. Shalaev, "Plasmon modes in metal nanowires and left-handed materials," *J. Nonlin. Opt. Phys. Mater.*, vol. 11, no. 3, pp. 65–74, 2002.

8. A. N. Lagarkov and A. K. Sarychev, "Electromagnetic properties of composites containing elongated conducting inclusions," *Phys. Rev. B*, vol. 53, pp. 6318–6336, March 1996.

9. L. V. Panina, A. N. Grigorenko, and D. P. Makhnovskiy, "Optomagnetic composite medium with conducting nanoelements," *Phys. Rev. B*, vol. 66, 155411, October 2002.

10. A. P. Vinogradov, L. V. Panina, and A. K. Sarychev, "Method for calculating the dielectric constant and magnetic permeability in percolation systems," *Sov. Phys. Dokl.*, vol. 34, no. 6, pp. 530–532, 1989.

11. A. N. Lagarkov, A. K. Sarychev, Y. R. Smychkovich, and A. P. Vinogradov, "Effective medium theory for microwave dielectric constant and magnetic permeability of conducting stick composites," *J. Electromagn. Waves Appl.*, vol. 6, no. 9, p. 1159, 1992.

12. D. Rousselle, A. Berthault, O. Acher, J. P. Bouchaud, and P. G. Zerah,"Effective medium at finite frequency: Theory and experiment," *J. Appl. Phys.*, vol. 74, no. 1, pp. 475–479, 1993.

13. A. N. Lagarkov, S. M. Matitsine, K. N. Rozanov, and A. K. Sarychev, "Dielectric properties of fiber-filled composites," *J. Appl. Phys.*, vol. 84, pp. 3806–3814, October 1998.

14. D. P. Makhnovskiy, L. V. Panina, D. J. Mapps, and A. K. Sarychev, "Effect of transition layers on the electromagnetic properties of composites containing conducting fibres," *Phys. Rev. B.*, vol. 64, 134205, September 11, 2001.

15. S. M. Matitsine, K. M. Hock, L. Liu, Y. B. Gan, A. N. Lagarkov, and K. N. Rozanov, "Shift of resonance frequency of long conducting fibers embedded in a composite," *J. Appl. Phys.*, vol. 94, pp. 1146–1154, July 15, 2003.

16. L. A. Vainshtein, *Electromagnetic Waves*, 2nd ed., Radio and Telecommunications, Moscow, 1988.

17. E. Hallen, *Electromagnetic Theory*, Chapman and Hall, London, 1962.

18. D. Landau and E. M. Lifshitz, *Electrodynamics of Continuous Media*, 2nd ed., Pergamon, Oxford, 1984.

19. J. A. Stratton, "Spheroidal functions," *Proc. Natl. Acad. Sci. USA*, vol. 21, pp. 51–56, 1935.

20. A. N. Lagarkov, S. M. Matitsine, K. N. Rozanov, and A. K. Sarychev, "Dielectric permittivity of fiber-filled composites: comparison of theory and experiment," *Physica A*, vol. 241, pp. 58–63, 1997.

21. C. A. Grimes, C. Mungle, D. Kouzoudis, S. Fang, and P. C. Eklund, "The 500 MHz to 5.50 GHz complex permittivity spectra of single-wall carbon nanotube-loaded polymer composites," *Chem. Phys. Lett.*, vol. 319, pp. 460–464, March 2000.

22. C. A. Grimes, E. C. Dickey, C. Mungle, K. G. Ong, and D. Qian, "Effect of purification of the electrical conductivity and complex permittivity of multiwall carbon nanotubes," *J. Appl. Phys.*, vol. 90, pp. 4134–4137, 2001.

23. D. J. Bergman and D. Stroud, "The physical properties of macroscopically inhomogeneous media," *Solid State Phys.*, vol. 46, pp. 148–270, 1992.

24. D. A. G. Bruggeman, "The calculation of various physical constants of heterogeneous substances. I. The dielectric constants and conductivities of mixtures composed of isotropic substances," *Ann. Phys. (Leipzig)*, vol. 24, pp. 636–664, 1935.

25. R. Landauer, "Electrical conductivity in inhomogeneous media," in *Electrical Transport and Optical Properties of Inhomogeneous Media*, J. C. Garland and D. B. Tanner, eds., AIP Conference Proceedings No. 40, AIP, New York, 1978, p. 2.

26. E. M. Purcell and C. R. Pennypacker, "Scattering and absorption of light by nonspherical dielectric grains," *Astrophys. J.*, vol. 186, p. 705, 1973.

27. B. T. Draine, "The discrete dipole approximation and its application to interstellar graphite grains," *Astrophys. J.*, vol. 333, p. 848, 1988.

28. M. Moskovits, private communication.

29. N. Yamamoto, K. Araya, M. Nakano, and F. J. Garcia de Abajo, OSA meeting 2002, Orlando, FL.

30. A. K. Sarychev and V. M. Shalaev, "Electromagnetic field fluctuations and optical nonlinearities in metal-dielectric composites," *Phys. Rep.*, vol. 333, p. 275, 2000.

31. J. A. Stratton, *Electromagnetic Theory*, McGraw-Hill, New York, 1941.

32. A. K. Sarychev, V. P. Drachev, H. K. Yuan, V. A. Podolskiy, and V. M. Shalaev, "Optical properties of metal nanowires," in *Nanotubes and Nanowires*, SPIE Proceedings Vol. 5219, 2003, pp. 92–98.

33. W. Gotschy, K. Vonmetz, A. Leither, and F. R. Aussenegg, "Thin films by regular patterns of metal nanoparticles: Tailoring the optical properties by nanodesign," *Appl. Phys. B*, vol. 63, no. 4, pp. 381–384, 1996.

34. J. D. Jackson, *Classical Electrodynamics*, John Wiley & Sons, New York, 1999.

9 An Overview of Salient Properties of Planar Guided-Wave Structures with Double-Negative (DNG) and Single-Negative (SNG) Layers

ANDREA ALÙ*,† and NADER ENGHETA*

*Department of Electrical and Systems Engineering
University of Pennsylvania
Philadelphia, PA 19104
United States

†Department of Applied Electronics
Università di Roma Tre
84-00146 Roma
Italy

In this chapter, we provide an overview of some of the results of our analytical studies of wave guiding properties of planar structures containing double-negative (DNG) and single-negative (SNG) metamaterials, in which, respectively, both or one of the material parameters (i.e., permittivity and permeability) possess negative real parts in a certain band of frequency. We have shown how guided modes in such structures exhibit unusual features. In particular, we discuss and review our theoretical results on planar waveguides with DNG and SNG metamaterials and their unconventional features, such as the possibility of guiding electromagnetic energy with lateral dimensions below the diffraction limit, large-aperture monomodal waveguiding using conjugate SNG bilayers, modal excitation by current sources and power flow distribution in the partially filled DNG waveguides, dispersion peculiarities of surface wave propagation along the DNG open slab waveguides, and anomalous contradirectional (backward) coupling between the DNG and dielectric open planar waveguides. Physical insights into our findings are also given.

9.1 INTRODUCTION

Since the pioneering work of Smith, Schultz, and their co-workers [1], in which they, inspired by the work of Pendry [2, 3], constructed a composite medium exhibiting negative refraction resulting from the negative real parts for its effective permittivity and permeability in the microwave regime, the interest in the fundamental aspects and the potential applications of these exotic metamaterials has grown considerably. The features of such a double-negative (DNG) medium (which is also referred to as a left-handed (LH) medium) had been postulated almost forty years ago by Veselago [4], but since then no sample of such a material had been found in nature or synthesized until the work reported in Ref. 1. Various properties of this class of metamaterials are now being studied by several groups worldwide, verifying their anomalous physical properties and salient features, and many ideas and suggestions for their potential applications have been proposed (see, e.g., Refs. 1–55). As one such idea, in some of our earlier works we theoretically suggested the possibility of having ultrathin, subwavelength cavity resonators in which a layer of the DNG medium is paired with a layer of conventional material (i.e., a "double-positive (DPS)" medium) [37–39]. Owing to the antiparallel nature of the phase velocity and Poynting vectors in a DNG slab, we theoretically found the possibility of resonant modes in electrically thin parallel-plate structures containing such DNG–DPS bilayer structures [37–40]. We then extended our work to fully analyze the guided modes in parallel-plate waveguides containing a pair of DNG and DPS slabs [41, 42]. Later in Ref. 43, we showed the effects of the anomalous mode coupling between DNG and DPS open waveguides located parallel to, and in proximity of, each other. Some other research groups have also explored certain aspects of waveguides involving DNG media (see, e.g., Refs. 49–53).

We have also been interested in exploring the properties of "single-negative (SNG)" materials in which *only one* of the material parameters, *not both*, has a negative real value. These SNG media also exhibit interesting properties when they are paired in a conjugate manner. These media include the epsilon-negative (ENG) media, in which the real part of permittivity is negative but the real permeability is positive, and the mu-negative (MNG) media, in which the real part of permeability is negative but the real permittivity is positive. The idea of constructing an effective DNG medium by having layers of SNG media has been explored by Fredkin and Ron [48]. We have analyzed in detail the wave reflection from and transmission through a pair of juxtaposed ENG and MNG slabs, revealing interesting properties such as resonances, transparency, anomalous tunneling, and zero reflection [47]. In our analysis, we also utilized appropriate distributed circuit elements in the transmission-line model for the pair of ENG–MNG layers, and we have theoretically explained and justified the unusual field behavior in these paired ENG–MNG structures. We have also shown that such lossless pairs may exhibit "interface resonance" phenomena, even though each slab alone does not manifest such an effect [47]. As a further contribution to the topic of wave interaction with SNG and DNG media, in our earlier work on metamaterial planar waveguides [42, 54] we studied the properties of guided modes in waveguides filled with a pair of SNG layers, showing possible unconventional features for such guided modes.

In this chapter, we review some of our results on guided modes in planar waveguides with DNG and/or SNG layers, which we have already reported elsewhere and the interested reader is referred to for detailed information, and in addition we present the analysis for certain other planar guided-wave structures containing these metamaterials. We discuss how their unconventional features may be utilized to overcome certain physical limitations present in conventional waveguides with standard DPS materials: for example, we note the possibility of guiding of waves in ultra-thin structures with lateral dimension below the diffraction limits. Other notable features are also discussed here.

Although the structures considered here are simple 2-D geometries, their waveguiding features indeed provide physical insights into exciting ideas and characteristics for other waveguide geometries filled with such metamaterials, with potential applications in the design of future devices and components for miniaturization of RF and optical interconnects.

In our analysis described in this chapter, all materials are assumed to be lossless, homogeneous, and isotropic, unless otherwise specified. Furthermore, it is worth noting that passive SNG and DNG metamaterials are inherently dispersive [4,5], and thus their material parameters may vary considerably with the frequency. Therefore, for the sake of simplicity, we fix the frequency of operation ω and we consider the values of permittivity and permeability of SNG, DNG, and DPS materials at this given frequency. All other parameters of the waveguides—for example, layer thicknesses and longitudinal wavenumbers—may vary.

9.2 PARALLEL-PLATE WAVEGUIDES WITH DNG AND SNG METAMATERIALS

In this section we briefly review our results revealing some of the anomalous features of parallel-plate waveguides partially or totally filled with layers of metamaterials. The interested reader is referred to our work for further detail [42]. The geometry is depicted in Fig. 9.1, where the region between two infinite perfectly electric conducting (PEC) plates separated by the distance $d = d_1 + d_2$ is filled with a pair of parallel layers made of any two of ENG, MNG, DNG, and DPS materials. A monochromatic time-harmonic excitation $e^{j\omega t}$ is assumed. The two slabs are characterized by their thicknesses d_1 and d_2 and their constitutive parameters ε_1, μ_1 and ε_2, μ_2, which are assumed real. In our general formulation, we do not assign any specific sign to these parameters. Later, however, we show how different signs will lead to different properties for guided modes. The Cartesian coordinate system (x, y, z) is shown in Fig. 9.1, and x is chosen as the direction of propagation of guided modes. As we have done in Ref. 42, satisfying the appropriate boundary conditions at $y = d_1$ and $y = -d_2$, the electric and magnetic field expressions for the TE^x mode

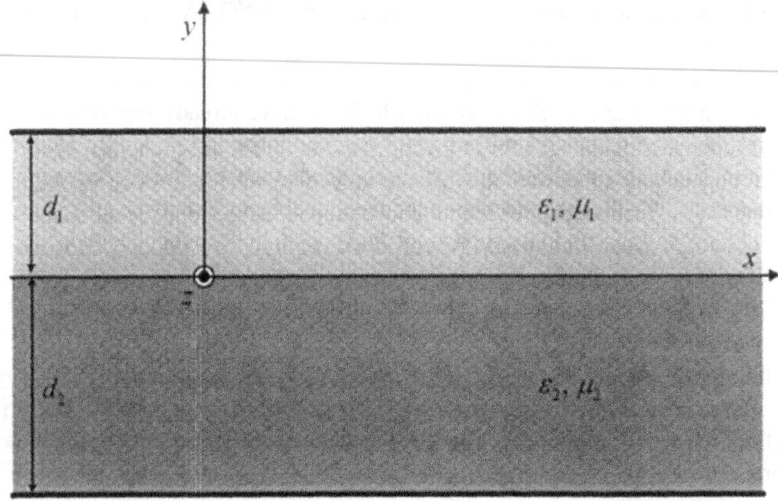

Fig. 9.1 Geometry of the parallel-plate waveguide filled with a pair of layers made of any two of epsilon-negative (ENG), mu-negative (MNG), double-negative (DNG), and double-positive (DPS) materials. After Ref. [42]. Copyright © 2004 IEEE.

are written as

$$E^{TE} = \hat{z}\, E_0^{TE}\, e^{-j\beta_{TE}x} \cdot \begin{cases} \sin\left(k_{t2}^{TE} d_2\right) \sin\left[k_{t1}^{TE}\left(d_1 - y\right)\right], & y > 0 \\ \sin\left(k_{t1}^{TE} d_1\right) \sin\left[k_{t2}^{TE}\left(y + d_2\right)\right], & y < 0 \end{cases} \tag{9.1}$$

$$H^{TE} = j\,\hat{x}\,\omega^{-1} E_0^{TE} e^{-j\beta_{TE}x} \cdot \begin{cases} -\mu_1^{-1} k_{t1}^{TE} \sin\left(k_{t2}^{TE} d_2\right) \cos\left[k_{t1}^{TE}\left(d_1 - y\right)\right] \\ \mu_2^{-1} k_{t2}^{TE} \sin\left(k_{t1}^{TE} d_1\right) \cos\left[k_{t2}^{TE}\left(y + d_2\right)\right] \end{cases}$$

$$-\hat{y}\,\omega^{-1}\beta_{TE} E_0^{TE} e^{-j\beta_{TE}x} \cdot \begin{cases} \mu_1^{-1} \sin\left(k_{t2}^{TE} d_2\right) \sin\left[k_{t1}^{TE}\left(d_1 - y\right)\right], & y > 0 \\ \mu_2^{-1} \sin\left(k_{t1}^{TE} d_1\right) \sin\left[k_{t2}^{TE}\left(y + d_2\right)\right], & y < 0 \end{cases} \tag{9.2}$$

where E_0^{TE} is the mode amplitude, determined by the excitation, and $k_{ti}^{TE} = \sqrt{k_i^2 - \beta_{TE}^2}$ with $k_i^2 = \omega^2 \mu_i \varepsilon_i$ for $i = 1, 2$. The corresponding expressions for the TM^x modes may be easily obtained through duality. For ENG and MNG slabs, where one of the material parameters is negative, we have $k_i^2 < 0$, and for propagating modes with real β the transverse wavenumber k_{ti} is always imaginary. However, for DPS and DNG slabs, $k_i^2 > 0$ and the transverse wavenumber k_{ti} may be real or imaginary, depending on the value of β. The field expressions in (9.1) and (9.2) and the corresponding expressions for the TM case, however, remain valid for any such cases. By applying the boundary conditions for the tangential components of the electric and magnetic fields at the interface $y = 0$, one finds the following two dispersion relations for the TE and TM modes, respectively:

$$\frac{\mu_1}{k_{t1}^{TE}} \tan\left(k_{t1}^{TE} d_1\right) = -\frac{\mu_2}{k_{t2}^{TE}} \tan\left(k_{t2}^{TE} d_2\right) \tag{9.3}$$

$$\frac{\varepsilon_1}{k_{t1}^{TM}} \cot\left(k_{t1}^{TM} d_1\right) = -\frac{\varepsilon_2}{k_{t2}^{TM}} \cot\left(k_{t2}^{TM} d_2\right) \tag{9.4}$$

Depending on the choice of material parameters, the above dispersion relations reveal interesting characteristics for the guided modes present in this waveguide. A complete analysis of these features of guided modes for different metamaterials may be found in Ref. 42. In the rest of this section, we will highlight and summarize some of the unusual properties of propagating guided modes in parallel-plate waveguides filled with pairs of ENG, MNG, and/or DNG slabs.

9.2.1 Large-Aperture Monomodal Waveguides with ENG–MNG Pairs

It is well known that in conventional waveguides filled with standard DPS materials a large aperture (i.e., the distance between the two plates, $d = d_1 + d_2$) provides an easier coupling of the incident plane-wave energy with the waveguide. However, at the same time this may present a disadvantage of exciting several propagating guided modes, which may cause signal dispersion and limitation on the communication performance. Therefore, usually the waveguide aperture is carefully chosen to allow the guide to operate above the dominant cutoff frequency (in order to have at least one mode propagating in the structure), but below the second- and higher-order cutoff frequencies (to avoid multimodal propagation). When the ENG–MNG pairs are chosen to fill the waveguide in Fig. 9.1, however, this limitation may be overcome. Even a wide aperture, in fact, may allow monomodal propagation inside such a waveguide, since (9.3) and (9.4) may admit only one single solution for real β, as we have studied in Ref. 42.

If a parallel-plate waveguide, in fact, is totally filled with only an ENG or MNG material, it cannot support any propagating mode (as is intuitively expected). However, pairing of the ENG–MNG materials allows an "interface resonance" at the interface between these two conjugate slabs, providing the possibility of propagating guided modes in such a waveguide [42, 54]. In particular, for an ENG–MNG waveguide, equations (9.3) and (9.4) may be solved for the thickness d_2 as a function of the other waveguide parameters, yielding

$$d_2^{TE} = \frac{\tanh^{-1}\left[\frac{|\mu_1|\sqrt{|k_2|^2+\beta_{TE}^2}}{|\mu_2|\sqrt{|k_1|^2+\beta_{TE}^2}} \tanh\left(\sqrt{|k_1|^2+\beta_{TE}^2}\, d_1\right)\right]}{\sqrt{|k_2|^2+\beta_{TE}^2}} \tag{9.5}$$

$$d_2^{TM} = \frac{\tanh^{-1}\left[\frac{|\varepsilon_2|\sqrt{|k_1|^2+\beta_{TM}^2}}{|\varepsilon_1|\sqrt{|k_2|^2+\beta_{TM}^2}}\tanh\left(\sqrt{|k_1|^2+\beta_{TM}^2}d_1\right)\right]}{\sqrt{|k_2|^2+\beta_{TM}^2}} \tag{9.6}$$

Owing to the single-valuedness of the hyperbolic tangent functions, these dispersion relations lead to single solutions, differently from the standard DPS case, for which the counterparts to equations (9.5) and (9.6) would have been multivalued. This implies that, if propagating modes can exist in such a waveguide (and this depends on the following conditions

$$\tanh\left(\sqrt{\beta_{TE}^2+|k_1|^2}d_1^{TE}\right) < \frac{|\mu_2|\sqrt{\beta_{TE}^2+|k_1|^2}}{|\mu_1|\sqrt{\beta_{TE}^2+|k_2|^2}} \tag{9.7}$$

$$\tanh\left(\sqrt{\beta_{TM}^2+|k_1|^2}d_1^{TM}\right) < \frac{|\varepsilon_1|\sqrt{\beta_{TM}^2+|k_2|^2}}{|\varepsilon_2|\sqrt{\beta_{TM}^2+|k_1|^2}} \tag{9.8}$$

derived in Ref. 42), these modes are of finite number, which does not increase if the waveguide thickness is increased. Several combinations of the waveguide parameters, in particular, show a single-solution for the dispersion relations (9.5) and (9.6), suggesting a monomodal propagation in such waveguides, independent of the total aperture size (e.g., even with a very large aperture). It should be noted that the field distribution is concentrated around the interface between the conjugate slabs (e.g., at the ENG–MNG interface) and decays towards the metallic walls, giving rise to a surface-plasmon-like propagation for these modes and thus justifying why an increase in $d = d_1 + d_2$ beyond a certain limit does not sensibly affect the modal properties of this waveguide. Figure 9.2 depicts the dispersion diagrams [i.e., equation (9.5) with d_1, d_2 and β_{TE} as parameters] for the single TE mode in two different ENG–MNG waveguides. In panel (a), the interface between the ENG–MNG media does not support a surface wave, whereas in the case shown in panel (b) it does, and the corresponding β_{TE} for the surface wave represents an asymptote for the required value of d_2 (since for this value of β_{TE} the second metallic plate may be infinitely far). In Fig. 9.2, the value of

$$d_2^* = \frac{1}{|k_2|}\tanh^{-1}\left(\sqrt{\frac{\mu_1\varepsilon_2}{\mu_2\varepsilon_1}}\right) \tag{9.9}$$

has been shown, which represents the particular case of a one-dimensional (1-D) cavity resonator with $\beta_{TE} = 0$ and d_1 being arbitrarily large. In this case, no other resonant mode is supported in this 1-D cavity, regardless of the thickness of the first slab, implying that an ultra-thin or very thick 1-D cavity with a single resonant mode is possible.

ENG-MNG

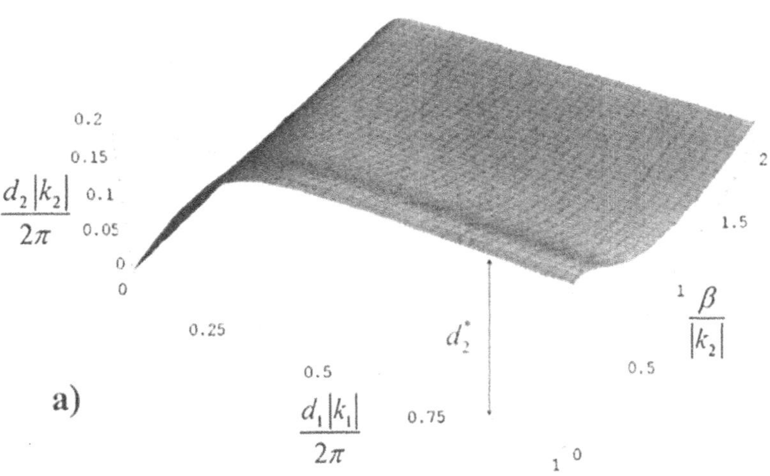

ENG-MNG

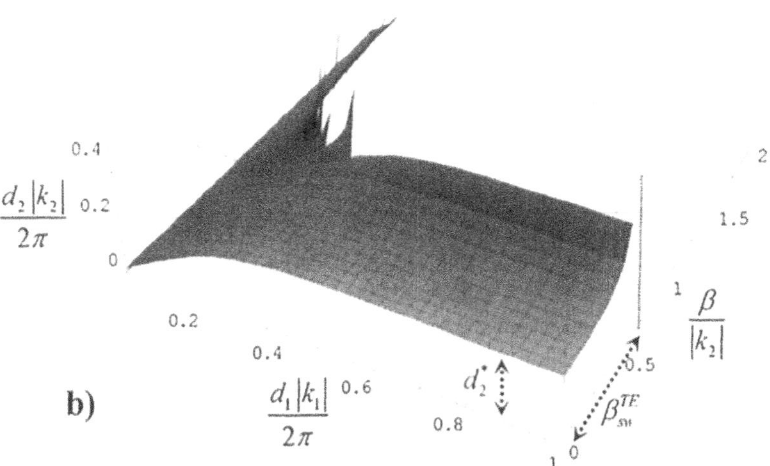

Fig. 9.2 Dispersion diagram for TE mode in an ENG–MNG waveguide, illustrating the relationship among normalized d_1 and d_2 and normalized real-valued β_{TE}, as described in (9.5), for two sets of material parameters for a pair of ENG–MNG slabs at a given frequency: (a) when $\varepsilon_1 = -2\varepsilon_0$, we have $\mu_1 = \mu_0$, $\varepsilon_2 = 2\varepsilon_0$, and $\mu_2 = -2\mu_0$; and (b) when $\varepsilon_1 = -5\varepsilon_0$, we have $\mu_1 = 2\mu_0$, $\varepsilon_2 = 2\varepsilon_0$, and $\mu_2 = -\mu_0$. The set of material parameters chosen in (a) does not allow a TE surface wave at the ENG–MNG interface, while the set chosen in (b) does. The value of d_2^* is given by (9.9). After Ref. [42]. Copyright © 2004 IEEE.

9.2.2 Waveguiding in Ultra-thin Structures with Lateral Dimension Below Diffraction Limits

In conventional waveguides, guided modes may not be supported with the lateral dimensions below a certain limit, generally close to $\lambda/2$. Specifically, in a closed waveguide with a metallic wall, reducing the cross-sectional size of the guide leads to cutting off the modes (and eventually the dominant mode), whereas in an open conventional dielectric waveguide, decreasing the size of the cross section results in broadening the lateral spread of the guided modes (and eventually leading to leakage of modes). This limitation may be removed by employing pairs of DNG, DPS, and/or SNG metamaterials filling the waveguide of Fig. 9.1, as it is described below. (For the open waveguides, this issue is discussed in Section 9.4.)

If the thicknesses $|k_1|\,d_1$ and $|k_2|\,d_2$ are assumed to be very small, equations (9.3) and (9.4) may be approximated, respectively, by

$$\gamma \simeq -\frac{\mu_2}{\mu_1} \tag{9.10}$$

$$\beta_{TM} \simeq \pm\omega\sqrt{\frac{\mu_1\gamma + \mu_2}{\gamma/\varepsilon_1 + 1/\varepsilon_2}} \tag{9.11}$$

where γ is shorthand for d_1/d_2 and obviously should always be a positive quantity. We note the fact that these approximate expressions are valid for thin waveguides loaded with any pair of slabs, since they have been obtained directly from the general dispersion relations (9.3) and (9.4). This point is physically justified considering the fact that in *thin* waveguides the *transverse* behavior of the field, which determines the possibility of a mode to propagate, is similar for DPS, DNG, and SNG materials, since the hyperbolic and trigonometric sinusoidal functions have somewhat similar behavior in the limit of small arguments.

For a thin waveguide filled with a pair of DPS–DPS layers (and similarly with a pair of ENG–ENG, DPS–ENG, MNG–MNG, DNG–MNG, or DNG–DNG layers), equation (9.10) may never be satisfied, because for these pairs $\mu_2/\mu_1 > 0$ and thus no TE mode may propagate in such a thin waveguide, as expected. This represents the diffraction limit mentioned above for standard waveguides. On the other hand, equation (9.11) will provide the approximate value for β_{TM} of the dominant TM mode, if β_{TM} turns out to be a real quantity for a given set of γ and material parameters. We note that β_{TM} depends on the ratio of layer thicknesses, not on the total thickness. Therefore, this TM mode has no cutoff thickness; that is, there is not a thickness below which this TM mode may not propagate. For a DPS–DPS or DNG–DNG thin waveguide, this TM mode exists for any ratio γ, and its β_{TM} is sandwiched between k_1 and k_2, which are effectively the two limits of equation (9.11) for $\gamma \to \infty$ and $\gamma \to 0$, respectively. This implies that the TM field distribution in the transverse section of a DPS–DPS or DNG–DNG thin waveguide has to be expressed using the exponential functions in one of the two slabs (in the one with smaller wavenumber) and the sinusoidal functions in the other slab. The allowable ranges of variation of β_{TM} in (9.11) in terms of γ are shown in Fig. 9.3 for various

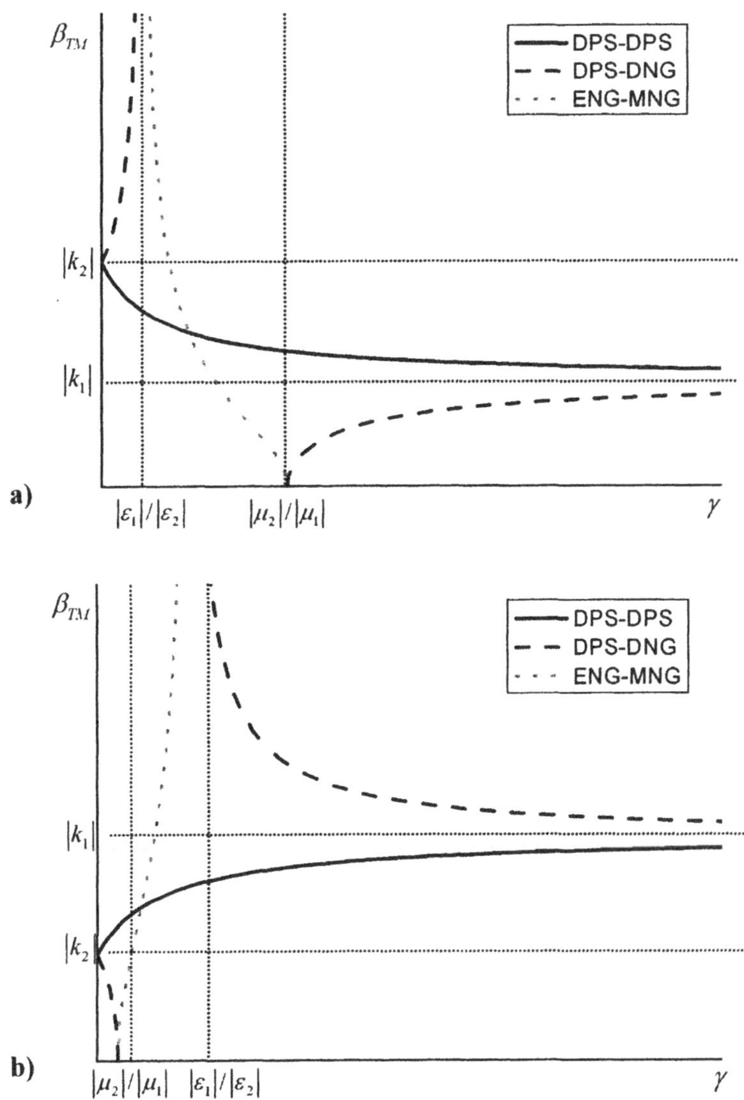

Fig. 9.3 β of the dominant TM mode for thin waveguides (with lateral thickness much less than the diffraction limit) filled with a pair of DPS–DPS, ENG–MNG, or DPS–DNG slabs, versus $\gamma = d_1/d_2$. (a) The material parameters are chosen such that $\varepsilon_1 = \pm 2\varepsilon_0$, $\mu = \pm\mu_0$, $\varepsilon_2 = \pm 3\varepsilon_0$, and $\mu = \pm\mu_0$ for which $|k_2| > |k_1|$. (b) The two slabs have been interchanged; that is, slab 1 and 2 in (a) are now slabs 2 and 1 in (b), respectively, thus $|k_2| < |k_1|$. Here we are concerned only with the positive real solutions for β_{TM}, but its negative real solutions are simply obtained by flipping its sign. After Ref. [42]. Copyright © 2004 IEEE.

pairs of slabs. As shown in this figure, the ENG–MNG pair behaves differently: The existence of a no-cutoff dominant TM mode is restricted to the waveguides with γ in the range between $|\varepsilon_1| / |\varepsilon_2|$ and $|\mu_2| / |\mu_1|$. However, its wavenumber β_{TM} is *not* restricted to any interval; that is, an ENG–MNG waveguide may have a dominant no-cutoff TM mode with β_{TM} ranging from zero to infinity (Fig. 9.3). The range of variation of β_{TM} in the thin DPS–DNG waveguide indeed differs from the ones in the thin ENG and in the standard DPS–DPS waveguides. Here β_{TM} may attain values only outside the interval between $|k_1|$ and $|k_2|$ (effectively "complementary" to the standard DPS–DPS case where β_{TM} is in this interval), and γ should also be outside the range between $-\mu_2/\mu_1$ and $-\varepsilon_1/\varepsilon_2$. The fact that thin waveguides loaded with "conjugate" pairs of metamaterials (e.g., DPS–DNG or ENG–MNG) may support nonlimited β_{TM} may offer interesting possibilities in designing very thin resonant cavities, as already proposed in Refs. 37–39, for which $\beta = 0$ when $\gamma = -\mu_2/\mu_1$, or for very thin waveguides having guided modes with high β, which may give rise to compact resonators and filters. A similar observation regarding the possibility of β_{TM} to be very large has also been made by Nefedov and Tretyakov [51].

Considering the TE case, a thin waveguide with a pair of ENG–MNG slabs (or also a pair of DNG–ENG, DPS–MNG, or DPS–DNG slabs) has $\mu_2/\mu_1 < 0$, and thus (9.10) may be satisfied for a certain value of γ. Equation (9.10) seems to be effectively independent of β_{TE}. However, we should note that in such a limit, the wavenumber β_{TE} of the guided mode may essentially attain any real value, as can be seen in Figs. 9.2a and 9.2b around the region where $|k_1| d_1 \to 0$ and $|k_2| d_2 \to 0$. In such a limit, no matter how thin these layers are (as long as they satisfy (9.10)), one (and only one) propagating mode may exist. In other words, this waveguide does not have a cutoff thickness for the TE modes, and thus they can support a guided mode, even though the lateral dimension can be well below the diffraction limit. This feature represents also a generalization of the analysis for the DPS–DNG thin cavity shown in Refs. 37–39. This may provide an interesting possibility for transporting RF or optical signals in ultra-thin guiding structures with cross-sectional dimensions below the diffraction limits. The case of open waveguides with such a property will be discussed in the next section.

9.2.3 Power Propagation in DPS–DNG Waveguides

In the previous section, we reviewed the field modal structure in the DPS–DNG and ENG–MNG planar waveguides and cavities, highlighting the possibility of modes with no cutoff thickness resulting in guided propagation in subwavelength waveguides and also monomodal waveguides with electrically large apertures. In this section, we approach the problem of the mode excitation in the DPS–DNG waveguides, in order to analyze how these peculiar modes are excited and understand how they carry power in these waveguides. (Parts of these results were first presented in a recent symposium [40].)

First, let us consider a planar current sheet at the interface between the two materials filling the waveguide—that is, on the $y = 0$ plane. Similar to what we have shown in Ref. 44 for the unbounded problem, we intuitively expect that the radiation from such a current source possess the typical backward behavior in the DNG slab. In other words, we expect to find two antiparallel power flows in the two slabs of the DPS–DNG waveguide. The solutions given in Ref. 44 may be easily extended to this case by augmenting the field expressions with the waves reflected from the parallel plates. An electric current source $\mathbf{J} = \hat{\mathbf{z}}\, J\, e^{-j\beta x} \delta\,(y)$, therefore, excites a TE electromagnetic mode in the waveguide with the following expressions:

$$\mathbf{E} = \hat{\mathbf{z}} \begin{cases} -\dfrac{j\omega\mu_0 J\, e^{-j\beta x} \frac{\sin[k_{t1}(d_1-y)]}{\sin[k_{t1}d_1]}}{D(\beta)}, & 0 \leq y < d_1 \\ -\dfrac{j\omega\mu_0 J\, e^{-j\beta x} \frac{\sin[k_{t2}(y+d_2)]}{\sin[k_{t2}d_2]}}{D(\beta)}, & -d_2 < y < 0 \end{cases} \tag{9.12}$$

$$\mathbf{H} = \hat{\mathbf{x}} \begin{cases} -\dfrac{J\, e^{-j\beta x} \frac{k_{t1}}{\mu_1} \frac{\cos[k_{t1}(d_1-y)]}{\sin[k_{t1}d_1]}}{D(\beta)} \\ \dfrac{J\, e^{-j\beta x} \frac{k_{t2}}{\mu_2} \frac{\cos[k_{t2}(y+d_2)]}{\sin[k_{t2}d_2]}}{D(\beta)} \end{cases} + \hat{\mathbf{y}} \begin{cases} \dfrac{j\, J\, e^{-j\beta x} \frac{\beta}{\mu_1} \frac{\sin[k_{t1}(d_1-y)]}{\sin[k_{t1}d_1]}}{D(\beta)}, & 0 \leq y < d_1 \\ \dfrac{j\, J\, e^{-j\beta x} \frac{\beta}{\mu_2} \frac{\sin[k_{t2}(y+d_2)]}{\sin[k_{t2}d_2]}}{D(\beta)}, & -d_2 < y < 0 \end{cases} \tag{9.13}$$

where $D\,(\beta) \equiv \frac{k_{t1}}{\mu_1} \cot\,[k_{t1}d_1] + \frac{k_{t2}}{\mu_2} \cot\,[k_{t2}d_2]$ is the common denominator. We note the analogy with equations (9.1)–(9.2) for the expression of the waveguide modes. All the field components share the same denominator, which represents the dispersion relation for these modes and is identically zero when (9.3) is satisfied. When this happens, in fact, the field amplitude diverges, since the structure is behaving like a resonator driven at the frequency of resonance, similar to what was obtained in Ref. 44 for the surface modes (i.e., surface waves) supported by a DPS–DNG interface.

The extension to an arbitrary distribution of current at the interface is of course straightforward: Assuming a sheet of electric current $\mathbf{J} = \hat{\mathbf{z}}\, J\,(x)\; \delta\,(y)$ (again exciting TE modes), this may be expanded using the standard Fourier transform:

$$\mathbf{J} = \hat{\mathbf{z}} \int_{-\infty}^{\infty} \tilde{J}\,(\beta)\, d\beta\, e^{-j\beta x}\, \delta\,(y) \tag{9.14}$$

where $\tilde{J}\,(\beta)$ is the Fourier transform of $J\,(x)$. Each infinitesimal contribution, $\tilde{J}\,(\beta)\, d\beta$, induces a TE mode in the waveguide, expressed by (9.12)–(9.13). Therefore, the total electric field in the waveguide may be expressed as

$$\mathbf{E} = \hat{\mathbf{z}} \begin{cases} -j\omega\mu_0 \int_{-\infty}^{\infty} \dfrac{\tilde{J}(\beta)\, e^{-j\beta x} \frac{\sin[k_{t1}(d_1-y)]}{\sin[k_{t1}d_1]}}{D(\beta)} d\beta, & 0 \leq y < d_1 \\ -\, j\omega\mu_0 \int_{-\infty}^{\infty} \dfrac{\tilde{J}(\beta)\, e^{-j\beta x} \frac{\sin[k_{t2}(y+d_2)]}{\sin[k_{t2}d_2]}}{D(\beta)} d\beta, & -d_2 < y < 0 \end{cases} \tag{9.15}$$

The magnetic field may obviously be written in a similar way and is not reported here. As expected, the integrand poles coincide again with the dispersion relations for the waveguide modes. For a sheet current source in a finite range $|x| < \hat{x}$, the evaluation of the field at a point sufficiently far from the source region is promptly performed by solving the residue problem for only the real roots of the denominator, which correspond to the propagating guided modes supported by the waveguide. For nondegenerate modes, these are single roots, and the integral may be substituted by the residue sum:

$$\int_{-\infty}^{\infty} \frac{N(\beta)}{D(\beta)}\, d\beta = 2\pi j \sum_{n=1}^{M} \frac{N(\beta_n)}{\left.\frac{\partial D(\beta)}{\partial \beta}\right|_{\beta=\beta_n}} \tag{9.16}$$

where β_n represent the M real roots of $D(\beta)$.

In the particular case of a line current at the origin, the current expression becomes $\mathbf{J} = \hat{\mathbf{z}}\, I\, \delta(x)\, \delta(y)$, and thus $\tilde{J}(\beta) = \frac{I}{2\pi}$. The electric field in the first slab may be expressed as

$$\begin{aligned} \mathbf{E} &= \hat{\mathbf{z}}\, I\omega\mu_0 \sum_{n=1}^{M} \frac{e^{-j\beta_n x} \frac{\sin[k_{t1n}(d_1-y)]}{\sin[k_{t1n}d_1]}}{\left.\frac{\partial}{\partial\beta}\left[\frac{k_{t1}}{\mu_1}\cot(k_{t1}d_1) + \frac{k_{t2}}{\mu_2}\cot(k_{t2}d_2)\right]\right|_{\beta=\beta_n}} \\ &= \hat{\mathbf{z}}\, I\omega\mu_0 \sum_{n=1}^{M} \frac{e^{-j\beta_n x} \frac{\sin[k_{t1n}(d_1-y]}{\sin(k_{t1n}d_1)}}{\beta_n \left[\frac{d_1 - \frac{\sin(2k_{t1n}d_1)}{2k_{t1n}}}{\mu_1 \sin^2(k_{t1n}d_1)} + \frac{d_2 - \frac{\sin(2k_{t2n}d_2)}{2k_{t2n}}}{\mu_2 \sin^2(k_{t2n}d_2)}\right]} \end{aligned} \tag{9.17}$$

which represents the explicit superposition of the M propagating guided modes in the waveguide with transverse wavenumbers $k_{t1n} = \sqrt{k_1^2 - \beta_n^2}$. The other field components may be easily obtained in a similar way.

The case of an arbitrary current $\mathbf{J}(x, y)$ in a closed region S (not necessarily distributed at the interface as supposed up to now) with finite volume may be studied by applying the conservation of energy. If the electric current is again supposed to be directed along $\hat{\mathbf{z}}$, exciting only TE modes, the total electric field in the waveguide may be expressed as a superposition of the infinite modes expressed by (9.1)–(9.2). Each of them is characterized by a longitudinal wavenumber β_n, which is a solution of the dispersion relation (9.3), being a real quantity for propagating guided modes and an imaginary quantity when the corresponding mode is evanescent. The amplitude of each mode E_{0n} is of course dependent on the source $\mathbf{J}(x, y)$. In order to relate these two quantities, we may apply the conservation of energy for the DPS–DNG waveguide.

Consider the expression of the Poynting vector inside the waveguide:

$$\mathbf{S}_n = \frac{|E_{0n}|^2\, e^{2x Im[\beta_n]}}{2\omega} \left\{\mathbf{S}_{nx}\hat{\mathbf{x}} + \mathbf{S}_{ny}\hat{\mathbf{y}}\right\}$$

where

$$\mathbf{S}_{nx} = \begin{cases} \frac{\beta_n^*}{\mu_1} \left|\sin\left(k_{tn1}\left(d_1 - y\right)\right)\right|^2 \left|\sin\left(k_{tn2} d_2\right)\right|^2, & 0 \leq y < d_1 \\ \frac{\beta_n^*}{\mu_2} \left|\sin\left(k_{tn1} d_1\right)\right|^2 \left|\sin\left(k_{tn2}\left(y + d_2\right)\right)\right|^2, & -d_2 < y < 0 \end{cases}$$

$$\mathbf{S}_{ny} = \begin{cases} \frac{j k_{tn1}^*}{\mu_1} \tan\left(k_{tn1}\left(d_1 - y\right)\right) \left|\cos\left(k_{tn1}\left(d_1 - y\right)\right)\right|^2 \left|\sin\left(k_{tn2} d_2\right)\right|^2 \\ -\frac{j k_{tn2}^*}{\mu_2} \tan\left(k_{tn2}\left(y + d_2\right)\right) \left|\sin\left(k_{tn1} d_1\right)\right|^2 \left|\cos\left(k_{tn2}\left(y + d_2\right)\right)\right|^2 \end{cases} \tag{9.18}$$

where * stands for complex conjugation. Considering real values for the β_ns, the above expression shows that for the DPS–DNG waveguide every mode carries a real power in the x-direction, just like the conventional case of DPS–DPS waveguides. The difference, however, is in the fact that the power flows in the two slabs are antiparallel in the DPS–DNG case, as intuitively expected. The total net power flowing out from a given cross section of the DPS–DNG waveguide section, therefore, leads to the algebraic sum of the two power flows—that is, the difference between the magnitudes of the two oppositely directed power flows. (This is analogous to the findings in Ref. 44.) Therefore, while in a standard (DPS–DPS) waveguide a mode with a positive real β_n (and with a positive phase velocity) carries power along the positive x-direction, in a DPS–DNG waveguide the phase velocity and the direction of power flow do not necessarily point in the same direction. In other words, a guided-mode's total net power flowing out of any cross section in the DPS–DNG waveguide may be either parallel or antiparallel with the phase velocity of that mode. Therefore, fixing the source region inside a DPS–DNG waveguide and considering the fact that for any guided mode its net power should flow away from the source, the sign of β_n may then be determined such that this net power condition is satisfied. Depending on the material parameters and slab thicknesses, the wavenumber β_n might be parallel or antiparallel with the direction of the net power flowing away from the source.

Following the previous considerations, in the DPS–DNG waveguide, like in any standard waveguide, the entire electric and magnetic fields may be decomposed into modes carrying power in the positive x-direction and modes carrying power in the negative x-direction. Due to the waveguide symmetry, such decomposition has the general form

$$\mathbf{E} = \sum_{n=1}^{M/2} a_n \mathbf{e_n} e^{-j\beta_n x} + \sum_{n=1}^{M/2} b_n \mathbf{e_n^*} e^{j\beta_n x}$$
$$\mathbf{H} = \sum_{n=1}^{M/2} a_n \mathbf{h_n} e^{-j\beta_n x} - \sum_{n=1}^{M/2} b_n \mathbf{h_n^*} e^{j\beta_n x} \tag{9.19}$$

where the β_n set is composed by those $M/2$ real solutions of (9.3) that each carries a positive net power flowing away from a closed mathematical surface containing S. Such β_ns are not all necessarily positive as we usually expect in a conventional DPS–DPS waveguide. The expressions for $\mathbf{e_n}$ and $\mathbf{h_n}$ in (9.19) may be easily deduced from (9.1)–(9.2).

For any two given modes in the DPS–DNG waveguide, the following general orthogonality relation is still valid:

$$\int_{-d_2}^{d_1} (\mathbf{e_n} \times \mathbf{h_m^*} + \mathbf{e_m^*} \times \mathbf{h_n}) \cdot \hat{\mathbf{x}}\, dy = 4\bar{P}_n \delta_{mn} \tag{9.20}$$

where $\bar{P}_n$ is the net power carried by the nth mode, normalized to the squared module of its amplitude, and δ_{mn} is the Kronecker delta function. As in a standard waveguide, the above relation guarantees that in DPS–DNG waveguides there is no power coupling between the two distinct (nondegenerate) modes.

The real power taken away from the current source by the nth mode may be calculated from (9.19) and, depending on the net-power-flow direction, is given by the upper or lower part of the following expression:

$$\iiint\limits_V \frac{1}{2}\mathrm{Re}\left[\mathbf{J}^* \cdot \left\{ \begin{array}{c} a_n \mathbf{e_n} e^{-j\beta_n x} \\ b_n \mathbf{e_n^*} e^{j\beta_n x} \end{array} \right.\right] dV \tag{9.21}$$

Owing to the conservation of energy, this quantity should be equal to the net real power of the nth mode flowing out of a closed mathematical surface containing S—that is, from two oppositely oriented cross sectional surfaces perpendicular to the x-axis. The real net power flowing out of each oriented section S for the nth mode is:

$$\iint\limits_S \frac{1}{2}\left\{ \begin{array}{c} |a_n|^2 \\ |b_n|^2 \end{array} \right. \mathrm{Re}\left[\mathbf{e_n} \times \mathbf{h_n^*}\right] \cdot d\mathbf{S} \tag{9.22}$$

and, by symmetry, $|a_n|^2 = |b_n|^2$. Therefore, by equating the two expressions, the values of the mode amplitude are given as

$$\begin{aligned} a_n &= \frac{\iiint_V \mathbf{J} \cdot \mathbf{e_n^*} e^{j\beta_n x} dV}{2\iint_S \mathrm{Re}[\mathbf{e_n} \times \mathbf{h_n^*}] \cdot d\mathbf{S}} \\ b_n &= \frac{\iiint_V \mathbf{J} \cdot \mathbf{e_n^*} e^{-j\beta_n x} dV}{2\iint_S \mathrm{Re}[\mathbf{e_n} \times \mathbf{h_n^*}] \cdot d\mathbf{S}} \end{aligned} \tag{9.23}$$

In the case of the line source $\mathbf{J} = \hat{\mathbf{z}} I \delta(x)\, \delta(y)$ placed at the origin, the field expression for every propagating mode coincides with the one in (9.17). In fact, for the electric field of the nth mode, we have

$$\iiint\limits_V \mathbf{J} \cdot \mathbf{e_n^*} e^{j\beta_n x} dV = \iiint\limits_V \mathbf{J} \cdot \mathbf{e_n} e^{-j\beta_n x} dV = I \sin(k_{tn2} d_2) \sin(k_{tn1} d_1)$$

$$
\begin{aligned}
2\iint_S \operatorname{Re}\left[\mathbf{e_n}\times\mathbf{h_n^*}\right]\cdot d\mathbf{S} &= 2\int_0^{d_1}\frac{\beta_n}{\omega\mu_0\mu_1}\sin^2\left(k_{tn2}d_2\right)\sin^2\left(k_{tn1}\left(d_1-y\right)\right)dy\\
&+ 2\int_{-d_2}^{0}\frac{\beta_n}{\omega\mu_0\mu_2}\sin^2\left(k_{tn1}d_1\right)\sin^2\left(k_{tn2}\left(d_2+y\right)\right)dy\\
&= \frac{\beta_n}{\omega\mu_0}\left(\frac{d_1-\frac{\sin(2k_{t1n}d_1)}{2k_{t1n}}}{\mu_1\sin^{-2}(k_{t2n}d_2)}+\frac{d_2-\frac{\sin(2k_{t2n}d_2)}{2k_{t2n}}}{\mu_2\sin^{-2}(k_{t1n}d_1)}\right)
\end{aligned}
\tag{9.24}
$$

We now focus on the denominator of the expression in (9.23). This denominator is proportional to the net power carried by the nth mode in the DPS–DNG waveguide. Unlike the case of the standard waveguide in which the net power for each mode with a non-zero real β is always positive, for the DPS–DNG waveguide such a net power may become zero due to the algebraic sum of the two oppositely-directed power flows in the DPS and DNG layers, implying that the denominators in (9.23) may go to zero under certain conditions. In these situations the field amplitude of such a mode becomes infinitely large. This is due to the fact that under such conditions the corresponding mode carries two oppositely directed power flows in the DNG and DPS slabs, equal in magnitude and opposite in direction, and therefore the waveguide is effectively behaving like a cavity resonator (although it is "open" ended). If the source drives a resonating mode in a cavity at its resonant frequency, it is known that in the ideal lossless scenario the induced field will diverge in the steady-state regime. In fact, we note that even in a conventional DPS-DPS waveguide whose dispersion relation has a solution for $\beta = 0$, the waveguide becomes a cavity resonator for $\beta = 0$, and an analogous phenomenon happens as the denominator of (9.23) goes to zero, as is clearly understandable.

Equation (9.24) reveals the conditions under which the net power may become zero. According to this relation, for a DPS–DPS waveguide, only when $\beta_n = 0$, a zero net power is achieved, which relates to the mode bouncing back and forth between the two parallel walls of the waveguide (i.e., a cavity resonator). However, when a DPS–DNG waveguide is considered, in addition to the case of $\beta_n = 0$, another condition can lead to zero net power and that is

$$
\frac{d_1-\frac{\sin(2k_{t1n}d_1)}{2k_{t1n}}}{|\mu_1|\sin^2\left(k_{t1n}d_1\right)}=\frac{d_2-\frac{\sin(2k_{t2n}d_2)}{2k_{t2n}}}{|\mu_2|\sin^2\left(k_{t2n}d_2\right)}
\tag{9.25}
$$

which can have solutions with $\beta_n \neq 0$. In order to understand this condition physically, let us consider a DPS–DNG monomodal waveguide, for which the only mode of propagation is a "forward" mode—that is, for which the phase flow is parallel to the energy flow. Such a mode carries its power in a unusual way: an observer located in the DPS slab sees a certain amount of power flowing away form the source, and he can measure such a power flow in the first slab, which is proportional to the left side of (9.25). The total power he measures, however, is *bigger* than the quantity actually taken away from the source. The exceeding part, however, cannot be used by the observer, since this power has to flow back to the source in the DNG slab to maintain equilibrium and energy conservation. This may appear to be counterintuitive, but it is all physically justifiable. In fact this description is very similar to what happens in a simple case of a normally incident plane wave which is partially reflected from a simple vacuum–dielectric interface. If an observer were able to distinguish between the incident and reflected power flows in this analogy, he would measure a higher value for the incident power than the net power actually flowing in the forward direction into the dielectric half space. If the reflecting surface is a perfect metal, all the radiation would be reflected back and the observer would not measure any net power flowing in one or another direction, even though each of the two power flows by itself carries a nonzero power.

In the DPS–DNG waveguide, the negative refraction [4] present at the interface between the two slabs effectively acts as a reflecting mechanism orthogonal to the interface, which partially (or totally) reflects the impinging power. From an electromagnetic point of view, the resonance condition, therefore, may be simply explained as the condition for which the *equivalent* orthogonal surface becomes totally reflecting, effectively transforming the waveguide into a cavity resonator. What we obtain is essentially a form of a standing wave, as in a usual cavity or in a Bragg reflector; however, unlike those cases, here the forward and backward waves are flowing in *spatially separated* regions in the two slabs. In Fig. 9.4, the electric and magnetic field variation along the cross section of a very thin DPS–DNG waveguide are shown. The waveguide total cross section is taken to be only one-hundredth of the wavelength in the first slab. Note that the ratio of the slab thicknesses satisfies (9.10) with a good approximation. Applying the previous formulas to this case, we obtain that if $P = P_1 - P_2$ (i.e., the net power carried by this mode, obtained as the difference between the power flows in the two slabs) we have $P_1 = 1.125P$ and $P_2 = 0.125P$. This conceptual example shows in principle the possibility of the energy transport by a very thin monomodal DPS–DNG waveguide with a very small electrical thickness while still supporting a mode carrying a nonzero net power.

There are other interesting issues related to the anomalous power propagation in DPS–DNG waveguides. A first topic concerns the relation between the power flows in the two slabs: We have seen that every mode in the DPS–DNG waveguide carries its power as two oppositely directed power flows in the two parallel slabs. Moreover, we have addressed the orthogonality properties of such modes, guaranteeing no power exchange between two distinct modes. Therefore, for each propagating mode, one would think that the portion of the power traveling away from the source in one of the two slabs may return back in the other slab. If true, this would imply that, in some section of the waveguide, a portion of the power from the first slab should cross the interface of the two slabs to revert its path. However, the Poynting vector expression (9.18) clearly reveals that *no real net power* crosses the interface between the two slabs, since every mode carries real power only along the longitudinal direction. In other words, it appears that the two antiparallel power flows travel separately in the two slabs, and the two powers do not vary in quantity along any homogeneous waveguide section. We may interpret the power propagation in such a waveguide as

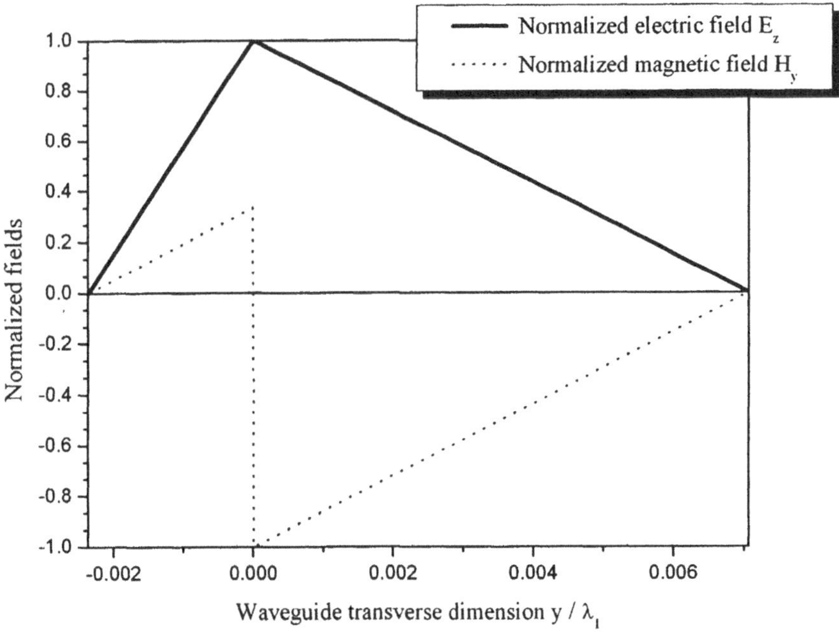

Fig. 9.4 Distribution of the normalized electric and magnetic field in a DPS–DNG thin waveguide with $\varepsilon_1 = 2\varepsilon_0$, $\mu_1 = \mu_0$, $\varepsilon_2 = -3\varepsilon_0$, $\mu_2 = -3\mu_0$, $\beta = 0.47k_1$.

sketched in Fig. 9.5: The portion of power which comes backward towards the source in the waveguide may be regarded as if exchanged in "loops" between the two slabs, each one-half of a longitudinal wavelength long. This sketch is well supported by the ray-theory approximation and the fact that at the DPS–DNG interface the refraction is negative. On the sections in which the loops cross the interface, however, transverse net power is zero, since in the lossless condition the two transverse power fluxes across the interface are equal and antiparallel, consistent with equation (9.18) in which no real net power crosses across the interface between the DPS and DNG slabs. When the loss is present, the field decays with $x = \hat{x}$ and a small transverse exchange of power is present, as can be verified by introducing a small loss in equation (9.18). In the lossless scenario, if no net power flows transversely, where, then, will the backward flow towards the source come from? In our analysis, we deal with an infinitely long DPS–DNG waveguide whose cross section does not vary with x, and has no interruptions or abruptions. In this case no transverse energy flow is required, since the modal structure is the same at every cross section. The issue arises in practical situations where a finite length of the waveguide is considered. The answer to our question about the transverse flow of power can be found exactly at the section where an interruption or an abruption is present. At such discontinuities, local reactive nonpropagating evanescent modes are excited in addition to the propagating guided modes, and they induce a *real* power flow in the *transverse* direction. In the

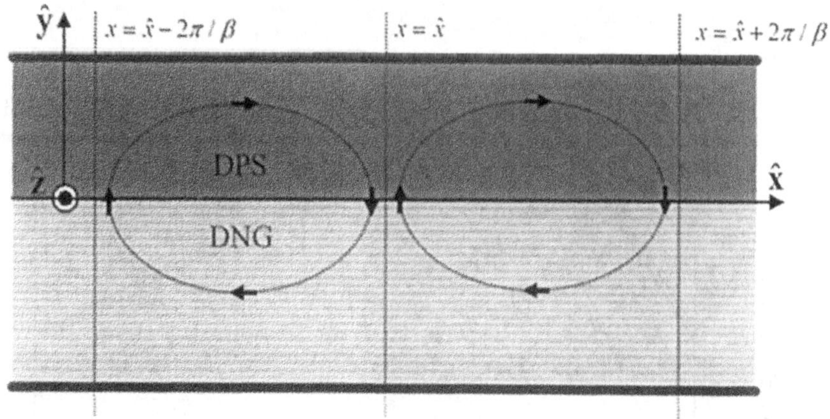

Fig. 9.5 Flow of energy in "loops." At a given section $x = \hat{x}$, the power flow in the transverse plane due to one loop is balanced by the flow due to the next one and thus no net transverse energy flow is present, if loss is neglected.

vicinity of these discontinuities, therefore, the exchange of power between the two slabs takes place and the backward flow is generated. Figure 9.6 shows the results of a mode-matching analysis of an abruption between a DPS–DNG waveguide (on the left) and an empty waveguide (on the right). The two waveguides support only one propagating mode each, whose longitudinal (real part of) Poynting vector distributions are described by the solid lines on the two sides of the abruption. The black arrows represent the local real part of the Poynting vector in each point of the waveguide. As may be clearly seen from the figure, far enough away from the abruption the arrows show the longitudinal power flow as given by the propagating modes only. In the DPS–DNG waveguide, in fact, they are oppositely directed and completely separated from each other. In the vicinity of the abruption, however, the distribution of the real part of the Poynting vectors (the black arrows) is rearranged by the presence of the evanescent non-propagating modes and they clearly reconstruct the transmitted mode (on the right of the abruption) and feed the backward flow (on the left of the abruption).

It should be mentioned that a similar behavior may be obtained in inhomogeneously filled circular waveguides loaded by ferrite rods, or in other nonhomogeneous circular waveguides in the neighborhood of the cutoff frequency, as shown by Clarricoats [56] and Waldron [57], in which standard dielectrics have been used. In these cases the supported modes carry power in such a way that in some regions of the waveguide cross section a negative power flux density is present (i.e., the power flow direction is antiparallel to the phase flow direction), while in the rest of the cross section the power flux density is positive. These "regions" of negative and positive power flux density may vary as the mode wavenumber changes in such inhomogeneously filled circular waveguides, whereas in our DPS–DNG planar waveguides, the separation between the two oppositely directed power flows is clearly distinguished.

9.3 OPEN SLAB WAVEGUIDES WITH DNG METAMATERIALS

In this section we analyze another guiding structure involving metamaterials, namely the case of an open planar waveguide consisting of a metamaterial slab supporting a surface-plasmon mode. (Some of our results for these guided-wave structures were first presented in Ref. 43.) The geometry is shown in Fig. 9.7, together with a Cartesian coordinate system for a DNG slab of thickness $2d$ and constitutive parameters ε and μ, surrounded by free space with parameters ε_0 and μ_0.

The TE spectrum of propagating modes in this open waveguide may be split into even and odd modes. The solution of the boundary-value problem yields the following expressions for the electric field of even and odd TE^x modes, respectively:

$$E_e^{TE} = \hat{z}\, E_{0e}^{TE}\, e^{-j\beta_{TE}x} \cdot \begin{cases} \cos\left(k_t^{TE} d\right) e^{-\sqrt{\beta_{TE}^2 - k_0^2}(y-d)}, & y > d \\ \cos\left(k_t^{TE} y\right), & |y| < d \\ \cos\left(k_t^{TE} d\right) e^{\sqrt{\beta_{TE}^2 - k_0^2}(y+d)}, & y < -d \end{cases} \tag{9.26}$$

$$E_o^{TE} = \hat{z}\, E_{0o}^{TE}\, e^{-j\beta_{TE}x} \cdot \begin{cases} \sin\left(k_t^{TE} d\right) e^{-\sqrt{\beta_{TE}^2 - k_0^2}(y-d)}, & y > d \\ \sin\left(k_t^{TE} y\right), & |y| < d \\ -\sin\left(k_t^{TE} d\right) e^{\sqrt{\beta_{TE}^2 - k_0^2}(y+d)}, & y < -d \end{cases} \tag{9.27}$$

where E_{0e}^{TE} and E_{0o}^{TE} are the mode amplitudes, and $k_t^{TE} = \sqrt{k^2 - \beta_{TE}^2}$ with $k^2 = \omega^2 \mu\, \varepsilon$. The corresponding expressions for the magnetic fields may be easily derived from the Maxwell equations:

$$\begin{aligned} H_e^{TE} = & \\ & +j\,\hat{x}\,\omega^{-1} E_{0e}^{TE} e^{-j\beta_{TE}x} \cdot \begin{cases} -\mu_0^{-1}\sqrt{\beta_{TE}^2 - k_0^2}\cos\left[k_t^{TE} d\right] e^{-\sqrt{\beta_{TE}^2 - k_0^2}(y-d)} \\ -\mu^{-1} k_t^{TE} \sin\left[k_t^{TE} y\right] \\ \mu_0^{-1}\sqrt{\beta_{TE}^2 - k_0^2}\cos\left[k_t^{TE} d\right] e^{\sqrt{\beta_{TE}^2 - k_0^2}(y+d)} \end{cases} \\ & -\hat{y}\omega^{-1}\beta_{TE} E_{0e}^{TE} e^{-j\beta_{TE}x} \cdot \begin{cases} \mu_0^{-1}\cos\left[k_t^{TE} d\right] e^{-\sqrt{\beta_{TE}^2 - k_0^2}(y-d)}, & y > d \\ \mu^{-1}\cos\left[k_t^{TE} y\right], & |y| < d \\ \mu_0^{-1}\cos\left[k_t^{TE} d\right] e^{\sqrt{\beta_{TE}^2 - k_0^2}(y+d)}, & y < -d \end{cases} \end{aligned} \tag{9.28}$$

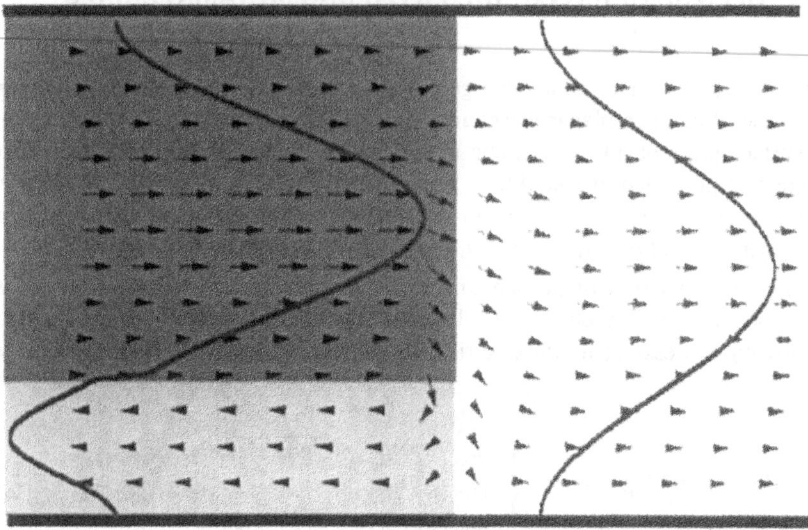

Fig. 9.6 Mode-matching analysis of an abruption. The black arrows represent the real part of the Poynting vector. On the left we have a DPS–DNG waveguide with one propagating mode with power distribution described by the solid line on the left. On the right there is an empty parallel-plate waveguide that supports again only one propagating mode with field distribution given by the solid line on the right.

$$H_o^{TE} =$$
$$+j\,\hat{x}\,\omega^{-1}E_{0o}^{TE}e^{-j\beta_{TE}x}\cdot\begin{cases} -\mu_0^{-1}\sqrt{\beta_{TE}^2-k_0^2}\sin\left[k_t^{TE}d\right]e^{-\sqrt{\beta_{TE}^2-k_0^2}(y-d)} \\ \mu^{-1}k_t^{TE}\cos\left[k_t^{TE}y\right] \\ -\mu_0^{-1}\sqrt{\beta_{TE}^2-k_0^2}\sin\left[k_t^{TE}d\right]e^{\sqrt{\beta_{TE}^2-k_0^2}(y+d)} \end{cases}$$
$$-\hat{y}\,\omega^{-1}\beta_{TE}E_{0o}^{TE}e^{-j\beta_{TE}x}\cdot\begin{cases} \mu_0^{-1}\sin\left[k_t^{TE}d\right]e^{-\sqrt{\beta_{TE}^2-k_0^2}(y-d)}, & y>d \\ \mu^{-1}\sin\left[k_t^{TE}y\right], & |y|<d \\ -\mu_0^{-1}\sin\left[k_t^{TE}d\right]e^{\sqrt{\beta_{TE}^2-k_0^2}(y+d)}, & y<-d \end{cases} \tag{9.29}$$

The expressions for the TM^x polarization may be obtained from duality and are not reported here. In the following we will focus on the TE polarization, but similar considerations may be made for the TM polarization.

Applying the proper boundary conditions for the magnetic fields at the three interfaces, we obtain the following dispersion relations for the even and odd TE

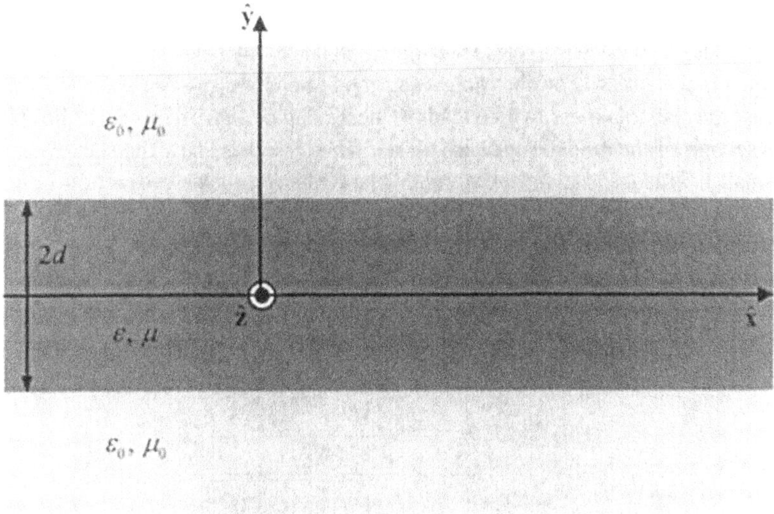

Fig. 9.7 Geometry of the open waveguide: a double-negative (DNG) slab surrounded by empty space.

guided modes, respectively:

$$\frac{\mu}{\sqrt{k^2-\beta_e^2}}\cot\left(\sqrt{k^2-\beta_e^2}d\right)=\frac{\mu_0}{\sqrt{\beta_e^2-k_0^2}} \tag{9.30}$$

$$\frac{\mu}{\sqrt{k^2-\beta_o^2}}\tan\left(\sqrt{k^2-\beta_o^2}d\right)=-\frac{\mu_0}{\sqrt{\beta_o^2-k_0^2}} \tag{9.31}$$

The analogous dispersions for the TM case may be promptly obtained substituting μ and μ_0 with ε and ε_0, respectively. These dispersion relations are valid for any complex value of ε and μ. In particular, in the limit of no loss (i.e., for real-valued constitutive parameters) these equations remain valid also when we consider negative parameters.

In the case of a standard DPS material, as is well known, propagating guided-mode solutions for the dispersion relations (9.30)–(9.31) are possible only when $k_0 < \beta < k$, yielding the real-valued expressions in both sides of the equations. In particular, the first even mode has no cutoff; that is, even when $d \to 0$, a solution for β_e does still exist, and $\beta_e \to k_0^+$. This implies that, although a surface wave is always supported by the DPS slab waveguide, conceptually even in the limit of the slab thickness tending to zero, when the slab thickness is too small the lateral distribution of the field of such a guided mode is widespread in the region surrounding the slab [as may be clearly verified from equations (9.26)–(9.29)], and essentially the mode becomes weakly guided. Therefore, when the guiding structure with the DPS material becomes very thin, the effective cross section of the guided mode becomes very large (in the limit of zero slab thickness the guided mode is simply a uniform plane wave). In other words, if even consider to reduce the slab thickness, the guided mode will travel with a transverse section much larger than the slab lateral dimension. (On the other hand, to confine the guided mode near the guiding slab, one needs to increase the slab thickness, which leads to increasing β_e. But in this case, the effective cross section of the guided mode is still not small, because now the slab thickness is not small.) This issue is indeed another manifestation of the diffraction limitation, which does not traditionally allow the signal transport in a guided structure thinner than a given dimension determined by the wavelength of operation.

The planar slab waveguide made of a DNG material may, in principle, overcome this limitation. In order to have a better insight into the differences between a DNG and a DPS slab waveguide, let us rewrite the dispersion relations (9.30) and (9.31) as

$$\tan(k_{te}d) = \frac{\mu\sqrt{(\Delta k\, d)^2 - (k_{te}d)^2}}{\mu_0\,(k_{te}d)} \tag{9.32}$$

$$-\cot(k_{to}d) = \frac{\mu\sqrt{(\Delta k\, d)^2 - (k_{to}d)^2}}{\mu_0\,(k_{to}d)} \tag{9.33}$$

where $k_{to} \equiv \sqrt{k^2 - \beta_o^2}$, $k_{te} \equiv \sqrt{k^2 - \beta_e^2}$, and $\Delta k^2 \equiv k^2 - k_0^2$. In this form, the two sides of the dispersion relations (9.32)–(9.33) may be easily plotted as a function of $(k_t d)$, with $(\Delta k\, d)$ as a varying parameter (Fig. 9.8), and thus the intersections of the two sides represent the solutions of (9.32) and (9.33)—that is, the propagating surface modes supported by the structure. In the figure, the circular dots indicate the propagation modes along a standard DPS slab, whereas the squares represent the modes in a DNG open slab waveguide with parameters ε and μ similar in magnitude, but opposite in sign, with those of the DPS slab. In the standard DPS case, as already anticipated, the first even mode (associated to the most left branch of the tangent in Fig. 9.8) does not show a cutoff thickness, since solutions are expected also when $(\Delta k\, d) \to 0$. In this case, however, the intersection is found for $(k_t d) \to 0$, that is, for $\beta_e \to k_0^+$. No odd modes, on the other hand, may propagate in very thin DPS open slab waveguides.

For the DNG case, the situation is different: The right-hand sides of the equations change sign and now the first odd mode has no cutoff thickness, even though the requirement of $(k_t d) \to 0$ is not required any more, but instead β_o increases as the slab thickness is reduced (and as a result the fields of such an odd mode are concentrated and confined more near the slab surface). This is a key advantage of such DNG open slab waveguides in design of very thin open waveguides with a concentrated cross section of guided mode. This may clearly overcome the usual diffraction limit for guided modes mentioned above, and it can provide a solution for transport of RF and optical energy in structures with small lateral dimension below the diffraction limit with possible applications to miniaturization of optical interconnects. Ideally in the lossless case, there is in principle no limitation on the compactness of such waveguides and confinement of the guided mode. However, in practice the loss is present and may limit the performance, and thus should be taken into consideration.

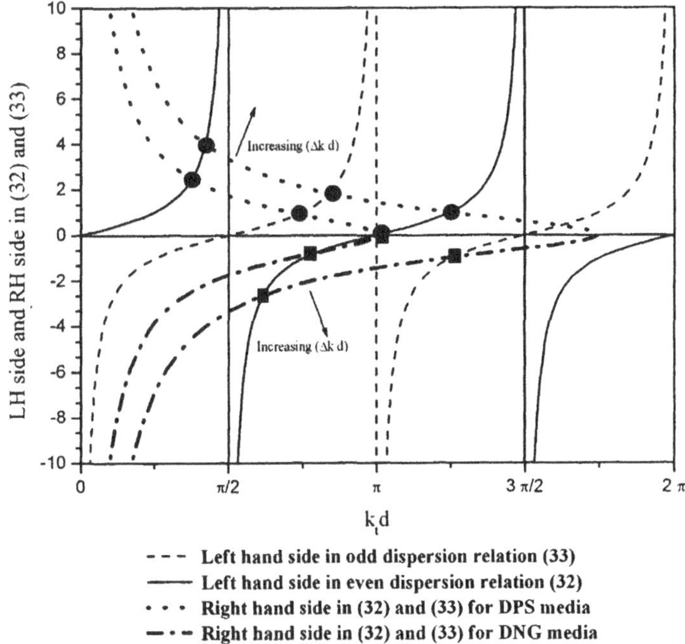

Fig. 9.8 Plots of the left- and right-hand sides of the dispersion relations (9.18) and (9.19) for DPS and DNG materials with $\mu = \pm\mu_0$.

Another peculiar distinction between the DNG and DPS cases may be noticed from Fig. 9.8: In the DPS case the derivatives of the two curves at the intersection points have always opposite signs, and due to the monotonic behavior of these curves, they will intercept only once per branch. As the slab thickness is decreased, the modes gradually disappear, as shown in Fig. 9.8. [As an example, let us take the second-order even mode in Fig. 9.8. When we reduce the slab thickness, the intersection of the two curves moves to the left and at the cutoff (i.e., at the minimum thickness for which this mode still propagates) both sides of the equation become identically zero. This point represents the cutoff dimension for this mode, and, as is evident from the dispersion relations, the corresponding β equals k_0.]

In the DNG case, on the other hand, the two sides of equations (9.32) and (9.33) share the same sign for their derivatives and therefore they may share more than one intersection point per branch. In the lower (i.e., negative) part of Fig. 9.8, in fact, it may be seen that the first even mode branch in the DNG case, now at the cutoff thickness of second-order even mode in the DPS case, has two distinct solutions. For that slab thickness, one of them does not correspond to $\beta = k_0$. However, if the slab thickness is decreased a little more, this even mode in the DNG case does not disappear and its β will anomalously *increase*, together with the field ratio confined inside the slab. The value of d representing the slab thickness for the second-order

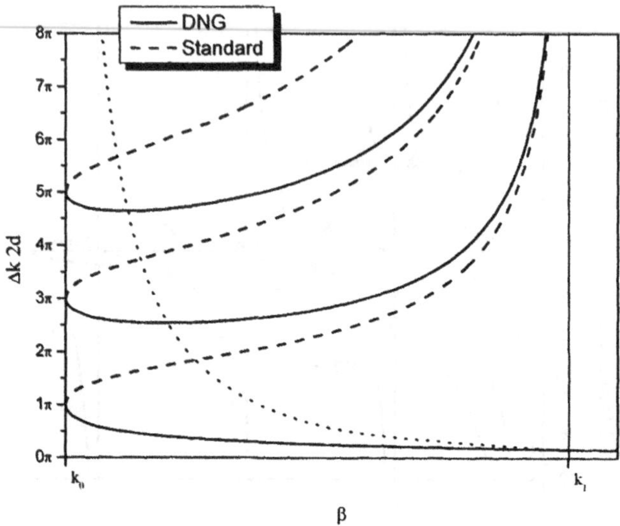

Fig. 9.9 Dispersion curves for odd modes in DNG and DPS slabs with $\mu = \pm 4\mu_0, \varepsilon = \pm 2\varepsilon_0$

mode cutoff in the DPS slab is *not* the cutoff dimension for the first even mode in the DNG slab. The *actual* cutoff dimension in this case, instead, is lower and it is achieved when the two curves in Fig. 9.8 are tangent to each other. The lower odd mode does not have a cutoff dimension, but the mode has a point with $\beta = k_0$, at which the mode is completely unbounded. Starting from this point and decreasing the slab thickness further, however, its β anomalously increases toward infinity (contrary to the DPS case), causing a more confined guided mode in a thinner DNG slab.

These features are clearly evident in the dispersion plots in Fig. 9.9, where the odd dispersion relation (9.33) is solved numerically and directly plotted as β_o versus the variable $(2\Delta k\, d)$. This figure shows the anomalous features of guided modes in the DNG open slab waveguide, most notably in the neighborhood of the cutoff (for $\beta \simeq k_0$) and for the lowest odd mode having β greater than k. In the region where β is near k_0, we may analytically study the behavior of odd and even modes by substituting $\beta \to k_0 + \delta\beta$, $\Delta k\, 2d \to m\pi + 2\,\delta\xi$ in the corresponding dispersion relations, where m is the integer index of the mode order (which is even when even modes are considered and odd for odd modes). The first-order approximation for both the dispersion relations yields:

$$\delta\xi = \frac{\mu}{\Delta k^2}\sqrt{2\,k_0\delta\beta} \tag{9.34}$$

Therefore, in the left part of Fig. 9.9 the curves are well approximated by square roots, for which the branch is determined by the sign of μ. For each DNG branch, as anticipated from the previous remarks on Fig. 9.8, there is an interval of values for d

below the "DPS cutoff" dimension for which two DNG modes with the same order are present with different β's. The DNG cutoff dimension is at the lower end of this interval, when the derivative of the dispersion curve is zero. These points all belong to the line:

$$d = \frac{-\mu_0 \mu \Delta k^2}{\sqrt{\beta^2 - k_0^2}\left[\beta^2\left(\mu^2 - \mu_0^2\right) + k^2\mu_0^2 - \mu^2 k_0^2\right]} \tag{9.35}$$

which has been shown as a dotted line on the plot.

From the above relation (9.35), one can see that for the DPS case the derivative of the dispersion curve will never be zero, since d should remain positive. In Fig. 9.9, the DPS curves are indeed monotonically growing. In the DNG case, on the other hand, the locus of the minima is represented by the intersections of (9.35) with the dispersion curves. In the special case of $\mu = -\mu_0$, (9.35) is interestingly simplified into $d = \frac{1}{\sqrt{\beta^2 - k_0^2}}$. The physical meaning of relation (9.35) and the particular behavior of the DNG plots in Fig. 9.9 will become clearer, once we discuss the characteristics of power flow in such guided-wave structures, as we show in the following.

In presence of an excitation (such as an impressed current source), the power carried by each mode may be evaluated using the Poynting Theorem, similarly to what was done in the previous section for the closed waveguide, exploiting the orthogonality relation between each pair of nondegenerate modes. In this case, the values of the amplitudes E_0 in the field expressions (9.26)–(9.29) can be expressed in terms of the current source as follows:

$$\begin{aligned} E_0^+ &= \frac{\iiint_V \mathbf{J}\cdot\mathbf{e}_\mathbf{n}^* e^{j\beta x}\, dV}{2\iint_S Re[\mathbf{e}_\mathbf{n}\times\mathbf{h}_\mathbf{n}^*]\cdot d\mathbf{S}} \\ E_0^- &= -\frac{\iiint_V \mathbf{J}\cdot\mathbf{e}_\mathbf{n} e^{-j\beta x}\, dV}{2\iint_S Re[\mathbf{e}_\mathbf{n}\times\mathbf{h}_\mathbf{n}^*]\cdot d\mathbf{S}} \end{aligned} \tag{9.36}$$

where V is the volume occupying by the impressed volume current source $\mathbf{J}$, $\mathbf{S}$ is a mathematical surface orthogonal to the slab axis, with its unit normal vector pointing outwards with respect to the source , $\mathbf{e_n}$ and $\mathbf{h_n}$ are the normalized transverse fields, proportional to (9.26)–(9.29), but carrying unit power. The superscripts for E_0 refer to the direction of power flow, i.e., E_0^+ and E_0^- are the amplitudes for the modes carrying power in the positive and negative x-direction, respectively.

It now becomes important to clarify which propagating modes we expect to carry power in the positive and in the negative direction. This may be achieved by analyzing the x component of the Poynting vector.

$$\mathbf{S_o}\cdot\hat{\mathbf{x}} = \frac{\beta}{2\omega}\left|E_{0o}\right|^2 \cdot \begin{cases} \mu_0^{-1}\sin^2(k_{to}d)e^{-2\sqrt{\beta_o^2-k_0^2}(y-d)} & y > d \\ \mu^{-1}\sin^2(k_{to}y) & |y| < d \\ \mu_0^{-1}\sin^2(k_{to}d)e^{2\sqrt{\beta_o^2-k_0^2}(y+d)} & y < -d \end{cases} \tag{9.37}$$

$$\mathbf{S_e}\cdot\hat{\mathbf{x}} = \frac{\beta}{2\omega}\left|E_{0e}\right|^2 \cdot \begin{cases} \mu_0^{-1}\sin^2(k_{te}d)e^{-2\sqrt{\beta_e^2-k_0^2}(y-d)} & y > d \\ \mu^{-1}\sin^2(k_{te}y) & |y| < d \\ \mu_0^{-1}\sin^2(k_{te}d)e^{2\sqrt{\beta_e^2-k_0^2}(y+d)} & y < -d \end{cases} \tag{9.38}$$

The total power flowing across a given cross section is obtained by the algebraic sum of two separate power flows: P_{empty}, the one in the empty space region $|y| > d$, and P_{slab}, the one flowing inside the slab. The two power flows are in general expressed as

$$\begin{aligned} P_{empty}^{odd} &= |E_{0o}|^2 \beta \omega^{-1} \mu_0^{-1} k_{to}^{-1} \sin^2(k_{to} d) \\ P_{slab}^{odd} &= |E_{0o}|^2 \beta \omega^{-1} \mu^{-1} \left(d - \frac{\sin(2k_{to}d)}{2k_{to}} \right) \end{aligned} \quad (9.39)$$

$$\begin{aligned} P_{empty}^{even} &= |E_{0e}|^2 \beta \omega^{-1} \mu_0^{-1} k_{te}^{-1} \cos^2(k_{te} d) \\ P_{slab}^{even} &= |E_{0e}|^2 \beta \omega^{-1} \mu^{-1} \left(d - \frac{\sin(2k_{to}d)}{2k_{to}} \right) \end{aligned} \quad (9.40)$$

In the DPS case, relations (9.39) and (9.40) reveals that for any propagating mode the power flow is parallel to the phase flow (which is defined by the sign of β). This means that each mode with positive β carries "positive" power, given by the sum of P_{empty} and P_{slab}, which flows to the right, away from the source. Conversely, a mode with a negative β carries a power to the left (negative sense of the x axis). This fact is reflected in the monotonic behavior of the DPS dispersion curves in Fig. 9.9, since the derivative $\frac{d(\Delta k\, 2d)}{d\beta}$, which is proportional to the group velocity $\frac{d\omega}{d\beta}$ for our monochromatic excitation, is always positive for positive β's and negative for negative β's.

In the DNG case, on the other hand, for every mode the signs of P_{empty} and P_{slab} are opposite, which implies that each mode carries two oppositely directed power flows, similar to what we have verified in a parallel-plate DPS–DNG waveguide in Ref. 38. The net power effectively carried by each mode is given by the algebraic sum of the two opposite power flows, which may be positive or negative, even if positive β are considered. In other words, we may find "backward" modes for which the direction of the net power flow is antiparallel with the direction of β. The more the guided mode is confined, the more the proportion of the power is concentrated inside the DNG slab, resulting in a backward mode. For high enough values of d in a DNG slab, most part of the power flow may be confined inside the slab and thus the direction of net power flow is opposite to the direction of β. A reduction in the slab dimension, in this case, is related to a reduction in the value of $|\beta|$ (as can be seen from the second odd mode in Fig. 9.9), and therefore the proportion of the power flow outside the slab is increased. Consequently, there is one value for the DNG slab thickness, for which $|P_{empty}|$ equals $|P_{slab}|$. In this case, no net power is actually extracted from the source, since all the power P_{empty} flowing away in one direction will come back as P_{slab} in the other direction. This special value of d does indeed satisfy relation (9.35) and it is represented by the minimum point in each dispersion curve. If one wants to decrease the value of $|\beta|$ even further (i.e., to obtain a less confined mode in the slab) one will have to *increase* the slab dimension, as seen from Fig. 9.9. This effect is physically related to the fact that at this special point there is a switch in the direction of net power flow, and the related group velocity $\frac{d\omega}{d\beta} \propto \frac{d(\Delta k\, 2d)}{d\beta}$ changes its sign. At this minimum, for which the group velocity goes to zero and no net power is carried by the mode, the structure effectively acts as an open "cavity resonator" and the field amplitude in (9.36) diverges, since no net power is effectively carried away from the source. (This is similar to exciting a closed cavity resonator at its resonant frequency.) The phenomenon is analogous to the case of parallel-plate DPS–DNG waveguide analyzed in the previous section. The dispersion relation plots for the odd modes in the DPS and DNG slabs, for both positive and negative β's, are shown in Fig. 9.10, underlining the direction of power flow according to the aforementioned arguments.

It is worth noting the importance of the first odd mode in the DNG slab. This mode, which is always a "forward" mode, has no cutoff thickness. Specifically, in the ideal lossless assumption, in principle there is no limit on the reduction of the slab thickness: As the thickness d is reduced further, the value of β is increased, the guided mode becomes more confined, and effective cross section of such modes is decreased. This feature, which is contrary to what we have in the standard DPS slab waveguide, can also be exhibited in the ENG and MNG slab waveguides that are effectively plasmonic waveguides. Therefore, the DNG slab waveguides, and also plasmonic waveguides, may potentially offer solutions to the problem of RF and optical energy transmission via guided structures with lateral cross section below the diffraction limits. Even if a limit is imposed due to the presence of inevitable ohmic loss in the slab material, we may still be able to overcome the diffraction limit by designing subwavelength slabs supporting a confined propagating beam. Finally, it is important to note that for such a thin DNG slab, the value of β can be much larger than the free space wavenumber k_0, implying very small *guide* wavelength $\lambda_g \equiv 2\pi/\beta \ll \lambda_o$. This may lead to exciting possibilities for design of ultra-thin cavities and devices using such DNG, ENG, MNG, or plasmonic slabs. This issue will be reported in a future paper.

9.4 THE CONTRADIRECTIONAL (BACKWARD) COUPLERS

In this section we describe the analysis of the coupling between two open slab waveguides; one made of conventional DPS materials and the other with a DNG metamaterial. (Some of the results of our analysis of the coupling phenomenon between the DPS and DNG slab waveguides were first presented in Ref. 43.) In the previous section, we described some of the unusual features of the DNG open slab waveguides, in particular the fact that such a waveguide supports guided modes in which the portion of the power flowing in the surrounding vacuum is antiparallel with respect to the portion flowing inside the DNG slab. Due to this anomalous property of the DNG open waveguide, one may intuitively expect that the coupling between such a DNG open waveguide and a standard DPS waveguide placed in its proximity could be "contradirectional" or "backward"; that is, if one of the two waveguides is excited to carry power in one direction, the second waveguide, through the coupling, might "redirect" back some of this power in the opposite direction. It is interesting to note

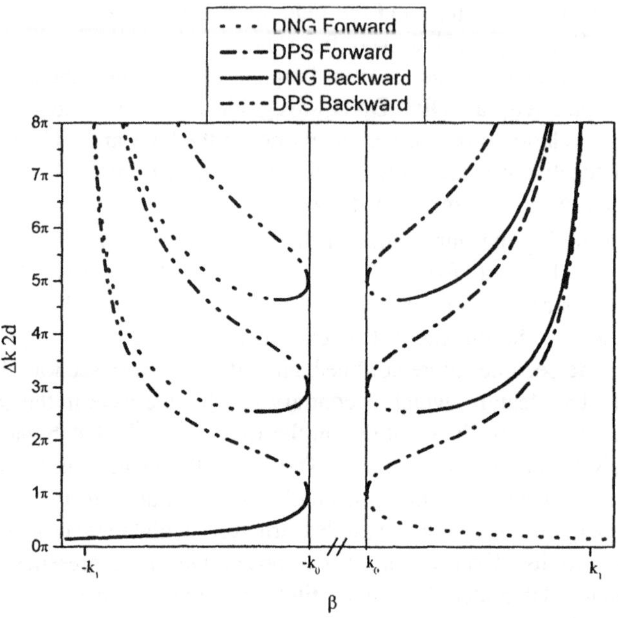

Fig. 9.10 Similar to Fig. 9.9, but emphasizing also the power flows, for positive and negative β.

that an analogous phenomenon has been observed and studied in the negative-index transmission-line couplers investigated by others [16, 19].

The coupling problem between the DNG and DPS open slab waveguides may be addressed by extending the analysis of surface-wave propagation in a planar dielectric slab, developed in the previous section, to the case in which the geometry is perturbed by the presence of a second slab. Clearly, the electromagnetic field distribution will be modified and part of the power flow guided by one slab may flow into the other one, owing to the coupling effect, as it happens between any two open waveguides placed close to each other.

The geometry of the problem is shown in Fig. 9.11, in which the two open slab waveguides separated and surrounded by a simple medium (e.g., free space) are considered. In this figure, a simple physical description of the contradirectional coupling of power flows in the two slabs is also given in terms of the phase and Poynting vectors.

Performing a rigorous modal analysis of the structure by starting from the Maxwell curl equations and imposing the suitable boundary conditions at all the interfaces, similar to what was done in the previous section, we obtain the following dispersion relation for the supported TE modes:

$$Disp_1\, Disp_2 = c_1 c_2 \tag{9.41}$$

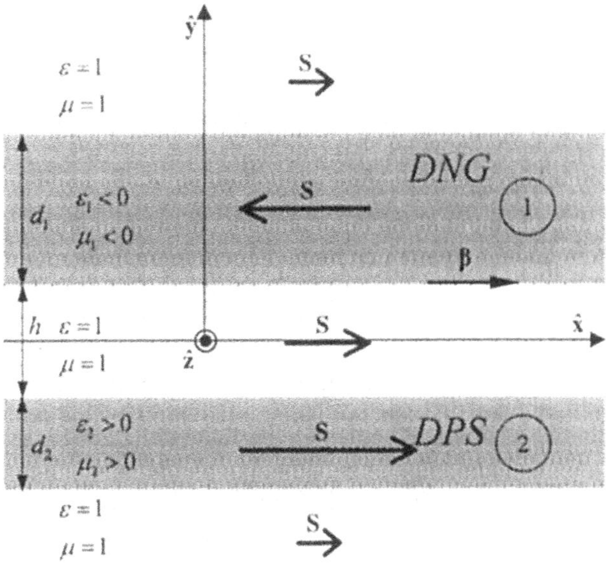

Fig. 9.11 The geometry of the contradirectional coupler with the DNG and DPS slab waveguides, with a sketch of the power flows for a given supported mode. After Ref. [43]. Copyright © 2002 IEEE.

where $Disp_i = 0$ (with $i = 1, 2$) is the modal dispersion relation of each slab alone (i.e., without coupling), represented by the product of the dispersion relations of even and odd TE modes in the isolated slab. These formulas are consistent with relations (9.30) and (9.31) (with a slight modification due to the different notation for the slab thickness):

$$Disp_i = \left(\sqrt{k_i^2 - \beta^2}\cot\left(\sqrt{k_i^2 - \beta^2}\frac{d_i}{2}\right) + \mu_i\sqrt{\beta^2 - k_0^2}\right) \cdot \left(\sqrt{k_i^2 - \beta^2}\tan\left(\sqrt{k_i^2 - \beta^2}\frac{d_i}{2}\right) - \mu_i\sqrt{\beta^2 - k_0^2}\right) \quad (9.42)$$

The coefficients on the right side of (9.41) are given by

$$c_i = \frac{1}{2}e^{-h\sqrt{\beta^2 - k_0^2}}\left[\beta^2\left(\mu_i^2 - \mu_0^2\right) + \mu_0^2 k_i^2 - \mu_i^2 k_0^2\right]\sin\left(\sqrt{k_i^2 - \beta^2}d_i\right) \quad (9.43)$$

and they take into account the coupling effect. We note that the dispersion relation (9.41) is valid for any complex ε_n and μ_n ($n = 1, 2$).The TM mode dispersion relation may be straightforwardly obtained using the duality principle. When the two waveguides are far apart (i.e., h is sufficiently large) the coupling term on the right side of the dispersion relation vanish, and equation (9.41) reduces into the dispersion relations for the two "decoupled" open waveguides, as expected. In this case, the modes in each waveguide are unperturbed, and thus there is no coupling present.

When h is reduced, however, the modes supported by each one of the two waveguides have field distributions that extend into the region occupied by the other waveguide. This can perturb their field distributions and their wavenumbers, and the new modes satisfy the exact dispersion relation (9.41) with a field distribution obtainable by solving the boundary value problem (their expressions are not reported here for sake of brevity).

If h is not too small, the presence of one waveguide will perturb only slightly the guided modes that are derived for each waveguide separately. For an illustrative example, Fig. 9.12 shows the real part of the longitudinal component of Poynting vector distribution for the geometry of Fig. 9.11, first using the modes obtained for the two "isolated" waveguides (left panel) in which they have wavenumbers $\beta_1^{no\ coupling}$ and $\beta_2^{no\ coupling}$, respectively, and then using the exact solutions of the boundary-value problem (right panel). From this figure, it is clear how the mode distribution is slightly changed by the presence of the coupling between the two waveguides. The two modes have been chosen to have similar β, and the two waveguides carry power in opposite directions. Clearly, following relation (9.27) the wavenumbers are also slightly modified by the coupling effect, and the new wavenumbers, taking the coupling into account, are $\beta_1^{coupling}$ and $\beta_2^{coupling}$. In the following we will show how such modifications may be explained and justified.

Assuming that the coupling is not very strong and that the mode distribution is not dramatically modified by such coupling, we may approach the problem using a perturbation technique [58]. Let the unperturbed geometry be defined by the following constitutive parameters:

$$\varepsilon^{(2)}(y) = \begin{cases} \varepsilon_2 & \text{in region 2} \\ \varepsilon_0 & \text{elsewhere} \end{cases}$$
$$\mu^{(2)}(y) = \begin{cases} \mu_2 & \text{in region 2} \\ \mu_0 & \text{elsewhere} \end{cases} \tag{9.44}$$

that is, the unperturbed case is represented by the presence of only the second waveguide surrounded by vacuum. The "perturbed" problem, characterized by $\varepsilon^{(1)}(y)$ and $\mu^{(1)}(y)$ as in Fig. 9.11, consists of the geometry of interest, with the two waveguides separated by the distance h. The perturbing functions $\Delta\varepsilon_1 = \varepsilon^{(1)} - \varepsilon^{(2)}$ and $\Delta\mu_1 = \mu^{(1)} - \mu^{(2)}$ are obviously different from zero only in region 1. (Analogously, we may define the perturbing functions $\Delta\varepsilon_2$ and $\Delta\mu_2$ obtained by considering the second waveguide as the perturbation. In such a case, $\Delta\varepsilon_2$ and $\Delta\mu_2$ are of course different from zero only in region 2.)

Since the field distribution is only slightly perturbed by the coupling, as shown in Fig. 9.12, we apply the standard perturbation technique [58] to approximate the perturbed fields as a superposition of the proper modes of the two open waveguides that are derived separately in the two unperturbed cases. In other words, the perturbed field $\{\mathbf{E}'(x,y), \mathbf{H}'(x,y)\}$ is approximated by a linear superposition of the unperturbed fields in each slab that are obtained by solving the exact dispersion relation (9.41) in the limit of $h \to \infty$.

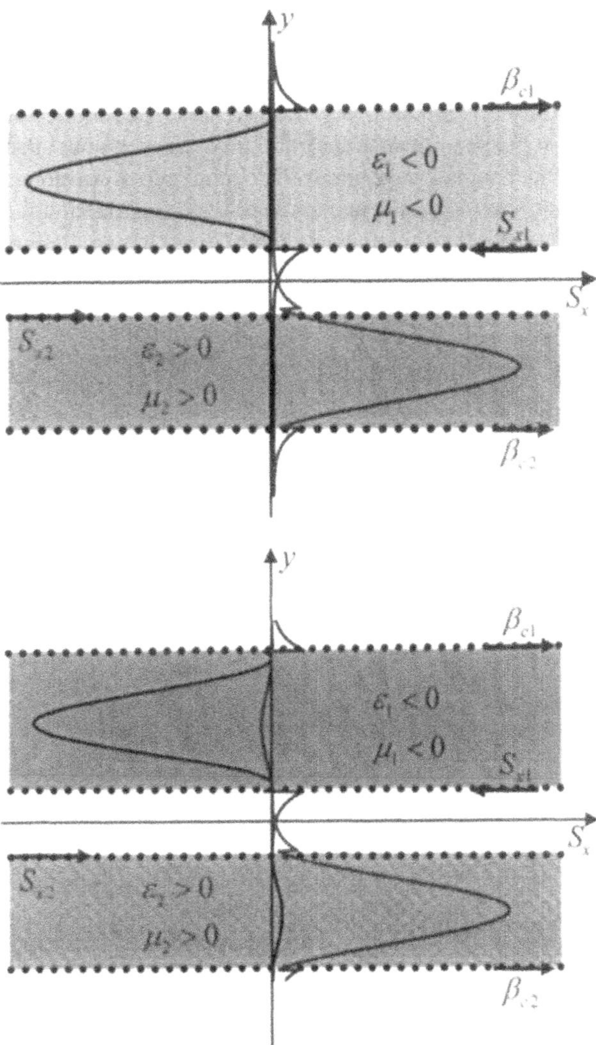

Fig. 9.12 Real part of the longitudinal component of Poynting vector distribution for the DPS and DNG waveguides, each supporting one mode. (a) Neglecting the coupling—that is, when the two waveguides are considered individually. (b) Solving exactly the boundary-value problem including the coupling effect.

If the two waveguides are monomodal and they support, respectively, the modes $\{\mathbf{e_1}(y), \mathbf{h_1}(y)\}$ and $\{\mathbf{e_2}(y), \mathbf{h_2}(y)\}$ [with field expressions given by (9.26)–(9.29)] the total perturbed fields may be expressed as

$$\left\{ \begin{array}{l} \mathbf{E}'(x,y) \\ \mathbf{H}'(x,y) \end{array} \right. \cong a_1(x) \left\{ \begin{array}{l} \mathbf{e_1}(y) \\ \mathbf{h_1}(y) \end{array} \right. e^{-j\beta_{n1}x} + a_2(x) \left\{ \begin{array}{l} \mathbf{e_2}(y) \\ \mathbf{h_2}(y) \end{array} \right. e^{-j\beta_{n2}x} \tag{9.45}$$

The amplitudes of the two modes vary with x, since, due to the coupling between the two modes, we expect a variation of the field and the power distribution as the modes travel in the coupled waveguides. The longitudinal wavenumbers β_1 and β_2 are the solutions of the respective unperturbed dispersion relations $Disp_1 = 0$ and $Disp_2 = 0$. The $\mathbf{e}$ and $\mathbf{h}$ expressions are normalized to carry unit power along the x-direction. In the case of a multimodal excitation, formula (9.45) may obviously be rewritten as

$$\left\{ \begin{array}{l} \mathbf{E}'(x,y) \\ \mathbf{H}'(x,y) \end{array} \right. \cong \sum_{i=1}^{N_1} a_{i1}(x) \left\{ \begin{array}{l} \mathbf{e_{i1}}(y) \\ \mathbf{h_{i1}}(y) \end{array} \right. e^{-j\beta_{i1}x} + \sum_{i=1}^{N_2} a_{i2}(x) \left\{ \begin{array}{l} \mathbf{e_{i2}}(y) \\ \mathbf{h_{i2}}(y) \end{array} \right. e^{-j\beta_{i2}x} \tag{9.46}$$

where N_i is the number of real solutions for propagating guided modes for $Disp_i = 0$ $(i = 1,\ 2)$.

Applying the Lorentz reciprocity theorem [58], one may write a set of N_1 equations (varying $m = 1 \ldots N_1$) for the unperturbed modes of the first waveguide:

$$\begin{aligned} &\frac{d}{dx} \int_{-\infty}^{\infty} \left[\mathbf{e}^*_{\mathbf{m1}}(y) \times \mathbf{A}(x,y) + \mathbf{B}(x,y) \times \mathbf{h}^*_{\mathbf{m1}}(y) \right] \cdot \hat{\mathbf{x}}\, dy \qquad (9.47) \\ &= -j\omega \int_{S_2} \varepsilon_0 \Delta\varepsilon_2 \mathbf{E}'(x,y) \cdot \mathbf{e}^*_{\mathbf{m1}}(y) e^{j\beta_{m1}x} + \mu_0 \Delta\mu_2 \mathbf{H}'(x,y) \cdot \mathbf{h}^*_{\mathbf{m1}}(y) e^{j\beta_{m1}x} dy \end{aligned}$$

where

$$\begin{aligned} \mathbf{A} &= \sum_{i=1}^{N_1} a_{i1}(x) \mathbf{h_{i1}}(y) e^{-j(\beta_{i1}-\beta_{m1})x} + \sum_{i=1}^{N_2} a_{i2}(x) \mathbf{h_{i2}}(y) e^{-j(\beta_{i2}-\beta_{m1})x} \\ \mathbf{B} &= \sum_{i=1}^{N_1} a_{i1}(x) \mathbf{e_{i1}}(y) e^{-j(\beta_{i1}-\beta_{m1})x} + \sum_{i=1}^{N_2} a_{i2}(x) \mathbf{e_{i2}}(y) e^{-j(\beta_{i2}-\beta_{m1})x} \end{aligned}$$

An analogous set of N_2 equations for the modes of the second waveguide is

$$\begin{aligned} &\frac{d}{dx} \int_{-\infty}^{\infty} \left[\mathbf{e}^*_{\mathbf{m2}}(y) \times \mathbf{A}(x,y) + \mathbf{B}(x,y) \times \mathbf{h}^*_{\mathbf{m2}}(y) \right] \cdot \hat{\mathbf{x}}\, dy \qquad (9.48) \\ &= -j\omega \int_{S_2} \varepsilon_0 \Delta\varepsilon_2 \mathbf{E}'(x,y) \cdot \mathbf{e}^*_{\mathbf{m2}}(y) e^{j\beta_{m1}x} + \mu_0 \Delta\mu_2 \mathbf{H}'(x,y) \cdot \mathbf{h}^*_{\mathbf{m2}}(y) e^{j\beta_{m1}x} dy \end{aligned}$$

where $\mathbf{A}$ and $\mathbf{B}$ are defined as in (9.47). Applying the mode orthogonality and considering the fact that for not-too-close waveguides we have

$$\int_{-\infty}^{\infty} \left(\mathbf{e_{in}} \times \mathbf{h_{km}}^* + \mathbf{e_{in}}^* \times \mathbf{h_{km}}\right) \cdot \hat{\mathbf{x}}\, dy \ll \int_{-\infty}^{\infty} \left(\mathbf{e_{in}} \times \mathbf{h_{in}}^* + \mathbf{e_{in}}^* \times \mathbf{h_{in}}\right) \cdot \hat{\mathbf{x}}\, dy \tag{9.49}$$

due to the decaying behavior of the modes outside the waveguide, (9.47) and (9.48) may be approximated by the coupling equations:

$$\frac{da_{m1}}{dx} = -j \sum_{k=1}^{N_2} C_{km1} a_{k2}(x) e^{-j(\beta_{k2}-\beta_{m1})x} \tag{9.50}$$

$$\frac{da_{m2}}{dx} = -j \sum_{k=1}^{N_1} C_{km2} a_{k1}(x) e^{-j(\beta_{k1}-\beta_{m2})x} \tag{9.51}$$

with

$$C_{km1} = \tfrac{\omega}{2} \int_{S_2} \left[\Delta\varepsilon_2 \mathbf{e}_{\mathbf{m1}}^*(y) \cdot \mathbf{e_{k2}}(y) + \Delta\mu_2 \mathbf{h}_{\mathbf{m1}}^*(y) \cdot \mathbf{h_{k2}}(y)\right] dy$$
$$C_{km2} = \tfrac{\omega}{2} \int_{S_1} \left[\Delta\varepsilon_1 \mathbf{e}_{\mathbf{m2}}^*(y) \cdot \mathbf{e_{k1}}(y) + \Delta\mu_1 \mathbf{h}_{\mathbf{m2}}^*(y) \cdot \mathbf{h_{k1}}(y)\right] dy$$

Equations (9.50) and (9.51) represent a set of $N_1 + N_2$ differential equations to be solved for the $a\,(x)$ unknowns.

The system admits solutions of the form $a_{mi} = a_{mi}^{(0)} e^{-j\gamma_{mi}x}$, where the coefficients γ satisfy the algebraic system:

$$\begin{aligned}
&\sum_{k=1}^{N_2} C_{k11} a_{k2}^{(0)} - \gamma_{11} a_{11}^{(0)} = 0 \\
&\vdots \\
&\sum_{k=1}^{N_2} C_{kN_11} a_{k2}^{(0)} - \gamma_{N_11} a_{N_11}^{(0)} = 0 \\
&\sum_{k=1}^{N_1} C_{k12} a_{k1}^{(0)} - \gamma_{12} a_{12}^{(0)} = 0 \\
&\vdots \\
&\sum_{k=1}^{N_1} C_{kN_22} a_{k1}^{(0)} - \gamma_{N_22} a_{N_22}^{(0)} = 0
\end{aligned} \tag{9.52}$$

and the relations $\gamma_{mi} + \beta_{mi} = K$, for $i = 1, 2$ and any m, with K an arbitrary parameter. Imposing the condition that the determinant of (9.52) is zero, together with the latter relations, gives a system to be solved for the γ_{mi}, after which the coefficients $a\,(x)$ may be determined with suitable boundary conditions.

In general, using the reciprocity theorem it may be shown that $C_{km1} = \pm C_{mk2}^*$. This represents a generalization of the well-known formula (with the plus sign) valid for a standard directional coupler with waveguides [58]. The sign here depends on the direction of the coupling: The plus is valid for any directional coupling for which the two modes are either both forward or both backward modes; and the minus for a contradirectional coupling for which the two modes carry their net power in opposite directions. We should also be reminded that a non-negligible coupling between the two modes with a sufficiently high value of the coupling coefficient C, is possible only when the two β's are close enough.

Let us now consider the coupling between two given modes with longitudinal wavenumbers β_{11} and β_{12} in the waveguide coupler in Fig. 9.13. At $x = 0$, only the second waveguide is excited. As is conventionally done, by solving the system (9.52) we obtain the amplitude coefficients:

$$\begin{aligned} a_{11}(x) &= Ae^{j\frac{\beta_{11}-\beta_{12}}{2}x}\sin(Sx) \\ a_{12}(x) &= \frac{A}{C_{111}}e^{j\frac{\beta_{12}-\beta_{11}}{2}x}\left[\frac{\beta_{12}-\beta_{11}}{2}\sin(Sx) + jS\cos(Sx)\right] \end{aligned} \tag{9.53}$$

with $S = \sqrt{\left(\frac{\beta_{11}-\beta_{12}}{2}\right)^2 + C_{111}C_{112}}$ and where A is related to the amplitude of the excitation.

When two conventional DPS slabs are considered, as in a standard waveguide directional coupler (Fig. 9.13a), the coupling of one waveguide with the other is "directional," since the two modes (which must have somewhat similar β) carry power in the same direction. In this case, $S = \sqrt{\left(\frac{\beta_{11}-\beta_{12}}{2}\right)^2 + |C_{111}|^2}$ and it is always a real quantity. The power carried by the two distinct modes is

$$\begin{aligned} P_1(x) &= |a_{11}(x)|^2 = |A|^2\sin^2(Sx) \\ P_2(x) &= |a_{12}(x)|^2 = \frac{|A|^2}{|C_{111}|^2}S^2\left[\left(\frac{\beta_{12}-\beta_{11}}{2S}\right)^2\sin^2(Sx) + \cos^2(Sx)\right] \end{aligned} \tag{9.54}$$

which shows the sinusoidal exchange of power, typical of a directional coupler [58]. In this case it is well known that the maximum power transferred from one mode to the other is given by

$$P_1\left(\frac{\pi}{2S}\right) = |A|^2 = P_2(0)\frac{|C_{111}|^2}{S^2} = P_2(0)\frac{1}{1+\left(\frac{\beta_{11}-\beta_{12}}{2|C_{111}|^2}\right)^2} \tag{9.55}$$

where $P_2(0) = \frac{|A|^2}{|C_{111}|^2}S^2$ is the power in the second mode at the beginning of the coupling (at $x = 0$). This value is obtained when the length of the coupler is chosen to be $L = \frac{\pi}{2S}$. These are well-known classic results and, of course, they correspond to the interference of two perturbed modes [solutions of (9.41)], which propagate in the whole structure and interfere along the coupler [43,58].

Let us now analyze the case depicted in Fig. 9.13b—that is, when the waveguide coupler consists of a DPS and a DNG slab. In this case, provided that one of the supported modes concentrate most of its power in the DNG slab (and not in its

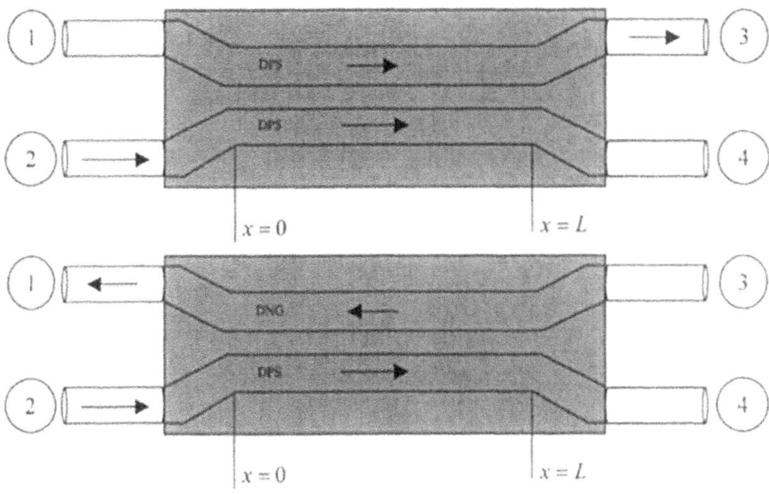

Fig. 9.13 (a) Standard directional coupler with DPS–DPS waveguides (similar to one in Ref. [58], page 226). (b) Contradirectional (backward) coupler formed by a DPS and a DNG slab.

surrounding region), a contradirectional coupling is expected, since the two nonperturbed modes, which must have similar β to allow a significant coupling, should support power flows with opposite directions in the two materials (as derived in the previous section) and therefore $S = \sqrt{\left(\frac{\beta_{11}-\beta_{12}}{2}\right)^2 - |C_{111}|^2}$. Notice that in this case, whatever power is coupled in the DNG slab, we expect to find it in port 1 and not in port 3, typical of a contradirectional coupling.

We now assume that the values of the two nonperturbed betas β_{11} and β_{12} are fixed. These values are close numbers and correspond to two nonperturbed modes, of which the one in the DNG is backward (since it concentrates its power flow inside the DNG slab). When the distance between the two waveguides is sufficiently large, the coupling is very weak between the two modes and we may assume that the corresponding coupling coefficient satisfies $|C_{111}|^2 < \left(\frac{\beta_{11}-\beta_{12}}{2}\right)^2$. In this case, S is a real quantity and the exchange of power remains sinusoidal as in a standard directional coupler. However, when we decrease the distance h between the two waveguides, the spatial period of the power exchange between the two waveguides *increases* (since C_{111} increases with the coupling), contrary to the case of two DPS waveguides in which such spatial period would decrease. The closer the waveguides are, the longer the coupler should be to achieve the maximum power transfer from port 2 to port 1. Even with this anomaly, reducing the distance between the two waveguides leads to the increase of the value of the total coupled power, since this quantity still follows relation (9.55). Eventually, as we decrease h further, we reach a point for which $S = 0$ and the spatial period of the sinusoidal exchange of coupled

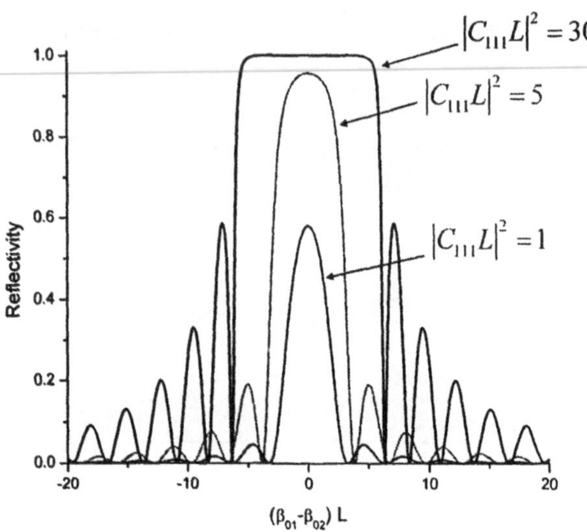

Fig. 9.14 Reflectivity of a contradirectional (backward) coupler for different values of its parameters. This may effectively be similar to the corresponding reflectivity of conventional grating (e.g., Ref. [58], page 240). However, here the "reflected" power is in a separate channel and is isolated from the "incident" one. In other words, their power flows are spatially localized in the two different waveguides.

power becomes infinite. At this point, in principle a "complete" exchange of power cannot be achieved with a finite coupler.

It is important to mention that these features may also be confirmed through solving the modes from the exact dispersion relation (9.41). In a standard waveguide directional coupler, when we fix the geometry of the two waveguides, and therefore fix the values of $\beta_1^{no\ coupling}$ and $\beta_2^{no\ coupling}$, as h is reduced the exact solutions of (9.41) $\beta_1^{coupling}$ and $\beta_2^{coupling}$ move farther from each other, and their interference spatial period consequently decreases. When instead we consider an antidirectional coupler with a DPS and a DNG, it may be verified that the two solutions $\beta_1^{coupling}$ and $\beta_2^{coupling}$ move closer as h is reduced, thus increasing the spatial period of their coupling. We get to a point (which is given in the approximate perturbation analysis by the condition $S = 0$) at which the two supported modes have the same beta (i.e., $\beta_1^{coupling} = \beta_2^{coupling}$), and the interference is no longer present (i.e., its period is infinite). By decreasing the distance h further, and consequently increasing the coupling coefficient C_{111}, an imaginary S is resulted in the perturbation approach—that is, an exponential variation (rather than a sinusoidal variation) for the power exchange, in which the power is "redirected" back to port 1 continuously and exponentially with a factor that increases as h decreases. In the exact approach, this is due to the fact that the two modes, shares the same real part, but starts having two oppositely-valued imaginary parts! The contradirectional (i.e., backward) coupler with a strong coupling, therefore, becomes somehow similar to a periodically corrugated waveguide (grating reflector [58]) in its stopband, but with the unusual feature that the "reflected" power is effectively flowing in a separate channel and is isolated from the "incident" one, which implies that their power flows are spatially localized in the two different waveguides. These results may be summarized in Fig. 9.14, where the reflectivity of the contradirectional (backward) coupler is plotted versus the difference between the two nonperturbed wavenumbers for different values of the coupling factor.

Acknowledgments

This work is supported in part by the Fields and Waves Laboratory, Department of Electrical and Systems Engineering, University of Pennsylvania. Andrea Alù has been supported by the scholarship "Isabella Sassi Bonadonna" from the Italian Electrical Association (AEI).

REFERENCES

1. R. A. Shelby, D. R. Smith, and S. Schultz, "Experimental verification of a negative index of refraction," *Science*, vol. 292, no. 5514, pp. 77–79, 2001.

2. J. B. Pendry, A. J. Holden, D. J. Robbins, and W. J. Stewart, "Magnetism from conductors and enhanced nonlinear phenomena," *IEEE Trans. Microwave Theory Tech.*, vol. 47, no. 11, pp. 2075–2081, November 1999.

3. J. B. Pendry, A. J. Holden, D. J. Robbins, and W. J. Stewart, "Low-frequency plasmons in thin wire structures," *J. Physics: Condensed Matter*, vol. 10, pp. 4785–4809, 1998.

4. V. G. Veselago, "The electrodynamics of substances with simultaneously negative values of ε and μ," *Sov. Phys. Usp.*, vol. 10, no. 4, pp. 509–514, 1968 [in Russian *Usp. Fiz. Nauk*, vol. 92, pp. 517–526, 1967].

5. R. W. Ziolkowski and E. Heyman, "Wave propagation in media having negative permittivity and permeability," *Phys. Rev. E.*, vol. 64, no. 5, 056625, 2001.

6. D. R. Smith, W. J. Padilla, D. C. Vier, S. C. Nemat-Nasser, and S. Schultz, "Composite medium with simultaneously negative permeability and permittivity," *Phys. Rev. Lett.*, vol 84, no. 18, pp. 4184–4187, 2000.

7. R. A. Shelby, D. R. Smith, S. C. Nemat-Nasser, and S. Schultz, "Microwave transmission through a two-dimensional, isotropic, left-handed metamaterial," *Appl. Phys. Lett.*, vol. 78, no. 4, pp. 489–491, 2001.

8. J. B. Pendry, "Negative refraction makes a perfect lens," *Phys. Rev. Lett.*, vol. 85, no. 18, pp. 3966–3969, 2000.

9. R. W. Ziolkowski, "Superluminal transmission of information through an electromagnetic metamaterial," *Phys. Rev. E.*, vol. 63, no. 4, 046604, April 2001.

10. R. W. Ziolkowski, "Pulsed and CW Gaussian beam interactions with double negative metamaterial slabs," *Opt. Express*, vol. 11, no. 7, pp. 662–681, April 7, 2003.

11. R. W. Ziolkowski and A. D. Kipple, "Application of double-negative materials to increase power radiated by electrically small antennas," *IEEE Trans. Antennas Propag.*, Special Issue on Metamaterials, vol. 51, no. 10, pp. 2626–2640, October 2003.

12. A. A. Oliner, "A periodic-structure negative-refractive-index medium without resonant elements," in *URSI Digest, 2002 IEEE AP-S International Symposium/USNC/URSI National Radio Science Meeting*, San Antonio (TX), June 16–21, 2002, p. 41.

13. A. A. Oliner, "A planar negative-refractive-index medium without resonant elements," in *Digest IEEE MTT International Microwave Symposium (IMS'03)*, Philadelphia, PA, June 8–13, 2003, pp. 191–194.

14. A. Grbic and G. V. Eleftheriades, "A backward-wave antenna based on negative refractive index L–C networks," in *Proceedings, 2002 IEEE AP-S Int. Symposium/USNC/URSI National Radio Science Meeting*, San Antonio (TX), June 16–21, 2002, vol. 4, pp. 340–343.

15. G. V. Eleftheriades, A. K. Iyer, and P. C. Kremer, "Planar negative refractive index media using periodically L–C loaded transmission lines," *IEEE Trans. Microwave Theory Tech.*, vol. 50, no. 12, pp. 2702–2712, December 2002.

16. R. Islam and G. V. Eleftheriades, "A planar metamaterial co-directional coupler that couples power backward," in *Digest IEEE MTT International Microwave Symposium (IMS'03)*, Philadelphia, PA, June 8–13, 2003, pp. 321–324.

17. C. Caloz, H. Okabe, T. Iwai, and T. Itoh, "Transmission line approach of left-handed materials," in *URSI Digest, 2002 IEEE AP-S International Symposium/USNC/URSI National Radio Science Meeting*, San Antonio (TX), June 16–21, 2002, p. 39.

18. L. Liu, C. Caloz, C.-C. Chang, and T. Itoh, "Forward coupling phenomena between artificial left-handed transmission lines," *J. Appl. Phys.*, vol. 92, no. 9, pp. 5560–5565, November 1, 2002.

19. C. Caloz, A. Sanada, L. Liu, and T. Itoh, "A broadband left-handed (LH) coupled-line backward coupler with arbitrary coupling levels," in *Digest IEEE MTT International Microwave Symposium (IMS'03)*, Philadelphia, PA, June 8–13, 2003, pp. 317–320.

20. Z. M. Zhang and C. J. Fu, "Unusual photon tunneling in the presence of a layer with a negative refractive index," *Appl. Phys. Lett.*, vol. 80, no. 6, pp. 1097–1099, February 11, 2002.

21. I. V. Lindell, S. A. Tretyakov, K. I. Nikoskinen, and S. Ilvonen, S. "BW media—media with negative parameters, capable of supporting backward waves," *Microwave Opt. Technol. Lett.*, vol. 31, no. 2, pp. 129–133, 2001.

22. S. A. Tretyakov, "Metamaterials with wideband negative permittivity and permeability," *Microwave Opt. Technol. Lett.*, vol. 31, no. 3, pp. 163–165, 2001.

23. K. G. Balmain, A. A. E. Luttgen, and P. C. Kremer, "Resonance cone formation, reflection, refraction and focusing in a planar anisotropic metamaterial," *IEEE Antennas Wireless Propag. Lett.*, vol. 1, pp. 146–149, 2002.

24. M. W. McCall, A. Lakhtakia, and W. S. Weiglhofer, "The negative index of refraction demystified," *Eur. J. Phys.*, vol. 23, pp. 353–359, 2002.

25. N. Garcia and M. Nieto-Vesperinas, "Left-handed materials do not make a perfect lens," *Phys. Rev. Lett.*, vol. 88, no. 20, 207403, 2002.

26. A. Lakhtakia, "Reversed circular dichroism of isotropic chiral mediums with negative permeability and permittivity," *Microwave Opt. Technol. Lett.*, vol. 33, no. 2, pp. 96–97, April 20, 2002.

27. A. N. Lagarkov and V. N. Kisel, "Electrodynamics properties of simple bodies made of materials with negative permeability and negative permittivity," *Doklady Physics*, vol. 46, no. 3, pp. 163–165, 2001 [in Russian *Dokl. Akad. Nauk*, vol. 377, no. 1, pp. 40–43, 2001].

28. M. W. Feise, P. J. Bevelacqua, and J. B. Schneider, "Effects of surface waves on behavior of perfect lenses," *Phys. Rev. B*, vol. 66, 035113, 2002.

29. R. Marques, F. Medina, and R. Rafii-El-Idrissi, (2002) "Role of bianisotropy in negative permeability and left-handed metamaterials," *Phys. Rev. B*, vol. 65, no. 14, 144440, 2002.

30. R. Marques, J. Martel, F. Mesa, and F. Medina, "A new 2-D isotropic left-handed metamaterial design: Theory and experiment," *Microwave Opt. Technol. Lett.*, vol. 36, pp. 405–408, December 2002.

31. S. Hrabar, Z. Eres, and J. Bartolic, "Capacitively loaded loop as basic element of negative permeability meta-material," in *Proceedings, 32nd European Microwave Conference 2002 (EuMC' 2002)*, Milan, Italy, September 24–26, 2002.

32. P. M. Valanju, R. M. Walser, and A. P. Valanju, "Wave refraction in negative-index media: Always positive and very inhomogeneous," *Phys. Rev. Lett*, vol. 88, no. 18, 012220, 2002.

33. D. R. Smith, D. Schurig, and J. B. Pendry, "Negative refraction of modulated electromagnetic waves," *Appl. Phys. Lett*, vol. 81, no. 15, pp. 2713–2715, October 7, 2002.

34. P. Gay-Balmaz and O. J. F. Martin, "Efficient isotropic magnetic resonators," *Appl. Phys. Lett.*, vol. 81, no. 5, pp. 939–941, July 29, 2002.

35. J. A. Kong, J. A., B.-I. Wu, and Y. Zhang, "A unique lateral displacement of a Gaussian beam transmitted through a slab with negative permittivity and permeability," *Microwave Opt. Technol. Lett.*, vol. 33, no. 2, pp. 136–139, 2002.

36. R. A. Silin and I. P. Chepurnykh, "On media with negative dispersion," *J. Commun. Technol. Electron.*, vol. 46, no. 10, pp. 1121–1125, 2001 [in Russian, *Radiotekhnika*, vol. 46, no. 10, pp. 1212–1217, 2001].

37. N. Engheta, "An idea for thin subwavelength cavity resonators using metamaterials with negative permittivity and permeability," *IEEE Antennas Wireless Propag. Lett.*, vol. 1, no. 1, pp. 10–13, 2002. Digital Object Identifier: 10.1109/LAWP.2002.802576.

38. N. Engheta, "Guided waves in paired dielectric-metamaterial with negative permittivity and permeability layers," in *URSI Digest, USNC-URSI National Radio Science Meeting*, Boulder, CO, January 9–12, 2002, p. 66.

39. N. Engheta, "Ideas for potential applications of metamaterials with negative permittivity and permeability," *Advances in Electromagnetics of Complex Media and Metamaterials*, NATO Science Series, S. Zouhdi, A. H. Sihvola, and M. Arsalane, eds., Kluwer Academic Publishers, Dordrecht, pp. 19–37, 2002.

40. A. Alù and N. Engheta, "Mode excitation by a line source in a parallel-plate waveguide filled with a pair of parallel double-negative and double-positive slabs," in *Proceedings, 2003 IEEE AP-S International Symposium*, Columbus, OH, June 22–27, 2003, Vol. 3, pp. 359–362.

41. A. Alù and N. Engheta, "Mono-modal waveguides filled with a pair of parallel epsilon-negative (ENG) and mu-negative (MNG) metamaterial layers," in *Digest IEEE MTT International Microwave Symposium (IMS'03)*, Philadelphia, PA, June 8–13, 2003, pp. 313–316.

42. A. Alù and N. Engheta, "Guided modes in a waveguide filled with a pair of single-negative (SNG), double-negative (DNG), and/or double-positive (DPS) layers," *IEEE Trans. Microwave Theory Tech.*, vol. MTT-52, no. 1, pp. 199–210, January 2004.

43. A. Alù and N. Engheta, "Anomalous mode coupling in guided-wave structures containing metamaterials with negative permittivity and permeability," in *Proceedings, 2002 IEEE—Nanotechnology*, Washington, DC, August 26–28, 2002, pp. 233–234.

44. A. Alù and N. Engheta, "Radiation from a traveling-wave current sheet at the interface between a conventional material and a material with negative permittivity and permeability," *Microwave Opt. Technol. Lett.*, vol. 35, no. 6, pp. 460–463, December 20, 2002.

45. N. Engheta, "Is Foster's reactance theorem satisfied in double-negative and single-negative media?" *Microwave Opt. Technol. Lett.*, vol. 39, no. 1, pp. 11–14, October 5, 2003.

46. N. Engheta, S. Nelatury, and A. Hoorfar, "The role of geometry of inclusions in forming metamaterials with negative permittivity and permeability" in *Proceedings, XXVII General Assembly of International Union of Radio Science (URSI GA'02)*, in Maastricht, The Netherlands, August 17–24, 2002. Paper number 1935 in the CD.

47. A. Alù and N. Engheta, "Pairing an epsilon-negative slab with a mu-negative slab: resonance, tunneling and transparency," *IEEE Trans. Antennas Propag.*, Special issue on "Metamaterials," vol. 51, no. 10, pp. 2558–2571, October 2003.

48. D. R. Fredkin and A. Ron, "Effective left-handed (negative index) composite material," *Appl. Phys. Lett.*, vol. 81, no. 10, pp. 1753–1755, September 2, 2002.

49. A. Topa, "Contradirectional interaction in a NRD waveguide coupler with a metamaterial slab," *XXVII General Assembly of International Union of Radio Science (URSI GA'02)*, in Maastricht, The Netherlands, August 17–24, 2002, paper number 1878 in the CD digest.

50. I. S. Nefedov and S. A. Tretyakov, "Theoretical study of waveguiding structures containing backward-wave materials," *XXVII General Assembly of International Union of Radio Science (URSI GA'02)*, Maastricht, The Netherlands, August 17–24, 2002, paper number 1074 in the CD digest.

51. I. S. Nefedov and S. A. Tretyakov, "Waveguide containing a backward-wave slab," e-print in arXiv:cond-mat/0211185 v1, at http://arxiv.org/pdf/cond-mat/0211185, November 10, 2002.

52. B.-I. Wu, T. M. Grzegorczyk, Y. Zhang, and J. A. Kong, "Guided modes with imaginary transverse wave number in a slab waveguide with negative permittivity and permeability," *J. Appl. Phys.*, vol. 93, no. 11, pp. 9386–9388, June 1, 2003.

53. C. Caloz, C.-C. Chang, and T. Itoh, "Full-wave verification of the fundamental properties of left-handed materials in waveguide configurations," *J. Appl. Phys.*, vol. 90, no. 11, pp. 5483–5486, December 2001.

54. A. Alù and N. Engheta, "Distributed-circuit-element description of guided-wave structures and cavities involving double-negative or single-negative media," in

Proceedings SPIE: Complex Mediums IV: Beyond Linear Isotropic Dielectrics, Vol. 5218, San Diego, CA, August 4–5, 2003, pp. 145–155.

55. A. Alù and N. Engheta, "Resonances in sub-wavelength cylindrical structures made of pairs of double-negative and double-negative or epsilon-negative and mu-negative coaxial shells," in *Proceedings of the International Conference in Electromagnetics and Advance Applications (ICEAA'03)* Torino, Italy, September 8–12, 2003, pp. 435–438.

56. P. J. B. Clarricoats, "Circular-waveguide backward-wave structures," *Proc. IEE*, vol. 110, no. 2, pp. 261–270, February 1963.

57. R. A. Waldron, "Theory and potential applications of backward waves in non periodic inhomogenoeus waveguides," *Proc. IEE*, vol. 111, no. 10, pp. 1659–1667, October 1964.

58. D. L. Lee, *Electromagnetic Principles of Integrated Optics*, John Wiley & Sons, New York, 1986.

10 Dispersion Engineering: The Use of Abnormal Velocities and Negative Index of Refraction to Control Dispersive Effects

MOHAMMAD MOJAHEDI and GEORGE V. ELEFTHERIADES

The Edward S. Rogers Sr. Department of Electrical and Computer Engineering
University of Toronto
Toronto, Ontario, M5S 3G4
Canada

10.1 INTRODUCTION

In recent years, two discoveries have led to the possibility of manufacturing materials and structures with dispersive behaviors previously thought unattainable. These are media with an effective negative index of refraction (NIR) also known as the left handed media (LHM) or negative-refractive-index metamaterials [1–4], and the possibility of measuring superluminal* or negative group velocities also generically referred to as abnormal group velocities [5–9]. In this manuscript we study the underlying physics and manifestations of these rather unusual behaviors, and we will see how a medium can be manufactured that simultaneously demonstrates both properties. But before proceeding, let us describe our motivation behind this approach and see how combining these two effects may lead to the design of new classes of materials with novel and counterintuitive dispersive effects, a subject we have termed "Dispersion Engineering."

The fact that any physically realizable medium must be dispersive is the consequence of the principle of causality, which demands that no effect precedes its cause [10]. In the simple case of one-dimensional analysis, the dispersive behavior of a medium can be described by the dependence of the propagation vector on frequency (or equally well the dependence of the frequency on the wavevector) according to

*The term *superluminal* implies group velocities in excess of the speed of light in vacuum.

Ref. 11

$$k(\omega) = k(\omega_0) + \left.\frac{dk}{d\omega}\right|_{\omega_0} (\omega - \omega_0) + \frac{1}{2}\left.\frac{d^2k}{d\omega^2}\right|_{\omega_0} (\omega - \omega_0)^2 + \cdots$$
$$= v_p^{-1}\omega_0 + v_g^{-1}(\omega - \omega_0) + \frac{1}{2}\psi(\omega - \omega_0)^2 + \cdots \quad (10.1)$$

In equation (10.1) the coefficients of expansion v_p, and v_g are the phase and group velocities, and ψ is the group velocity dispersion (GVD) given by

$$\text{GVD} = \frac{d^2k}{d\omega^2} = -\frac{1}{v_g^2}\left(\frac{dv_g}{d\omega}\right) \quad (10.2)$$

With the photonic dispersion relation $k(\omega) = \omega\, n_p(\omega)/c$, the dispersive effects of equation (10.1) can be also described in terms of the phase index (commonly referred to as the index of refraction, n_p) and its higher-order derivatives according to

$$k(\omega) = \frac{\omega}{c} n_p(\omega)$$
$$= \left.\frac{\omega_0}{c} n_p\right|_{\omega_0} + \left.\frac{n_g}{c}\right|_{\omega_0} (\omega - \omega_0) + \frac{1}{2c}\left.\left[2\frac{dn_p}{d\omega} + \omega\frac{d^2n_p}{d\omega^2}\right]\right|_{\omega_0} (\omega - \omega_0)^2 + \cdots$$
$$(10.3)$$

In equation (10.3) the second coefficient of expansion is the group delay which is related to the phase index by

$$n_g = n_p + \omega\, dn_p(\omega)/d\omega \quad (10.4)$$

The relations between the phase index and phase velocity, group index and group velocity are then as follows:

$$v_p = \frac{c}{n_p(\omega)} \quad (10.5)$$
$$v_g = \frac{c}{n_g(\omega)} \quad (10.6)$$

So far, in discussing the dispersive effects signified by phase velocity, group velocity, group velocity dispersion, and so on, we have made an implicit assumption that the medium under consideration has a non-negligible spatial extent, or more rigorously $L > \lambda$, where L is the physical length of the one dimensional medium and λ is the wavelength of the excitation. However, the aforementioned dispersive effects can be formulated in a more general way that is equally applicable to both spatially extended ($L > \lambda$) or spatially negligible systems ($L < \lambda$). This formulation relies on the notion of transfer function (impulse response) which can be used to relate the input and the output. The transfer function, also sometimes referred to as the system response or network function, is a complex quantity given by

$$T(\omega) = |T(\omega)| \exp[j\,\phi(\omega)] \quad (10.7)$$

With a relatively constant value for the transmission function magnitude, or equally well a sufficiently narrowband excitation, the phase of the transfer function can be expanded in a Taylor series according to

$$\phi(\omega) = \phi(\omega_0) + \left.\frac{\partial \phi}{\partial \omega}\right|_{\omega_0} (\omega - \omega_0) + \frac{1}{2} \left.\frac{\partial^2 \phi}{\partial \omega^2}\right|_{\omega_0} (\omega - \omega_0)^2 + \cdots \tag{10.8}$$

or

$$\phi(\omega) = -\tau_p \omega_0 - \tau_g (\omega - \omega_0) - \frac{1}{2}\mathrm{GDD}(\omega - \omega_0)^2 + \cdots \tag{10.9}$$

where the phase delay (τ_p), group delay (τ_g), and group delay dispersion (GDD) are given by

$$\tau_p = -\left.\frac{\phi}{\omega}\right|_{\omega_0} \tag{10.10}$$

$$\tau_g = -\left.\frac{\partial \phi}{\partial \omega}\right|_{\omega_0} \tag{10.11}$$

$$GDD = -\left.\frac{\partial^2 \phi}{\partial \omega^2}\right|_{\omega_0} \tag{10.12}$$

The connections between the phase and group delays are applicable to both spatially extended and spatially negligible systems [equations (10.11) and (10.12)], and the phase and group velocity (applicable to a spatially extended system, $\lambda > L$) is then as follows:

$$v_p = \frac{c}{n_p(\omega)} = \frac{L}{\tau_p} \tag{10.13}$$

$$v_g = \frac{c}{n_g(\omega)} = \frac{L}{\tau_g} \tag{10.14}$$

In fact, equation (10.14) can be obtained using more rigorous arguments based on the Fourier transform theorem [12], or more intuitively by considering the wave propagation through a slab of thickness L , matched to its surrounding media (i.e., no reflection at the interfaces), having the transmission function $T = \exp(j\phi) = \exp[-j\, k(\omega)\, L]$. Finally, two more points are worth mentioning. First, from equation (10.14) it is clear that for a physical system of length L the sign of the group velocity and group delay are the same. Second, when discussing a spatially extended system, the fundamental requirements of causality, also referred to as "primitive causality," must be augmented with relativistic causality, also referred to as "macroscopic" or "Einstein causality" [13]. We note that the concept of primitive causality is more general than macroscopic causality since it does not rely on the existence of a finite speed (c) for propagation of the "cause."

Now, the notion of "Dispersion Engineering," alluded to earlier, reflects our desire to synthesize and control various dispersive effects, and in particular their associated signs, as manifested by the phase delay, group delay, group delay dispersion, and so on [see equation (10.9) or similarly equations (10.1) or (10.3)]. In the remaining parts of this manuscript, we focus only on the first two terms of the expansion—that is, the phase and group delays or equally well the phase and group velocities—and leave the consideration of the group delay dispersion to later times.

From a physical point of view, the phase and group delays are the delays encountered by sinusoidal time harmonics and the wave packet envelope (composed of such harmonics) as they propagate through the media, respectively. Under usual propagation conditions, both the phase and group velocities (read phase and group delays) are positive, indicating the fact that both the sinusoidal time harmonics and the pulse envelope move away from the source.

In the case of negative phase but positive group velocity (read negative phase but positive group delays) the sinusoidal time harmonics move toward the source while the wave packet envelope moves away from the source. This phenomenon is sometimes referred to as backward-wave propagation [14] and is the signature of the LHM studies so far [4, 15–17].

More interestingly, under some conditions, it is also possible to observe positive phase but negative group velocities (read positive phase but negative group delays) [18–21]. Under these circumstances, for a finite-length medium exhibiting such abnormal behavior and illuminated by a source outside, the observer will note that the sinusoidal time harmonics move away from the source but the wave packet envelope (inside the medium) moves toward the source. Stated otherwise, our observer will note that the peak (envelope) of a well-behaved pulse will emerge from the medium prior to the peak of the incident pulse entering it. This counterintuitive behavior is a subclass of the so-called "abnormal group velocities," which will be revisited shortly in the next section.

Finally, by combining the backward waves and abnormal group velocities, one can synthesize a medium with simultaneous negative phase and group velocities (read negative phase and group delays.) Under this condition, both the sinusoidal time harmonics and the wave packet envelope move toward the source. This case is of particular interest to us and will be investigated theoretically and experimentally in the following sections.

In Section 10.2 we revisit the concept of abnormal group velocities (superluminal or negative) and will provide a short overview of the field. We begin our study of the NRI media exhibiting both negative phase and negative group velocities with the simple case of slab having a Lorentzian magnetic and electric response. This simple case will set the stage for more detailed analysis of NRI media realized by periodically loading a transmission line which exhibits both negative phase and group delays. The theoretical and experimental studies of such media can be found in Section 10.4. For the sake of brevity, the results of similar observations in an LHM consisting of split-ring resonator (SRR) and strip wires will be presented elsewhere, whereas an interested reader may consult Ref. 22 for a brief description of this situation. Finally, we summarize our work in Section 10.5.

10.2 ABNORMAL GROUP VELOCITY

Soon after Einstein's formulation of special relativity in 1905, the question of wave propagation in a medium with Lorentz–Lorenz dispersion captured the attention of the researchers of the time [23]. It was known, at least theoretically, that within the region of anomalous dispersion for such medium the group velocity—describing the velocity by which the peak and hence the envelope of a well-behaved wave packet travels—can exceed the speed of light in vacuum; that is, it becomes superluminal. On its face value, this theoretical possibility was in contrast with the requirements of relativistic causality as formulated by Einstein. Sommerfeld and then his postdoctoral fellow, Brillouin, undertook the analysis of this problem, the result of which was published in 1914 and later republished and expanded in Ref. 23. The two authors defined or clarified many velocity terms such as phase, group, energy, "signal,"† and most importantly the first and second forerunner (precursor) velocities. One important aspect of their work was establishing the fact that the velocity of the earliest field oscillations known as the front will never exceed the speed of light in vacuum; and in fact, under all circumstances it remains exactly luminal. While this seminal work confirmed the compatibility of the relativistic causality and wave propagation in a Lorentzian medium, Brillouin along with many others considered the superluminal or negative group velocities as unphysical, and perhaps a mere mathematical consideration [24–26].

Decades later, in 1970, Garret and McCumber revisited the same problem and concluded that under certain easily satisfied conditions, superluminal or negative group velocities may be observed and are therefore physical [27]. In their work, they considered the propagation of Gaussian packets and showed that these could travel at abnormal group velocities without significant distortion of the pulse shape even though the pulse was attenuated.

Chu and Wong (1982) were the first to experimentally demonstrate the existence of abnormal group velocities for picosecond laser pulses propagating through the excitonic absorption line of a GaP:N sample [28]. Since then, abnormal group velocities have been measured in various structures including photonic crystals, undersized waveguides, misaligned horn antennas, side-by-side prisms, and inverted media [6, 8, 20, 21, 29–33].

Figure 10.1 summarizes the results. Consider a medium of length L, excited from left by a smoothly varying pulse (such as a Gaussian or Sinc). The pulse on the right is the transmitted pulse (output). Under normal propagation conditions the medium group index [see equation (10.4)] is greater than one, which implies a group delay larger than L/c and a group velocity less than c. In other words, the output pulse peak (envelope) is delayed as compared to the input. On the other hand, for

†The "signal" velocity as defined by Sommerfeld and Brillouin is approximately the velocity of the half-maximum point, and by their own admission is an arbitrary construct. In other words, in light of new observations of abnormal velocities, the so-called "signal" velocity should not be confused with genuine information velocity, which is constrained by relativistic causality.

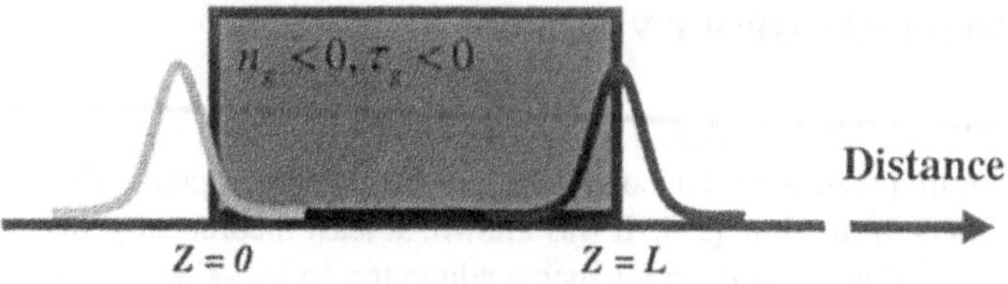

Fig. 10.1 A well-behaved wave packet incident from left and traveling through a medium of length L.

$0 < n_g < 1$, the group velocity is larger than c and the group delay is less than L/c. This is the case of superluminal group velocity. When the group index approaches zero ($n_g \to 0$), the group velocity grows unbounded ($v_g \to \infty$) and the group delay approaches zero ($\tau_g \to 0$). To the observer, the peak of the output pulse in Fig. 10.1 appears at the right interface ($z = L$) at the same time the peak of the input wave packet is at the left interface ($z = 0$). Now, if we continue with our analysis and consider a negative group index ($n_g < 0$), then both the group velocity and group delay are also negative ($v_g < 0, \tau_g < 0$). Under this condition, the observer notes that the peak of the output leaves the medium prior to the peak of the input entering it, as it is depicted in Fig. 10.1.

While Fig. 10.1 depicts the pulse propagation in space, better insight may be obtained by considering the behavior in time. Figure 10.2 shows the abnormal wave propagation in the time domain. The picture is equally applicable to both cases of superluminal or negative group velocities. Whereas in the later case the input and output are to be understood as the input to and output from the medium of length L (depicted in Fig. 10.1), in the former case the input is to be understood as a wave packet traveling a distance L through vacuum and the output is a pulse traveling through a medium of the same length with $0 < n_g < 1$. There are two points worth emphasizing.

First, for the passive medium, the superluminal or negative group velocity is accompanied by attenuation of the pulse. In other words, for such a medium, the wave packets depicted in Fig. 10.2 are considered to be normalized to their respective maximum values. However, such attenuation is not a necessary condition for all cases of abnormal group velocities. In fact, it has been shown that an inverted medium (a medium with gain) can display abnormal group velocities without attenuation [18–21, 34].

Second, as Fig. 10.2 shows and as discussed in more detail in Sections 10.4.4.2 and 10.4.4.3, while the peak of the output precedes the input peak, the earliest part of the output pulse—presented by a discontinuity in the envelope or higher-order derivatives of the envelope—is retarded with respect to the input. In other words, while the peak of the output is advanced in time, its front is not. In this sense, the "genuine information" conveyed by our electromagnetic pulse (for example, a "1" or

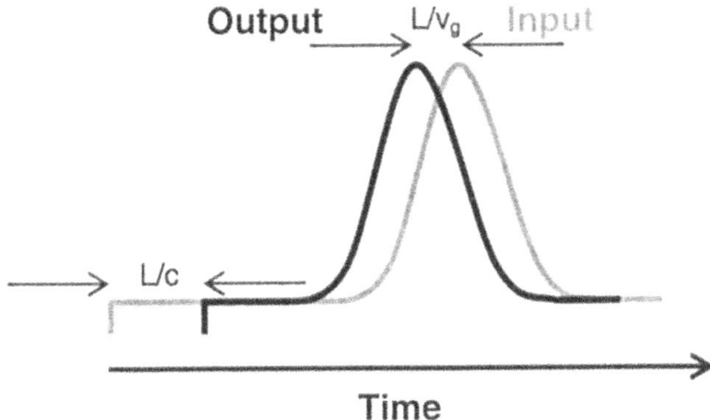

Fig. 10.2 The input and output wave packets. The output pulse travels with superluminal or negative group velocities, while its front (the discontinuity) propagates at luminal speed.

a "0") is carried by the points of nonanalyticity (discontinuities) [8,9], and therefore from the theoretical point of view there is no contradiction between the behavior depicted in Fig. 10.2 and the requirements of causality. It must be added that for a Lorentzian medium, the oscillation frequency of the points of nonanalyticity (front) is extremely high while their associated amplitude is very low; hence, from a practical point of view, the detection of the front and precursor fields may not be the most convenient scheme for routine detection of signals.

10.3 WAVE PROPAGATION IN A SLAB WITH NEGATIVE INDEX OF REFRACTION

A brief review of current literature shows that the meaning of negative group velocity and its connection with LHM is mired by misunderstandings and misconceptions [2,4,35,36]. The subject of negative group velocity in such media is of particular interest, since most theoretical and experimental studies presented so far only consider the case of antiparallel phase and group velocities (backward waves) for which the group velocity is positive and points away from the radiating source, while the phase velocity is negative and points toward the source [3,4,15,36]. To begin our discussion of negative group velocity and group delay in LHM, we start with the simple case of a slab with simultaneous negative permittivity and permeability.

The medium is characterized by [3]

$$\epsilon_{eff} = 1 - \frac{\omega_{ep}^2 - \omega_{eo}^2}{\omega^2 - \omega_{eo}^2 - j\gamma_e\omega} \tag{10.15}$$

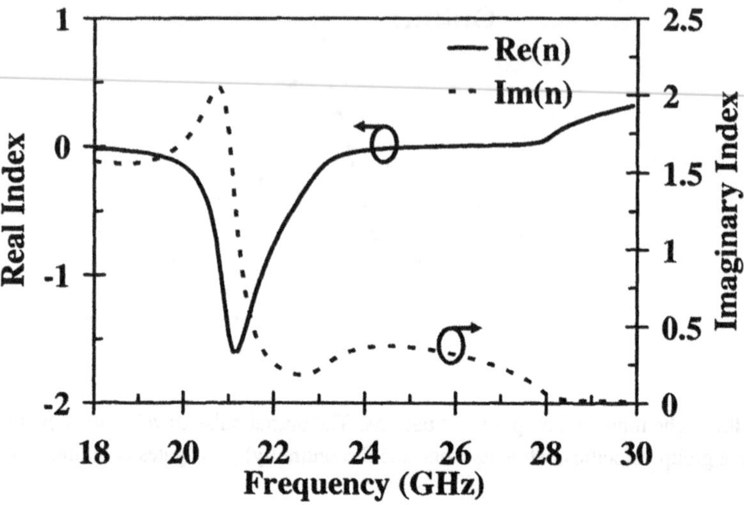

Fig. 10.3 The real and imaginary parts of the index of refraction. $\omega_{eo} = 0$ GHz, $\omega_{ep} = 2\pi \times 28$ GHz, $\omega_{mo} = 2\pi \times 21$ GHz, $\omega_{mp} = 2\pi \times 23$ GHz. $\gamma_e = 1.6 \times 10^9 s^{-1}$, and $\gamma_m = 4 \times 10^9 s^{-1}$.

and by

$$\mu_{eff} = 1 - \frac{\omega_{mp}^2 - \omega_{mo}^2}{\omega^2 - \omega_{mo}^2 - j\gamma_m \omega} \tag{10.16}$$

where ω_{ep}, ω_{mp} are the electric and magnetic plasma frequencies and ω_{eo}, ω_{mo} are the electric and magnetic resonance frequencies respectively. The γ_e and γ_m are the phenomenological electric and magnetic damping constants. In regions for which the real parts of the effective permeability and permittivity are both negative, the index of refraction is also negative [4]. Figure 10.3 shows the real and imaginary parts of the effective index calculated from $n_{eff} = \sqrt{\epsilon_{eff}}\sqrt{\mu_{eff}}$.

Figure 10.4 shows the geometry of the problem under consideration. A slab of thickness L with dispersion characteristics depicted in Fig. 10.3 is irradiated by plane waves from a source located to its left at negative z-values. The transmission coefficient (magnitude and phase) can then be calculated according to

$$T(\omega) = \frac{t_{12}t_{23}e^{-jk_2L}}{1 + r_{12}r_{23}e^{-j2k_2L}} = |T(\omega)|\, e^{j\phi} \tag{10.17}$$

where $t_{i,j}$ and $r_{i,j}$ are the Fresnel transmission and reflection coefficients corresponding to the slab boundaries and k_2 is given by

$$k_2 = \frac{2\pi}{\lambda} n_2 \cos\theta_2 \tag{10.18}$$

In the following we assume that the NIR medium is surrounded by vacuum ($n_1 = n_3 = 1$) and is illuminated at normal incidence ($\theta_1 = 0$).

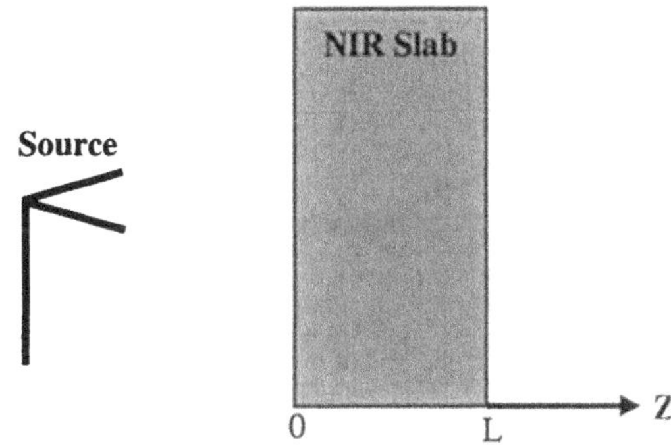

Fig. 10.4 A slab with NIR irradiated by a source to its left.

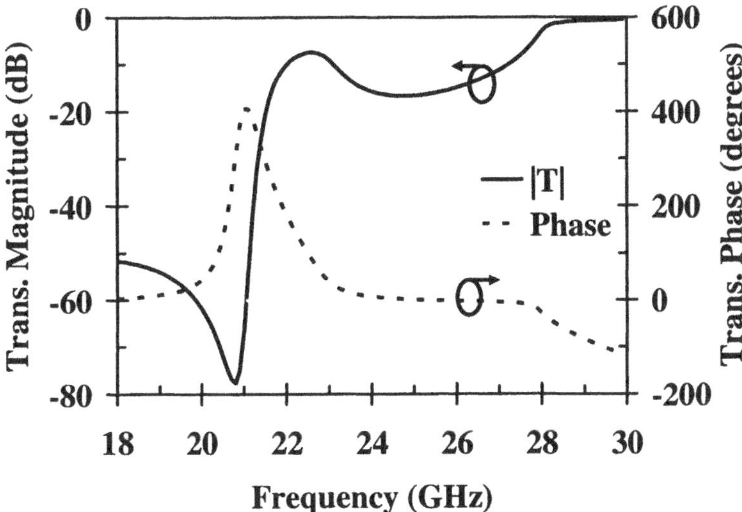

Fig. 10.5 Transmission magnitude and phase for a 1-cm-thick section of an NIR medium. All material parameters are the same as in Fig. 10.3.

Figure 10.5 shows the transmission function (magnitude and phase) for a left-handed slab, 1 cm thick. Note that in the vicinity of minimal transmission, corresponding to the region of anomalous dispersion, the slope of the transmission phase changes sign, implying a change of the sign for the group delay and group velocity.

The group delay and the real part of the index are plotted in Fig. 10.6. From the figure it is evident that group delay and hence the group velocity are negative within the region of anomalous dispersion and are positive away from it. Note that the real part of the index is negative from 18 to 25.4 GHz, while at the frequency $\nu_t = 21.2$ GHz the group delay changes sign from negative to positive.

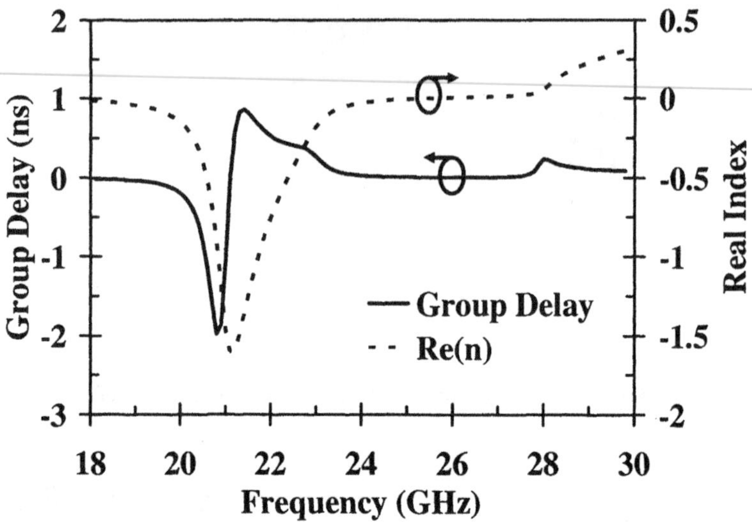

Fig. 10.6 Group delay and real part of the index for a 1-cm-thick section of NIR medium. All material parameters are the same as in Fig. 10.3.

This implies that for frequencies greater than ν_t the group velocity is positive, whereas the phase velocity remains negative, corresponding to the backward-wave propagation discussed earlier. The fact that the group velocity is positive for frequencies greater than ν_t can also be seen from the behavior of the index of refraction in Fig. 10.6. In this frequency range, $\omega\, dn/d\omega$ is positive and larger than n, indicating a positive value for the group velocity calculated from*

$$v_g = \frac{c}{n + \omega\, dn/d\omega} = \frac{c}{n_g} \tag{10.19}$$

where n_g is the group index. The existence of regions of negative group/negative phase velocities, and positive group/negative phase velocities (backward waves) for the above case can also be verified using full-wave simulations.

From the above discussions it is clear that LHM, similar to right-handed media (RHM), possesses an anomalous dispersion region in which the group velocity is negative. However, the LHM anomalous dispersion region differs from that of RHM in at least two respects. First, in the case of LHM, the negative group velocity is also accompanied with a negative phase velocity. Second, at the minimal dispersion point ($dn/d\omega = 0$) or frequency interval for which $dn/d\omega \approx 0$ the LHM exhibits a group

*Equation (10.19) assumes perfect matching between the slab and the surrounding media, that is, $r_{12} = r_{23} = 0$. The effects of mismatches (interfaces) which produce negligible positive delays will not alter the conclusions presented above. Note that equation (10.11), which is used to plot the group delay in Fig. 10.6, takes into account the positive delays associated with the interfaces.

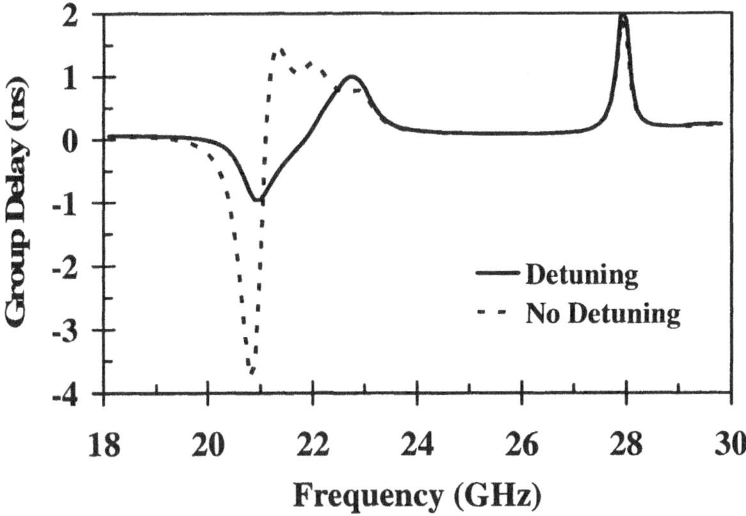

Fig. 10.7 Group delay for detuned (solid curve) and non-detuned (dashed curve) 1DPC with 8 LHM slabs separated by air.

velocity given by

$$v_g \approx v_p = \frac{c}{n} < 0 \tag{10.20}$$

which is negative, in contrast to the case of LHM.

Finally, note that negative refractive index is an artificial dispersion in which the characteristics of the underlying subwavelength unit cell control the overall dispersive behavior. It is then possible to slightly vary the frequency response of each unit cell (detuning) in order to broaden the frequency range over which negative phase and group delays are exhibited. As a proof of concept, Fig. 10.7 shows the group delays for a one-dimensional photonic crystal (1DPC) consisting of 8 RHM slabs separated by air. For the detuned structure (solid curve), the magnetic resonance of each slab is increased by 1 GHz as compared to the previous layer—starting with $\omega_{mo} = 21$ GHz for the first slab—while keeping all other parameters the same as before. The figure also shows the group delay for the 1DPC without detuning (dashed curve). As a result of detuning, the negative group delay bandwidth in Fig. 10.7 has increased from 1.7 GHz to 2 GHz (an increase of 18%) while the absolute value of negative group delay has decreased (from -3.7 ns to -0.97 ns). The above tradeoff between the bandwidth and the amount of negative delay is a fundamental design constraint.

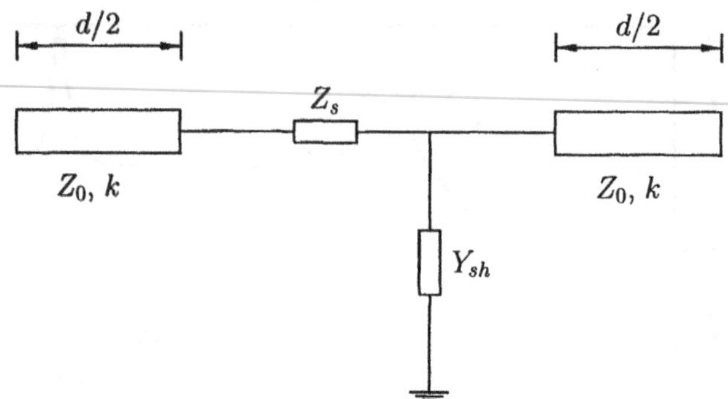

Fig. 10.8 A unit cell of a transmission line loaded with lumped series impedance Z_s and shunt admittance Y_{sh}.

10.4 PERIODICALLY LOADED TRANSMISSION LINE WITH AN EFFECTIVE NEGATIVE INDEX OF REFRACTION AND NEGATIVE GROUP INDEX

In the previous section we discuss the ideal, but not practical, case of a slab or multiple slabs having an effective negative index of refraction. Here, we concentrate on the actual structures that are manufactured to exhibit such responses. We begin with the general theory of the periodically loaded transmission line (PLTL) and show how such a device can be modified in order to exhibit combinations of positive or negative phase and group delays.

10.4.1 General Theory of PLTL Exhibiting Negative Phase Delay

Figure 10.8 shows the unit cell of a transmission line repeatedly loaded with lumped series impedance (Z_s) and shunt inductance (Y_{sh}). This periodic structure can be considered as an effective medium, provided that the dimensions of the unit cell are small as compared to the excitation wavelength. The study of such PLTL with the help of a dispersion diagram is a subject well examined in electromagnetic theory [37,38] and will be used in our analysis.

The loading elements Z_s, and Y_{sh} can be chosen such that the overall result is a medium exhibiting backward-wave propagation—that is, a medium with an effective negative index of refraction. This approach in designing a medium with backward-wave propagation (negative phase velocity) is well-described in other chapters of this book and will not be repeated here. It suffices to say that a two-dimensional version of such a PLTL has been used to demonstrate focusing of a radiating cylindrical source [15].

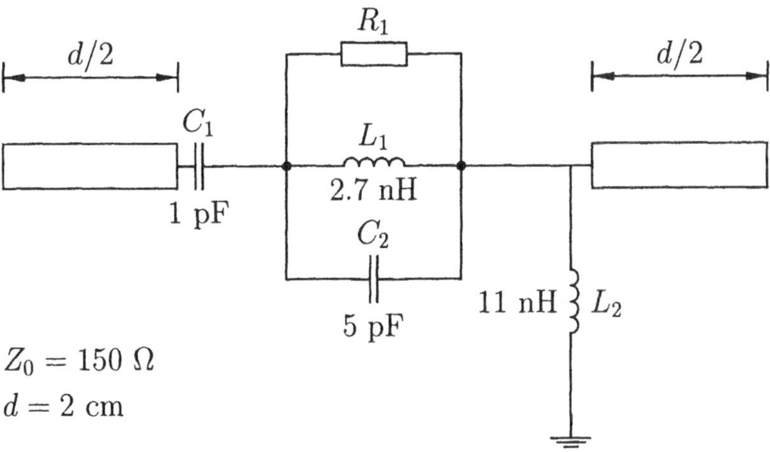

Fig. 10.9 Unit cell of the proposed loaded transmission line which exhibits negative refractive index as well as negative group delay. Typical component values are also shown. After Ref. [41]. Copyright © 2004 IEEE.

10.4.2 Frequency Domain Simulations

For the PLTL discussed in Section 10.4.1 the loading elements Z_s and Y_{sh} can be chosen such that the resulting structure exhibits both negative group delay (negative group velocity) and negative phase delay (negative phase velocity). Figure 10.9 shows the unit cell of such a PLTL [39]. Using the ABCD transmission matrix, the complex propagation constant (γ) of the periodic structure is given by

$$\cosh \gamma d = \cos[(\alpha + j\beta)d] = \cos kd + j\frac{(Z_s + Y_{sh}Z_0^2)}{2Z_0}\sin kd + \frac{Z_s Y_{sh}}{2}\cos kd \tag{10.21}$$

Here, α and β are the attenuation and phase constants of the periodically loaded medium, whereas k, Z_0, and d are the propagation constant, the characteristic impedance, and the length of the unit cell for the unloaded line, respectively.

In order for our PLTL to exhibit a region of anomalous dispersion with negative group delay in addition to negative phase delay (an effective negative index), the line is loaded in series with capacitor C_s and an $R_rL_rC_r$ resonator, and in shunt with an inductor L_{sh} [40]. The series impedance (Z_s) and shunt admittance (Y_{sh}) of (10.21) are then given by

$$Z_s = \frac{1}{j\omega C_s} - \frac{j\omega\frac{1}{C_r}}{\omega^2 - j\omega\frac{1}{R_rC_r} - \frac{1}{L_rC_r}} \tag{10.22}$$

$$Y_{sh} = \frac{1}{j\omega L_{sh}} \tag{10.23}$$

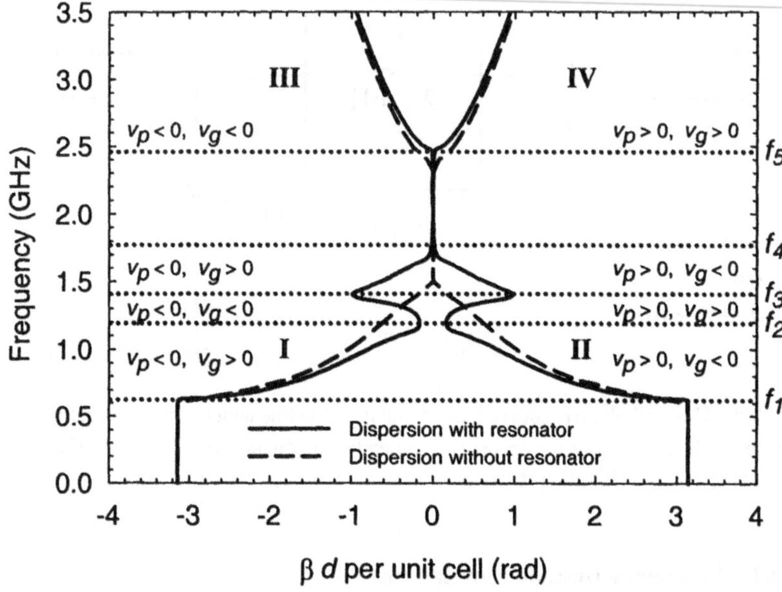

Fig. 10.10 Dispersion diagram (solid curve) of the proposed transmission line medium exhibiting simultaneous negative refractive index and negative group velocity in the first passband. Dotted curve shows the dispersion of the periodic medium without the $R_r L_r C_r$ resonators. After Ref. [40]. Copyright © 2003 IEEE.

Note that the resonant frequency of the parallel $R_r L_r C_r$ resonator, $f_0 = 1/2\pi\sqrt{L_r C_r}$, is also approximately the center frequency of the region of anomalous dispersion.

Figure 10.10 (solid curve) shows the dispersion diagram of the proposed periodic structure. The component values used to produce the curves are indicated in Fig. 10.9. The characteristic impedance of the unloaded line used in the simulation is 150 Ω and the length of the unit cell (d) is 2 cm. The first passband extends from frequency f_1 to f_4 which also spans the region of anomalous dispersion ($f_2 < f < f_3$). The second stopband ($f_4 < f < f_5$) and second passband ($f > f_5$), along with the appropriate signs for the phase and group velocities in each branch are also shown.

Figure 10.10 indicates that within the first passband ($f_1 < f < f_4$) branches marked I and II can describe the wave propagation in our PLTL. A question then can be asked. Which of the two branches correctly describes the wave propagation through the structure? To answer this, we may consider the following. First, the dashed curve in Fig. 10.10 shows the dispersion relation for a transmission line without the $R_r L_r C_r$ resonant circuit. As stated in Section 10.4.1 and described in Ref. 15, such a transmission line has been shown to exhibit an equivalent negative index of refraction—that is, a negative value for βd—which designates the branch

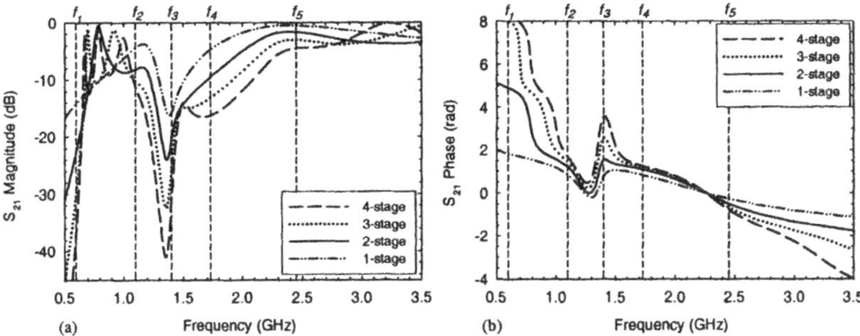

Fig. 10.11 (a) Calculated $|S_{21}|$ for the PLTL of Fig. 10.9 with different number of stages. (b) Calculated unwrapped S_{21} phase for the same transmission line with different number of stages. After Ref. [40]. Copyright © 2003 IEEE.

I as the appropriate choice. We may now consider the presence of the $R_r L_r C_r$ resonator as a perturbation to the previously studied case, and as such we once again must choose branch I as our dispersion curve. Second, the difference in the insertion phase for two transmission lines with different lengths can be used to deduce the proper branch. This point is discussed in the next section, and again it is seen that branch I correctly describes the wave propagation in our PLTL. Finally, we should note that within the second passband ($f > f_5$), the dispersive behavior depicted by the branch IV properly describes the wave propagation for our PLTL.

At this point, a few remarks regarding the relative signs of the phase and group velocities are in order. As Fig. 10.10 shows, for branch I and within the frequency range $f_1 < f < f_2$ the phase and group velocities are antiparallel (have opposite signs). This traditionally describes backward-wave propagation [14] and is the regime under which the theoretical and experimental work in Refs. 3, 15, and 36 were carried out. The frequency range $f_2 < f < f_3$ of branch I corresponds to the region of anomalous dispersion for which the phase and group velocities are parallel and are both negative. This frequency interval designates a band for which the term negative group velocity can be correctly used in connection with NRI metamaterials. The frequency range $f_3 < f < f_4$ is once again the region of backward-wave propagation, whereas for $f > f_5$ in branch IV the PLTL behaves as a normal medium with parallel and positive phase and group velocities.

To properly describe the wave propagation through a finite length PLTL, a unit cell of which was shown in Fig. 10.9, we will use the scattering matrix formulation. In order to closely emulate the experimental results of the next section, we will monotonically increase the number of unit cells from one to four and terminate the transmission line with a 50-Ω impedance.

Figures 10.11a and 10.11b show the S_{21} (transmission function) magnitude and phase for the PLTL as the number of unit cells is increased. The frequency bands corresponding to the first and second passbands and stopbands along with the region of anomalous dispersion are also displayed. As expected, within the region of

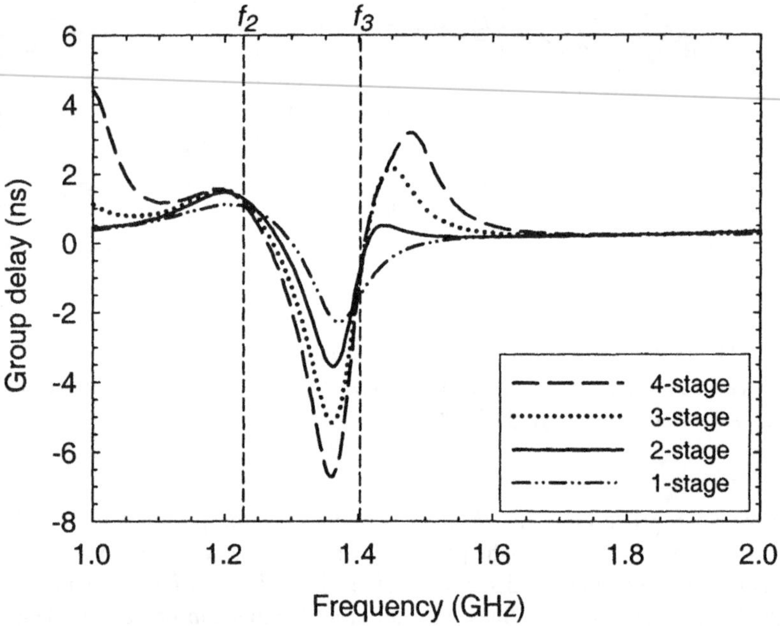

Fig. 10.12 Calculated group delay for the PLTL of Fig. 10.9 with 1, 2, 3, and 4 stages. This figure only shows the region of anomalous dispersion and its vicinity. After Ref. [40]. Copyright © 2003 IEEE.

anomalous dispersion ($f_2 < f < f_3$), the transmission magnitude is minimal, and it is within this frequency band that the negative group delay is to be observed [18]. Figure 10.11b shows the unwrapped transmission phase for the same range of frequencies. From the figure it is clear that within the region of anomalous dispersion ($f_2 < f < f_3$) the derivative of the phase function (ϕ) reverses its sign, hence implying the existence of a negative group delay and group velocity.

The fact that our PLTL, within the frequency bands $f_1 < f < f_4$, exhibits an equivalent negative index of refraction can also be verified from Fig. 10.11b. Assuming an unbounded medium (i.e., neglecting the mismatches*), the difference between the insertion phases of two PLTLs of lengths d_1 and d_2 is given by

$$\Delta\phi = \phi_2 - \phi_1 = \frac{\omega\, n(\omega)}{c}(d_2 - d_1) \tag{10.24}$$

Note that for $d_2 > d_1$ and normal media ($n > 0$), the difference in the insertion phase calculated from (10.24) is negative ($\Delta\phi < 0$), whereas from Fig. 10.11b, in the frequency band $0.5 < f < 2.3$ GHz, it is positive, indicating an equivalent negative

*Including effects of the boundaries (mismatches) only complicates the calculations but will not change the final conclusions.

refractive index. Interestingly, as Fig. 10.11b implies, for the second passband ($f > 2.3$ GHz) $\Delta\phi$ is negative, implying a normal transmission line operation. Finally, we expect that as the number of stages increases, the finite-length PLTL more closely approximates the dispersion characteristics of infinitely long PLTL depicted in Fig. 10.10.

Figure 10.12 shows the calculated group delay [equation (10.11)] for our PLTL with 1, 2, 3, and 4 unit cells. In accordance with the results for an infinitely long PLTL depicted in Fig. 10.10, it is seen that for a finite-length PLTL, the group delay is negative within the frequency band $f_2 < f < f_3$ and is positive away from the anomalous dispersion region. It must be noted that as the length of the finite-length PLTL is increased, the amount of negative delay (in absolute value sense) is also increased. In other words, longer transmission lines produce more time advances (negative delays) as compared to shorter lines, however, at the cost of reducing the transmitted signal amplitude. In the next section a frequency-domain setup is used to verify these theoretical predictions.

10.4.3 Frequency Domain Measurements

To verify our theoretical predictions, a coplanar waveguide (CPW), printed on Rogers 5880 substrate with dielectric constant of 2.2 and thickness of 0.381 mm, was designed. The CPW line was periodically loaded with surface-mounted chips of size 1.5 mm by 0.5 mm, such that one unit cell was approximately 2 cm long. To perform the experiment, PLTLs with 1, 2, 3, and 4 unit cells were fabricated. Figure 10.13 shows a PLTL with 3 stages. The device was connected to a vector network analyzer (HP-8722C), and in order to measure the transmission function (S_{21}) a full two-port calibration was performed.

The magnitude and phase of S_{21} are displayed in Figs. 10.14a and 10.14b respectively. The stopbands, passbands, and anomalous dispersion bands are also shown. Figure 10.14a clearly indicates that, in accordance with the theoretical predictions of the previous section, as the number of unit cells is increased, the magnitude of the insertion loss also increases. Furthermore, as discussed earlier, Fig. 10.14b shows that in the frequency band $f_1 < f < f_4$ the phase differences ($\Delta\phi$) between two PLTLs of different lengths ($d_2 > d_1$) are positive, implying that the PLTLs exhibit an effective negative index of refraction. On the other hand, for $f > f_4$, $\Delta\phi$ is negative, indicating a normal transmission line behavior. Finally, in Fig. 10.14b the region of anomalous dispersion ($f_2 < f < f_3$) can be identified by the reversal of the slope.

While the overall agreement between the theoretical predictions of Figs. 10.11a and 10.11b and the experimental results of Figs. 10.14a and 10.14b is good, in general, a shift of 50 to 80 MHz can be detected. For example, the experimental value for the center frequency of the region of anomalous dispersion is 1.29 GHz, whereas the theoretically predicated value is approximately 1.37 GHz. Moreover, around the resonances, more losses are predicted by the simulations as compared to the experimental results.

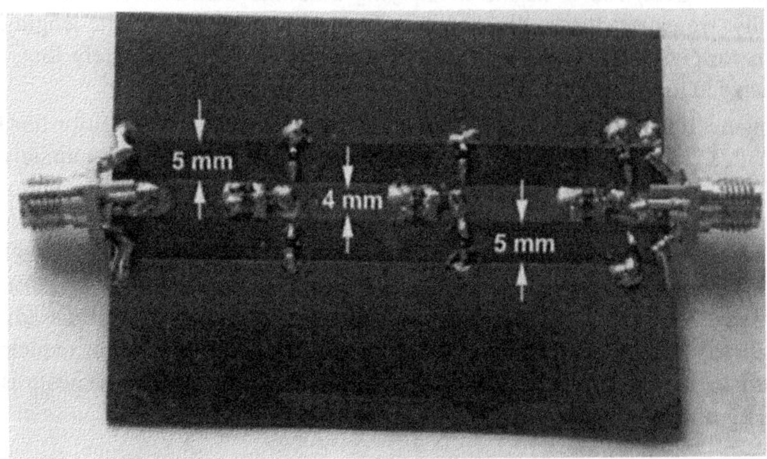

Fig. 10.13 PLTL with 3 stages. The board is Rogers 5880 with a substrate thickness of 0.381 mm, a relative permittivity of 2.2, a loss tangent of 0.0009, and volume and surface resistivities of 2×10^7 MΩ·cm and 3×10^8 MΩ, respectively. The copper cladding thickness is 17 μm. The center conductor of the waveguide has a width of 4 mm, and the slots have a width of 5 mm. After Ref. [41]. Copyright © 2004 IEEE.

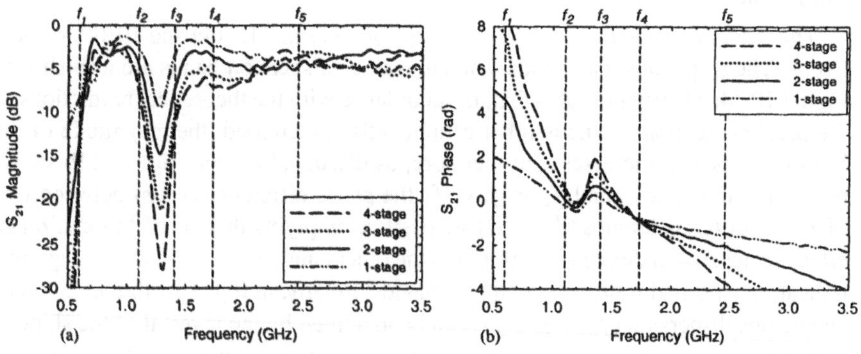

Fig. 10.14 (a) Measured S_{21} magnitudes of the PLTLs with one, two, three, and four unit cells. (b) Measured unwrapped S_{21} phases of the same PLTLs. After Ref. [40]. Copyright © 2003 IEEE.

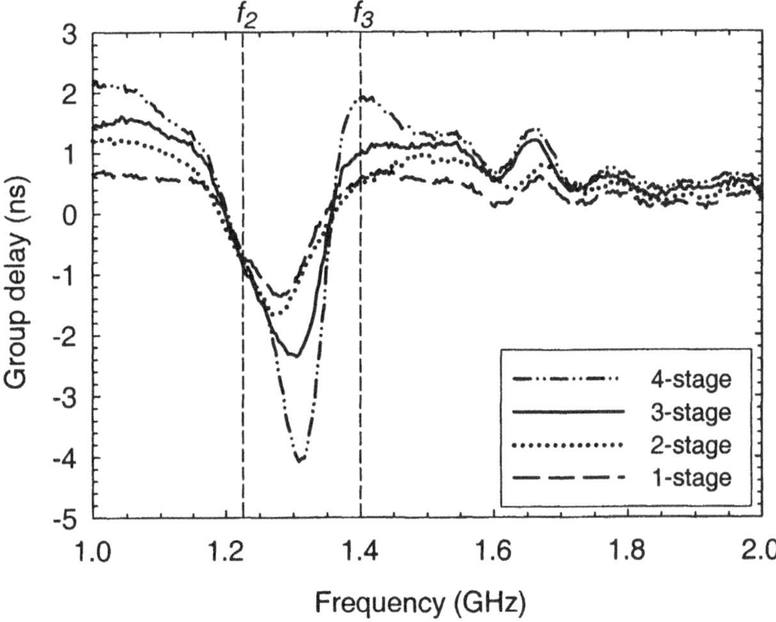

Fig. 10.15 Measured group delay for the PLTL with 1, 2, 3, and 4 unit cells. Larger negative delays (in an absolute sense) are measured for longer transmission lines. After Ref. [40]. Copyright © 2003 IEEE.

These discrepancies can be accounted for by considering a few factors. First, in all of our simulations we have used the nominal values associated with the surface-mount lumped elements provided by the manufacturer. Our experience has shown that in many cases, in part due to the embedded parasitics, the actual measured values can be significantly different. Second, in our simulations the resistance and conductance associated with the inductor L_r and the capacitor C_r have been ignored. The effect of this series resistance for the inductor and conductance for the capacitor is to reduce the overall impedance of the parallel $R_r L_r C_r$ resonant circuit, hence reducing the theoretically predicted insertion losses. Third, for the PLTLs with more than one stage, the resonant frequency for each stage is slightly different from the others due to variations in the component values. This nonhomogeneity was not taken into account in our theoretical model, and in practice it broadens the anomalous dispersion region, thus reducing the overall measured insertion losses in addition to decreasing the slope of the phase within this region.

The group delay for each truncated PLTL is shown in Fig. 10.15. The frequency band of interest is the anomalous dispersion region ($f_2 < f < f_3$) in which the group delay is more negative for longer transmission lines. The measured maximum group delay for the four-stage PLTL is approximately -4 ns compared to -7 ns obtained from the simulations. This difference is attributed to the decrease in slope of the transmission phase as discussed above.

10.4.4 Time Domain Simulations

In our discussion of Section 10.1 we observed that the phase delay is the delay associated with underlying sinusoidal harmonics, whereas the group delay is the delay of the pulse envelope. In a system supporting negative phase delay the output sinusoidal harmonics lead the input, while in a system with negative group delay, the peak of the output wave packet precedes the peak of the input wave packet. In this section, we theoretically and experimentally study these effects directly in the time domain [41]. Since the concept of negative phase delay (phase lead) is well understood within standard circuit analysis, we spend most of our time describing the negative group delay for structures supporting both behaviors. We start our discussion with simulating a PLTL which exhibits negative or positive phase or group delays depending on the frequency of operation and show how despite the counterintuitive shift of the pulse envelope to earlier times, Einstein's causality is not violated.

10.4.4.1 Negative Group Delay In order to study time-domain behavior of our PLTL, three loaded CPW transmission lines with unit cells depicted in Fig. 10.9 were considered. The only difference between the unit cell studied here and the one discussed in Sections 10.4.2 and 10.4.3 is that the value of R_r was reduced from 300 Ω to 150 Ω. The total lengths of the lines with one, two, and three unit cells were 2, 4, and 6 cm, respectively. The transmission lines were excited with Gaussian pulses of temporal length 30 ns, modulated at the resonance frequency of the series $R_rL_rC_r$ loading element (1.3 GHz). Using the specification sheet for the Rogers 5880 samples with conductor thickness of 17 μm, the substrate and conductor losses were included in our analysis, whereas the lumped components used in the simulation were assumed to be ideal.

The simulations were performed using Agilent's Advanced Design System (ADS), where Fig. 10.16 shows the calculated voltage waveforms at the input and output of the loaded lines. The peaks of all three output pulses *precede* the input peaks by an amount proportional to the length of the line. In other words, since the longer lines have more unit cells, they generate a larger negative group delay. This negative delay is mostly due to the series $R_rL_rC_r$ resonator and thus resonant absorption losses are also introduced, as indicated by the drop in magnitude of the output voltage waveforms. For example, in the case of the 2-cm transmission line, a negative delay of −0.89 ns is predicted while the output voltage peak is approximately 15 percent of the input. Note that some of the predicted losses are due to mismatched impedances between the loaded transmission line (150 Ω) and the source (50 Ω).

10.4.4.2 Luminal Front Velocity Figure 10.16 shows that the pulse peak and envelope propagate with a negative group delay and consequently the pulse travels

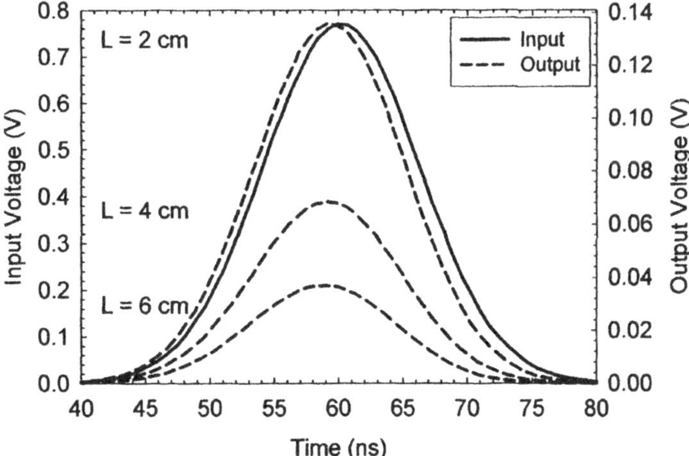

Fig. 10.16 Time-domain simulations showing negative group delay for the 2-cm, 4-cm, and 6-cm transmission lines, with delays of -0.89 ns, -1.17 ns, and -1.53 ns, respectively. After Ref. [41]. Copyright © 2004 IEEE.

with a negative group velocity. Contrary to the traditional point of view, negative and superluminal group velocities are therefore physical and measurable and do not contradict the requirements of relativistic causality. While a rigorous analysis of this point can be found elsewhere [8], a short justification can be provided by considering the following: Every causal signal has a starting point in time, before which the signal does not exist. This starting point is marked by a discontinuity in the pulse envelope or higher-order derivatives of the envelope, at which point the pulse is no longer analytic. These points of nonanalyticity are the conveyers of *genuine information* and can be shown to propagate at exactly the speed of light c under all circumstances [8, 9, 23] and thereby fulfill the requirements of the relativistic causality. In short, for a smoothly varying pulse, presented by an analytical function, there is no more information in the pulse peak than in its earliest parts.

The propagation of these discontinuities can be examined using time-domain simulations as shown in Fig. 10.17. The discontinuities in the pulse waveform were established by introducing a "turning-on" point, commonly referred to as the front. The propagation of the front through the PLTLs of different lengths, having negative group delays, can be seen by examining the first 0.3 ns of the pulse evolution, shown on a logarithmic scale in Fig. 10.17. The output pulse fronts for the three structures all suffer the expected *positive luminal delays* with respect to the input fronts, given by L/c, where L is the length of the transmission line. Thus the simulations show that causality is preserved as seen by the fact that discontinuities in the pulse travel at exactly the speed of light in vacuum.

While the simulations indicate the causal propagation of information by the points of nonanalyticity, the amplitudes associated with these fronts are particularly small,

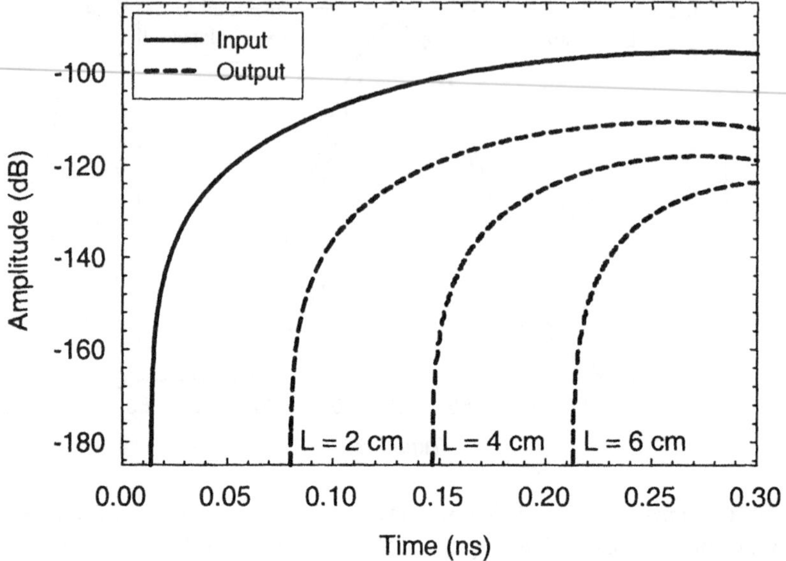

Fig. 10.17 Time-domain simulations of the modulated pulse fronts, plotted on a logarithmic (dB) scale. The input pulse front always precedes the output front by a time equal to L/c, where L is the length of the line. On the other hand (see Fig. 10.16), the output peak precedes the input peak. After Ref. [41]. Copyright © 2004 IEEE.

making their experimental detection a challenging task. This difficulty is one of the reasons that we practically detect a "signal" by observing its maximum or half-maximum points, which in turn can be made to propagate superluminally or with negative velocities.

10.4.4.3 Physical Mechanism Underlying Negative Group Delay The mechanism behind the pulse advancement can be also explained in terms of pulse reshaping. We can study the time evolution of a pulse by considering the spatiotemporal voltage distributions of the individual Fourier components making up the pulse. The system under study, shown schematically in Fig. 10.18, consists of two sections of regular transmission line occupying the regions $z < 0$ and $z > a$, surrounding a PLTL section of length a. The PLTL is assumed to be a transmission line of length 2 cm, having a dispersive behavior determined by the dispersion relation (10.21) and operated within the anomalous dispersion band. That is, the PLTL exhibits both negative refractive index and negative group velocity properties.

Consider a modulated Gaussian pulse, with center frequency in the anomalous dispersion band, excited on the $z < 0$ transmission line segment. By Fourier analysis, this waveform can be decomposed into many single-frequency sinusoidal components.

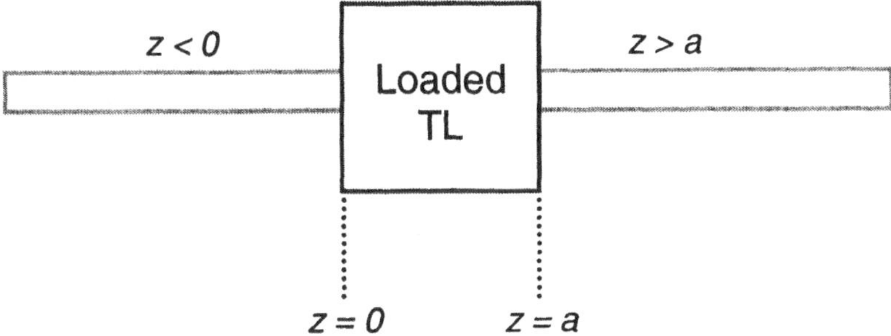

Fig. 10.18 A transmission line consisting of two regular lines ($z < 0$ and $z > a$) and a PLTL ($0 < z < a$) used for the simulations that explain the mechanism behind negative group delay. After Ref. [41]. Copyright © 2004 IEEE.

The peak of the pulse is formed at the position where these individual frequency components interfere constructively, and the nulls of the pulse are formed where these components interfere destructively.

The space- and time-dependent voltage distribution $V_n(z, t)$ for the nth spectral component of the Gaussian pulse is given by

$$V_n(z,t) = \begin{cases} G_n \cos(\omega_n t - k_n z), & z < 0 \\ G_n e^{-\alpha_n z} \cos(\omega_n t - \beta_n z), & 0 < z < a \\ G_n e^{-\alpha_n a} \cos(\omega_n t - k_n[z - a] - \beta_n a), & z > a \end{cases} \quad (10.25)$$

Here ω_n and G_n are the frequency and amplitude of the nth harmonic, and k_n is the propagation constant on the regular transmission line, in the regions $z < 0$ and $z > a$. In the PLTL section $0 < z < a$, the propagation and the attenuation constants of the nth harmonic are β_n and α_n, respectively, calculated from the dispersion relation (10.21). Note that, according to (10.25), the peak of the pulse strikes the interface $z = 0$ at $t = 0$.

Figure 10.19a displays three spectral components of a Gaussian pulse with frequencies in the anomalous dispersion band at the instant $t = -13$ ns, calculated from (10.25). In addition to the underlying harmonics, Fig. 10.19a also displays the pulse envelope, so that the peak location can be clearly identified. It is evident from the figure that the frequency components add up in phase and a peak is formed in the $z < 0$ section of the transmission line.

As time progresses, the pulse propagates along the transmission line and the early part of the pulse encounters the PLTL section. By virtue of the phase compensation caused by the anomalous dispersion, the negative group delay transmission line rearranges the relative phases of the individual frequency components. Since the phase response of the line is approximately linear and the magnitude response is approximately flat over the bandwidth of the Gaussian pulse, the frequency components add up to produce a close copy of the original pulse, in the region $z > a$. This output pulse appears at $t = -0.5$ ns, before the input peak reaches the first interface, as

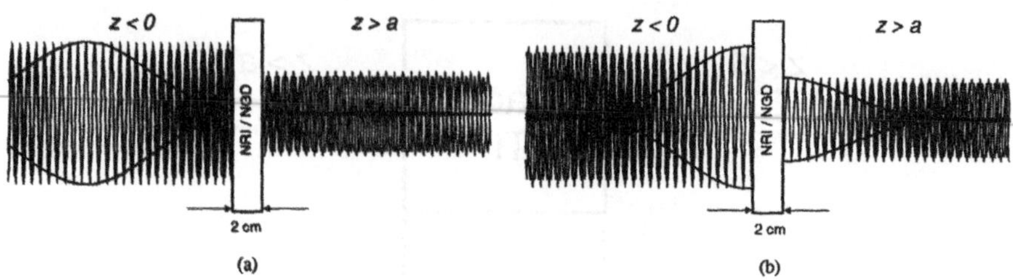

Fig. 10.19 Simulations illustrating the pulse-reshaping mechanism which underlies the negative group delay. (a) Three main frequency components of the Gaussian pulse and the resulting pulse envelope 13 ns before the input peak reaches the loaded transmission line interface. (b) The same three frequency components 0.5 ns before the input peak reaches the interface; at this point a peak has been already formed at the output. After Ref. [41]. Copyright © 2004 IEEE.

shown in Fig. 10.19b. Note that the output pulse amplitude is reduced in magnitude relative to the input pulse, though the envelope retains its basic shape. Figure 10.19b thus shows that the peak of the output pulse appears at the output terminal 0.5 ns before the input peak reaches the input terminal. Note that the effects of reflections from the interfaces in these simulations have been ignored. These reflections produce standing waves in the $0 < z < a$ section, and thus cause a further reduction in the transmitted pulse amplitude; however, they do not affect the location of the pulse peak.

10.4.5 Time-Domain Measurements

To verify our theoretical predictions, coplanar waveguides with 1, 2, and 3 unit cells depicted in Fig. 10.9 were manufactured (recall that in our time-domain analysis the value of R_r is 150 Ω). The experimental setup used to measure the group delay is schematically shown in Fig. 10.20. A baseband Gaussian pulse of temporal width 40 ns was created with a Tektronix AWG2041 arbitrary waveform generator (ARB) and was modulated with a Rohde & Schwartz SMV03 vector signal generator at frequencies between 1.1 and 1.5 GHz. The modulated signal was then divided by a 1×2 splitter. Any discrepancy in length between the two cables joining the splitter to the oscilloscope will introduce an inherent delay between the two paths, thereby affecting the accuracy of the final group delay measurements. Therefore, both outputs of the splitter were initially connected to the Channels 1 and 3 of an Agilent 54846 Infiniium oscilloscope (bandwidth 2.25 GHz) for a calibration measurement. The delay was measured on the Infiniium scope and electronically equalized to 0 ± 0.1 ns, using the oscilloscope internal functions. After this calibration step, the CPW was inserted into the Channel 3 cable, as indicated in Fig. 10.20. In this way, both the input and output signal of the PLTL were simultaneously recorded on the oscilloscope.

Figure 10.21a shows the behavior of the 3-stage loaded transmission line operated at 1.11 GHz, in the band of positive group delay—that is, away from the anomalous

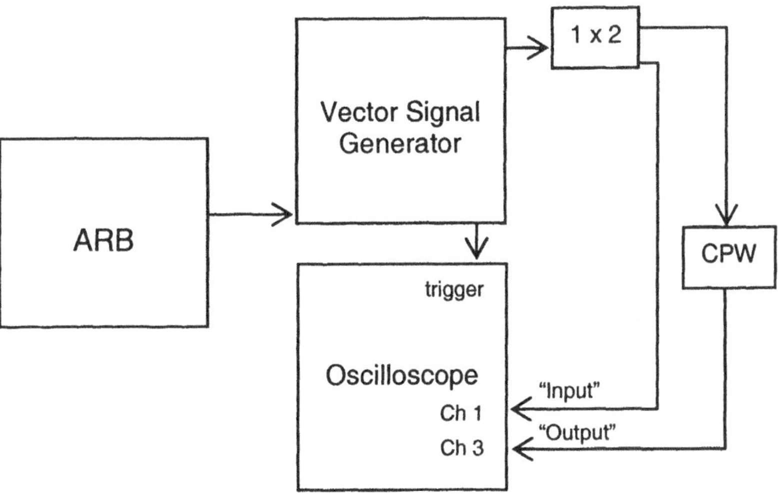

Fig. 10.20 Schematic diagram of the experimental setup used to measure negative group delay in the time domain. After Ref. [41]. Copyright © 2004 IEEE.

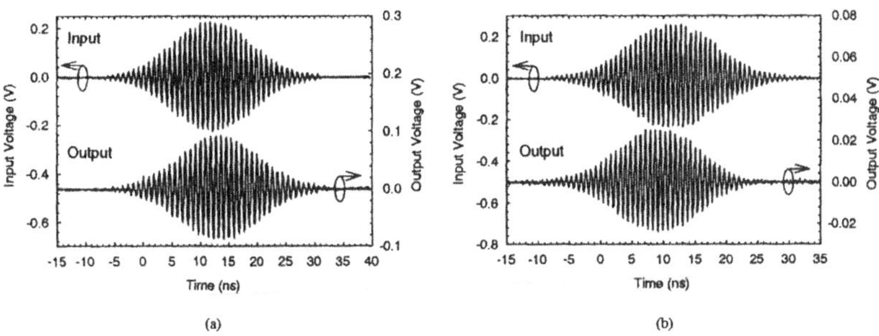

Fig. 10.21 Time-domain experimental results for the 3-stage negative delay circuit at two frequencies. (a) Positive delay at a center frequency of 1.11 GHz, 160 MHz below resonance. (b) Negative group delay at the resonance frequency of 1.27 GHz. After Ref. [41]. Copyright © 2004 IEEE.

dispersion band. For this case a positive group delay of approximately +1.5 ns, due to propagation along the 6-cm line, was observed. Under normal conditions, therefore, the peak of the output pulse appears at a later time than the peak of the input pulse. In contrast, Fig. 10.21b shows the input and output pulses when the PLTL is operated within the anomalous dispersion band, at the resonance frequency of 1.27 GHz, where a negative group delay of −3.1 ns was measured. Note that in Fig. 10.21b the output peak precedes the input peak; this unusual outcome is the meaning of negative group delay.

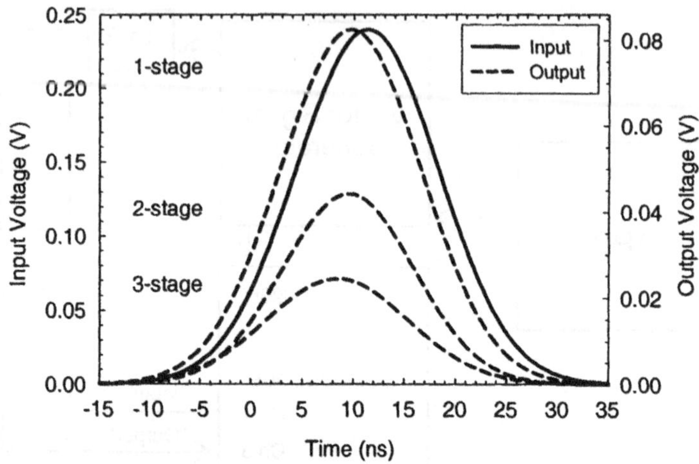

Fig. 10.22 Experimental results showing extracted pulse envelopes for the 1-, 2- and 3-stage transmission lines, with delays of -1.6 ns, -1.9 ns, and -3.1 ns, respectively. After Ref. [41]. Copyright © 2004 IEEE.

Figure 10.22 shows the measured input pulse (solid curve) and output pulses (dashed curves) at the point of maximum negative group delay, approximately 1.27 GHz, for the 1-, 2-, and 3-stage circuits. For clarity, only the pulse envelopes are shown. These curves are the experimental validation of Fig. 10.16. The envelopes were extracted from the raw data by fitting a three-parameter Gaussian curve. The peak arrival times were acquired from the Gaussian fit parameters to within ± 0.2 ns. At the $R_r L_r C_r$ resonance frequency, the 1-, 2- and 3-stage circuits exhibit group delays of -1.6 ns, -1.9 ns and -3.1 ns, respectively. Note that, as expected, the greatest negative delay and the greatest attenuation are found for the longest transmission line, and the least negative delay and least attenuation are found for the shortest line.

In comparing Figs. 10.16 and 10.22 the trend that longer lines have greater negative delay and greater insertion loss is common to both simulation and experiment; however, there are also some discrepancies. For example, there is generally less attenuation and more pulse advancement in the experiments. These discrepancies are due to the differences between the components used in the simulations and those in the actual devices. First, nominal values for the components were used in the simulations. In practice, however, the components have manufacturer stated* tolerances of $\pm 5\%$. By including these tolerances in our simulations, we found that the discrepancies between the measured and calculated group delay can be reduced by half. The group delay is particularly sensitive to changes in the resistor or capacitor in the $R_r L_r C_r$ resonator, and variation in these component values will affect the

*As alluded to in Section 10.4.3, the actual tolerance may well be above the manufacturer stated values.

slope of the transmission phase, thus altering the amount of negative group delay. Second, and more importantly, the simulations use ideal component models, and thus the self-resonant behaviors of the capacitors and inductors were not included. In practice, the self-resonances can change the overall impedance of the $R_rL_rC_r$ resonator, thereby altering the device attenuation and negative group delay. These two effects may be included in the simulations if measured S-parameters are used for each component, a tedious but effective method of improving the agreement between experiment and simulation.

10.5 CONCLUSIONS

In this chapter we have studied the dynamics of wave propagation in a general medium having both an effective negative refractive index and a negative group velocity. Our study was motivated by our desire to control the dispersive effects such as phase velocity, group velocity, and group velocity dispersion in general, and the associated signs of the first two effects in particular.

We began our studies by formulating equivalent ways of describing the aforementioned dispersive effects in terms of the phase index and its higher-order derivatives, or more generally by formulating these effects in terms of various delays such as phase delay, group delay, group delay dispersion, and so on. Our attention has been focused on the phase and group delays and their associated signs, leaving the remaining dispersive terms for later considerations.

We then proceeded to discuss the concept of abnormal group velocities for which the group velocity can become superluminal (exceeding the speed of light in vacuum) or negative, without violating the principles of relativistic causality. This then served as a conduit to bring together the two notions of negative phase and negative group velocities (negative phase and negative group delays) in structures that can exhibit both behaviors in addition to normal wave propagation (positive phase and group velocities) or backward waves (negative phase but positive group velocities.)

We continued our analysis with the case of a single slab possessing Lorentzian electric and magnetic responses. This case was chosen for both its generality and simplicity. We theoretically showed that such a medium can support both negative phase and group velocities.

We then considered a practical periodically loaded transmission line (PLTL) to demonstrate some of the above theoretical considerations. A CPW transmission line was periodically loaded with series capacitor and shunt inductor in addition to a resonant RLC circuit such that the overall transmission line exhibited positive or negative phase velocity and positive or negative group velocity depending of the exact combination. Assuming an infinitely long PLTL, the structure was studied using periodic analysis. For a finite-length structure, the scattering matrix formulation was used to calculate the medium's response. The theoretical findings were further confirmed using frequency-domain measurements. We also performed theoretical and experimental studies of our PLTL in the time domain. It was observed that such a medium can be made to operate with both negative phase and group delays (negative phase and group velocities) for which the output peak envelope precedes the input peak.

Acknowledgments

The authors would like to thank their students who have contributed to the present work, in particular Suzanne J. Erickson, Omar F. Siddiqui, Jonathan Woodley, and Mark Wheeler. This work was supported by the Natural Sciences and Engineering Research Council of Canada under Grant RGPIN 249531-02, by the Photonics Research Ontario under Project 03-26, and with the funds provided by the Canada Foundation for Innovation and Ontario Innovation Trust.

REFERENCES

1. V. G. Veselago, "Properties of materials having simultaneously negative values of the dielectric ϵ and the magnetic μ susceptibilities," *Sov. Phys. Solid State*, vol. 8, pp. 2854–2856, 1967.

2. V. G. Veselago, "The electrodynamics of substances with simultaneously negative values of ϵ and μ," *Sov. Phys. Usp.*, vol. 10, no. 4, pp. 509–514, January–February 1968.

3. R. A. Shelby, D. R. Smith, and S. Schultz, "Experimental verification of a negative index of refraction," *Science*, vol. 292, pp. 77–79, 2001.

4. D. R. Smith, W. J. Padilla, D. C. Vier, S. C. Nemat-Nasser, and S. Schultz, "Composite medium with simultaneously negative permeability and permittivity" *Phys. Rev. Lett.*, vol. 84, pp. 4184–4187, 2000.

5. R. Y. Chiao and A. M. Steinberg, "Tunneling Times and Superluminality," *Progress in Optics*, vol. 37, pp. 345–405, 1997.

6. A. M. Steinberg, P. G. Kwiat, and R. Y. Chiao, "Measurement of the single-photon tunneling time," *Phys. Rev. Lett.*, vol. 71, pp. 708–711, 1993.

7. M. Mojahedi, E. Schamiloglu, K. Agi, and K. J. Malloy, "Frequency domain detection of superluminal group velocities in a distributed Bragg reflector" *IEEE J. Quantum Electron.*, vol. 36, pp. 418–424, 2000.

8. M. Mojahedi, E. Schamiloglu, F. Hegeler, and K. J. Malloy, "Time-domain detection of superluminal group velocity for single microwave pulses," *Phys. Rev. E*, vol. 62, pp. 5758–5766, 2000.

9. M. Mojahedi, K. J. Malloy, G. V. Eleftheriades, J. Woodley, and R. Y. Chiao, "Abnormal wave propagation in passive media," *IEEE J. Sel. Top. Quantum Electron.*, vol. 9, pp. 30–39, 2003.

10. G. B. Arfken, *Mathematical Methods for Physicists*, 3rd ed., Academic Press, Orlando, FL, 1985.

11. J. D. Jackson, *Classical Electrodynamics*, 3rd ed., John Wiley & Sons, New York, 1998.

12. A. Papoulis, *The Fourier Integral and its Applications*, McGraw-Hill, New York, 1962.

13. H. M. Nussenzveig, *Causality and Dispersion Relations*, Academic Press, New York, 1972.

14. S. Ramo, J. R. Whinnery, and T. Van Duzer, *Fields and Waves in Communication Electronics*, 3rd ed., John Wiley & Sons, New York, 1994.

15. G. V. Eleftheriades, A. K. Iyer, and P. C. Kremer, "Planar negative refractive index media using periodically L–C loaded transmission lines," *IEEE Trans. Microwave Theory Tech.*, vol. 50, pp. 2702–2712, December 2002.

16. M. Notomi, "Theory of light propagation in strongly modulated photonic crystals: Refractionlike behavior in the vicinity of the photonic band gap," *Phys. Rev. B*, vol. 62, pp. 10696-10705, 2000.

17. R. Marques, J. Martel, F. Mesa, and F. Medina, "Left-handed-media simulation and transmission of EM waves in subwavelength split-ring-resonator-loaded metallic waveguides," *Phys. Rev. Lett.*, vol. 89, 183901, 2002.

18. E. L. Bolda, R. Y. Chiao, and J. C. Garrison, "Two theorems for the group velocity in dispersive media," *Phys. Rev. A*, vol. 48, pp. 3890–3894, 1993.

19. E. L. Bolda, J. C. Garrison, and R. Y. Chiao, "Optical pulse-propagation at negative group velocities due to a nearby gain line," *Phys. Rev. A*, vol. 49, pp. 2938–2947, 1994.

20. L. J. Wang, A. Kuzmich, and A. Dogariu, "Gain-assisted superluminal light propagation," *Nature*, vol. 406, pp. 277–279, 2000.

21. M. D. Stenner, D. J. Gauthier, and M. A. Neifeld, "The speed of information in a 'fast-light' optical medium," *Nature*, vol. 425, pp. 695–698, 2003.

22. J. Woodley and M. Mojahedi, "Negative group velocity in left-handed materials," presented at 2003 *IEEE International Symposium on Antennas and Propagation: URSI North American Radio Science Meeting*, Columbus, OH, USA, 2003.

23. L. Brillouin, *Wave Propagation and Group Velocity*, Academic Press, New York, 1960.

24. L. Brillouin, *Wave Propagation in Periodic Structures: Electric Filters and Crystal Lattices*, 1st ed., McGraw-Hill, New York, 1946.

25. M. Born and E. Wolf, *Principles of Optics*, 4th ed., Pergamon Press, New York, 1970.

26. L. D. Landau, E. M. Lifshittz, and L. P. Pitaevski, *Electrodynamics of Continuous Media*, 2nd rev. and enl. ed., Pergamon Press, New York, 1984.

27. C. G. B. Garrett and D. E. McCumber, "Propagation of a Gaussian light pulse through an anomalous dispersion medium," *Phys. Rev. A*, vol. 1, pp. 305–313, 1970.

28. S. Chu and S. Wong, "Linear pulse-propagation in an absorbing medium," *Phys. Rev. Lett.*, vol. 48, pp. 738–741, 1982.

29. A. Ranfagni, D. Mugnai, P. Fabeni, and G. P. Pazzi, "Delay-time measurements in narrowed wave-guides as a test of tunneling," *Appl. Phys. Lett.*, vol. 58, pp. 774–776, 1991.

30. A. Enders and G. Nimtz, "Photonic-tunneling experiments," *Phys. Rev. B*, vol. 47, pp. 9605–9609, 1993.

31. A. Ranfagni, P. Fabeni, G. P. Pazzi, and D. Mugnai, "Anomalous pulse delay in microwave propagation: A plausible connection to the tunneling time," *Phys. Rev. E*, vol. 48, pp. 1453–1460, 1993.

32. P. Balcou and L. Dutriaux, "Dual optical tunneling times in frustrated total internal-reflection," *Phys. Rev. Lett.*, vol. 78, pp. 851–854, 1997.

33. D. Mugnai, A. Ranfagni, and L. Ronchi, "The question of tunneling time duration: A new experimental test at microwave scale," *Phys. Lett. A*, vol. 247, pp. 281–286, 1998.

34. R. Y. Chiao and J. Boyce, "Superluminality; parelectricity; and Earnshaws theorem in media with inverted populations," *Phys. Rev. Lett.*, vol. 73, pp. 3383–3386, 1994.

35. R. A. Shelby, D. R. Smith, S. C. Nemat-Nasser, and S. Schultz, "Microwave transmission through a two-dimensional, isotropic, left-handed metamaterial," *Appl. Phys. Lett.*, vol. 78, pp. 489–91, 2001.

36. D. R. Smith and N. Kroll, "Negative refractive index in left-handed materials," *Phys. Rev. Lett.*, vol. 85, pp. 2933–2936, October 2000.

37. D. M. Pozar, *Microwave Engineering*, 2nd ed., John Wiley & Sons, New York, 1998.

38. R. E. Collin, *Foundations for Microwave Engineering*, 2nd ed., McGraw-Hill, New York, 1992.

39. O. Siddiqui, M. Mojahedi, S. Erickson, and G. V. Eleftheriades, "Periodically loaded transmission line with effective negative refractive index and negative group velocity," presented at 2003 *IEEE International Symposium on Antennas and Propagation: URSI North American Radio Science Meeting*, Columbus, OH, USA, 2003.

40. O. F. Siddiqui, M. Mojahedi, and G. V. Eleftheriades, "Periodically loaded transmission line with effective negative refractive index and negative group velocity," *IEEE Trans. Antennas Propag.*, vol. 51, pp. 2619–2625, 2003.

41. O. F. Siddiqui, S. J. Erickson, G. V. Eleftheriades, and M. Mojahedi, "Time-domain measurement of negative group delay in negative-refractive-index transmission-line metamaterials," *IEEE Trans. Microwave Theory Tech.*, vol. 52, pp. 1449–1454, 2004.

Index

U

V

W

Y

Printed and bound by CPI Group (UK) Ltd, Croydon, CR0 4YY
16/04/2025
14658345-0005